FROM ARABIA TO THE PACIFIC

Drawing upon invasion biology and the latest archaeological, skeletal and environment evidence, *From Arabia to the Pacific* documents the migration of humans into Asia, and explains why we were so successful as a colonising species.

The colonisation of Asia by our species was one of the most momentous events in human evolution. Starting around or before 100,000 years ago, humans began to disperse out of Africa and into the Arabian Peninsula, and then across southern Asia through India, Southeast Asia and south China. They learnt to build boats and sail to the islands of Southeast Asia, from which they reached Australia by 50,000 years ago. Around that time, humans also dispersed from the Levant through Iran, Central Asia, southern Siberia, Mongolia, the Tibetan Plateau, north China and the Japanese islands, and they also colonised Siberia as far north as the Arctic Ocean. By 30,000 years ago, humans had colonised the whole of Asia from Arabia to the Pacific, and from the Arctic to the Indian Ocean as well as the European Peninsula. In doing so, we replaced all other types of humans such as Neandertals and ended five million years of human diversity.

Using interdisciplinary source material, *From Arabia to the Pacific* charts this process and draws conclusions as to the factors that made it possible. It will be invaluable to scholars of prehistory, and archaeologists and anthropologists interested in how the human species moved out of Africa and spread throughout Asia.

Robin Dennell is Emeritus Research Professor at Exeter University, UK. In his early career, he was primarily interested in the Neolithic of Europe and Southwest Asia. During 1981–1999, his main research was on the Palaeolithic and Pleistocene of Pakistan. In 2003, he was awarded a three-year British Academy Research Professorship to write *The Palaeolithic Settlement of Asia* (2009), the first overview of the Asian Early Palaeolithic and Pleistocene. Since 2005, he has conducted research with Chinese colleagues into the Pleistocene and Palaeolithic of China. He was elected a Fellow of the British Academy in 2012.

FROM ARABIA TO THE PACIFIC

How Our Species Colonised Asia

Robin Dennell

LONDON AND NEW YORK

First published 2020
by Routledge
2 Park Square, Milton Park, Abingdon, Oxon OX14 4RN

and by Routledge
52 Vanderbilt Avenue, New York, NY 10017

Routledge is an imprint of the Taylor & Francis Group, an informa business

British Library Cataloguing-in-Publication Data
A catalogue record for this book is available from the British Library

Library of Congress Cataloging-in-Publication Data
A catalog record has been requested for this book

ISBN: 978-0-367-48239-8 (hbk)
ISBN: 978-0-367-48241-1 (pbk)
ISBN: 978-1-003-03878-8 (ebk)

Typeset in Bembo
by Integra Software Services Pvt. Ltd.

For all those – past, present and future – interested in Asia and the history of our species

CONTENTS

List of figures *ix*
List of tables *xiii*
Preface *xiv*
Acknowledgements *xvii*
List of abbreviations *xix*

1 Invasion biology and the colonisation of Asia 1

2 The African background: hominins to humans 28

3 The climatic and environmental background to the human colonisation of Asia 48

PART 1
Prologue: the southern dispersal across Asia **71**

4 Arabia to the Thar Desert 77

5 The Oriental Realm of South Asia 106

6 Sunda and mainland Southeast Asia 134

7 Wallacea and Sahul 166

PART 2
Prologue: the northern dispersal across Asia **195**

8 Southwest Asia: from the Levant to Iran 203

9 Central Asia, southern Siberia and Mongolia 248

10 China 281

11 Humans on the edge of Asia: the Arctic, Korean Peninsula and the Japanese islands 317

12 How, when and why did our species succeed in colonising Asia? 346

General index *357*
Site index *362*

FIGURES

1.1	Point and range distribution	3
1.2	Range expansion and range shift	5
1.3	Range and metapopulation distribution	7
1.4	Jump dispersal	12
1.5	Coastal dispersal as a "string of pearls" or as a broad front	14
1.6	Colonisation through assimilation	15
1.7	Colonisation through replacement	16
1.8	Learning a new landscape in occupied and unoccupied lands	18
1.9	Responses of moose to auditory predator cues among predator-naïve and -savvy herds	20
1.10	Limitations of the fossil record	22
2.1	Jebel Irhoud skull compared with a Neandertal and late *Homo sapiens*	29
2.2	Behavioural innovations of the African Middle Stone Age	33
2.3	The "green Sahara"	34
3.1	Marine isotope stages (MIS) of the last 550,000 years and their subdivisions	49
3.2	The marine isotope stages (MIS 1–MIS 6) of the last 150,000 years	50
3.3	Rainfall variability over the last 6,000 years in a semi-arid landscape of northeast Asia	51
3.4	Biogeographic realms of Asia during MIS 3 and MIS 4	52
3.5	The principal mountain areas of Asia	55
3.6	The principal desert areas of Asia	56
3.7	Convergent and divergent corridors	57
3.8	Winter temperatures in northern Asia	60
3.9	Wind chill chart	61
3.10	Plant versus hide and fur clothing in different environments	62
3.11	Habitat loss and fragmentation	64

3.12 Sink and source populations 66
4.1 Present-day rainfall levels of the Arabian Peninsula 79
4.2 Palaeodrainage of Arabia 81
4.3 Comparison of palaeo-environmental records and dated Middle Palaeolithic sites in Arabia 82
4.4 Estimated annual rainfall in the Arabian Peninsula in the last interglacial 83
4.5 A selection of Middle Palaeolithic artefacts from Arabia 85
4.6 Levallois and Nubian methods of core reduction 88
4.7 Assemblage C from Jebel Faya 89
4.8 Middle Palaeolithic artefacts from Jebel Qattar 1, Nefud Desert 91
4.9 Ancient and modern drainage systems of Southwest Asia and land exposed during the Last Glacial Maximum 96
4.10 Potential corridors for dispersal across Iran 97
4.11 New evidence for point technologies from the Thar Desert 99
5.1 Principal features and Late Palaeolithic sites of India 107
5.2 Indian vegetation in MIS 5 109
5.3 Pollen frequencies in core SK-128A-31 from the Indian Ocean west of India 110
5.4 South Asia, showing reconstructed vegetation zones for ca. 30 ka ago and location of microlithic sites 111
5.5 South Asian monsoon strength, hominin fossils and key archaeological sites and industries 114
5.6 Blades, microblades and cores from Jwalapuram 9, Jerreru Valley 120
5.7 Blades from site 55, Riwat, Pakistan 123
5.8 Plan of site 55, Riwat, Pakistan 124
5.9 Microliths from Batadomba lena, Sri Lanka 125
5.10 Fauna from Fa Hien lena 126
6.1 Coastlines in Southeast Asia when sea levels were 40 m below present levels 136
6.2 The principal rivers on Sunda and adjacent mainland Southeast Asia when sea levels were 75 m below the present 137
6.3 Vegetation zones and major rivers on Sunda during the Last Glacial Maximum (LGM) 138
6.4 Vegetation of Southeast Asia in MIS 5 139
6.5 Vegetation of Southeast Asia in MIS 4 140
6.6 The consequences of rises in sea level on animal populations 142
6.7 The cave of Tam Pa Ling, Laos 147
6.8 Niah Cave, Borneo 150
6.9 Palawan today and during the Last Glacial Maximum (LGM) 153
7.1 Wallacea, with sea levels at –50 m, and major early human sites 167
7.2 The Philippines, when sea levels were 125 m lower than now 168
7.3 Sea currents in Wallacea 170
7.4 Sea levels and uplift rates in Wallacea 175
7.5 Problems in sampling rock shelter occupation records in Wallacea 180

7.6 The broken, but elaborate, 35,000-year-old projectile point from Matja Kuru 2, Timor Leste 181
7.7 Early human sites on Pleistocene Sahul 186
8.1 Middle and Upper Palaeolithic sites in the Levant 204
8.2 Recent excavations at the Upper Palaeolithic site of Manot, Israel 214
8.3 Initial Upper Palaeolithic (IUP) retouched tools from Üçağızlı, Turkey 216
8.4 Modern vegetational zones of Israel 219
8.5 The climatic record from the Dead Sea 220
8.6 The speleothem record from Manot cave, Israel 221
8.7 The speleothem record from Karaca Cave, Turkey 226
8.8 Palaeolithic sites in Iran and the Zagros Mountains 228
8.9 Early Upper Palaeolithic stone tools from Yafteh, western Iran 231
9.1 Location of Middle and Upper Palaeolithic sites in Central Asia 250
9.2 Location of sites in the Altai Mountains 253
9.3 Dating the appearance of *H. sapiens* at Denisova in Siberia 257
9.4 The location of the main stratified sites with early blade assemblages in Transbaikalia and Mongolia 262
9.5 The environmental record of Lake Kotokel over the last 46,000 years 263
9.6 Personal ornaments in Initial Upper Palaeolithic (IUP) sites in Siberia 266
9.7 Location of Upper Palaeolithic sites in the Tolbor Valley, Mongolia 269
9.8 Initial Upper Palaeolithic (IUP) artefacts from Tolbor 16 273
9.9 Climatic and environmental history of northern Mongolia since 40 ka 274
10.1 Fossil hominin and early Upper Palaeolithic sites in China and neighbouring countries 283
10.2 The Chinese speleothem record of climatic change over the last 70,000 years 285
10.3 Estimated rainfall in north China since the last interglacial 286
10.4 Estimated rainfall over the last 140,000 years in the Loess Plateau 287
10.5 Changes in vegetation in Inner Mongolia over the last 140,000 years 288
10.6 The coastal shelf of east China 289
10.7 Model of immigration into China in warm periods 291
10.8 Model of immigration into China in cold periods 292
10.9 Shuidonggou 1 and 2 300
10.10 Blade production techniques at Shuidonggou (SDG) 1, lower cultural layer 301
11.1 Stratigraphic units and palaeo-environments of Arctic Siberia and Western Beringia 319
11.2 Palaeolithic sites in the Arctic 320
11.3 Ivory and bone tools from the Yana RHS site 321

11.4 Coastlines of Japanese islands during the Last Glacial Maximum (LGM) 328
11.5 Main vegetation zones in Japan today and during the Last Glacial Maximum (LGM) 329
11.6 Possible colonisation routes to the Japanese islands 330
11.7 Examples of the major stone tool types from early Upper Palaeolithic assemblages in Japan 331
11.8 Summary of innovations in the Late Palaeolithic of Japan and north China 332
11.9 Earliest archaeological dates from the Ryuku Islands 334
11.10 The sequence of Sakitari Cave, Okinawa 335
11.11 The Kelp Highway and relevant archaeological sites along the Pacific Rim 338
12.1 Developmental rates in apes and humans 353

TABLES

6.1 Human skeletal remains from Southeast Asia and Sunda 145
7.1 Palaeolithic sites in Wallacea and Sahul with C14 dates 177
7.2 Intermittent occupation records from Wallacea 179
8.1 The fossil hominin evidence from the Levant 205
8.2 Dates (ka) for Levantine hominin skeletal remains 207
8.3 Radiocarbon dates of Upper Palaeolithic sites in the Zagros 229
8.4 The fauna from Wezmeh Cave, western Iran 230
8.5 Prey taken at Zagros Caves 235
9.1 Fauna associated with Middle and Upper Palaeolithic sites in Central Asia 252
9.2 Fauna of the Altai in the Late Pleistocene 255
9.3 Summary of the Early Upper Palaeolithic in the Trans-Baikal 264
9.4 Initial Upper Palaeolithic sites in Siberia and Mongolia with personal ornaments 265
9.5 Upper Palaeolithic dates from sites in Mongolia 271
10.1 Dates for Shuidonggou (SDG) 1 and SDG 2 302
11.1 Radiocarbon-dated Late Palaeolithic sites from the Korean Peninsula 324
12.1 Examples from Palaeolithic Asia of imagination, ingenuity and inventiveness that are unique to our species and not found among the previous inhabitants 353

PREFACE

This book focuses on the first appearance of our species, *Homo sapiens*, in Asia. This was one of the most momentous events in human evolution, and one which resulted in our species being the only hominin in the whole of Africa, Asia and Europe. In some regions of Asia, when this happened can be indicated with some precision – for example, the main Japanese islands were first colonised around 38,000 years ago. In other areas, our first appearance is known to within a few millennia: for example, from Iran through southern Siberia to Mongolia and north China, this likely happened between 50,000 and 45,000 years ago. In other regions such as India, we have very little idea of when our ancestors first appeared because of a lack of well-dated skeletal material and a shortage of well-dated archaeological assemblages. An additional complication is that the dispersal of our species across Asia was unlikely to have been a single event. Instead it is much more likely to have been composite, with at least two main dispersals across southern and continental Asia, and in Southwest Asia – and perhaps further east – there may well have been several episodes of dispersal, depending upon the prevailing climate and availability of water. I try to show what we think we know, and what we know we don't know about this momentous colonisation of the largest continent by our species. As I hope to show, this is a story of immense ingenuity, inventiveness and adaptation that amply demonstrates our talents as a coloniser and as an invasive species. In terms of time frames, my coverage ends around 30,000 years ago, by which time humans were present from Tasmania to the Arctic, from the Atlantic and Mediterranean to the Pacific, and were even on the Tibetan Plateau. In geographical terms, I define Asia as the area east of the Mediterranean and Ural Mountains, but I exclude the Caucasus region on the grounds that it can be better treated as part of Europe.

My main sources are the stones and bones – the bread and butter of Palaeolithic archaeology – evidence about the climate and environment, and of course the skeletal evidence for our species and those whom they replaced. I also use ancient DNA (aDNA) where it is available, and I expect this source to increase greatly in importance in the new few years. What I have not used is the enormous literature derived from the genetic analysis of living human populations. There are two main reasons for this decision. The first is simply one of space as there is a limit to what can be crammed into a book of this length. The second is that there has been a tendency in recent years for the "traditional" sources of evidence – the stone tools, associated animal bones, and skeletal evidence – to be downplayed in a favour of a framework established by genetic analyses of modern populations. These typically state that people with a particular genetic make-up dispersed from area A to area B at such and such a time, and attach to that narrative whichever pieces of skeletal data or archaeological evidence that are consistent with that scenario. They are often shown as maps with arrows denoting the direction of dispersal, with in most cases, a total disregard for the landscapes through which people moved. The timing of such dispersals is usually derived from estimates of rates of genetic change, and should be seen as hypotheses waiting to be tested, and not taken (as often happens) at face value. Additionally, these scenarios of when and by which routes humans dispersed take no account of the what people did in the landscapes that they colonised, how they adapted to them, what choices they made, or their climatic and environmental circumstances. In other words, they have rarely told me about the topics that interest me most. So, in defence of my own profession as primarily a Palaeolithic archaeologist, I try to show as best I can what we know about our colonisation of Asia from the evidence they left behind over 30,000 years ago rather than from the genetic make-up of people who may or may not be their direct descendants.

My approach is primarily biogeographical. When our species colonised Asia, it adapted to an immense variety of environments, from the deserts of Southwest and Central Asia, to the cold steppes of Mongolia and north China, to the high Arctic, and the rainforests of South and Southeast Asia. Each had its own characteristic fauna, flora, climate, relief and environmental challenges. The easiest way to summarise these is to follow the lead of Alfred Wallace (1823–1913) – the father of biogeography – and break Asia into three major biological realms of the African-Arabian, the Oriental and the Palearctic. These realms were never static, and their boundaries changed in keeping with the numerous climatic shifts of the last glacial cycle. I find these categories useful as a way of circumscribing the opportunities and challenges that each presented to both their resident species and a dispersive (or invasive) species such as ourselves. Because the term "invasive" is emotive, for the most part in this book, I prefer to talk in terms of human dispersal into a region. That way, our dispersal can be seen in the same light as the dispersal of mice and mammoths. I do, however, refer to our talents as an invasive species at the beginning and at the end of the book when I try to explain the reasons behind our success as a coloniser.

The structure of the book is as follows: in Chapter 1, I discuss how we, as *Homo sapiens*, have been supremely successful as an invasive species, and how invasion (or dispersal) biology can offer insights into how we managed to colonise the entire planet. In Chapter 2, I discuss the African background to the colonisation of Asia. I focus on two recent developments: first, recent evidence that the earliest members of our species are now dated to more than 300,000 years old; and second, the evidence for a "green Sahara" that contained numerous lakes and rivers in interglacial times, and provided corridors between sub-Saharan and North Africa as well as a spring board for human groups to enter Arabia and the Levant. Before turning to the Asian story of colonisation, I discuss in Chapter 3 the Asian climatic and environmental background, and argue that a southern and northern dispersal was inevitable given basic topographic realities of continental Asia. Chapter 4 discusses the evidence from the Arabian Peninsula but also the eastward extension of arid and semi-arid landscapes of southern Iran, Pakistan and the Thar Desert. India and Sri Lanka are discussed in Chapter 5; unfortunately, Myanmar remains a major gap in our knowledge of what would have been both a barrier and a corridor into Southeast Asia. In Chapter 6, I discuss Sundaland, or the landmass of mainland Southeast Asia, and the islands of Sumatra, Java and Borneo that would have been conjoined when sea levels fell by 50 m and more. This leads in Chapter 7 to Wallacea and Sahel, or the principal islands east of the Wallace Line (Flores, Sulawesi, Timor and the Philippines) and the conjoined landmass of New Guinea, Australia and Tasmania. Here, I depart from normal practice in regarding Wallacea as a biogeographical province in its own right and not simply part of Island Southeast Asia. The second part of the book turns to the northern dispersal of our species: in Chapter 8, from the Levant to the Zagros Mountains and interior of Iran, and in Chapter 9, Central Asia, Siberia and Mongolia. Chapter 10 covers China (including the Tibetan Plateau), and Chapter 11 ends with the colonisation of the Arctic, the Korean Peninsula and the Japanese islands. I end in Chapter 12 with a brief consideration of what it means to be "human", and why prehistory is written by us rather than Neandertals.

Exeter and Lemnos, Greece,
October 2019

ACKNOWLEDGEMENTS

My first debt of gratitude is to the Leverhulme Trust for awarding me an emeritus grant (EM-2016-05) that funded my travel and other costs to the meeting of the Asian Palaeolithic Association at Denisova, Siberia, in July 2018; the Indo-Pacific Association meeting in Hue, Vietnam, in September 2018; the Institute of Vertebrate Palaeontology and Palaeontology, Beijing, in January 2019; the INQUA meeting in Dublin, in July 2019; as well as a research trip to Israel in May 2019. I sincerely thank Chris Bae, Katerina Douka and Mike Petraglia for inviting me in 2016 to a Wenner-Gren symposium in Cintra, Portugal, on the human colonisation of Asia in the Late Pleistocene, where I first explored some of the ideas about invasive biology that I develop in this book. I am also extremely grateful to the Asian Palaeolithic Association for inviting me to their meeting at Denisova, and to Professor Yoshi Nishiaki for inviting me to a workshop on early *Homo sapiens* in Asia in Kyoto in December 2018. Other meetings before I started on this book have also been enormously influential: these include a meeting organised by Huw Groucutt and James Blinkhorn in Oxford in 2012 on deserts and the Middle Palaeolithic; a symposium organised by Nicci Boivin, Mike Petraglia and Remy Crassard and funded by the Ffyssen Foundation in Paris, 2015 on human dispersal; and an invitation to speak at the China Archaeological Congress, Zhengzhou, in May 2016. I thank also Maria Martinón-Torres for inviting me to give a seminar on the dispersal of *Homo sapiens* across Asia at UCL, London, in February 2017 and Martin Porr for his invitation to me to give the Hilgendorf Lecture in Tübingen in May 2017. Martin Porr and Jaqueline Matthews are also thanked for inviting me to their workshop in 2015 on decolonisation and human origins in Perth, Australia. I owe a huge amount to my friend and colleague Nuria Sanz, formerly director of the UNESCO HEADS (Human evolution, adaptation, dispersal and social development) programme, at which as a member of the scientific committee, I learnt much relevant to this book at meetings in Jeongok, South Korea (2012, on Asia), Tübingen (2013, on Eurasia), Puebla, Mexico, (2013, on the Americas), Xalapa, Mexico (2015, on rainforests), and Mexico City (2017, on primatology); those meetings were always a pleasure and an education. Thanks are also due to Anna Belfer-Cohen, Adam Brumm, Iain Davidson, Anatoly Derevianko, Jo Kaminga, Ofer Marder, James O'Connell, Gonen Sharon and Li Feng for their advice and insights.

There are also two groups of people to thank. I am especially grateful to the following for reading chapters of this book and preventing me from making huge errors of omission and commission: Huw Groucutt and Eleanor Scerri (Chapters 2 and 4 on Africa and Arabia); James Blinkhorn and Parth Chauhan (Chapter 4, and Chapter 5 on India); Julien Louys (Chapter 6 on Southeast Asia and Chapter 7 on Wallacea); Mae Goder-Goldberg (Chapter 8 on the Levant); Saman Heydari-Guran and Elham Ghasidian (Chapter 8 on Iran); Evgenny Rybin, Irnina Khatsenovich and Masima Izuho (Chapter 9 on Siberia and Mongolia); Gao Xing (Chapter 10 on China), and Masima Izuho (Chapter 11 on Korea and Japan). Needless to say, the errors are all mine.

The second group is larger and more amorphous but includes all those I met at the meetings mentioned above who talked about the Palaeolithic, human evolution, the Pleistocene, their own research and the usual hot conference gossip.

Finally, I am immensely grateful to my wife Linda Hurcombe, not least for her insights into perishable material culture and for reading the entire text, and also my son Patrick, who will doubtless be horrified to learn that he influenced my thoughts in Chapter 12.

ABBREVIATIONS

BP	before present, or rather, before 1950 because of the amount of contamination in atmospheric radioactivity that was subsequently caused by nuclear weapon testing.
MIS	Marine isotope stage – a period of the earth's climate determined by isotopic analysis of microscopic shellfish (foraminifera) in marine sediments.
ka	a thousand years.
ka cal BP	calibrated thousands of years before present. Radiocarbon years are not equivalent to calendar years so need to be calibrated, or converted into calendar years.
Hominin	hominins are those precursors of *Homo sapiens* that are distinct from the lineages of our closest relatives, the bonobo, chimpanzee and gorilla. For the purposes of this book, hominins refer to those members of the genus *Homo* that inhabited Asia before our species: *Homo neanderthalensis*, *H. luzonensis*, *H. floresiensis* and *H. erectus*.

1

INVASION BIOLOGY AND THE COLONISATION OF ASIA

Introduction

As Pat Shipman (2015) has pointed out, our species – *Homo sapiens* – is the ultimate invasive species, as well as the main omission on the list of invasive species in the 1992 Convention on Biological Diversity (Davis 2009: 4). When our species first emerged in Africa around 300,000 years ago (see Chapter 2), there were several species of our relatives outside Africa. By 30,000 years ago, these were all extinct, and we were the only member of our own genus *Homo* worldwide. This expansion of our species from Africa across ultimately the rest of the world is usually treated as an example of dispersal, in other words, as the expansion of a species beyond its original range. "Dispersal" is a neutral term, whereas "invasion" is used in a pejorative sense to denote the negative consequences of a dispersal upon indigenous species, either directly – by eliminating them as competitors – or indirectly – by, for example, out-competing them for resources, or by disrupting the ways they used their environment. As an invasive species, we have been very good at doing all three: to quote John Shea (2011: 28), "'Plays well with others' is not something one is likely to see on *H. sapiens*' evolutionary report card".

As an invasive[1] species, we exhibited three prime features. First, we colonised all areas that were already inhabited; second, we colonised new habitats that had never been inhabited, such as Australia, the Japanese islands, the Americas, the Tibetan Plateau, and new environments such as rainforests and the Arctic. Thirdly, we eliminated all indigenous hominin species, albeit with some evidence of interbreeding with Neandertals and Denisovans (see Chapter 9). These three features were unique in the evolutionary history of our lineage – for at least four million years, hominins had been a diverse family but were now reduced to just one species. As Ian Tattersall (2000: 56) remarked, "Once we were not alone"; now we are most definitely on our own.

Biologists have paid considerable attention to invasive species, by which they mean ones that impact negatively on us, either directly (such as pathogens) or indirectly (for example, pests that attack our crops and livestock) (see e.g. Clobert et al. 2001; Davis 2009; Eatherley 2019; Elton 1958; MacDonald 2003; Shigesada and Kawasaki 1997; Thompson 2014). These studies have been at a variety of scales, from continental (as with the invasive behaviour of grey squirrels against native red ones in Europe, or the cane toad in Australia) to regional (as with monitoring the invasive behaviour of plants, birds or insects) or laboratory-sized experiments (as with watching invasive species of bacteria). Nearly all are short-term, as is the nature of research in ecology, and few span more than a few decades. As such, they may seem inapplicable to studying at continental or even regional scales how our species invaded Eurasia and other regions over periods of several tens of millennia. Nevertheless, some of the basic ideas are valuable in providing a framework by which we might begin to understand how we were so dramatically successful as an invasive species (see e.g. Smith 2013: 74–76 for their application in an Australian context). These ideas are particularly useful within the wider field of population biology when it comes to studying how species react to climate change – obviously a major concern these days – and how invasive species manage to out-compete and sometimes replace indigenous species.

When considering the behaviour of a colonising species such as *Homo sapiens*, we need first to consider the populations that were colonised.

The colonised: the indigenous inhabitants of Asia

As noted above, there were several types of indigenous hominins in Asia at the time of contact with our species. These formed discrete (allopatric) populations that did not overlap much: Neandertals across continental Eurasia (Chapters 8 and 9), *H. erectus* in East and Southeast Asia (Chapters 6 and 10), *H. floresiensis* known (so far) only from the island of Flores, *H. luzonensis* in the Philippines (see Chapter 7) and probably a different type in South Asia (see Chapter 5). A biological population is "all individuals of a given species in a prescribed area" (MacDonald 2003: 15). Depending upon the choice of scale, this can be a region, country or a continent. Here, we are primarily concerned with biological populations of different species over their total range in Asia. Biologists (including palaeoanthropologists) can show the distribution of a population in a number of ways. The first is by a point distribution – in other words, to show on a map where a species has been observed (see Figure 1.1). The main advantage of this type of map is that it can show where a species is most often observed, and where it is rarely seen. (The equivalent in human origins research is a map showing the location of key fossil specimens and Palaeolithic sites). Its main drawback is that the level of observation may not be uniform across the area enclosed by the map: for example, some areas may contain more records simply because there are more observers. Point distributions therefore have an inherent degree of uncertainty over whether gaps in distribution or areas

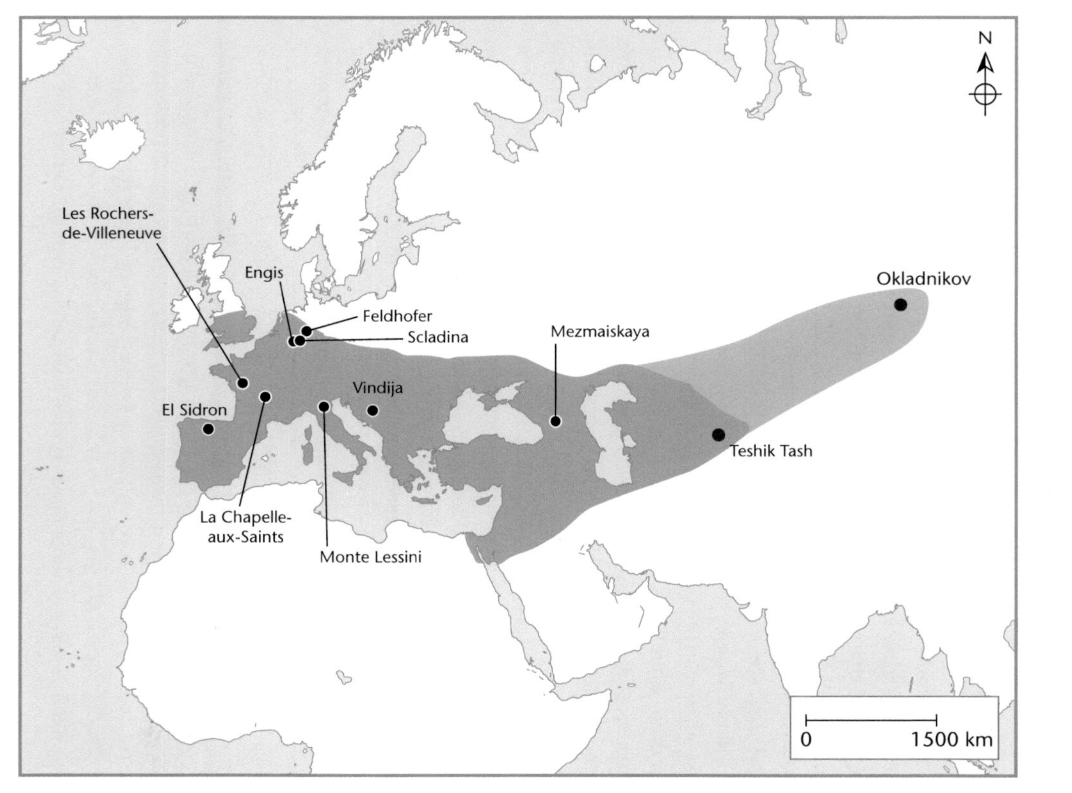

FIGURE 1.1 Point and range distribution

Here, a point distribution shows where Neandertal fossils were found; the range distribution shows the total area over which they have been found.

Source: Krause et al. 2007, Fig. 1.

showing few records are real or an artefact of inadequate observation. These uncertainties also exist in palaeoanthropology. For example, some areas have been investigated very intensively for over a century (such as western Europe), whereas others (such as Central Asia) have seen far less fieldwork. The fact that there are many more examples of Neandertals in western Europe than in Central Asia may therefore indicate different population densities, or may merely be a product of different research histories. A second way of showing the distribution of a population is to show its total range (see Figure 1.1). This gets around the problem of uneven coverage because areas with numerous examples of a species are treated the same way as ones with few records. The main drawback of this type of map is that it does not show where population levels are highest, and it presents the impression that population density is even across the entire range. In palaeoanthropology, maps showing the total range of a species such as Neandertals provide a rough indication of the maximum total area they may have inhabited during the 200,000 years or so that they existed, and when climatic conditions were at their most favourable. What they won't show is how their distribution may have changed over time, or their distribution when climatic conditions were at their worst.

Despite these limitations, maps showing the total range of a species are an essential first step in tackling palaeodemography. If conducted as an integrative exercise, the limit of the total range can be discussed in terms of the prevailing climate, relief, resources, seasonality, the behavioural competence and ecological tolerance of a species, and perhaps other limiting factors such as diseases. The limits of the range might be defined by establishing how and why populations declined to the point at which extinction outweighed recruitment and they were no longer viable. For example, the northern limit of the Neandertal range might have been defined by winter temperatures, length of daylight in winter, scarcity of fuel or a combination of these. The southern limit of the Neandertal range has attracted less attention, but might have been set by summer temperatures, competition with other hominins, or an unwillingness to leave the Eurasian Palearctic Faunal Realm (see Chapter 3).

To go further, we need clarity on two issues. The first is the extent to which behaviour changed over time in an evolving species of hominin; for examples, Neandertals 50 ka ago may have been more competent than their counterparts at 100 ka or 250 ka, and able to inhabit a wider range of environments. Likewise, our own species 40 ka ago was more competent than 60 ka earlier. Second, we need clarity on their range under a particular type of climate. When climatic conditions were at their most severe (as in a glacial maximum or its equivalent in non-glaciated regions), hominins such as Neandertals would have been restricted to a few glacial refugia. The maximum range should occur in the optimal parts of a glacial cycle as an example of range *extension* or range *expansion*; that is to say, the northern boundary of their range expanded northwards, and they simply expanded into its northern parts (Figure 1.2A). However, as a complicating factor, the expansion of a hominin range during an interglacial

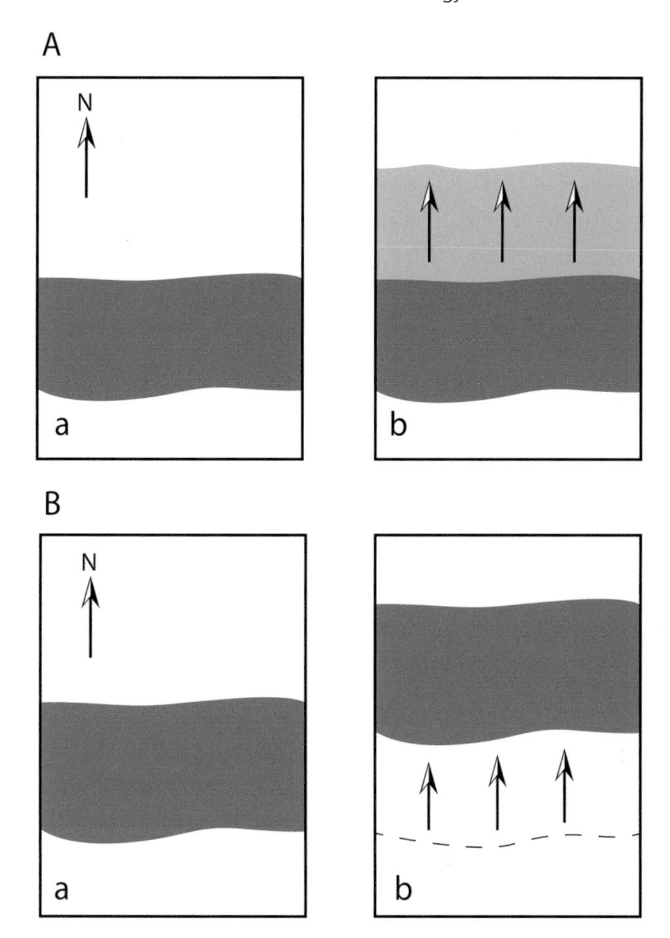

FIGURE 1.2 Range expansion and range shift

These diagrams use as an example of a population in a northerly region in response to climatic amelioration. Figure (A) shows range expansion: (a) shows part of the range of a species in a cold period. With climatic amelioration (b), groups are able to expand northwards, and the area of the total range now increases. Note that the southern boundary of the range remains unchanged. (A similar example at a smaller scale would be the expansion of range into higher altitudes of mountain ranges when climate improved.)

Figure B shows range shift: (a) shows part of the range of a species in a cold period. With climatic amelioration (b), the entire range moves northwards but remains the same size.

Source: The author.

might also be an example of a *shift* in the environmental range; in other words, both the southern and northern boundaries of their preferred habitat shifted northwards (see Figure 1.2B), and the hominins simply stayed within it without any population increase or change in their survival strategies (see Roebroeks 2006 for a discussion of these points).

Determining the extent to which behaviour changed over time and how hominins responded to a *range extension* – usually defined by the movement northwards of the northern range boundary – or a *range shift* – when both the southern and northern boundaries moved northwards in good times but southwards during climatic downturns – requires detailed, well-dated archaeological evidence that can be integrated with local climatic records. Unfortunately, that quality of evidence is rarely available for most of continental Asia, so it is rarely possible to explore these issues.

These problems aside, we need at this point is to consider how hominin populations in Asia were structured at the time of contact with our species.

Population structure

Hominin populations at a species level in Asia in the Late Middle Pleistocene (ca. 300–125,000 years ago) were likely structured in the sense that gene flow (in other words, sexual relations between its members) was not random or unimpeded across its total range. As with many mammals, we are not panmictic, meaning that there are no genetic or behavioural restrictions upon mating. Instead, most gene flow and social interaction would have occurred within sub-groups of the total population, and these were primarily defined by where they lived. These sub-groups would have interacted with each other, thereby preventing speciation through isolation. There is even likely to have been some gene flow between different hominin species, given that different species (and even genera) such as lions and tigers can potentially interbreed[2] to produce ligers and tions (there are also beefalos, or the offspring of American bison and domestic cows, and even zonkeys, or the progeny of zebras and donkeys). We will see later (Chapter 9) evidence that *H. sapiens* interbred with Neandertals and Denisovans, and perhaps even with east Asian *H. erectus* (Chapter 10). Overall, however, most interactions would have taken place within sub-groups of a species. These sub-groups are called metapopulations.

Metapopulations and palaeodemes

The sub-sets that biologists call metapopulations are groups of spatially separated populations that are interdependent to a greater or lesser degree. (As an example, think of groups of farms and villages in a region that are linked to each other through kinship, economy, transport and business.) Figure 1.3 shows how metapopulations might be distributed across the total range occupied by a species. In this figure, each metapopulation contains several small groups whose membership is largely constant from year to year, and whose monthly and annual movements are usually over the same territory (as with hunting bands, or subsistence farmers). These small groups are the basic subsistence units that created the archaeological record in caves and open-air locations investigated by Palaeolithic archaeologists.

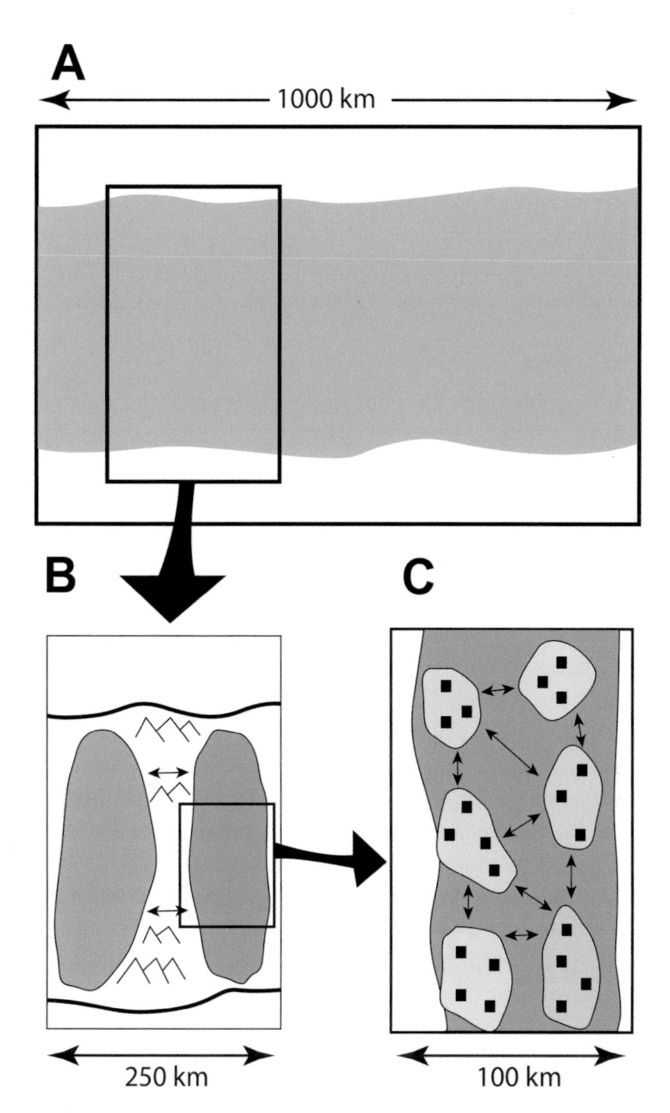

FIGURE 1.3 Range and metapopulation distribution

Within the total range of a species (A), there would have been spatially separated but inter-dependent populations (B). At a smaller scale, each metapopulation would have comprised further sub-sets of subsistence groups (C) whose membership was largely constant from year to year, and whose monthly and annual movements were usually over the same territory. These groups would form networks within which mates, resources, and information could be exchanged. The distances shown in these figures is intended only to show relative differences in the scale of analysis.

Source: The author.

These metapopulations may each have had their own cultural traditions, language, customs and so on, and interacted in a wide variety of ways with neighbouring and more distant metapopulations. Within a metapopulation, some groups may have had frequent contact with others, but some only rarely. Biologists often refer to metapopulations as loose or tight. Loose ones are those with weak links between groups: an example is subpopulations of arctic foxes inhabiting different islands. Others interact frequently and form tight metapopulations, such as birds nesting and living in different but nearby woodlands (MacDonald 2003: 15). Humans are a prime example of a species that is extremely good at forming tight metapopulations (or networks) that maintain cohesion through kinship, ideology or other forms of corporate identity. Hunters and foragers, for example, typically live in groups within an area as part of a network that maintains contact with other groups in order to share information, exchange and obtain scarce resources (such as furs, obsidian or ochre), and mating partners. Within metapopulations, there is thus gene flow and information exchange between its constituent sub-groups. Networks have attracted a great deal of study by archaeologists at levels from inter-group interactions to large-scale trade networks (see e.g., Knappett 2011, 2013), and there are obvious commonalities between archaeological networks and demographic and biological metapopulations.

Metapopulations of a species are usually spatially discrete and found within specific areas. They can be isolated from other metapopulations by distance, by barriers (such as a mountain range), or by habitat: for example, metapopulations in rainforest might not often interact with those in neighbouring woodlands. Usually, however, metapopulations do interact with other metapopulations. Most of these interactions would have occurred on their edges, either where they overlapped or were close (see Figure 1.3) and connected by corridors. There will therefore be some degree of gene flow between metapopulations as well as within them. An important type of interaction between human metapopulations that might be detected archaeologically concerns long-distance exchange networks, whereby groups in one metapopulation can obtain rare or valued materials from several hundred kilometres away: examples are the distribution of Japanese obsidian on the Asian mainland over distances of 1,000 km after 30 ka (see e.g. Ikeya 2015; Kuzmin 2006; and Chapter 11); another is the presence of anthraxolite (a soft coal-like type of rock) and amber that came from 600 km away at the 27,000-year-old Yana RHS site on the Arctic Ocean (Pitulko et al. 2004; and Chapter 11). Over these distances, it is possible that items were moved between different metapopulations, and not within a single large one. Another indicator of interactions between metapopulations might be the distribution of rare and highly specific items such as ostrich eggshell beads over enormous areas of north China, Mongolia and southern Siberia (see Chapters 9 and 10).

Palaeodemes in human evolutionary studies are the equivalent of metapopulations. Clark Howell (1996, 1999) stressed the importance of the palaeodeme as "the basic stuff of the hominin fossil record" (Howell 1996: 9) as a unit of analysis in palaeoanthropology. A deme is "a communal interbreeding population within a species … distinguishable by reproductive (genetic), geographic, and ecological (habitat) parameters". Together, demes (sometimes as isolated populations) constitute subspecies, or the "aggregate of local populations of a species inhabiting a geographic subdivision of the range of the species" (ibid.: 8–9). In areas that contain a detailed skeletal record, it might be possible to demonstrate local palaeodemes: we will see in the next chapter how African populations were probably structured, and in Chapter 10, how the same may have been true of populations in China.

Core areas, backwaters and cul-de-sacs

Because population densities were never even across the entire range occupied by a hominin species, some parts of its range are likely to have been core areas where resources were most abundant and predictable. The East African Rift Valley and southwest France are two examples of probable core areas. In most of Asia, it is currently difficult to identify core areas because so little fieldwork has taken place. We can sometimes make an informed guess by considering the local climate and likely resources, but should not make the mistake of assuming that an area in which some fieldwork has taken place was by default a "centre" or a core area of settlement.

Backwaters and cul-de-sacs

Some populations may have remained isolated over long periods of time, and thereby lived in regions that might be regarded as backwaters or cul-de-sacs. They might therefore have formed relic populations that either died out, or much later mixed with immigrant groups after a long period of isolation, in much the same way as a rock pool when the tide comes in. Populations that were isolated for long periods provide one explanation for the persistence of "primitive" traits in a palaeodeme or metapopulation: we will see later (Chapter 10) some possible candidates from southwest China. We should, however, be cautious in dismissing a region as a backwater or cul-de-sac without strong reasons. Hallam Movius (1948: 411), for example, famously (or infamously) dismissed the whole of southern and eastern Asia as "a region of cultural retardation". That type of sweeping judgement about a region that was based on a small amount of imperfect evidence can easily become self-fulfilling because it provides no incentive to investigate it further (see Dennell 2014, 2016).

The colonisers

The colonising populations – or rather, metapopulations – of *Homo sapiens* that entered Asia probably came from several source populations in east and northeast Africa, and at different times (see Chapter 2). Some of these dispersals were probably unsuccessful and short-lived, but others were more persistent. Whether successful or not, these initial pioneers were probably part of structured populations composed of sub-groups, or metapopulations that formed networks that allowed the sharing of information and the exchange of mating partners.

The behaviour of metapopulations is crucial when considering colonisation. First, "At any given latitude or effective temperature the structure of the subsistence resource base can vary widely in terms of predictability, patchiness and density in space and time" (Ambrose and Lorenz 1990: 9). Consequently, dispersal into a diverse environment is more effective if these differences are dealt with by a metapopulation formed of several integrated groups than a single, nucleated population. Second, tight metapopulations have a major advantage over discrete, nucleated populations, in that a large area can be occupied in a short space of time providing that groups maintain sufficient connectivity with each other to ensure that each (or most) survive. A third point is that dispersal takes place at the edge of the inhabited range, and thus in the least familiarised part of the landscape. A metapopulation can move into new territory on a trial and error basis: some groups might flourish, and may thus be able to recruit more members, or form other groups that could disperse further over new territory; if some groups fail, they might be replaced by the immigration of other groups. Colonisation is clearly successful if more groups survive than fail within a metapopulation, and often, initial attempts at colonisation fail because too few groups survive (as with, for example, some of the earliest European settlements in the Americas). Collectively, a species organised into tight metapopulations with a high degree of connectivity between groups could also acquire far more knowledge of their environment as an information network than a group could obtain on its own. Colonisation by a network of groups into areas already occupied by other hominins (such as Neandertals) would have an advantage over the indigenous groups if the networks of the immigrant population were more effective. Connectivity within a metapopulation is also crucial in maintaining cultural diversity. As Hovers and Belfer-Cohen 2013: 347) point out:

> gains in technological diversity (i.e., incorporation of inventions as society-wide innovations) are not dependent only on population size. Even if a new behavior is beneficial and population size is above 250 individuals, it will not spread and cultural diversity will not increase if levels of interconnectedness (e.g., the number of linkages between individuals in networks of different mating behaviors) are low.

The colonisers who entered Asia encountered an enormous diversity of landscapes that varied greatly in animals and plants throughout the year, and/or from year to year. In some areas, resources might have been abundant year-round, predictable from season to season or year to year, and easily accessible without having to travel long distances. In contrast, other environments were ones of low abundance and high risk: semi-arid and arid landscapes, for example, are ones where the risk of drought is high, and survival may depend on a few widely spaced sources of water.

The crucial difference between the colonisation of Asia by our species and the colonisation of Australia and the Americas is that Asia was already inhabited by several types of indigenous residents. Incoming groups of *Homo sapiens* had therefore to deal with not only a wide variety of landscapes, but also ones that were already occupied. How then might our species have colonised Asia?

Seven ways to colonise a continent

There is no single method of colonising a continent, especially if much of it is already inhabited by indigenous populations, but at least seven methods may have been used by our species when colonising Asia.

Possible ways of colonising

1 Avoidance

The path of least resistance is for a colonising population to occupy areas that have no indigenous residents. Coastlines, for example, might have been uninhabited when our species entered Asia, and some researchers have proposed that our species dispersed rapidly across southern Asia by taking a coastal route. Another environment that might have been uninhabited before *H. sapiens* is rainforest, and that possibility is explored in Chapters 5 and 6.

2 Co-existence

In areas of low productivity, and/or with low densities of indigenous inhabitants, colonisers might simply have been able to co-exist in the same region with little animosity from or interaction with the local residents. One possible example that we will meet in Chapter 9 is the co-existence of Neandertals and Denisovans in the Altai Mountains of Siberia. As seen by the end of this book, it is perhaps debateable whether our species maintained this type of peaceful co-existence over several millennia.

3 Jump dispersals: the cautious and the bold

In Asia, the clearest example of jump dispersals involves the colonisation of islands, whereby colonists "jumped" across open water to a new area (Figure 1.4). Jump dispersals also happened when people dispersed bypassed unproductive parts of a coastline, or "jumped" between water sources when crossing deserts. A different type of jump dispersal has been proposed by American palaeoanthropologist David

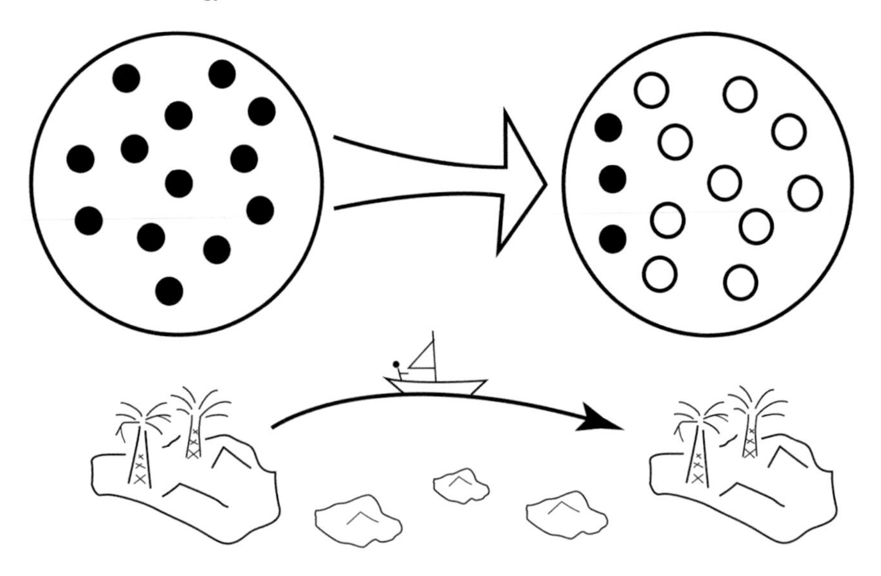

FIGURE 1.4 Jump dispersal

This diagram shows bold colonisation by jump dispersal. Here, some groups (A, black circles) at the edge of an island metapopulation take the risk of jumping past a series of unproductive islands to find a better island (B) than their present location. Although the risk of failure is high, success means that a new region can be colonised by descendant groups (white circles).

Source: The author.

Meltzer (2009: 234–238), who has suggested two ways by which the first colonists in North America might have dealt with landscapes that have high or low productivity. He names these as the cautious and the bold. *Cautious* groups were ones that remained anchored to patches of high, year-round productivity. Over time, some may have sought similar locations nearby and developed those in turn, but generally, dispersal rates were slow: such groups "moved only slowly across the latitudes and longitudes, each daughter colony being spawned by the overflow of a saturated estate" (Beaton 1991: 220–221). An example in Asia of such a process might be the northern Levant, with its rich coastal plain and inland lakes. Rather than envisaging human settlement there after 100–125 ka as a "failed dispersal" (Shea 2008), it might instead be an example of how cautious colonisers made the most of a rich environment, beyond which resources in the deserts and semi-arid plains to the east were more scattered, less predictable, harder to obtain, and a disincentive to dispersal (see Chapter 8).

In contrast, *bold* colonists are ones that move when productivity declines. These colonists inhabit areas that will not support long-term residence of the kind favoured by cautious colonists, so they have a high incentive to move, and especially to "jump" across areas that are low in productivity (such as steppes or deserts). During sudden environmental downturns, for example, water resources might diminish, and game become less plentiful; onward movement to new areas is risky, but nevertheless less risky than clinging to a declining resource base. Bold colonists tend to use mobility to explore and become familiar with large areas (Veth 2005); as example, judging from where different types of stone were obtained, the occupants of Puritjarra 35 ka in Western Australia were probably familiar with ca. 10,000 sq km of territory (Smith 2013: 90). Bold colonists also have a characteristic method of expansion "in which migrants generate new satellite colonies rather than simply expanding the occupied area from its periphery" (Smith 2013: 75). They thus form a metapopulation of spatially separate groups, some of which may fail, but which can be replaced by colonists, in the same way as in source-sink models (see Chapter 3). An Asian example might be the colonisation of the North China Plain, where resources were scarce and often unpredictable, winters harsh, and few areas allowed intensive long-term occupation. Bold colonists are also ones that seek out naïve prey that is unfamiliar with humans and thus easier to hunt; when the animals become more wary and vigilant, bold colonists will move to new areas where the prey is naïve (see below).[3]

Meltzer's suggestions about cautious and bold colonists are interesting and applicable to a variety of Asian landscapes. Their inherent weakness in an Asian context is that both cautious and bold colonists would likely encounter indigenous populations, and thus the problem has to be faced of how the colonisers displaced or replaced the indigenous populations. Two scenarios are likely.

4 Coastal dispersal

Some researchers (e.g. Stringer 2000; Mellars 2006; Mellars et al. 2013) have suggested that humans took a coastal route out of Africa along southern Asia towards Australia as a way of explaining what appeared to be a rapid expansion of the human range after 60 ka. There are three main ways by which this may have happened. One is by staying on land and dispersing along a coast. This type of expansion can be envisaged as a "string of pearls", of groups strung out along a length of coastline (see Figure 1.5A). An inherent weakness of this type of dispersal is that groups are limited in their choice of mates to the group either ahead or behind them, and those at the front can choose only from the group immediately to their rear. A more plausible method of coastal dispersal is along a broad front that might include a coastline but only as a minor component of a dispersal that was largely inland (Figure 1.5B). A more effective way is by a series of jump dispersals (see above), in which colonists bypassed unproductive sections of a coast by using boats. This is not without risk, especially if groups become separated from each other and face a shortage of prospective mates. Additionally, coastal environments need to be well established and stable for a coastal dispersal to be worthwhile, and this is unlikely to happen when sea levels were rising or falling, as was often the case in the last glacial cycle (Westley and Dix 2006).

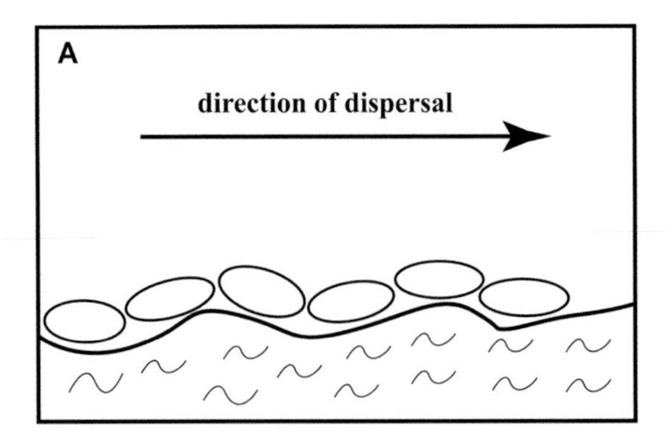

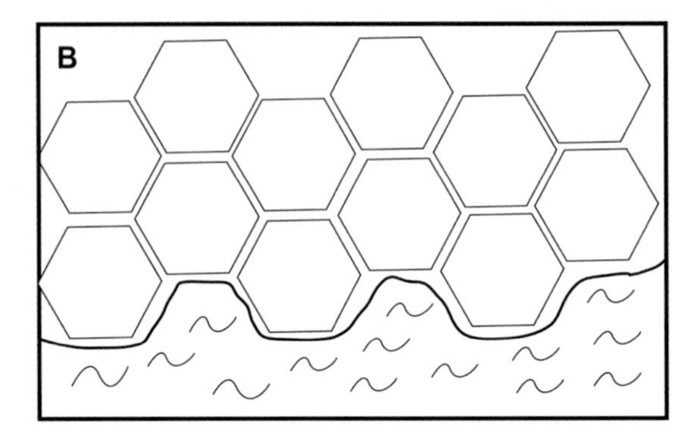

FIGURE 1.5 Coastal dispersal as a "string of pearls" or as a broad front

A: Here, a colonising population is dispersing along a coastline. Groups in the middle can choose mates from in front or behind it, but the leading group is restricted to the group behind it.

B: Here, a colonising population is dispersing along a coastline but also inland. Each coastal group can recruit mates from at least three neighbouring groups, and those inland have access to five. This method of dispersal would have been far more effective than one that restricted itself to a land-based route along the coast.

Source: The author.

5 Colonisation through assimilation of the indigenous population

In this scenario, part of a metapopulation began to invade an area that was occupied by a different type of hominin. The invasive metapopulation then proceeded to assimilate the females of reproductive age, thus degrading the previous viability of the indigenous population (see Figure 1.6). This type of scenario is indicated by evidence of gene flow from Neandertals and Denisovans into *Homo sapiens* outside Africa. As we shall see in Chapter 9, there may have been an "interbreeding bonanza" (Callaway 2016) between *H. sapiens*, Neandertals and Denisovans. Although we may never know the nature of these encounters between *H. sapiens* and Neandertals, it is likely that interbreeding, whether consensual or forcible, was

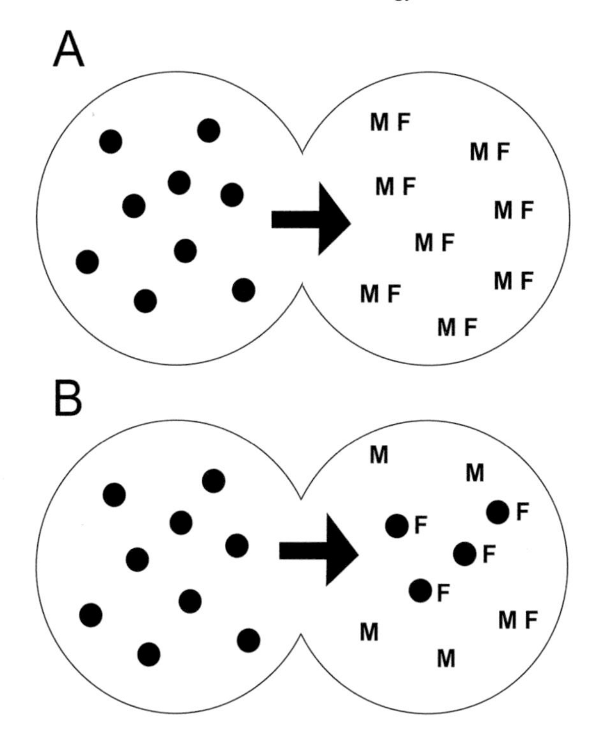

FIGURE 1.6 Colonisation through assimilation

In this scenario, part of a metapopulation (A, black circles) begins to invade an area occupied by a different type of hominin, shown as MF, with M = males and F = females. The invasive metapopulation then proceeds to assimilate the females of reproductive age (B), thus degrading the previous viability of the indigenous population.

Source: The author.

largely between invasive male *H. sapiens* and indigenous female Neandertals and Denisovans. The long-term consequences of these encounters may have been negative (such as male hybrid infertility [Sankararaman et al. 2014]), or positive if leading to "hybrid vigour", enhanced immunity (see Stewart and Stringer 2012), and the acquisition of beneficial genes such as those that facilitated life at high altitudes such as the Tibetan Plateau (Huerta-Sánchez and Casey 2015) or in cold environments. Colonisation through assimilation also helps explain the evidence for hybridisation in the East Asian skeletal evidence for *H. sapiens* in the last glacial cycle (see Chapter 10).

6 Colonisation through population replacement

In this scenario, incoming colonising groups simply replaced the indigenous inhabitants (see Figure 1.7). This process may have been violent but may also have occurred because the incoming population out-competed the local population for key

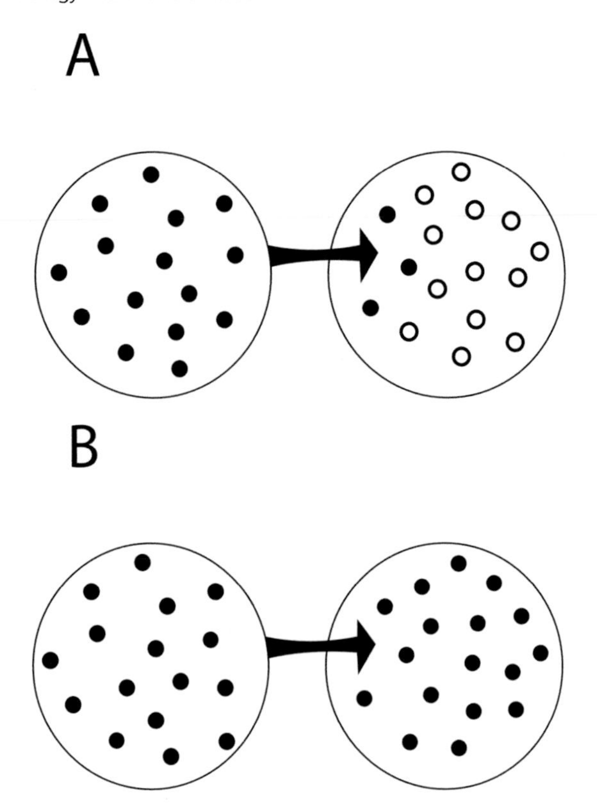

FIGURE 1.7 Colonisation through replacement

Here, a metapopulation (A, black circles) begins to invade an area already occupied by other groups (open circles). In B, the indigenous occupants are replaced. This scenario is one explanation for the replacement of Neandertals by *H. sapiens* in western Eurasia, or the replacement of *H. floresiensis* by *H. sapiens* on Flores, Indonesia.

Source: The author.

resources and locations and destroyed their connectivity between groups. Some bird populations compete for the best nesting sites rather than for resources, and humans may have done the equivalent by appropriating the best caves. Until recently, colonisation by total replacement was the dominant model for explaining the expansion of *H. sapiens* across Eurasia. The discovery that all Eurasians contain some Neandertal DNA (Green et al. 2008, 2010) has muted this type of explanation, and most researchers are now sympathetic to models that allow for some degree of assimilation of indigenous Eurasians by immigrant populations of *Homo sapiens*.

7 Colonisation via contagion

Another possible way in which our species may have been able to colonise areas that were already inhabited is through the inadvertent transmission of diseases that were both lethal and infectious for the indigenous population (Bar-Yosef and Belfer-Cohen 2001). There are numerous examples from recent colonial history in which indigenous American and Australian populations were decimated by the introduction of smallpox, measles, influenza and other diseases from Europeans, and it is not impossible that indigenous Eurasians were highly susceptible to diseases that had an African origin. However, Asia has its own considerable set of diseases that might have been equally lethal to any colonising population that originated in Africa, and these might have delayed colonisation until the colonisers had acquired some degree of immunity. Diseases might therefore have accelerated or impeded the rate of colonisation, depending upon whether they originated amongst the colonisers or the colonised.

Landscapes and learning

Learning a new landscape is a key aspect of colonisation (see for example Rockman and Steele 2003; and Roebroeks 2003), and would have been a key feature of the colonisation of Asia by our species. During this process, humans would have had to learn how to survive and adapt in environments as different as the rain forests of Southeast Asia and the Arctic. Some landscapes would have been easy to colonise because the resources – animal and plant foods, flakeable stone, plants that could be used for their fibres or as medicine, potential fuel resources, etc. – are already known to the coloniser from prior experience. By far the easiest landscapes to colonise are those that through range extension (see above) become larger as a result of climate change. As example, during the transition to an interglacial, groups living near the margins of an area of semi-arid grassland might find that an adjacent area of desert became semi-arid grassland. Apart from having to learn about its topography, water supplies, sources of stone and location of food resources, no change of behaviour would be needed to colonise it. The Arabian Peninsula provides one example of how groups already resident in semi-arid grasslands in north and northeast Africa could have easily moved into Arabia without having to change their behaviour when it too became grassland as a result of climate change. Another example might be the Thar Desert of northwest India when it became less arid and better watered in the last interglacial (see Chapter 4).

At the other extreme are landscapes in which none of the food and other resources are familiar. The hardest landscapes to colonise in Asia were probably the rainforests of South and Southeast Asia, and the landscapes of the Arctic. For humans who were familiar with grasslands and open woodlands, there is almost nothing familiar about rainforests. The plant foods are not obvious; some are toxic if eaten raw and require processing by leaching and rinsing; the plants that could be used for cordage, containers or clothing are unfamiliar; there are no herd animals, and some of the potential prey such as monkeys live in the tree canopy and thus require new methods of hunting. There is also a new range of poisonous insects, reptiles and diseases to learn about, and a new range of medicines to develop. In the

Arctic, there would have been a very steep and unforgiving learning curve in how to survive the bitterly cold conditions of winter. First attempts are unlikely to have been successful, and we might expect a high rate of failure.

A crucial factor for a colonising population is whether the area is already inhabited. If it is, landscape learning should be much easier.[4] At the very least, incoming groups could copy what the local population was eating and using; if relations were cordial, then they would have the equivalent of an on-site tutorial (see Figure 1.8). Genetic and skeletal evidence of hybridisation (as between our species and Neandertals and Denisovans – Chapter 9) provides a good indication that the newcomers were learning from the local population – if they were exchanging mates (and genes), it would surely follow that they were also exchanging knowledge. We can expect that across much of Asia, incoming humans would have been able to take advantage of indigenous knowledge. It is also likely that later colonising groups would take advantage of the knowledge of the initial colonising population.

We can only speculate over how long it would have taken to learn a new landscape. Blanton (2003) suggests that it took the English settlers at Jamestown, Virginia 20 years to sense local climate patterns, whilst Tolan-Smith (2003) thought it might have taken a couple of thousand years for groups to adapt to post-glacial Britain. Clearly, much depends upon the scale and nature of the environment. Landscapes of high biodiversity – such as rainforests – would presumably require a much longer learning process than those of low biodiversity.

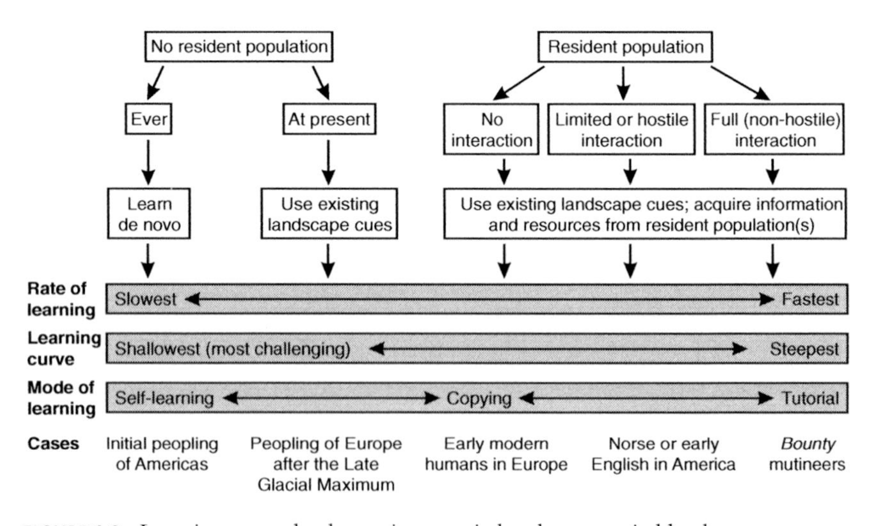

FIGURE 1.8 Learning a new landscape in occupied and unoccupied lands

This figure shows the number and complexity of issues involved in learning a new landscape. As shown, the main issue is whether a colonising population is entered a new landscape that is empty, or already inhabited.

Source: Meltzer 2003, Fig. 13.1.

Meltzer (2003) makes the important point that learning is not only a lengthy process but one that is gender-linked. Men hunting large animals, for example, would need to cover a large area and learn the necessary "macroenvironmental" skills for wayfinding and navigation, whereas females (and children) who gathered would need to learn at a smaller and more detailed scale about the location and seasonality of plant resources and small game.

In general terms, the human colonisation of Asia would likely have been a series of gradual steps from familiar to less familiar landscapes, rather than a sudden leap into landscapes that were totally unfamiliar. On those grounds, the rainforests of South and Southeast Asia would probably have been the last environments in those regions to colonise, and only after humans had learnt how to colonise adjacent types of vegetation. In Central and northern Asia, the learning process might have proceeded from learning how to survive, for example, in an area with one month of temperatures as low as –20° C. to surviving in another area with two months of temperatures as low as at –40° C. (see Chapters 3 and 11). Likewise, learning how to live in a winter landscape of snow and ice is not an overnight process for those who have never encountered a snowfall. We shall see in later chapters how and what humans had to learn when colonising these new environments.

Invasive species and naïve faunas

One fascinating aspect of the colonisation of some parts of Asia (as well as the Americas and Australia) that I explored in previous papers (Dennell 2017a, 2017b) is that humans sometimes encountered prey species that had never seen a human. The prey, in other words, was naïve. In contrast, prey animals in areas inhabited by humans are vigilant, or "savvy", because they realise that humans pose a real and immediate threat. As examples, Canadian elk (red deer in European terms, or *Cervus elaphus*) are extremely vigilant to the presence of humans near roads and frequency of traffic (Ciuti et al. 2012). Another study showed (Thurfjell et al. 2017) that female elk that are more than 9–10 years old in two areas of Canada are almost invulnerable to hunting by humans. This is because they are savvy to the point that they are aware of the hunting season, and even of the type of weapons that are being used. In the hunting season, they stay clear of roads to avoid being seen by hunters in vehicles; if hunters are using bow and arrow, the elk even retreat to rocky terrain where it is difficult to get a clear shot.

Naïve prey face a steep learning curve in survival when confronted with an invasive predator such as wolves or humans. Although there are no regions left where the indigenous large-mammal fauna is unaware of the potential or actual threat posed by humans, there are examples from North America and northern Scandinavia of how recent faunas have reacted to the re-introduction of predators such as bear and wolf as conservation measures. Berger and colleagues (2001) tested "naïve faunas" for their vigilance to predator cues versus the responses in "predator-savvy" populations (see Figure 1.9). They found that predator-savvy

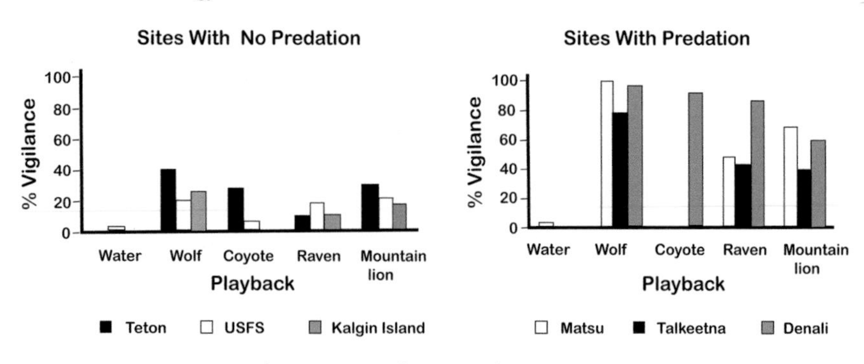

FIGURE 1.9 Responses of moose to auditory predator cues among predator-naïve and -savvy herds

Moose in areas where there was no predation were far less vigilant in their responses to the sounds of wolf, coyote and ravens (notable scavengers of predator kills) and even mountain lion than those in areas of predation. The sound of running water was used as a control in both groups.

The place names (Teton, Kalgin Island etc.) are nature reserves where moose were observed.

Source: Berger et al. 2001, Fig. 1.

moose were far more alert than naïve ones to the sound and scent of bears and wolves, and to raven calls (because these birds scavenge the carcasses of moose that were killed by bears) (Berger et al. 2001: 1037).

Berger and colleagues (2001) also investigated whether lack of vigilance in naïve moose led to greater mortality through predation by bears along the colonising front than in the centre. They found that both male and female bears consumed twice the rate of moose carcasses along the dispersal front than away from it; that male bears fed about two-thirds more often along the front than away, and most significantly, predation was far more successful at the edge than in the centre of the moose populations. As they summarise, "Our findings indicate that naïve individuals are i) conspicuously lacking in astuteness and ii) experiencing a blitzkrieg" (Berger et al. 2001: 1038).

Berger et al. (2001) also showed that prey extinction is avoided because the prey populations quickly learn the importance of vigilance, and thereby become predator-savvy. Significantly, moose changed from being predator-naïve to -savvy in a single generation: they learnt to re-establish a "landscape of fear" (Laundré 2001).

Re-colonisation and the loss of vigilance

Loss of vigilance amongst prey populations may also have been a major factor affecting the way hominins (and later, humans) re-colonised regions that had been depopulated during glacial periods. Remarkably, Berger et al. (2001) note that loss of vigilance occurred in less than 100 years, and in one case, between

40 and 75 years. Repeatedly through the Pleistocene, especially across the northern limits of the hominin range, from north China through Central Asia to west and northwest Europe, human populations declined with each return to cold, dry (i.e. glacial) conditions, especially when winters were sub-freezing for prolonged periods. In those areas where hominins either declined in number or disappeared, much of the resident fauna would have remained, or been replaced by, animals better adapted to the cold. Those animals that remained would have lost their vigilance regarding hominins, in the same way that moose in Sweden lost their vigilance against wolves when these became locally extinct. When hominins were able to move northwards again when the climate improved, they would have encountered a naïve fauna, and thus re-colonisation would have been easier – not just because of climate change, but because the prey was initially easier to kill. During the initial process of colonisation, they would have been at an advantage if they kept moving into areas where the prey fauna was still naïve. A good example of where this process was likely repeated at the onset of each interstadial and interglacial is northwest Europe, and the clearest and best documented example of this process, is probably the region's recolonisation in the final part of the last glacial period, during which humans moved back into the depopulated areas of northern Europe and ultimately colonised Britain, Scandinavia and the Baltic region.

When considering the dispersal of our own species in the last 60,000 years into regions such as Australia, Japan, the Philippines, Tibet, the Americas and the Arctic, they also entered environments containing naïve faunas that had never previously encountered humans. Americanists have been aware of this point, and unlike their counterparts in the Old World, have emphasised how the naivety of the indigenous fauna at the time of contact with humans would have facilitated their rapid movement across both North and South America (see e.g. Meltzer 2009; Dillehay 2000). In the case of Japan, with its endemic island fauna, the first colonists would have found hunting far easier than on the mainland (Chapter 11). The same is likely true of Australia, and it is not surprising that there has been active debate over the role of humans in megafaunal extinctions in Japan and Australia (see e.g. Iwase et al. 2012; Saltré et al. 2016).

The fossil record: first and last appearances

The human skeletal record is crucial for establishing when our species first appeared in a region outside Africa, and when the indigenous inhabitants last died out. In most textbooks on human evolution, maps showing the expansion of our species out of Africa rely very heavily upon the earliest fossil examples of our species in a region. Yet we need to recognise the limitations of this record. An obvious one is dating – some specimens turn out to be much older or younger than first thought, and others are found in deposits that are extremely hard to date precisely. But more importantly, we will never identify the precise moment when a group of colonists first arrived, nor will we ever detect the point at which a population dwindled to zero and became locally extinct (see Figure 1.10). Instead, the earliest fossil example of a species in a new area records the point when it is sufficiently numerous that some of its members may become part of the fossil record. In poorly documented areas – like most of Asia – we then have to add other limitations, such as the chances that the skull and dentition are preserved (as these are the easiest to identify to species level), the chances of it being discovered in its original context by qualified researchers, and the chances that the context can be securely dated. Similarly, the last fossil example of a species such as *Homo erectus* in a region denotes when it was still sufficiently numerous that some of that population would end up as part of the fossil record.

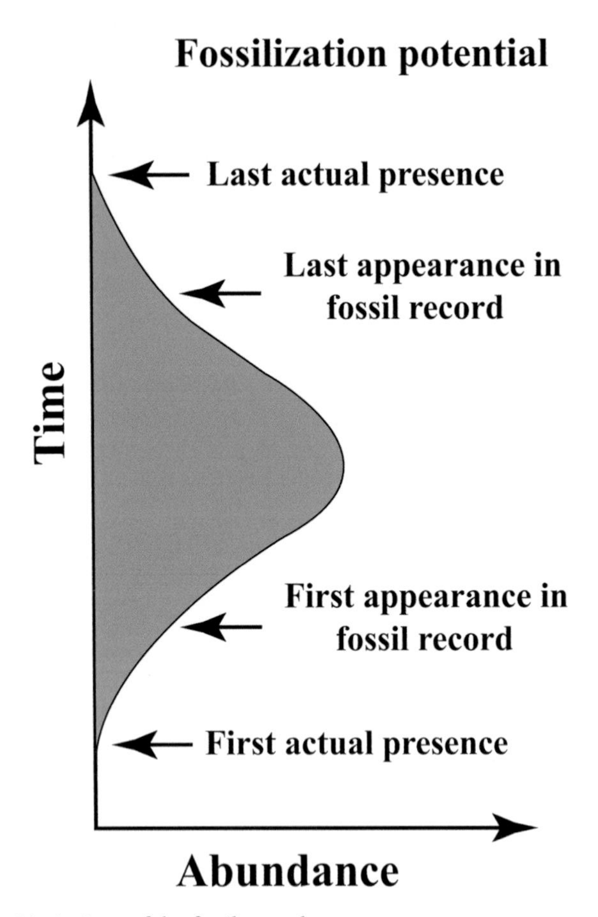

FIGURE 1.10 Limitations of the fossil record

The fossil record never demonstrates the first and last actual appearance of a taxon. Instead, its first and last appearance of a taxon will always underestimate its first and last actual appearance.

Source: The author.

We therefore need to distinguish between the first *recorded* presence and the first *actual* presence of a species; and between the last *recorded* presence and its last *actual* presence of its predecessor. In a sub-region such as western Europe, which has (certainly by comparison with Asia) a detailed fossil record and reasonably good proxy indicators of *H. sapiens* (such as blade assemblages and bone points), we can narrow down the gap between the first recorded presence and the first actual presence for *H. sapiens*, and the last recorded and last actual presence of Neandertals to a few thousand years. In most of Asia, there is not so much a gap between these points as a vast chasm. As an example, in South Asia, the last (in fact, only) recorded fossil presence of an indigenous inhabitant may be as old as 230 ka, and the earliest skeletal example of our species dates from only ca. 36 ka (see Chapter 5). The essential point here is that the fossil skeletal record will always understate when a species first appears, and it will always understate the age of the last appearance of its predecessor.

Because stone tools are more durable that a human skeleton, and far more common, artefacts are usually the main guide to when hominins first appeared in an unoccupied territory, whether Britain, Australia, the Americas or Japan. Unfortunately, stone artefacts offer little guidance about their makers or subsequent population movements. First, there is usually no clear correlation between a stone tool assemblage and a hominin species (with a possible exception for Neandertals and Mousterian assemblages in Europe and Asia outside the Levant); and second, the stone artefacts used by early *Homo sapiens* and the indigenous groups that they replaced are often too similar for them to be used as clear indicators of their makers.

The take-home message is that in poorly documented regions, the first recorded presence of *H. sapiens* is likely to seriously underestimate how long it has been resident, and the last recorded presence of its indigenous predecessor is likely to be far earlier than its last actual presence. This problem is unavoidable, and will accompany us throughout our exploration of the Asian evidence.

Summary

Invasion (or dispersal) biology offers a useful way of studying how humans might have colonised Asia and replaced all indigenous inhabitants. Both the colonising and colonised populations can be envisaged as networks of interacting local groups, or metapopulations, and depending upon local circumstances, the immigrant metapopulations of our species had various options about how they might interact with local groups, or disperse into and across different types of environment. An important part of the colonising process involved learning about new landscapes and their resources. In many parts of Asia, this included the opportunities offered by a prey fauna that was naïve in that it had never encountered humans, or had forgotten that they posed a threat. As a final point, the limitations of the human skeletal record are discussed in terms of its inability to show the first appearance of an immigrant population, and the last appearance of the indigenous inhabitants. At this point, we can consider the African background to the colonisation of Asia by *H. sapiens*.

Notes

1 An invasive species has also been referred to a "weed species" (Cachel and Harris 1998), meaning that it can flourish in disturbed or unstable habitats. Because humans are so good at colonising any environment, whether unstable or not, I prefer to use the term "invasive species".
2 Most of these hybrid species are produced under artificial conditions, often for the simple reason that the parent species rarely meet in the wild. The offspring are also often sterile, as with mules, the product of female horses and male donkeys.
3 Spikins (2015) points out that jump dispersals by bold colonists may be forced rather than voluntary. In metapopulations with strict norms of behaviour, transgressors may be forcibly expelled. Although difficult to demonstrate from Palaeolithic evidence, her suggestion is worth noting when considering instances of short-term human presence in areas such as offshore islands or the arctic that were far from the main areas of settlement.
4 There are examples where the colonising population stubbornly refused to learn from the indigenous population: one clear example are the first English settlers at Jamestown, Virginia, who displayed "a lethal combination of ethnocentrism, ignorance, and misplaced priorities in their interaction with the environment", to the bewilderment of the local Indian population (see Blanton 2003: 190).

References

Ambrose, S.H. and Lorenz, K.G. (1990) Social and ecological models of the Middle Stone Age in southern Africa. In P. Mellars (ed.) *The Emergence of Modern Humans*, Edinburgh: Edinburgh University Press, pp. 3–33.

Bar-Yosef, O. and Belfer-Cohen, A. (2001) From Africa to Eurasia – early dispersals. *Quaternary International*, 75: 19–28.

Beaton, J. (1991) Colonizing continents: some problems from Australia and the Americas. In T.D. Dillehay and D.J. Meltzer (eds) *The First Americans: Search and Research*, Baton Rouge, Tuscon: CRC Press, pp. 209–230.

Berger, J., Swenson, J.E. and Persson, I.-L. (2001) Recolonizing carnivores and naïve prey: conservation lessons from Pleistocene extinctions. *Science*, 291: 1036–1039.

Blanton, D.B. (2003) The weather is fine, wish you were here, because I'm the last one alive: "learning" the environment in the English New World colonies. In M. Rockman and J. Steele (eds) *Colonization of Unfamiliar Landscapes: The Archaeology of Adaptation*, London and New York: Routledge, pp. 190–200.

Cachel, S. and Harris, J.W.K. (1998) The lifeways of *Homo erectus* inferred from archaeology and evolutionary ecology: a perspective from East Africa. In M. Petraglia and R. Korisettar (eds) *Early Human Behavior in Global Context: The Rise and Diversity of the Lower Palaeolithic Record*, London: Routledge, pp. 108–132.

Callaway, E. (2016) Evidence mounts for interbreeding bonanza in ancient human species. *Nature*, doi:10.1038/nature.2016.19394.

Clobert, J., Danchin, E., Dhont, A.A. and Nichols, J.D. (2001) *Dispersal*. Oxford: Oxford University Press.

Ciuti S., Northrup, J.M., Muhly, T.B., Simi, S., Musiani, M. et al. (2012) Effects of humans on behaviour of wildlife exceed those of natural predators in a landscape of fear. *PLoS ONE*, 7(11): e50611, doi:10.1371/journal.pone.0050611.

Davis, M.A. (2009) *Invasion Biology*. Oxford: Oxford University Press.

Dennell, R.W. (2014) East Asia and human evolution: from cradle of mankind to cul-de-sac. In R.W. Dennell and M. Porr (eds) *Southern Asia, Australia and the Search for Human Origins*, Cambridge: Cambridge University Press, pp. 8–20.

Dennell, R.W. (2016) Life without the Movius Line. *Quaternary International*, 400: 14–22.

Dennell, R.W. (2017a) Pleistocene hominin dispersals, naïve faunas and social networks. In N. Boivin, R. Crassard and M. Petraglia (eds) *Human Dispersal and Species Movement from Prehistory to the Present*, Cambridge: Cambridge University Press, pp. 62–89.

Dennell, R.W. (2017b) Hominin dispersals and naïve faunas. In B. Bajd (ed.) *Palaeoanthropology: Recent Advances and Future Prospects*, Ljubljana: University of Ljubljana, pp. 91–102.

Dillehay, T. (2000) *The Settlement of the Americas*. New York: Basic Books.

Eatherley, D. (2019) *Invasive Aliens: Plants and Animals from Over There That Are Over Here*. London: Harper Collins.

Elton, C.S. (1958) *The Ecology of Invasions by Animals and Plants*. London: Methuen.

Green, R.E., Krause, J., Briggs, A.W., Maricic, T., Stenzel, U. et al. (2010) A draft sequence of Neandertal genome. *Science*, 328: 710–722.

Green, R.E., Malspinas, A.-S., Krause, J., Briggs, A.W., Johnson, P.L.F. et al. (2008) A complete Neandertal mitochondrial genome sequence determined by high-throughput sequencing. *Cell*, 134: 416–426.

Hovers, E. and Belfer-Cohen, A. (2013) On variability and complexity lessons from the Levantine Middle Paleolithic record. *Current Anthropology*, 54(S8): 337–357.

Howell, F.C. (1996) Thoughts on the study and interpretation of the human fossil record. In W.E. Meikle, F.C. Howell and N.G. Jablonski (eds) *Contemporary Issues in Human Evolution*, California Academy of Sciences Memoir 21: 1–45.

Howell, F.C. (1999) Paleo-demes, species clades, and extinctions in the Pleistocene hominin record. *Journal of Anthropological Research*, 55: 191–243.

Huerta-Sánchez, E. and Casey, F.P. (2015) Archaic inheritance: supporting high-altitude life in Tibet. *Journal of Applied Physiology*, 119(10): 1129–1134, doi:10.1152/japplphysiol.00322.2015.

Ikeya, N. (2015) Maritime transport of obsidian in Japan during the Upper Paleolithic. In Y. Kaifu, T. Goebel, H. Sato and A. Ono (eds) *Emergence and Diversity of Modern Human Behavior in Paleolithic Asia*, College Station: Texas A&M University Press, pp. 362–375.

Iwase, A., Hashizume, J., Izuho, M., Takahashi, K. and Sato, H. (2012) Timing of megafaunal extinction in the Late Pleistocene on the Japanese archipelago. *Quaternary International*, 255: 114–124.

Knappett, C. (2011) *An Archaeology of Interaction: Network Perspectives on Material Culture and Society*. Oxford: Oxford University Press.

Knappett, C. (ed.) (2013) *Network Analysis in Archaeology: New Approaches to Regional Interaction*. Oxford: Oxford University Press.

Krause, J., Orlando, L., Serre, D., Viola, B., Prüfer, K., Richards, M.P., Hublin, J.-J., Hänni, C., Derevianko, A.P. and Pääbo, S. (2007) Neanderthals in Central Asia and Siberia. *Nature*, 449: 902–904.

Kuzmin, Y.V. (2006) Recent studies of obsidian exchange networks in prehistoric Northeast Asia. In D.E. Dumond and R.L. Bland (eds) *Archaeology in Northeast Asia: On the Pathway to Bering Strait*, Oregon: Museum of Natural History, pp. 61–72.

Laundré, J.W. (2001) Wolves, elk, and bison: re-establishing the "landscape of fear" in Yellowstone National Park, U.S.A. *Canadian Journal of Zoology*, 79(8): 1401–1409.

MacDonald. G.M. (2003) *Biogeography: Space, Time and Life*. New York: John Wiley and Sons.

Mellars, P. (2006) Why did modern human populations disperse from Africa ca. 60,000 years ago? A new model. *Proceedings of the National Academy of Sciences USA*, 103(25): 9381–9386.

Mellars, P., Gori, K.C., Carr, M., Soares, P.A. and Richards, M.B. (2013) Genetic and archaeological perspectives on the initial modern human colonization of southern Asia. *Proceedings of the National Academy of Sciences USA*, 110: 10699–10704.

Meltzer, D. (2003) Lessons in landscape learning. In M. Rockman and J. Steele (eds) *Colonization of Unfamiliar Landscapes: The Archaeology of Adaptation*, London and New York: Routledge, pp. 222–241.

Meltzer, D. (2009) *First Peoples in a New World: Colonizing Ice Age America*. Berkeley: University of California Press.

Movius, H.L. (1948) The lower Palaeolithic cultures of southern and eastern Asia. *Transactions of the American Philosophical Society*, 38(4): 329–420.

Pitulko, V.V., Nikolsky, P.A., Girya, E.Y., Basilyan, A.E., Tumskoy, V.E. et al. (2004) The Yana RHS site: humans in the Arctic before the Last Glacial Maximum. *Science*, 303: 52–56.

Rockman, M. and Steele, J. (eds) (2003) *Colonization of Unfamiliar Landscapes: The Archaeology of Adaptation*. London and New York: Routledge.

Roebroeks, W. (2003) Landscape learning and the earliest peopling of Europe. In M. Rockman and J. Steele (eds) *Colonization of Unfamiliar Landscapes: The Archaeology of Adaptation*, London and New York: Routledge, pp. 99–115.

Roebroeks, W. (2006) The human colonisation of Europe: where are we? *Journal of Quaternary Science*, 21(5): 425–435.

Saltré, F., Rodríguez-Rey, M., Brook, B.W., Johnson, C.N., Turney, C.S.M. et al. (2016) Climate change not to blame for late Quaternary megafauna extinctions in Australia. *Nature Communications*, 7: Article number: 10511, doi:10.1038/ncomms10511.

Sankararaman, S., Mallick, S., Dannemann, M., Prüfer, K., Kelso, J., Pääbo, S., Patterson, N. and Reich, D. (2014) The genomic landscape of Neanderthal ancestry in present-day humans. *Nature*, 507: 354–357.

Shea, J.J. (2008) Transitions or turnovers? Climatically-forced extinctions of Homo sapiens and Neanderthals in the East Mediterranean Levant. *Quaternary Science*, 27: 2253–2270.

Shea, J.J. (2011) *Homo sapiens* is as *Homo sapiens* was: behavioral variability versus "Behavioral Modernity" in Paleolithic Archaeology. *Current Anthropology*, 52(1): 1–35.

Shigesada, N. and Kawasaki, K. (1997) *Biological Invasions: Theory and Practice*. Oxford: Oxford University Press.

Shipman, P. (2015) *The Invaders: How Humans and Their Dogs Drove Neandertals to Extinction*. Cambridge, Mass.: Harvard University Press.

Smith, M. (2013) *The Archaeology of Australia's Deserts*. Cambridge: Cambridge University Press.

Spikins, P. (2015) The geography of trust and betrayal: moral disputes and Late Pleistocene dispersal. *Open Quaternary* (orchid.org/0000-0002-9174-5168), http://eprints.whiterose.ac.uk/92253.

Stewart, J.R. and Stringer, C.B. (2012) Human evolution out of Africa: the role of refugia and climate change. *Science*, 335: 1317–1321.

Stringer, C.B. (2000) Coasting out of Africa. *Nature*, 405: 24–25.

Tattersall, I. (2000) Once we were not alone. *Scientific American*, 282: 56–62.
Thompson, K. (2014) *Where Do Camels Belong? The Story and Science of Invasive Species*. Vancouver: Greystone Books.
Thurfjell, H., Ciuti, S. and Boyce, M.S. (2017) Learning from the mistakes of others: how female elk (*Cervus elaphus*) adjust behaviour with age to avoid hunters. *PLoS ONE*, 12(6): e0178082.
Tolan-Smith, C. (2003) The social context of landscape learning and the late glacial-early postglacial recolonization of the British Isles. In M. Rockman and J. Steele (eds) *Colonization of Unfamiliar Landscapes: The Archaeology of Adaptation*, London and New York: Routledge, pp. 116–129.
Veth, P. (2005) Cycles of aridity and human mobility risk minimization among Late Pleistocene desert foragers of the Western Desert, Australia. In P. Veth, M. Smith and P. Hiscock (eds) *Desert Peoples: Archaeological Perspectives*, Oxford: Blackwell Publishing, pp. 100–115.
Westley, K. and Dix, J. (2006) Coastal environments and their role in prehistoric migrations. *Journal of Maritime Archaeology*, 1: 9–28.

2

THE AFRICAN BACKGROUND

Hominins to humans

Introduction

First, the African skeletal, archaeological and environmental records that are relevant to the dispersal of our species into Asia are examined.

African evidence for the origins of *Homo sapiens*

Genetic data indicate that our species diverged from our sister species between 400 ka and 700 ka, so its roots are clearly deep in the Middle Pleistocene (Stringer 2016). The earliest skeletal indications of *H. sapiens* now extend back to ca. 300,000 years (300 ka), as a result of re-investigations of the cave of Jebel Irhoud, Morocco. This cave was excavated in the 1960s, when several hominin remains were discovered, associated with Middle Palaeolithic – now called Middle Stone Age – stone tools. These are dominated by Levallois technology, include a high proportion of retouched tools, especially pointed forms and convergent scrapers, and lack Acheulean or Aterian elements (Richter et al. 2017). At the time, the skeletal remains were thought to be ca. 40 ka old and have Neandertal characteristics. Later, these were re-dated to ca. 160 ka, and in the light of further discoveries of skeletal evidence, regarded as more similar to our species than to Neandertals. Importantly, analysis of the teeth in the mandible of a young individual showed that it had developed at the same rate as a living human, and therefore at a slower rate than earlier types of hominin (Smith et al. 2007). The reason why this is important is that it indicates a longer and slower rate of development in childhood, which increased the length of childhood dependency upon the parents, but also allowed for the acquisition of complex skills whilst the brain was still developing. (I return to this point at the end of the book.) Recently, the cave has been re-investigated, and a comprehensive dating programme has established that layer 7, which contained Middle Stone Age artefacts, some important new fossil specimens and probably the skeletal remains found in the 1960s, has an age range of 280–350 ka – almost twice as old as previously thought (Richter et al. 2017). The latest assessment is that the face of the Jebel Irhoud individuals is essentially identical to that in living humans (and can thus be distinguished from other types of hominins) but the brain case retains the primitive elongated shape and low height, compared with the globular shape of humans today (Hublin et al. 2017; Stringer and Galway-Witham 2017; Neubauer et al. 2018; see Figure 2.1 and Chapter 12). Nevertheless, these individuals from Jebel Irhoud can be classed as an early form of *Homo sapiens*. Furthermore, the crania are similar to a partial cranium from Florisbad, South Africa, which is dated to ca. 260 ka (Grün et al. 1996) and which some researchers regard as an early representative of our species (Stringer 2016). This implies that individuals like those from Jebel Irhoud and Florisbad were widespread across the whole of Africa by 260 ka. The possibility that our species may have been pan-African by 260 ka is supported by a recent analysis of the DNA from a juvenile boy from Ballito Bay, KwaZulu-Natal, South Africa, who died ca. 2,000 years ago (Schlebusch et al. 2017). His DNA contained no admixture from Bantu Africans or Europeans, and Schlebusch and his colleagues estimated the population divergence time between this boy and other groups as beyond 260 ka – close to the age of the Florisbad and Jebel Irhoud specimens.

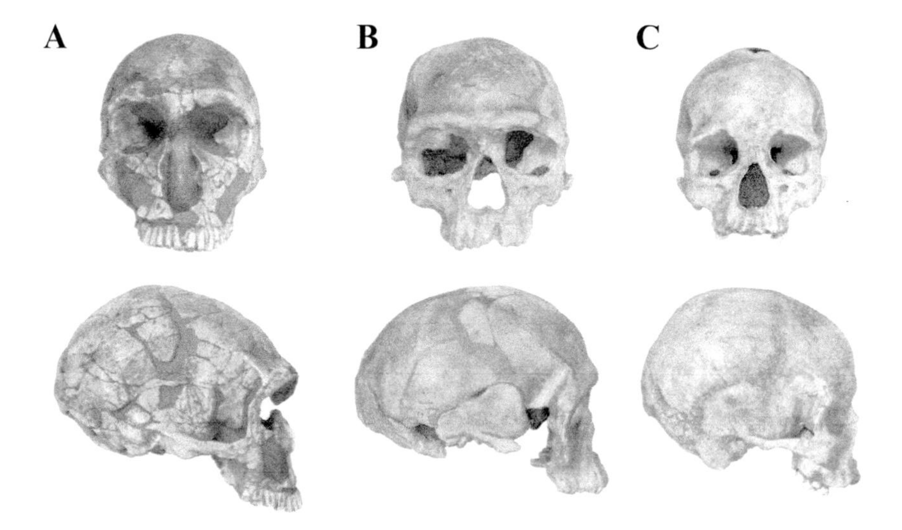

FIGURE 2.1 Jebel Irhoud skull compared with a Neandertal and late *Homo sapiens*

A. An approximately 60,000–40,000-year-old skull from La Ferrassie, France, is an example of a late Neandertal.

B. The facial shape of a Jebel Irhoud fossil previously discovered at the site shows similarities to the structure of more-modern humans, such as the presence of delicate cheekbones. However, the shape of the braincase (the section of the skull enclosing the brain) is archaic in form, and has an elongated shape that is less globular than the structure of more-modern *H. sapiens*.

C. An approximately 20,000-year-old *H. sapiens* fossil.

Source: Modified from Stringer and Galway-Smith 2017, Fig. 1.

Before this radical re-assessment of Jebel Irhoud, the earliest evidence of our species came from Northeast rather than Northwest Africa. Crania from Herto, Middle Awash Valley, Ethiopia, were dated at ca. 160 ka and classed as *Homo sapiens idaltu* (meaning "root") (White et al. 2003). There are also two crania from the Omo Valley, also in Ethiopia, that were dated at ca. 195 ka (McDougall et al. 2005). Of these, Omo 2 has a large but not globular braincase (Neubauer et al. 2018). Other specimens that can be included as early African examples of *H. sapiens* are from Ngaloba (= Laetoli specimen #18) and Eyasi, Tanzania, and Eliye Springs, Kenya (Stringer 2016). At the time, these discoveries supported the idea that our species originated in Northeast Africa, but the new re-appraisals of Jebel Ihoud and Florisbad suggest that this may no longer be the case. In fact, there may not have been a single place of origin of our species: instead, it might have emerged gradually from exchanges with several populations. If so, the search for a "centre" of origin for our species is likely to prove futile. (However, we should also note that the African fossil skeletal record for the period 400–200 ka is poorly dated, and too meagre to allow us to identify specific regions in which *Homo sapiens* may have first developed, especially as vast swathes of Africa – for example, West and Central Africa – have no Middle Pleistocene fossil record.)

"Archaic" versus "modern" Homo sapiens

Until recently, some researchers classified the earliest specimens of *H. sapiens* as "archaic", in contrast to later ones that looked "anatomically modern". (We will see in Chapter 10 that these terms have also been applied to Chinese specimens.) Others have objected on the grounds that these terms of "archaic" and "modern" are not applied to other taxa; for example, no palaeontologist would think it necessary or useful to describe a rodent specimen as "archaic" or "anatomically modern" *Rattus rattus*. The new specimens from Jebel Irhoud and its dating call into question the value of these terms. Because the face could be described as "anatomically modern" but the cranium as "archaic", the specimens clearly exhibit a mixture of features, and we should perhaps think about variability within and between different populations and not whether one specimen is more "primitive" than others. For the sake of clarity, we might distinguish early specimens from later ones on the basis of their dating, but without imposing value-laden terms such as "archaic", "primitive" or "transitional" upon a poorly documented fossil skeletal record. This is borne out by the features of the Late Pleistocene cranium from Iwo Eleru, Nigeria.

Iwo Eleru

Iwo Eleru is a rock shelter in southwest Nigeria that contained a burial that was ca. 11.7–16.3 ka old. What is peculiar about the cranium is that it is more like early examples of *Homo sapiens* such as Omo Kibish II or Qafzeh (Israel) than other Late Pleistocene African individuals. As with the Jebel Irhoud specimens, the Iwo Eleru cranium shows a combination of "modern" features (such as a low supraorbital projection) and "archaic" ones such as a high cranial length (Harvati et al. 2011). This implies that population structure in Africa may be very ancient and that our species emerged from a complex evolutionary process. This specimen also illustrates the futility of trying to classify specimens as either "archaic" or "modern".

Broken Hill and Homo naledi

Another recent discovery is that African hominin populations were surprisingly diverse around the time that some were emerging as *Homo sapiens*. Recent research suggests that the partial skeleton from Broken Hill (Kabwe) in Zambia that is attributed to *H. heidelbergensis* may be only 250–300 ka in age (Buck and Stringer 2014). If so, it overlaps with the earliest representatives of our own species. A completely different discovery of roughly the same age comes from the Rising Star cave, South Africa, where ca. 1,500 bones and teeth have been retrieved from the Naledi and Dinaledi Chambers, deep underground and accessible only via narrow passageways accessible only to experienced (and slender) cavers. These skeletal remains indicate a small type of hominin, now called *H. naledi*, that has features shared by early types of hominin such as *H. habilis* and *A. africanus*, as well as later ones (Berger et al. 2017; Hawks et al. 2017), and a cranial capacity of only ca. 620 cc. There is as yet no evidence that it made stone tools. Surprisingly, *Homo naledi* appears to be much younger than might be assumed from its appearance and is "only" 236–335 ka old (Dirks et al. 2017) – and is thus a contemporary of the earliest members of our own species. Whether they interbred is unclear but perhaps unlikely. Here, its main significance is that the background to our evolution is more complex than thought only a few years ago.

Population structure

The African fossil record for human evolution between 400 ka and 200 ka is poor but shows a considerable degree of diversity, although this may be exaggerated by uncertainties over dating and stratigraphic context. Genetic data also indicate a complex history that may include admixture. It is therefore less certain that our species originated from a single population in a single place of origin, and more likely that we originated and diversified within spatially discrete populations that were connected by sporadic gene flow – in other words, from populations that were structured into metapopulations or palaeodemes (see Chapter 1). (The Iwo Eleru specimen mentioned above provides one example of this.) Phillip Gunz and colleagues (2009) have suggested that these may have been subdivided over time and space, whereby "transient populations are connected by migration, subject to extinction and rebirth by colonization, as well as to fluctuation in local size" (Gunz et al. 2009: 6094). This type of patterning has been called "African multi-regionalism" by

Chris Stringer (2016) as a way of emphasising the importance of inter-connected sub-populations. Recently, Eleanor Scerri and colleagues (including this author) have argued further that our species evolved in Africa in subdivided populations and not from a single continental one that was panmictic, in other words, one in which all sexually active individuals had equal chances of mating (Scerri et al. 2018).This assessment has major implications on how we view the subsequent colonisation of Asia by our species: separate metapopulations might have split or merged, whereas others could have left Africa at different times and by different routes (see Garcea 2012). (Interpretations from genetic data of past demography are also affected if populations were structured and not panmictic [see Mazet et al. 2016].) At this point, we can consider what the archaeological record can inform us about our origins, and about the way these early populations of our species behaved.

The African Middle Stone Age

Jebel Irhoud is unusual among the sites for which there is skeletal evidence for early *Homo sapiens* in that it also contained associated archaeological evidence. This came from layers 6 and 7 from the new excavations, which were dated by thermoluminescence to 302 ± 32 ka and 315 ± 34 ka respectively, with a 95% probability that the layer 7 artefacts are 247–385 ka in age. The stone artefacts were dominated by Levallois technology. Acheulean types of bifaces and Aterian tanged artefacts were absent. These results are consistent with those for other early MSA assemblages in Africa, such as those in East and southern Africa (Richter et al. 2017). The earliest MSA assemblages date back ca. 315 ka at Jebel Irhoud and c. 280 ka at Kapthurin, East Africa and Florisbad, South Africa; data at present do not indicate whether the African MSA pre-dates *Homo sapiens*, or whether it was used only by our species. Its broad characteristics are summarised by Barham and Mitchell (2008: 218–259), and also by McBrearty and Brooks (2000). Their summary diagram of the earliest dates at which some of its elements appear is shown in Figure 2.2. Leaving aside the thorny question of whether the first indication of any of these innovations means that they were used locally, regionally, or at a continental sale, and whether they were used consistently after their first appearance or were short-lived, the value of this figure here is in showing the range of skills deployed by *H. sapiens* after 300 ka.

Our attention now shifts to the archaeological record of Africa north of the Sahara, but first we need to look at the corridors that linked North and sub-Saharan Africa.

A "green Sahara": corridors and barriers

The Sahara is currently a massive barrier that separates North from sub-Saharan Africa. Recent research has produced dramatic evidence that the Sahara has not always been a barrier or as arid as today, but was often crossed by rivers and

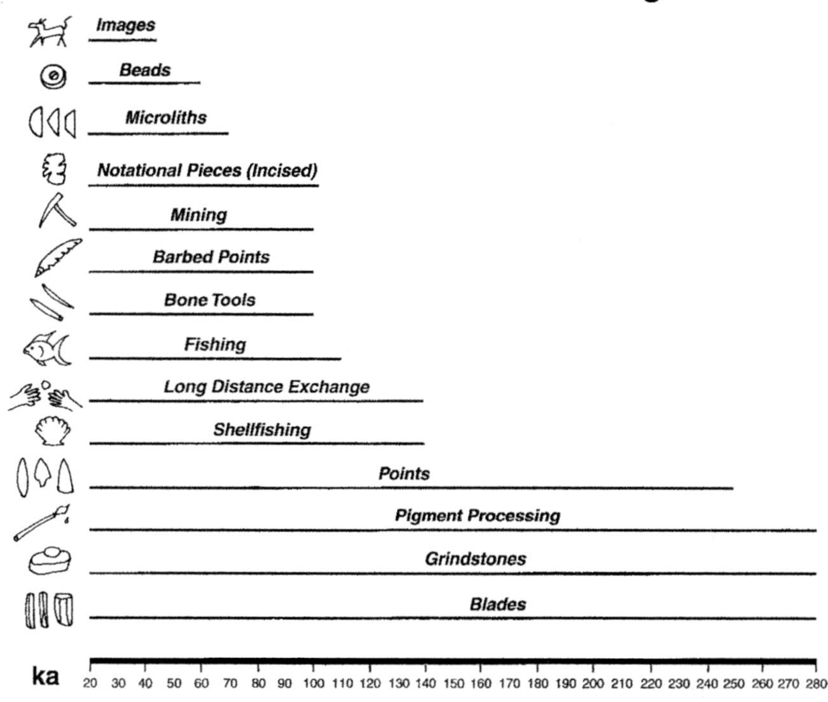

FIGURE 2.2 Behavioural innovations of the African Middle Stone Age
Source: McBrearty and Brooks 2000, Fig. 13.

contained lakes covering thousands of square kilometres (see Figure 2.3). According to Larrasoaña and colleagues (2013), there have been numerous "green Sahara periods" (GSPs) over the last eight million years. These were not trivial episodes: if, as the authors suggest, we assume a generation length (from birth to parenthood) of 15 years and optimal "green" episodes of 4,000–8,000 years, "hundreds of generations of hominins could have thrived in North Africa during each GSP" (Larrasoaña et al. 2013: 9). (This is not to say that the entire Sahara would have been lush green grassland: in "green" periods, lakes formed and rivers were activated but most of the landscape would have been semi-arid; see Scerri et al. 2014.) Larrasoaña and colleagues (2013) note "Subsequent GSP terminations would have caused major population collapses and displacements, and habitat fragmentation with associated potential for isolating small residual populations. These conditions would be conducive to genetic drift and strong selection pressures, possibly to the point of speciation or extinction" (Larrasoaña et al. 2013: 9). Significantly in the context of when humans might have left Africa, the Sahara was probably hyper-arid and depopulated in MIS 6 (Scerri 2017; Scerri et al. 2014). Regarding the last 500,000 years, at least five humid

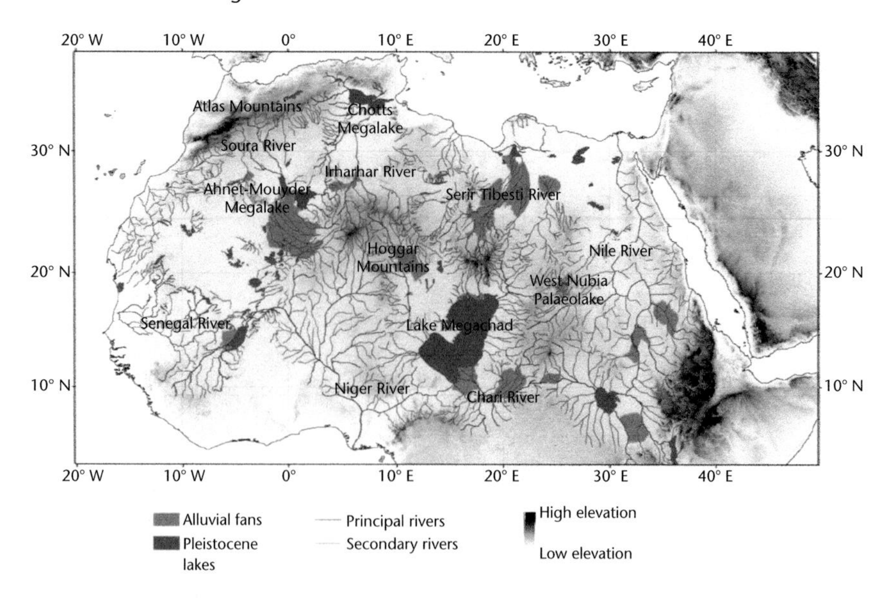

FIGURE 2.3 The "green Sahara"

This shows the topography of North Africa showing the major rivers and lakes that were active in the Sahara during the last interglacial. Under these conditions, there were no major barriers preventing *H. sapiens* (and other animals) from dispersing into Southwest Asia.

Source: Scerri et al. 2014, Fig. 3.

phases can be recognised in the Fazzan Basin, Libya, with the oldest tentatively correlated with marine isotope stage (MIS) 11 (ca. 390–425 ka). The Fazzan Basin also contains evidence of a palaeolake Lake MegaFazzan covering at least 76,250 km^2 (Armitage et al. 2007; Osborne et al. 2008), or more than twice the size of Belgium. Several studies identify major humid phases between 150 ka and 50 ka. Blome and colleagues (2012: 574) report wet conditions between 135 ka and 115 ka in the Egyptian Sahara and across North Africa that was followed by aridity before another wet period from 100 ka to 75 ka. Casteñada et al. (2009: 20159) note wet periods between 50 ka and 45 ka during MIS 3 and between 120–110 ka during MIS 5, and also make the point that these wet phases "coincide with major human migration events out of sub-Saharan Africa". A "green" period that is especially marked occurred during the last interglacial (MIS 5e). Osborne and colleagues (2008: 16446) note from data in the Libyan Sahara "an uninterrupted humid connection between the southern Sahara region and the Mediterranean at ca. 120 ka". More recently, Coulthard et al. (2013) identify three major riverine corridors across the Sahara during the last interglacial. The longest is the Irharhar in the western Sahara that was probably strongly seasonal in its flow. In the eastern Sahara 2,000 km to the west, the Sahabi and Kufrah systems were close to perennial. All three systems would have facilitated movement across the Sahara. In a major synthesis based on ground data and satellite imagery, Drake and colleagues (2011) showed how the Sahara in the last interglacial was crossed by four major drainage systems (including the Nile) and included lakes as large as Lake Megachad (the enlarged ancient version of modern Lake Chad), which in the early Holocene covered ca. 360,000 sq km (an area larger than the British Isles at 315,000 sq km), and which may have been of comparable size in the last interglacial. As Drake et al. (2011) and Scerri et al. (2014) note, these corridors could have been used by hominins to cross the Sahara during the last interglacial (MIS 5e). Because Arabia was also green at this time (see Chapter 3), it would also have been possible for them to enter Southwest Asia. Recently, El-Shenawy et al. (2018) identified from the analysis of a cave speleothem from Wadi Sannur, Egypt, three periods of widespread rainfall at in the Sahara 335 ± 12 ka, 219.4 ± 7.3 ka, and 128.5 ± 1.1 ka, and Lamb and colleagues (2018) identified four wet periods from a core at Lake Tana, Ethiopia, at >150–144 ka (MIS 6), 125–93 ka (MIS 5e–5c), 82–73 ka (MIS 5a), and 42–34 ka (MIS 3). There were clearly several "windows of opportunity" for humans to disperse into the Sahara and leave Africa before 60,000 years ago; we shall see in the following chapters the evidence from Asia that they may have used some of those opportunities.

Stone tool assemblages from the Sahara and North Africa

Our understanding of the Palaeolithic of the Sahara and North Africa has changed as much in recent years as our knowledge of past climatic change in that region. Briefly, the pre-Upper Palaeolithic of the Sahara and North Africa comprises assemblages classified as Aterian (defined largely by tanged points), Mousterian and Nubian (defined by a type of Levallois core reduction for producing pointed flakes), with the cave site of the Haua Fteah in Libya (see below) adding a pre-Aurignacian and a Levallois-Mousterian which were defined by Charles McBurney (1967). The most interesting of these, and one of the most widely distributed across the Sahara west of the Nile is the Aterian. In the last century, this was often regarded as a Late Middle Palaeolithic entity that was considered as a late variant of a North African Mousterian. This is no longer the case. Richter and his team (2010) have shown that the Aterian is not a chronostratigraphic unit; in other words, confined to a particular timeframe. At the cave of Ifri n'Ammar, Morocco, tanged items and personal ornaments are found in a layer dated to 83.3 ± 5.6 ka and above a layer lacking tanged items and dated to 130.0 ± 7.8 ka. However, the underlying layer, dated to 145 ± 9 ka, contains tanged objects (so far, their earliest dated appearance), whilst the underlying layer, dated to 171 ± 12 ka, lacks them. (These age estimates are weighted averages, and not as precise as they might appear; see Richter et al. 2010, Tables 2–5.) These results indicate that tanged stone tools are not confined to one particular period; Scerri (2013a) reached a similar conclusion in her recent analysis of Aterian assemblages.

The distinctiveness of the Aterian from other types of lithic assemblage is also questionable. Dibble and his team, drawing upon information from the excavation of Grotte de Contrebandiers, Morocco and other sites, showed that "it is not possible to distinguish the Aterian and Maghrebian Mousterian on the basis of anything other than the presence or absence of the tanged types"; consequently "the Aterian should no more be viewed as a type of Mousterian with tanged pieces than the Maghrebian Mousterian should be considered as a type of Aterian without those elements" (Dibble et al. 2013: 207). They furthermore propose that in the light of discoveries made since Aterian and North African Mousterian sites were identified, these should be regarded as Maghrebian variants of the African Middle Stone Age, and not as part of a European and Southwest Asian Middle Palaeolithic. This proposal is sensible as Aterian assemblages are associated with the remains of *H. sapiens*, whereas the Middle Palaeolithic in Europe and Southwest Asia is overwhelmingly (but not exclusively – see Chapter 8) associated with Neandertals.

In recent reviews, Scerri (2013b, 2017) has substantially revised assessments of the lithic assemblages from the Sahara and North Africa that are classified variously as Aterian, Mousterian, and as part of the Nubian Complex. This is a difficult undertaking because much of the material was collected by different teams, of different nationalities and with different intellectual traditions, and often working under different colonial administrations. Additionally, lithic industries were defined by different criteria: for example, the Aterian by the presence of finished tools (which often formed a small part of an assemblage), by reduction techniques (as with "Nubian" Levallois cores that are also found outside Nubia, or the lower Nile), or by the type of retouch and similarities with Levantine assemblages (as with some North African assemblages). What she shows is that Aterian assemblages are not uniform across the Sahara; instead, those in the eastern Sahara have more in common with Nubian assemblages in Egypt and the Sudan than with Aterian assemblages in the western Sahara. As for the importance of tanged artefacts, it seems that these are likely to have been a common response to hafting different types of stone artefacts onto a wooden shaft, rather than a distinctive marker of a cultural group. Scerri (2013a: 4244) also notes the suggestion by Wadley and colleagues (2009) that MSA groups in southern Africa used compound resins to create a less brittle glue that lessened the chances of hafted tools shattering when under pressure. A similar technology may have been used by North African MSA populations for hafting tanged artefacts.

The important point that emerges from Scerri's (2013b, 2017; Scerri et al. 2014) syntheses of the Saharan Middle Stone Age assemblages, whether Aterian, Mousterian or Nubian, is the way they can be mapped onto a "green Sahara" (see Figure 2.3): as she points out, these populations had a general sub-Saharan background when they first colonised (and re-colonised) the lake and river systems that crossed the Sahara in moist periods, and then developed their own regional variants.

Because even small reductions in rainfall could have endangered these water supplies, these groups always lived in areas with a high degree of risk: even in moist periods, there was the danger that the rains might fail. When the climate became more arid, they retreated or became locally extinct; in moist episodes they might re-colonise and recombine, but they were at all times tethered to water supplies. They also formed potential source populations that allowed "the possibility of structured groups both to expand and 'push' less structured groups into new territories. In both cases such 'pushes' may have extended out of Africa" (Scerri 2013b: 127). As suggested in Chapter 1, such groups might have been "bold" colonists who were prepared to undertake jump dispersals to find more attractive areas if local resources were diminishing.

The Haua Fteah

The Haua Fteah is an enormous cave with an interior width of 80 m that lies ca. 1 km from the Mediterranean in Cyrenaica, Libya, and on the northern edge of a plateau known as the Gebel Akhdar that lies north of the Sahara. Excavations in the 1950s showed that it contains a 14 m sequence of deposits that span the Middle Palaeolithic to the Neolithic (McBurney 1967). Recent re-examination of the sequence using dating and palaeoenvironmental techniques that were not available in the 1950s has shown that the earliest artefacts at the base of the deposits are probably last interglacial in age (Barker et al. 2012). McBurney (1967) classified the base of the artefact sequence from the bottom of the Deep Sounding to layer XXXV (Douka et al. 2014) as pre-Aurignacian (denoting the presence of blades, burins and points); followed by ones he called Levallois-Mousterian (layers XXV–XXXIV). Neither is found at other sites in North Africa. There are no clear indications of tanged artefacts, although the cave is so large that the excavated area might not have sampled the entire range of assemblages in the cave deposits. The Levallois-Mousterian assemblages were followed by an Upper Palaeolithic series of assemblages that McBurney called the Dabban (layers XVI–XXIV), after the site of ed-Dabba, also in Libya. Recent re-dating of the Haua Fteah sequence indicate that the "pre-Aurignacian" is likely 70.5–80.3 ka; the Levallois-Mousterian layers date from 48.7 to 68.1 ka; and the Dabban from 18.1 to 40 ka (Jacobs et al. 2017). Two mandibles of *Homo sapiens* were found in layer XXXIII and are now dated to 65–73 ka (all dates at a confidence level of 68.2%; Douka et al. 2014). Current re-assessments of the artefact sequence suggest that the Levallois-Mousterian assemblages have closer similarities to assemblages in the Levant or Egypt than the Sahara, and the pre-Aurignacian (which is not very abundant) at the base of the sequence could be simply a variant of the overlying Levallois-Mousterian assemblages (Reynolds 2013). The significance of the Haua Fteah is that it implies that the population history of the Gebel Akhdar is different from that in the Sahara and may be connected with the Levant or Egypt.

Implications of recent African discoveries for dispersals out of Africa

Recent research in this region has important implications on the timing and origin of human dispersals out of Africa into Southwest Asia. One is the apparent contradiction of the emerging picture from Jebel Irhoud that on the one hand, these "pre-modern" forms of *Homo sapiens* are supposed to be pan-African and distributed over the African continent from north to south, yet on the other hand, current skeletal evidence from Asia implies that they never ventured outside Africa until the last interglacial, when *Homo sapiens* is recorded for the first time in the Levant. (Or possibly 177,000–194,000 years ago if the dating of the new find from Misliya Cave [Hershkovitz et al. 2018] or even 210 ka if the new find from Apidima, Greece [Harvati et al. 2019; see Chapter 8] is robust.) Why it appears to have fastidiously avoided Arabia and the Levant for at least 100,000 or perhaps almost 200,000 years is currently inexplicable. It cannot be for environmental reasons, as the Sahara was not always a barrier, nor was Arabia always the inhospitable desert that it is today (see Chapter 4). We know that during the last interglacial, both were "green", with several rivers that provided corridors across both land masses, as well as numerous lakes, and these conditions likely prevailed in the previous interglacials MIS 7 (ca. 200 ka) and MIS 9 (ca. 300–337 ka) (see Larrasoaña et al. 2013). We may yet find that the earliest dispersals of *H. sapiens* out of Africa into the Arabian Peninsula occurred before the last interglacial (MIS 5e) and perhaps as early as MIS 7 (190–230 ka) or even MIS 9 (300–337 ka). Unfortunately, the African evidence for the MSA does not provide any obvious indication of when our species might have left Africa: to quote Chris Tyron and Tyler Faith (2013: 249–50):

> The observed variability of the eastern African MSA record reduces its utility in identifying any sort of archaeological marker for dispersals 'out of Africa'. Rather, it represents the long-term outcome of a series of local adaptations made by Middle and Upper Pleistocene populations that included *Homo sapiens*.

Recent re-evaluation of the Saharan evidence implies that there were several potential source populations with a Middle Stone Age technology that could have spread into the Arabian Peninsula and perhaps the Levant during moist episodes from MIS 9 onwards. The earliest dispersals of our species from Africa are therefore likely embedded in a Middle Stone Age or Middle Palaeolithic context. As Stringer and Gallway-Witham (2017: 213) point out, "corridors on the African periphery might have periodically linked northern Africa and western Asia". We shall return to this possibility in Chapter 3.

The modernity debate, and the origins of "modern" behaviour

No aspect of palaeoanthropology has generated more debate, disagreement and passion than that concerning the origin and unique nature of our behaviour.

The African MSA, and the origins of "modern" behaviour

It is only members of our species that are referred to as "human": males and females become men and women, and immature individuals become children. At what point can or should we use the term "human" instead of "hominin" when describing their skeletal remains? And at what point can we regard their behaviour as "modern" in the implied sense that their cognitive abilities were the same as ours? This is a notoriously difficult question, and one for which there is no clear answer. As Ian Tattersall (2003: 233) remarked, "humanity is to scientists much as pornography is to prosecutors: they know it when they see it, even if they cannot define it".

To take the first of these questions about the skeletal evidence, we have seen that the 300-ka-old individuals from Jebel Irhoud had a "modern" type of face, but a braincase that was elongate and low, and thus more like earlier types of hominin than our own. It is only with the emergence of a rounded braincase after ca. 100,000 years ago (Neubauer et al. 2018) that the cranium appears modern. Hublin and colleagues (2017: 291) suggest that this reshaping of the internal braincase shape "occurred together with genetic changes affecting brain connectivity, organization and development". It is perhaps at this point, when those stages were complete, that "hominins" became "human" in terms of their morphology. Let us now take the second question and assess archaeological evidence for whether the behaviour of *Homo sapiens* during the African Middle Stone Age was "modern".

In a landmark paper, McBrearty and Brookes (2000) argued that "modern" human behaviour originated in the African MSA and did so in a gradual manner over a 200,000-year period. For them there was no eureka moment, no switch that suddenly transformed behaviour from "primitive" to "modern". Instead, they argued, skills emerged gradually, as shown in Figure 2.2. Their argument was in head-on opposition to the theory advocated by Mellars (1989) that there had been an "Upper Palaeolithic Revolution" around 40 ka in Europe. For him, the appearance at or shortly after 40 ka in Europe of blade assemblages, elaborate human burials with grave goods, the use of bone and ivory tools, pendants and other forms of personal ornamentation, art, whether portable (as with figurines) or on cave walls, and musical instruments all showed a "modern" type of behaviour that contrasted sharply with that of Neandertals.

These researchers are major figures because they addressed fundamental aspects of our evolution at a continental level: there are surely few more important questions to ask in palaeoanthropology than "when and how did we become human?" They also demonstrated the potential of archaeological evidence to illuminate human behaviour, and to do more than demonstrate the regional affinities of different artefacts, or issues of chronology. They have often been criticised for relying on "shopping lists" of traits that they consider important indicators of behaviour. Inevitably, given their research experience, these lists were Afrocentric in the case of McBrearty and Brooks, and Eurocentric in the case of Mellars. This criticism is unfair in that no one could be expected to be equally familiar with both continents (and also with Asia and Australia), and inappropriate because any attempt to define a species' behaviour will invoke as many lines of evidence as possible, and these can inevitably be presented as a list. The more appropriate criticism is whether the lists of McBrearty and Brooks, and Mellars successfully identify a suite of behavioural traits that are unique to *Homo sapiens*.

Early Homo sapiens *and Neandertals*

What has emerged in the last few years is that the behavioural skills of *Homo sapiens* in the African Middle Stone Age overlapped considerably with those of Neandertals in Europe (see e.g. Roebroeks and Soressi 2016; Villa and Roebroeks 2014). For example, the MSA inhabitants of Pinnacle Point, South Africa, used fire to modify the flaking properties of silcrete – a technique which requires careful control of temperature and length of treatment (Brown et al. 2009) – yet this is easily matched by the abilities of Neandertals to make adhesives from birch resin (Cârciumaru et al. 2012; Koller et al. 2001), under conditions at least as precise as those used at Pinnacle Point. Likewise, the use of shell beads in the African MSA that is often cited as evidence of symbolism (Bouzouggar et al. 2007; d'Errico et al. 2005) can be matched by similar Neandertal examples from the Cueva de los Aviones (Zilhão et al. 2010) and the use of eagle talons and feathers for ornamentation at Krapina (see e.g. Radovčić et al. 2015), Gorhams Cave, Gibraltar (Finlayson et al. 2012), and other European sites (Finlayson and Finlayson 2016; Peresani et al. 2011; Romandini et al. 2014). Neandertals also used string or cordage (Hardy et al. 2013) and could also (although not often) catch small game such as rabbits (Pelletier et al. 2019). As another example, the simple 75–100 ka engravings on slabs of ochre at Blombos (Henshilwood et al. 2011) can be matched by a similar type of linear engraving at Gorham's Cave, Gibraltar (Rodríguez-Vidal et al. 2014), dated at a minimum of 39 ka (Rodríguez-Vidal et al. 2014).The ochre abstract drawing on a 73,000-year-old slab from Blombos Cave, South Africa (Henshilwood et al. 2018) might be the earliest example of art, but is matched by recent dating of cave art in Spain which shows that Neandertals might also have produced rock art 60,000 years ago (Hoffman et al. 2018a, but see Aubert et al. 2018 for a dissenting view and Hoffman et al. 2018b for a rebuttal). MSA *Homo sapiens* exploited marine resources at Pinnacle Point (Jeradino and Marean 2010), but so did Neandertals in southern Iberia (Cortés-Sánchez et al. 2011). Ochre provides another example: it is evidenced as a possible colourant at South African sites such as Blombos Cave at 100 ka (Henshilwood et al. 2011), but also at Maastricht-Belvedere, the Netherlands, 200–250 ka, where large amounts were probably imported over distances of several kilometres (Roebroeks et al. 2012). Artefacts made on blades are not a specific African innovation, as they were abundant at Qesem, Israel, in a context dating to ca. 400 ka (Shimelmitz et al. 2011). We shall see in Chapter 9 evidence that our species may have learnt the techniques for working hide from Neandertals (Soressi et al.

2013). These overlaps all suggest that the cognitive abilities of *Homo sapiens* in the African MSA and European Neandertals developed along very similar lines up to when Neandertals became extinct ca. 40,000 years ago: different types of large-brained hominins in Africa and Europe were able to undertake many of the same tasks by the Late Middle Pleistocene, and appear to have had similar skill sets (see e.g. Zilhão 2012): as he points out, some modern humans were "behaviorally Neandertal" and some Neandertals were "behaviorally modern" (Zilhão 2006: 191). At present, it seems that early *Homo sapiens* in Africa had no obvious competitive advantage over their Neandertal contemporaries – at least in terms of their material culture. As argued during this book, if our species had an advantage over their contemporaries, it lay more in their colonising and problem-solving abilities than in their material outputs.

The debate over why and how our species replaced Neandertals is predominantly a Eurocentric one. This debate is important, but is only a small part of the story about the success of *H. sapiens* as a colonising species. As I show by the last chapter, most of this story took place in Asia.

The Australian evidence

To compound further the problems of identifying specifically "modern human" traits from archaeological evidence, Habgood and Franklin (2008) undertook the same type of analysis as McBrearty and Brookes did for Africa. The Australian case is particularly interesting because the colonists were almost certainly *Homo sapiens* who arrived ca. 50–55 ka by boat or raft across tens of kilometres of open sea (see O'Connell et al. 2018; and Chapter 7). What Habgood and Franklin showed was that the first Australians brought with them an exceedingly simple lithic technology but gradually innovated, and independently invented grinding stones, ground-stone axes, the use of ochre, ornaments, art and microliths: in other words, they did not arrive with a "package" of "modern" human behaviour that could be identified from artefactual evidence; instead, it was the mental aptitude for technology inside their brains that enabled them to innovate, and adapt to their circumstances. If this is the case, we should not expect to find a close correspondence between "modern" behaviour and sets of artefacts in the Palaeolithic. As Iain Davidson (2013: 14) has pointed out, "recent attempts to show that the early Australians lacked the 'full package' of modern behavior miss the fundamental point that modern behaviour will be patterned, but that there will not be many requirements about what those patterns will be". It is the capacity to pattern behaviour that matters, not whether it is shown in neat packages of material culture.

The "modernity debate" and Asia

How should these debates over the origin and definition of "modern human behaviour" influence assessments of the evidence from Asia after it was colonised by *Homo sapiens*? Should we try to cherry-pick instances of "modern" behaviour in the Asian Palaeolithic? So far, no one has attempted to identify "modernity" in the Asian Palaeolithic, perhaps unsurprisingly given the size and diversity of Asia. Additionally, as we shall see in the following chapters, the archaeological evidence from across much of Palaeolithic Asia is so poor that it is often hard enough to infer which type of hominin was responsible for it, let alone decide whether its makers behaved in a "modern" or "archaic" manner, however the behaviour of either type might be defined. In my mind, there are two good reasons why the search for "modernity" from artefactual records is flawed. The first is that indicators of "modern" behaviour that are devised for Africa, Europe, or Australia have so far failed to identify any trait or set of traits that is unique to our species, universal in occurrence, and identifiable in the archaeological record. There is no reason to expect any different from Asia. In any case, we need to be aware of the dangers of regionalism. As Shea (2011: 9) points out:

> If, for example, the first archaeologists had been Polynesians, the important hallmarks of behavioral modernity as they conceptualized it might include ocean-going watercraft, celestial navigation skills, pelagic fishing, hunting marine mammals, horticulture, domesticated pigs and dogs, ceramics, edge-ground stone axes, monumental architecture, and feather cloaks … Our Polynesian prehistorians would probably regard carved antler tools, cave art, and prismatic blade production as quaint local phenomena of no obvious evolutionary significance.

The second and more important reason is that pigeon-holing archaeological features as "modern" or "archaic" can easily distract us from thinking about the underlying adaptability and variability that produced that particular set of features. To quote Shea (2011: 15) again, "Calling a particular behavior 'modern' clarifies neither the occurrence of evidence for that behavior in a given context nor the sources of its variability. It merely adds a layer of interpretation between ourselves and observations of a particular phenomenon." Instead, he argues that it is better to examine the costs, benefits, risks and consequences of behaving in a particular way. In an Asian context, that approach has much to offer when considering different type of climates and landscapes that humans encountered when colonising Asia. If we can understand better these factors, we might then be in a better position to decide how and why *Homo sapiens* differed from its contemporaries, whether or not we decide arbitrarily to class them as "modern" or "archaic". What I hope will become clearer by the end of this book is that the Asian record can illuminate some of the key features that made us different from Neandertals and the other indigenous populations in Asia prior to their extinction.

References

Armitage, S.J., Drake, N.A., Stokes, S., El-Hawat, A., Salem, M.J., White, K., Turner, P. and McLaren, S.J. (2007) Multiple phases of North African humidity recorded in lacustrine sediments from the Fazzan Basin, Libyan Sahara. *Quaternary Geochronology*, 2: 181–186.

Aubert, M., Brumm, A. and Huntley, J. (2018) Early dates for "Neanderthal cave art" may be wrong. *Journal of Human Evolution*, 125: 215–217.

Barham, L. and Mitchell, P. (2008) *The First Africans: African Archaeology from the Earliest Toolmakers to Most Recent Foragers*. Cambridge: Cambridge University Press.

Barker, G., Bennett, P., Farr, L., Hill, E., Hunt, C. et al. (2012) The Cyrenaican Prehistory Project 2012: the fifth season of investigations of the Haua Fteah cave. *Libyan Studies*, 43: 115–136.

Berger, L.R., Hawks, J., Dirks, H.G.M., Elliot, M. and Roberts, E.M. (2017) *Homo naledi* and Pleistocene hominin evolution in subequatorial Africa. *eLife*, 6: e24234, doi:10.7554/eLife.24234.

Blome, M.W., Cohen, A.S, Tryon, C.A., Brooks, A.S. and Russell, J. (2012) The environmental context for the origins of modern human diversity: a synthesis of regional variability in African climate 150,000–30,000 years ago. *Journal of Human Evolution*, 62(5): 563–592.

Bouzouggar, A., Barton, N., Vanhaeren, M., d'Errico, F., Collcutt, S. et al. (2007) 82,000-year-old shell beads from North Africa and implications for the origins of modern human behavior. *Proceedings of the National Academy of Sciences USA*, 104: 9964–9969.

Brown, K.S., Marean, C.W., Herries, A.R., Jacobs, Z., Tribolo, C. et al. (2009) Fire as an engineering tool of early modern humans. *Science*, 325: 859–862.

Buck, L.T. and Stringer, C.B. (2014) *Homo heidelbergensis*. *Current Biology*, 24: R214–215.

Cârciumaru, M., Ion, R.-M., Niţtu, E.-C. and Ştefânescu, R. (2012) New evidence of adhesive as hafting material on Middle and Upper Palaeolithic artefacts from Gura Cheii Râşsnov Cave (Romania). *Journal of Archaeological Science*, 39: 1942–1950.

Casteñada, I.S., Mulitz, S., Schefuß, E., Lopes dos Santos, R.A., Sinninghe Damste, J.S. and Schouten, S. (2009) Wet phases in the Sahara/ Sahel region and human migration patterns in North Africa. *Proceedings of the National Academy of Sciences USA*, 106: 20159–20163.

Cortés-Sánchez, M., Morales-Muñiz, A., Simón-Vallejo, M.D., Lozano-Francisco, M.C., Vera-Peláez, J.L., Finlayson, C. et al. (2011) Earliest known use of marine resources by Neanderthals. *PLoS ONE*, 6(9): e24026, https://doi.org/10.1371/journal.pone.0024026.

Coulthard, T.J., Ramirez, J.A., Barton, N., Rogerson, M. and Brücher, T. (2013) Were rivers flowing across the Sahara during the last Interglacial? Implications for human migration through Africa. *PLoS One*, 8(9): e74834, doi:10.1371/journal.pone.0074834.

Davidson, I. (2013) Guest Editorial: peopling the last new worlds: the first colonisation of Sahul and the Americas. *Quaternary International*, 285, 1–29.

d'Errico, F., Henshilwood, C., Vanhaeren and Niekerk, K. van (2005) *Nassarius kraussianus* shell beads from Blombos Cave: evidence for symbolic behaviour in the Middle Stone Age. *Journal of Human Evolution*, 48(1): 3–24.

Dibble, H.L., Aldeias, V., Jacobs, Z., Olszewski, D.I., Rezek, Z. et al. (2013) On the industrial attributions of the Aterian and Mousterian of the Maghreb. *Journal of Human Evolution*, 64: 194–210.

Dirks, P.H.G.M., Roberts, E.M., Hilbert-Wolf, H., Kramers, J.D., Hawks, J. et al. (2017) The age of *Homo naledi* and associated sediments in the Rising Star Cave, South Africa. *eLife*, 6: e24231, doi:10.7554/eLife.24231.

Douka, K., Jacobs, Z., Lane, C., Grün, R., Farr, L. et al. (2014) The chronostratigraphy of the Haua Fteah cave (Cyrenaica, northeast Libya). *Journal of Human Evolution*, 66: 39–63.

Drake, N.A., Blench, R.M., Armitage, S.J., Bristow, C.S. and White, K.H. (2011) Ancient watercourses and biogeography of the Sahara explain the peopling of the desert. *Proceedings of the National Academy of Sciences USA*, 108: 458–462.

El-Shenawy, M.I., Kim, S.-T., Schwarcz, H., Asmeron, Y. and Polyak, V.J. (2018) Speleothem evidence for the greening of the Sahara and its implications for the early human dispersal out of sub-Saharan Africa. *Quaternary Science Reviews*, 188: 67–76.

Finlayson, C., Brown, K., Blasco, R., Rosell, J., Negro, J.J. et al. (2012) Birds of a feather: Neanderthal exploitation of raptors and corvids. *PLoS One*, 7(9): e45927.

Finlayson, S. and Finlayson, C. (2016) The birdmen of the Pleistocene: on the relationship between Neanderthals and scavenging birds. *Quaternary International*, 421: 78–84.

Garcea, E. (2012) Successes and failures of human dispersals from North Africa. *Quaternary International*, 270: 119–128.

Grün, R., Brink, J.S., Spooner, N.A., Taylor, L., Stringer, C.B. et al. (1996) Direct dating of Florisbad hominid. *Nature*, 382: 500–501.

Gunz, P., Bookstein, F.L., Mitteroecker, P., Stadlmayr, A.Q., Seidler, H. and Weber, G. W. (2009) Early modern human diversity suggests subdivided population structure and a complex out-of-Africa scenario. *Proceedings of the National Academy of Sciences USA*, 106: 6094–6098.

Habgood, P.J. and Franklin, N.R. (2008) The revolution that didn't arrive: a review of Pleistocene Sahul. *Journal of Human Evolution*, 55: 187–222.

Hardy, B.L., Moncel, M.-H., Daujeard, C., Fernandes, P., Béarez, P. et al. (2013) Impossible Neanderthals? Making string, throwing projectiles and catching small game during Marine Isotope Stage 4 (Abri du Maras, France). *Quaternary Science Reviews*, 82: 23–40.

Harvati, K., Röding, C., Bosman, A.M., Karakostis, F.A., Grün, R. et al. (2019) Apidima Cave fossils provide earliest evidence of *Homo sapiens* in Eurasia. *Nature*, 571: 500–504, doi:10.1038/s41586-019-1376-z.

Harvati, K., Stringer, C., Grün, R., Aubert, M., Allsworth-Jones, P. and Folorunso, C.A. (2011) The Later Stone Age calvaria from Iwo Eleru, Nigeria: morphology and chronology. *PLoS ONE*, 6(9): e24024, doi:10.1371/journal.pone.0024024.

Hawks, J., Elliott, M., Schmid, P., Churchill, S.E., de Ruiter, D.J. et al. (2017) New fossil remains of *Homo naledi* from the Lesedi Chamber, South Africa. *eLife*, 6: e24232, doi:10.7554/eLife.24232.

Henshilwood, C., d'Errico, F., Niekerk, K.L. van, Coquinot, Y., Jacobs, Z., Lauritzen, S.-E., Menu, M. and García-Moreno, R. (2011) A 100,000-year-old ochre-processing workshop at Blombos Cave, South Africa. *Science*, 334: 219–222, doi:10.1126/science.1211535.

Henshilwood, C.S., d'Errico, F., Niekerk, K.L. van, Dayet, L., Queffelec, A. and Pollarolo, L. (2018) An abstract drawing from the 73,000-year-old levels at Blombos Cave, South Africa. *Nature*, 562: 115–118.

Hershkovitz, I., Weber, G.W., Quam, R., Duval, M., Grün, R. et al. (2018) The earliest modern humans outside Africa. *Science*, 359: 456–459.

Hoffman, D.L., Standish, C.D., García-Diez, M., Pettitt, P.B., Milton, J.A. et al. (2018a) U-Th dating of carbonate crusts reveals Neandertal origin of Iberian cave art. *Science*, 359: 912–915.

Hoffman, D.L., Standish, C.D., Garcia-Diez, M., Pettitt, P.B., Milton, J.A. et al. (2018b) Response to Aubert et al.'s reply "Early dates for 'Neanderthal cave art' may be wrong" [*Journal of Human Evolution*, 125: 215–217]. *Journal of Human Evolution*, 135 (2019) 102644: 1–5.

Hublin, J.-J., Ben-Neer, A., Bailey, S., Freidline, S.E., Neubauer, S. et al. (2017) New fossils from Jebel Irhoud, Morocco and the pan-African origin of *Homo sapiens*. *Nature*, 546: 289–292.

Jacobs, Z., Li, B., Farr, L., Hill, E., Hunt, C. et al. (2017) The chronostratigraphy of the Haua Fteah cave (Cyrenaica, northeast Libya) – optical dating of early human occupation during Marine Isotope Stages 4, 5 and 6. *Journal of Human Evolution*, 105: 69–88.

Jeradino, A. and Marean, C.W. (2010) Shellfish gathering, marine paleoecology and modern human behavior: perspectives from cave PP13B, Pinnacle Point, South Africa. *Journal of Human Evolution*, 59: 412–424.

Koller, J., Baumer, U. and Mania, D. (2001) Hi-tech in the Middle Palaeolithic: Neandertal-manufactured pitch identified. *European Journal of Archaeology*, 4(3): 385–397.

Lamb, H.F., Bates, C.R., Bryant, C.L., Davies, S.J., Huws, D.G., Marshall, M.H., Roberts, H. M. and Toland, H. (2018) 150,000-year palaeoclimate record from northern Ethiopia supports early, multiple dispersals of modern humans from Africa. *Scientific Reports*, 8: 1077, doi:10.1038/s41598-018-19601-w.

Larrasoaña, J.C., Roberts, A.P. and Rohling, E.J. (2013) Dynamics of Green Sahara periods and their role in hominin evolution. *PLoS One*, 8(10): e76514.

McBrearty, S. and Brooks, A.S. (2000) The revolution that wasn't: a new interpretation of the origin of modern human behaviour. *Journal of Human Evolution*, 39(5): 453–563.

McBurney, C.B.M. (1967) *The Haua Fteah in Cyrenaica and the Stone Age of the South-East Mediterranean*. Cambridge: Cambridge University Press.

McDougall, I., Brown, F.H. and Fleagle, J.G., (2005) Stratigraphic placement and age of modern humans from Kibish, Ethiopia. *Nature*, 433: 733–736.

Mazet, O., Rodríguez, W., Grusea, S., Boitard, S. and Chikhi, L. (2016) On the importance of being structured: instantaneous coalescence rates and human evolution – lessons for ancestral population size inference? *Heredity*, 116: 362–371.

Mellars, P. (1989) Major issues in the emergence of modern humans. *Current Anthropology*, 30(3): 349–385.

Neubauer, S., Hublin, J.-J. and Gunz, P. (2018) The evolution of modern human brain shape. *Science Advances*, 4: eaao5961.

O'Connell, J.F., Allen, J., Williams, M.A.J., Williams, A.N., Turney, C.S.M., Spooner, N., Kamminga, J., Brown, G. and Cooper, A. (2018) When did *Homo sapiens* first reach Southeast Asia and Sahul? *Proceedings of the National Academy of Sciences USA*, 115(34): 8482–8490, doi:10.1073/pnas.1808385115.

Osborne, A.H., Vance, D., Rohling, E.J, Barton, N., Rogerson, M. and Fello, N. (2008) A humid corridor across the Sahara for the migration of early modern humans out of Africa 120,000 years ago. *Proceedings of the National Academy of Sciences USA*, 105: 16444–16447, doi:10.1073/pnas.0804472105.

Pelletier, M., Desclaux, E., Brugal, J.-P. and Texier, P.-J. (2019) The exploitation of rabbits for food and pelts by last interglacial Neandertals. *Quaternary Science Reviews*, 224: 105972, doi:10.1016/j.quascirev.2019.105972.

Peresani, M., Flore, I., Galaa, M., Romandini, M. and Tagliacozzo, A. (2011) Late Neandertals and the intentional removal of feathers as evidenced from bird bone taphonomy at Fumane Cave 44 ky B.P. *Proceedings of the National Academy of Sciences USA*, 108: 3888–3893.

Radovčić, D., Sršen, A.O., Radovčić, J. and Frayer, D.W. (2015) Evidence for Neandertal jewelry: modified white-tailed eagle claws at Krapina. *PLoS ONE*, 10(3): e0119802, doi:10.1371/journal.pone.0119802.

Reynolds, T. (2013) The Middle Palaeolithic of Cyrenaica: is there an Aterian at the Haua Fteah and does it matter? *Quaternary International*, 300: 171–181.

Richter, D., Grün, R., Joannes-Boyau, R., Steele, T.E., Amani, F. et al. (2017) The age of the hominin fossils from Jebel Irhoud, Morocco, and the origins of the Middle Stone Age. *Nature*, 546: 293–296.

Richter, D., Moser, J., Nami, M., Eiwanger, J. and Mikdad, A. (2010) New chronometric data from Ifri n'Ammar (Morocco) and the chronostratigraphy of the Middle Palaeolithic in the Western Maghreb. *Journal of Human Evolution*, 59: 672–679.

Rodríguez-Vidal, J., d'Errico, F., Giles Pacheco, F., Blasco, R., Rosell, J. et al. (2014) A rock engraving made by Neanderthals in Gibraltar. *Proceedings of the National Academy of Sciences USA*, 111: 13301–13306.

Roebroeks, W., Sier, M., Nielsen, T.K., de Loecker, D., Parés, J.P., Arps, C.E.S. and Mücher, H.J. (2012) Use of red ochre by early Neandertals. *Proceedings of the National Academy of Sciences USA*, 109(6): 1889–1894, doi:10.1073/pnas.1112261109.

Roebroeks, W. and Soressi, M. (2016) Neandertals revised. *Proceedings of the National Academy of Sciences USA*, 113: 6372–6379.

Romandini, M., Peresani, M., Laroulandie, V., Metz, L., Pastoor, A., Vaqueros, M. and Slimak, L. (2014) Convergent evidence of eagle talons used by late Neanderthals in Europe: a further assessment on symbolism. *PLoS One*, 9(7): e101278.

Scerri, E.M.L. (2013a) On the spatial and technological organisation of hafting modifications in the North African Middle Stone Age. *Journal of Archaeological Science*, 40: 4234–4248.

Scerri, E.M.L. (2013b) The Aterian and its place in the North African Middle Stone Age. *Quaternary International*, 300: 111–130.

Scerri, E.M.L. (2017) The North African Middle Stone Age and its place in recent human evolution. *Evolutionary Anthropology*, 26(3): 119–135.

Scerri. E.M.L., Drake, N.A., Jennings, R. and Groucutt, H.S. (2014) Earliest evidence for the structure of *Homo sapiens* populations in Africa. *Quaternary Science Reviews*, 101: 207–216.

Scerri, E.M.L., Thomas, M.G., Manica, A., Gunz, P., Stock, J. et al. (2018) Did our species evolve in subdivided populations across Africa, and why does it matter? *Trends in Ecology and Evolution*, 33(8): 582–594.

Schlebusch, C.S. Malmström, H., Günther, T., Sjödin, P., Coutinho, A. et al. (2017) Ancient genomes from southern Africa pushes modern human divergence beyond 260,000 years ago. *bioRxi*, doi:10.1101/145409.

Shea, J.J. (2011) *Homo sapiens* is as *Homo sapiens* was: behavioral variability versus "Behavioral Modernity" in paleolithic archaeology. *Current Anthropology*, 52(1): 1–35.

Shimelmitz, R., Barkai, R. and Gopher, V. (2011) Systematic blade production at late Lower Paleolithic (400–200 kyr) Qesem Cave, Israel. *Journal of Human Evolution*, 61: 458–479.

Smith, T.M., Tafforeau, P., Reid, D.J., Grun, R., Eggins, S., Boutakiout, M. and Hublin, J.-J. (2007) Earliest evidence of modern human life history in North African early *Homo sapiens*. *Proceedings of the National Academy of Sciences USA*, 104(15): 6128–6133.

Soressi, M., McPherron, S.P., Lenoir, M., Dogand, T., Goldberg, P. et al. (2013) Neandertals made the first specialized bone tools in Europe. *Proceedings of the National Academy of Sciences USA*, 110: 14186–14190.

Stringer, C. (2016) The origin and evolution of *Homo sapiens*. *Philosophical Transactions of the Royal Society Series B*, 371: 20150237, doi:10.1098/rstb.2015.0237.

Stringer, C. and Galway-Witham, J. (2017) On the origin of our species. *Nature*, 546: 212–213.

Tattersall, I. (2003) Response to three roots of human recency molecular anthropology, the refigured Acheulean, and the UNESCO response to Auschwitz by R.N. Proctor. *Current Anthropology*, 44(2): 213–239.

Tyron, C.A. and Faith, J.T. (2013) Variability in the Middle Stone Age of Eastern Africa. *Current Anthropology*, 54(Supplement 8): S234–S254.

Villa, P. and Roebroeks, W. (2014) Neandertal demise: an archaeological analysis of the modern human superiority complex. *PLoS One*, 9(4): e96424, doi:10.1371/journal.pone.0096424.

Wadley, L., Hodgskiss, T. and Grant, M. (2009) Implications for complex cognition from the hafting of tools with compound adhesives in the Middle Stone Age, South Africa. *Proceedings of the National Academy of Sciences USA*, 106: 9590–9594.

White, T.D., Asfaw, B., DeGusta, D., Tilbert, H., Richards, G.D., Suwa, G. and Howell, F.C. (2003) Pleistocene *Homo sapiens* from Middle Awash, Ethiopia. *Nature*, 423: 742–747.

Zilhão, J. (2006) Neandertals and moderns mixed, and it matters. *Evolutionary Anthropology*, 15: 183–195.

Zilhão, J. (2012) Personal ornaments and symbolism among the Neanderthals. *Developments in Quaternary Science*, 16: 35–49.

Zilhão, J., Angelucci, D., Badal-García, E., d'Errico, F., Daniel, F. et al. (2010) Symbolic use of marine shells and mineral pigments by Iberian Neandertals. *Proceedings of the National Academy of Sciences USA*, 107(3): 1023–1028.

3

THE CLIMATIC AND ENVIRONMENTAL BACKGROUND TO THE HUMAN COLONISATION OF ASIA

Introduction: Asian climate

There are three main climate systems in Asia. The first affects western Asia and is driven by westerly winds from the Mediterranean as well as the Black and Caspian Seas across Southwest and Central Asia. Here, most rainfall falls in winter and spring. Inland from the coast of northern Israel, Lebanon, and western Turkey (where rainfall can reach 1,000 mm), rainfall totals decrease dramatically, and almost all of the Arabian Peninsula and much of Syria, Iraq, Iran (particularly on the Iranian Plateau) and Central Asia is semidesert or desert, with rainfall as low as <50 mm/year. Exceptions where rainfall is higher are the Caucasian, Taurus, and Zagros Mountains of the Caucasus region, eastern Turkey, and western Iran, respectively, and the southern coasts of the Black and Caspian Seas. The second and more important Asian weather system is the monsoon. This has two components – the Indian and East Asian monsoons. These operate as vast conveyor belts, transferring cold, dry air from northern Asia to the equatorial parts of the Indian and Pacific Oceans in the winter monsoon, and then warm, moist air back across the south and east parts of the Asian landmass in the summer monsoon. A distinctive feature of the monsoon over an area much larger than that affected by its summer rainfall is the alternation of prevailing winds,[1] with nearly opposite directions between the winter and summer seasons. In summer, southerly and southwest winds carry huge amounts of moisture from the Indian Ocean over the Indian subcontinent and Myanmar. The East Asian monsoon originates in the South China Sea and western Pacific Ocean and carries rain to Southeast Asia and China. At its southern limit, it interacts with the Australian monsoon, which is chiefly active over northern Australia (Liu and Ding 1998). There is also an African monsoon that affects the southern part of the Arabian Peninsula (Chapter 4). The third weather system affects Siberia and the Arctic, and is strongly continental, with short, often hot summers and long, bitterly cold winters.

Climatically, Asia has some of the greatest contrasts on the planet. A ridiculous 26,461 mm (86.75 feet) of rain fell at Cherrapunji in northeast India in 1860–1861, including 9,300 mm (30 feet) in one month (Guinness World Records 2004: 68), which qualifies it as the wettest place ever recorded,[2] yet several of Asia's deserts – such as the Rub'al Khali of Arabia and the Taklamakan of north China – are among the most arid and hottest parts of the world. At Verkhoyansk, eastern Siberia, winter temperatures can fall to –68° C. (–90° F.), thus making it the coldest part of the northern hemisphere; because summer temperatures there can exceed 37° C., its annual temperature range of 105° C. is the widest in the world (Guinness World Records 2004: 68). Siberia, covering one-twelfth of the earth's land surface, has permafrost up to 1.5 km deep (Tumel 2002: 149).

Asian climate, 300,000–30,000 years ago

As the origin of our species in Africa now extends beyond 300,000 years ago (see previous chapter), we can use the last half million years as the baseline of a framework of climatic change (see Figure 3.1). The climatic stages of these records are defined worldwide by the marine isotope stages (MIS) derived from analyses of marine sediment cores, supplemented by loess and palaeosol sequences in China, cores from the Greenland and Antarctic ice-sheets as well as analyses of pollen profiles, microfauna, cave spleothems and other sources. For those unfamiliar with these, cold stages have even numbers (e.g. MIS 2, MIS 4, etc.) and warm episodes have odd numbers (e.g. MIS 1 [the present interglacial], MIS 3, etc.), and each of these can be further subdivided. For our purposes, MIS 6 was the penultimate cold phase (equivalent to a glaciation in northern Europe and North America), MIS 5 for the last interglacial (which is also subdivided into sub-stages 5e, the earliest, to 5a), and MIS 4 and 3 occurred between 75,000 and 25,000 years ago during the last glacial cycle. The Last Glacial Maximum is defined by MIS 2, when ice sheets were at their largest between ca. 21,000 and 18,000 years ago, and sea levels as much as 100–120 m below the present, thus exposing a vast amount of land (see Chapter 6). The dates of these stages vary by region and choice of data set, but approximate

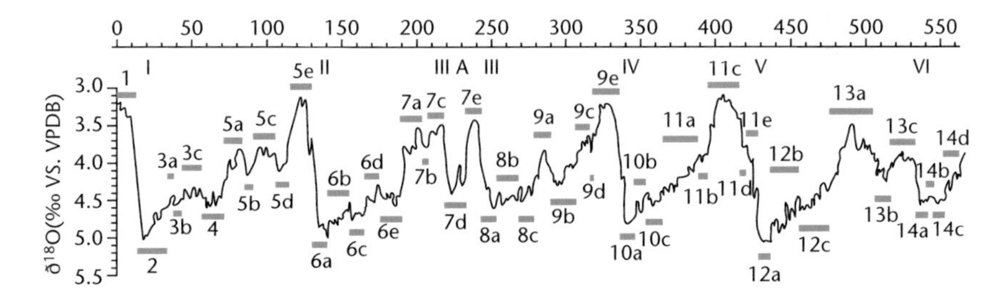

FIGURE 3.1 Marine isotope stages (MIS) of the last 550,000 years and their subdivisions Even numbers (MIS 2, MIS 4, etc.) indicate cold, dry periods and odd numbers (MIS 3, MIS 5, etc.) indicate warmer and moister ones.

Source: Gibbard and Lewin 2016, Fig. 1b.

ones are shown in Figure 3.2. For most of the time that humans were dispersing across and colonising Asia, the climate was not only colder and drier than today, but unstable. The key variable across most of Asia in the last glacial cycle was rainfall, particularly in regions that were arid or semi-arid. In South, Southeast and East Asia, the critical factors were the relative strengths of the winter and summer monsoons, as these determined the amount of rainfall and the boundary between the two. In East Asia (Chapter 10), the summer monsoon was able to penetrate further north in warm, moist periods, but retreated in colder and drier episodes when the winter monsoon was dominant.

There were also frequent climatic oscillations that operated on time-scales of a few centuries or even decades (and thus within the experience of a human generation). The greatest variability in rainfall occurs in arid and semi-arid regions, with variations of up to 50% from year to year. Figure 3.3 shows a 6,000-year record from the Holocene of northeast China as an example of rainfall variability from an arid and semi-arid region of Asia. Because of the variable nature of rainfall in these regions, they were high-risk for human residents and colonisers. Seasonal scarcity in resource availability is not necessarily a problem if it is predictable, but "Unpredictable variation in

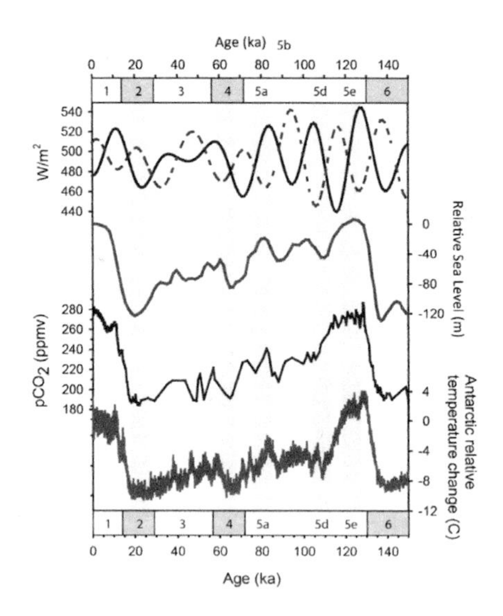

FIGURE 3.2 The marine isotope stages (MIS 1–MIS 6) of the last 150,000 years

The general trends are that MIS 2 (the Last Glacial Maximum) and MIS 6 (the penultimate glaciation) were extremely cold; the last interglacial (MIS 5e) was as warm as the present MIS 1; in MIS 3 and MIS 4 (when most of the expansion of our species across Asia occurred), the earth's climate was variable but usually cooler than at present.

Key: A (top): June insolation at 60° N. (unbroken line) and December insolation at 60° S. (dashed line); (B) relative sea level reconstructed from benthic foraminiferal oxygen isotope records; (C) atmospheric carbon dioxide concentration and (D) (bottom) ice core temperatures reconstructed from Antarctica (EPICA Dome C, grey). Marine isotope stages (MIS) 5e, 5d, 5a, 4, 3, 2 and 1 are indicated at the top and bottom of the figure; shading indicates the 10,000-yr time periods over which characteristics are averaged.

Source: Kohfeld and Chase 2017, Fig. 1.

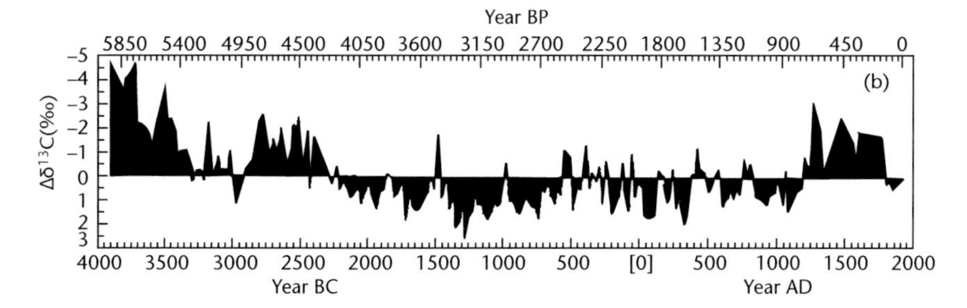

FIGURE 3.3 Rainfall variability over the last 6,000 years in a semi-arid landscape of northeast Asia

Source: Hong et al. 2011, Fig. 1.

the distribution of resources … constitutes ecological risk since it means the outcome of foraging behaviour is uncertain" (Burke et al. 2017: 717). We shall see later how humans might have responded to this type of risk.

In addition, there were cold, short and sudden episodes in the last glacial cycle known as Heinrich Events that are denoted by increased calving of icebergs in the North Atlantic and short-lived but extensive advances of pack ice. The ones that probably most affected human dispersal across continental Asia were Heinrich Events 4, 5 and 6 (ca. 38 ka, 48 ka, ~60 ka respectively) (Blunier and Brook 2001); these are detectable in the east Mediterranean (Bartov et al. 2003), the Arabian Sea (Schulz et al. 1998), southern (Wang et al. 2001) and central China (Porter and An 1995) as well as Greenland (Hemming 2004). Their effects would have been most strongly felt at the northern limits of the human range, north of 40° N. (roughly from Beijing to Tbilisi, Georgia, and Rome), and in semi-arid environments that were vulnerable to reductions in rainfall and winter temperature.

Asian biogeography

In 1876, Alfred Wallace – the father of biogeography – divided the world into biogeographic realms based on the types of animals and plants in each region: an Ethiopian Realm for Africa, a Palearctic Realm for continental Eurasia, an Oriental Realm for South and Southeast Asia, and an Australian Realm. Since then, there have been numerous variants of his scheme, and the latest is by Holt and colleagues (2013), who were able to draw upon a vastly larger dataset than was at Wallace's disposal. Because my main interests are humans and the animals that they hunted over tens of millennia during the last 100,000–200,000 years, my preferred scheme is shown in Figure 3.4. Following Holt et al. (2013), I recognise an Afro-Arabian Realm that includes the Arabian Peninsula and southern Iran and Pakistan as far east as the Thar Desert of northwest India (see Chapter 4). The Oriental Realm encompasses monsoonal South and Southeast Asia (Chapters 5 and 6), but I treat Wallacea as a separate Realm in its own

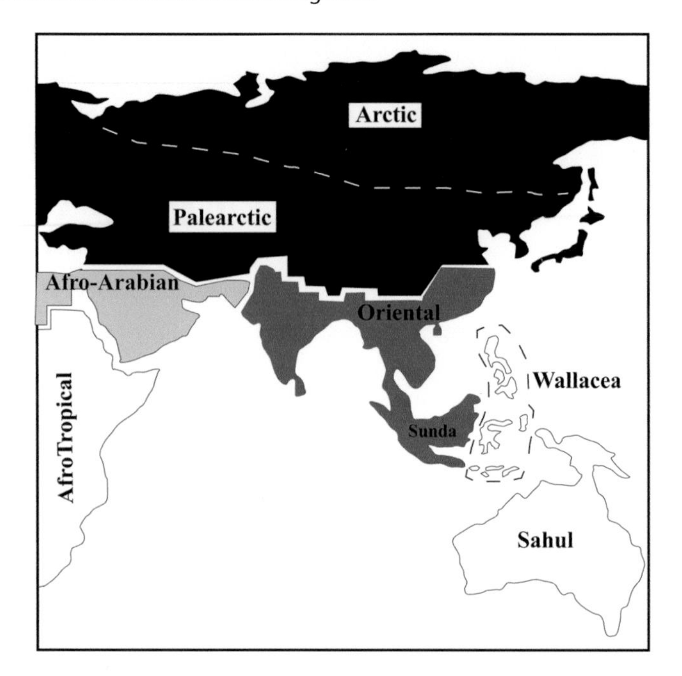

FIGURE 3.4 Biogeographic realms of Asia during MIS 3 and MIS 4

Sea levels are shown at 40–60 m below present levels for the Arabian-Persian Gulf, Sunda and Sahul.

Source: Redrawn and modified from Holt et al. 2013, Fig. 1.

right and not as simply part of island Southeast Asia (see Chapter 7). Because the boundary between the Palearctic and Oriental Realms in East Asia frequently shifted during the Pleistocene (Chapter 10), I ignore the Sino-Japanese Realm of Holt et al. (2013). The Palearctic Region covers the whole of Europe and continental Asia (Chapter 9), with an Arctic sub-Realm across northern Siberia where annual and winter temperatures are much lower (Chapter 11). The boundaries of these realms – and particularly the southern edge of the Palearctic Realm – fluctuated considerably during the last glacial cycle.

Geographic considerations

Regarding human dispersal across Asia, simple maps showing known sites and dates mask the most important topographic and climatic factors that most likely affected the rate and direction of dispersal.

Topographic factors

In idealised circumstances, there are no impediments to a colonising species dispersing into a new environment, and occupying the entire space at a uniform rate. In reality, all environments offer different degrees of "friction" that impede, direct and modify the process of colonisation. With the colonisation of Asia by our species, the humans that dispersed across Asia and ultimately to North America and Australia had to traverse and inhabit an immense variety of landscapes. Some were occupied more or less continuously after colonisation, others less so, and some rarely if at all. Following Smith's (2013) discussion of the colonisation of Australia, we can begin by recognising three types of terrain relevant to a dispersing species: refugia, barriers and corridors.

Refugia, barriers and corridors

The key point about these categories is that they are not immutable but dependent upon two factors. The first is the prevailing climate, as under some climatic circumstances, barriers could become corridors (or vice versa). The second and arguably more important is that through changes in behaviour, *H. sapiens* was able to find refugia in climatic downturns in areas that earlier hominins would have found uninhabitable, and colonise or use as corridors parts of the Asian landscape that had previously been barriers.

Refugia

Refugia are areas that could still be occupied at times of maximum environmental stress, and without them, a species would become regionally extinct because it had nowhere else to live. Refugia are also important as genetic bottlenecks (Bennett and Provan 2008): all species pass through bottlenecks but emerge from refugia. Strictly speaking, we should distinguish between glacial and interglacial refugia (Stewart et al. 2010), but in most biogeographic discussions, the term "refugia" denotes only glacial ones. Glacial refugia are areas where the climate was sufficiently mild, and where there were sufficient resources for humans to survive climatic downturns such as Heinrich Events and the Last Glacial Maximum during MIS 2. The main glacial refugia in Asia (as in Europe) for hominins (including *H. sapiens*) would have been along the southern edge of the maximum human range. Although current data do not permit the identification of Asian glacial refugia, some of the most likely are the Levant, the Caspian foreshore, the Ganges floodplain and Deccan peninsula in India; the Yangtze Valley and parts of south China; and the Sunda Shelf of island Southeast Asia (Dennell 2009; Louys and Turner 2012; Rabett 2012 for Southeast Asia). All these areas probably served as refugia for earlier hominins, and their settlement records should indicate (when better known) occupation during glacial maxima such as the LGM (Last Glacial Maximum), when conditions were at their most severe. Two other areas that may have been used as glacial refugia after 40 ka are a conjoined Hokkaido–Sakhalin Peninsula in northeast Asia (Izuho 2014; see Chapter 11), and perhaps also "Greater Beringia" between 30 and 16 ka (Mulligan and Kitchen 2014). For animals such as arctic lemming, arctic fox and musk-ox, interglacial refugia are in the Arctic regions of Siberia and North

America. There would also have been interglacial refugia for some plants and animals in high mountain regions, and also islands at times of high sea level. This last possibility may have been important in Southeast Asia where an enormous area of land was drowned in the last interglacial when sea levels were at their highest (Chapter 6).

Our species widened its range of glacial refugia in two ways. One was by adapting to tropical rainforest, which *H. sapiens* was the first hominin species to colonise (Roberts and Petraglia 2015). At present, the earliest examples are from Sri Lanka and Southeast Asia (see Chapters 5 and 6), where rainforests were continuously occupied after 45 ka in Sri Lanka and much earlier in Southeast Asia, and were thus glacial refugia. The second was at the northern limits of the hominin range where *H. sapiens* (and perhaps Neandertals) were able to survive climatic downturns by, for example, devising effective insulation in clothing and shelters, or by overcoming winter scarcity through food storage. As a consequence, *H. sapiens* may have been able to create refugia in northern areas that would have been previously uninhabitable. Beeton and colleagues (2013), for example, suggest that parts of Central Asia may have been a refugium throughout the last glaciation despite its harsh winter conditions. This possibility raises serious issues when considering the dispersal of *H. sapiens* across the northern part of the human range in Asia because of uncertainties over whether a site or group of sites in a cold period (such as a Heinrich Event) indicates a successful adaptation in a refugium, or a short-term failure by a population that foundered. At present, our information "is not sufficiently detailed to determine what sites can serve as evidence for survival through major natural calamities and what sites are those of people who perished" (Bar-Yosef 2017: 66).

One measure of the effectiveness of a refugium is the ease with which species can disperse from it when conditions improve. Dispersal from the Levant either north or eastwards is easy, for example, relative to the Indian sub-continent, where dispersal outwards is largely blocked by the Himalayas and Karakorum mountains to the north, the deserts of Baluchistan to the west, and the mountains of northern Myanmar to the east, in much the same way as dispersal from an Italian refugium was blocked by the Alps (Hewitt 1999).

Barriers

The most obvious permanent barriers to human settlement in Asia were high mountain ranges and deserts. The principal areas of high ground in Asia are shown in Figure 3.5. The Anatolian Plateau, with its harsh winters, would also have been a barrier for much of MIS 4–MIS 2 (Kuhn 2010). The height and extent of the Tibetan Plateau and adjoining Hindu Kush–Karakorum–Himalayan ranges to the south, and the Pamirs and Tienshan to the north, made it inevitable that human expansion across continental Asia would split around them into a northern and southern route. This split would probably have occurred further west because of the deserts of the Iranian Plateau and Central Asia. To the north, it would probably have been easier for humans to disperse north of the Pamirs into southern Siberia, as suggested by Goebel (2015), than southwards across the edges of the Taklamakan desert between the Tien Shan mountains and the northern edge of the Tibetan Plateau. (Recent evidence, however, suggests that some movement occurred across this desert barrier; see Chapter 10.) To the south, the salt and sand deserts of Central Iran (the Dasht-i-Kavir and Dasht-i-Lut) and the southern Zagros Mountains would have forced humans to a corridor region along the coasts of the Arabian/Persian Gulf and Indian Ocean (Chapter 4). The principal Asian deserts (see Figure 3.6) nowadays cover over six million sq. km but would have been larger during cold, arid periods. Landforms vary from extensive dune fields (e.g. the Thar of northwest India and the Badan Jarain of north China), salt flats and playas (e.g. the Dasht-i-Lut of Iran) or stony pavements (e.g. the Gobi in China). Asian deserts are also classed as either hot or cold, depending upon their average winter temperatures: the cold Asian deserts are those in Central Asia, north China and Mongolia where winter temperatures are sub-freezing for more than two months (Dennell 2013). (Temperatures as low as –54° C. have been recorded in the Turfan Basin of northwest China.) All in all, the combination of mountain, desert and harsh climate across Southwest and Central Asia would have presented enormous challenges to human colonisation (not least in developing warm winter clothing; see below) and inevitably have made a northern dispersal across Asia a much harder proposition than one along its southern flanks.

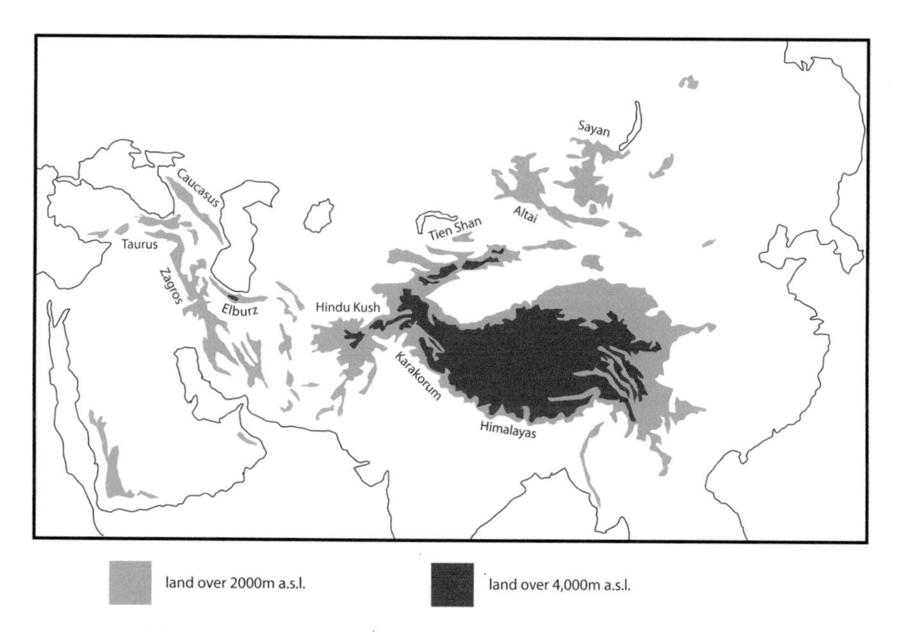

FIGURE 3.5 The principal mountain areas of Asia
Source: The author.

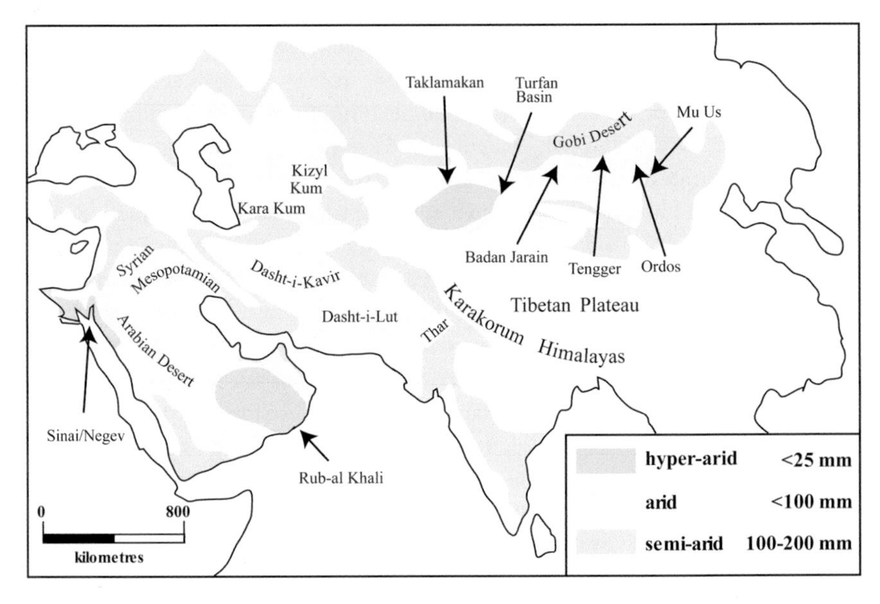

FIGURE 3.6 The principal desert areas of Asia
Source: Dennell 2013, Fig. 1.

Corridors

Permanent corridors were probably low mountain passes, and the lower parts of major river systems and their tributaries. The most useful rivers for a species dispersing longitudinally in Asia were those trending west–east, such as the extinct ones in Arabia (see Breeze et al. 2015), the Ganges and Narmada in India, and the Yellow and Yangtze rivers in China. In Siberia, northward-flowing rivers provided major highways to the Arctic, either in summer by boat or in winter as ice highways (Chapter 11).

Corridors are important in several ways besides indicating the routes that people may have taken. First, they are critical for maintaining connectivity between groups and metapopulations. At a local, short-term level, corridors facilitate exchange networks between groups, whether for goods, mates or information. At a long-term, regional level, corridors are essential for allowing gene flow between metapopulations and preventing isolation. Secondly, they may be major areas of settlement in their own right. For example, the western Zagros Mountains are corridors between present-day Iraq and the Iranian Plateau, and north–south between northwest Iran and the Arabian/Persian Gulf, but they are and have been important areas of settlement in their own right. The same is likely true of Beringia and the major river valleys of East and Southeast Asia, and Siberia. Thirdly, corridors are especially important when they diverge or converge. When they diverge, they open up new possibilities for a colonising population – as when, for example, a west–east corridor diverges into a corridor running south to north. They thus provide a way of linking regions – we need to remember that regions are rarely isolated from each other, but connected through corridors. Corridors that converge can also be important contact zones between different populations, as indicated in Figure 3.7. We shall encounter some possible examples of convergent and divergent corridors in Iran and Pakistan in Chapter 4.

In continental Asia, barriers and corridors were sometimes interchangeable depending upon the prevailing climate. For example, in climatic downturns, lowered snowlines would have made some mountain passes unusable as corridors, or usable for a shorter period each year. This was probably important in Southwest Asia, where

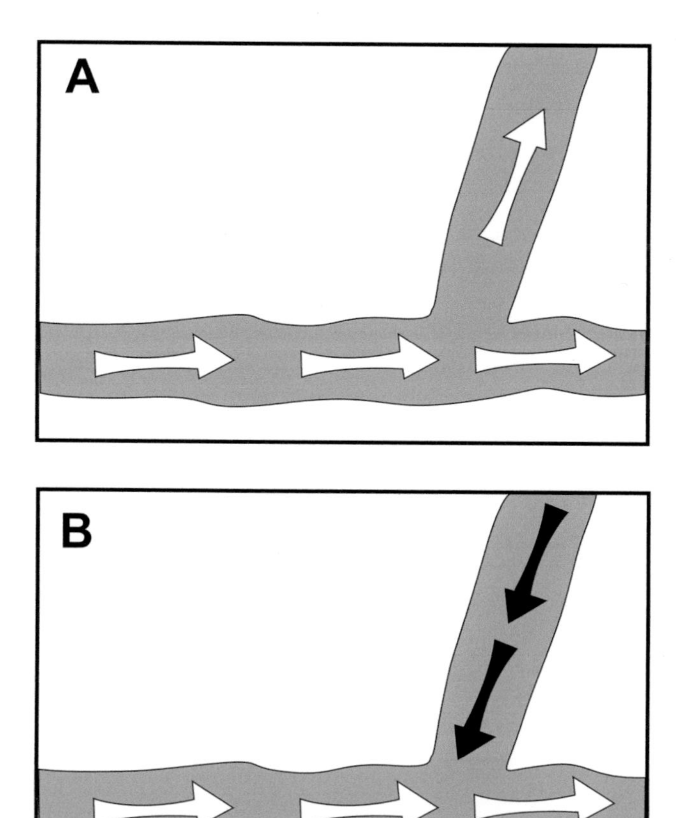

FIGURE 3.7 Convergent and divergent corridors

A. A corridor diverges, and provides a second potential route for a dispersing population.

B. Two corridors converge that are used by two populations. This type of convergence was probably important in providing contact zones (such as the Indus Valley) where two populations – such as *H. sapiens* and Neandertals – would have met at the junction of their respective corridors.

Source: The author.

summer snowlines were depressed by up to 1,800 m in the Zagros Mountains (see Chapter 8) with the consequence of some habitat loss and fragmentation (see below), and greater restrictions on movement. Similarly, deserts such as the Arabian and the Thar were likely barriers during arid periods, but potentially corridors in moister episodes such as MIS 5 and parts of MIS 3 and MIS 4 when there were active lakes and rivers (Blinkhorn et al. 2013; Rosenberg et al. 2011; see Chapter 4); the same is likely true of other Asian deserts such as those of Central Asia, and north and northwest China (Chapters 9 and 10). Dispersal across continental Asia would have been harder when the climate became colder and drier because of the depression of snowlines and the expansion of deserts. These conditions created additional hazards for colonisers that were not encountered across southern Asia.

Navigability (wayfinding) and ease of movement

From a coloniser's viewpoint, two important factors are navigability and ease of movement (Meltzer 2009: 221–224). Areas offering easy movement and abundant landmarks (such as along major rivers or broad inter-montane valleys) are easier to colonise than those with few landmarks and difficult to traverse, such as mountain ranges, dense rainforest or sand deserts. Coastlines with long sandy beaches are easier to traverse than those with numerous inlets and steep cliffs. At smaller scales, sightlines and local landmarks are important aids to wayfinding (see Guiducci and Burke [2016] for a Palaeolithic example).

Constraints on dispersal along a southern and a northern route

Two major constraints affected dispersal across southern and continental Asia. In southern Asia between the Arabian Peninsula and Island Southeast Asia (ISEA), sea levels and rainfall had very different consequences.

Southern Asia

In southern Asia, the last interglacial presents a paradox in that it simultaneously facilitated and impeded dispersal from Arabia to Southeast Asia. The climatic and environmental changes that occurred between MIS 5e and MIS 3 had very different consequences in southern and continental Asia. Across southern Asia, the key variables were rainfall and sea level. Increased rainfall in MIS 5 and parts of MIS 3 was critical in creating a "green Arabia" with extensive lakes, rivers and grasslands, in opening a corridor between Arabia and northwest India, and in greening the Thar Desert (see Chapter 4). These developments probably offset the advantages of low sea levels in MIS 4 that led to narrowing of the Bab el Mandab Strait and the exposure of the present-day Arabian/Persian Gulf. In Southeast Asia, both sea levels and rainfall were critical. In MIS 5, higher rainfall led to an expansion of rainforest, and high sea levels inundated several million square kilometres of land (Voris 2000), and turned the massive Sunda Shelf into a mass of islands, thus impeding overland dispersal but possibly incentivising seafaring (see Chapter 6). The expansion of rainforest and rise in sea levels combined created barriers to colonisation. The fragmentation of this landscape also had major consequences on the connectivity between human (and other animal) populations and their viability. In MIS 4, rainfall decreased and sea levels fell, both of which aided dispersal: first, the replacement of tropical forest by sub-tropical woodland and grassland created a savannah corridor (Louys and Turner 2012) through a newly emerged Sunda Shelf (Chapter 6). There was thus a see-saw effect across southern Asia, with higher levels of rainfall in MIS 5 and the moister parts of MIS 3 favouring dispersal in Arabia, but lower rainfall and lower sea levels in Southeast Asia favouring dispersal in MIS 4 or the drier parts of MIS 3. On these grounds, we might predict evidence of dispersal from Arabia to northwest India and probably into India by MIS 5, but less evidence of dispersal into ISEA and mainland Southeast Asia until MIS 5b, MIS 5d and MIS 4, when the vegetation became less dense and sea levels fell.

Central and northern Asia

An inescapable reality in Central Asia, Mongolia, north China and Siberia is the brutality of the winters (see Figure 3.8). Average January temperatures along the northern route are consistently and often substantially sub-freezing (see Chapters 9–11). In Siberia, the coldest place is Oymyakon, with a record low of –68° C. and an *average* annual temperature of –15° C. (and no record of above-freezing temperatures between 25 October and 17 March). In contrast, most of the inhabitants of southern Asia, from Arabia to Indonesia, are unlikely ever to have endured sub-freezing temperatures. Before going any further, we need to consider briefly three potentially lethal effects of sub-freezing temperatures: frostbite, hypothermia and wind chill.

Frostbite

Frostbite occurs when flesh is exposed to sub-freezing temperatures. Typically, one's extremities are affected first – ears, fingers, tips of noses and toes. The first phase of frostbite is frost nip, and the skin goes cold, numb and white, often with a tingling sensation. This can be reversed by gentle warming. The second intermediate phase is more serious as it results in tissue damage: blisters develop, and the underlying tissue feels numb and frozen. The third advanced stage is known as deep frost bite, and can affect nerves, tendons, muscle and bones. When the skin thaws, blood-filled blisters develop and these turn into thick black scabs. Affected tissue may die (tissue necrosis) and amputation is normally needed to prevent the spread of gangrene (see e.g. www.nhs.uk/conditions/frost bite/symptoms/).

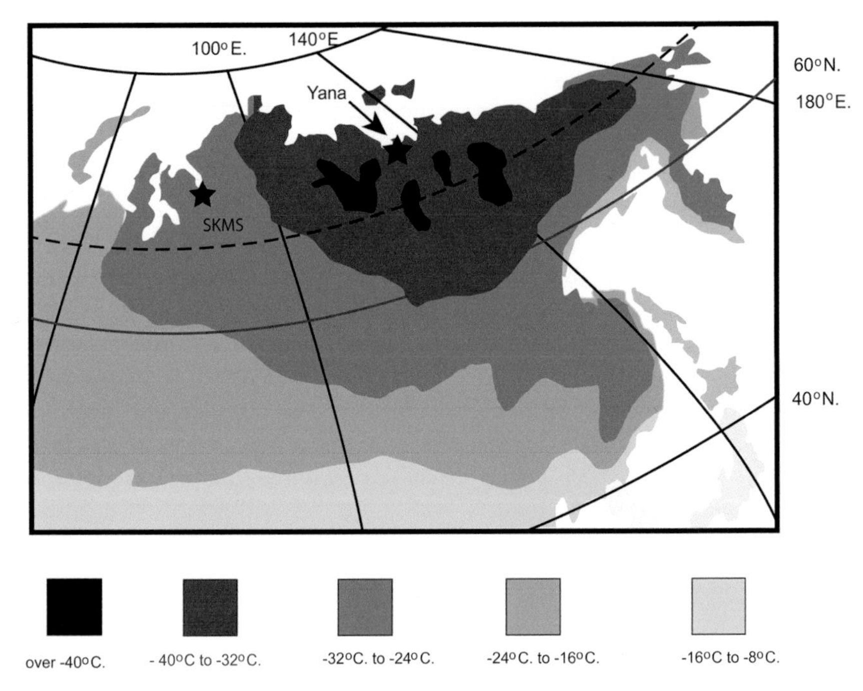

FIGURE 3.8 Winter temperatures in northern Asia

January average temperatures in northern Siberia and the location of Yana and the Sopochnaya Karga mammoth site (SKMS). The dashed line indicates the Arctic Circle.

Source: Redrawn from the *Encyclopedia of World Geography* 1974.

Hypothermia

Hypothermia is also potentially lethal, and sets in when the body's temperature falls from its normal temperature of ca. 37° C. Early symptoms include intense shivering, slurred speech, and loss of co-ordination and concentration. If not rectified, the victim can lapse into semi-consciousness and then a coma when the body temperature has fallen to ca. 30° C. In extreme cases, and for reasons poorly understood, victims have been known to remove all their clothing and then crawl into a small enclosed space before dying. Death normally occurs when the body temperature has fallen to 23° C. (see www.watchandride.com/blog/snowboard-optimisation/hypothermia/).

Wind chill

Wind chill is an especially lethal way of accelerating heat loss from the body. As shown in Figure 3.9, a moderate wind of only 15 mph (24 km/hr) at 0° F. (–17° C.) produces a wind chill temperature of –19° F. (–28° C.). At this temperature, frostbite will start to develop in 30 minutes. If the air temperature then fell

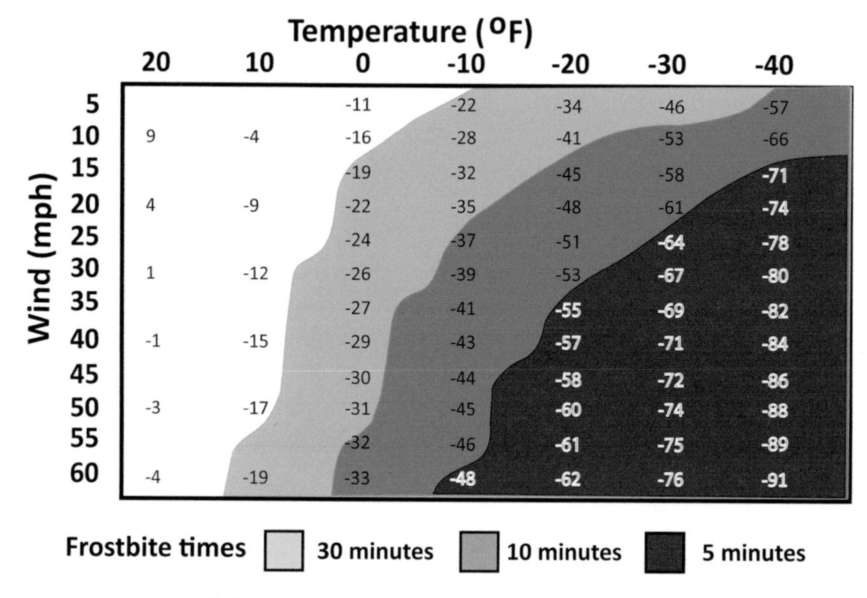

FIGURE 3.9 Wind chill chart

Source: Based on wind chill chart in https://en.wikipedia.org/wiki/Wind_chill.

to –20° F (–29° C.) and wind speed increased to 20 mph (32 km/hr), the wind chill factor would increase to –48° F. (–44° C.), and frostbite would develop in only ten minutes.

The sub-freezing winters of Central and northern Asia required a totally different range of adaptations from those needed in southern Asia. Essential requirements for surviving bitter, cold winters and avoiding frostbite and hypothermia were adequate winter dwellings, and control of fire for heat, cooking and light (see Hosfield 2016 for a discussion of these factors in a European Early and Middle Pleistocene context). The most essential requirement would have been warm, wind-proof clothing (see Gilligan 2010; Hurcombe 2014). Iain Davidson (2013) has made a perceptive contribution in noting how the proportion of the body covered by clothing and the materials used in clothing in ethnographic examples vary by latitude from the equator to the Arctic (Figure 3.10). In the tropics, clothing can even be ignored, but when worn, most ethnographic examples are made from plants. In continental Asia, most of the body had to be covered, and the main material is hide, and in the Arctic, fur. The division between the regions where clothing was based on plants or hides matches very closely the division of a southern from a northern dispersal across continental Asia, and lies roughly along the southern boundary of the Palearctic Faunal Realm (see Figure 3.4). Converting animal skin into warm, insulated and wind-proof clothing made from hides to avoid frostbite and hypothermia required nothing less than a revolution in technology. We will explore this in Chapter 9 when discussing Central Asia and Siberia.

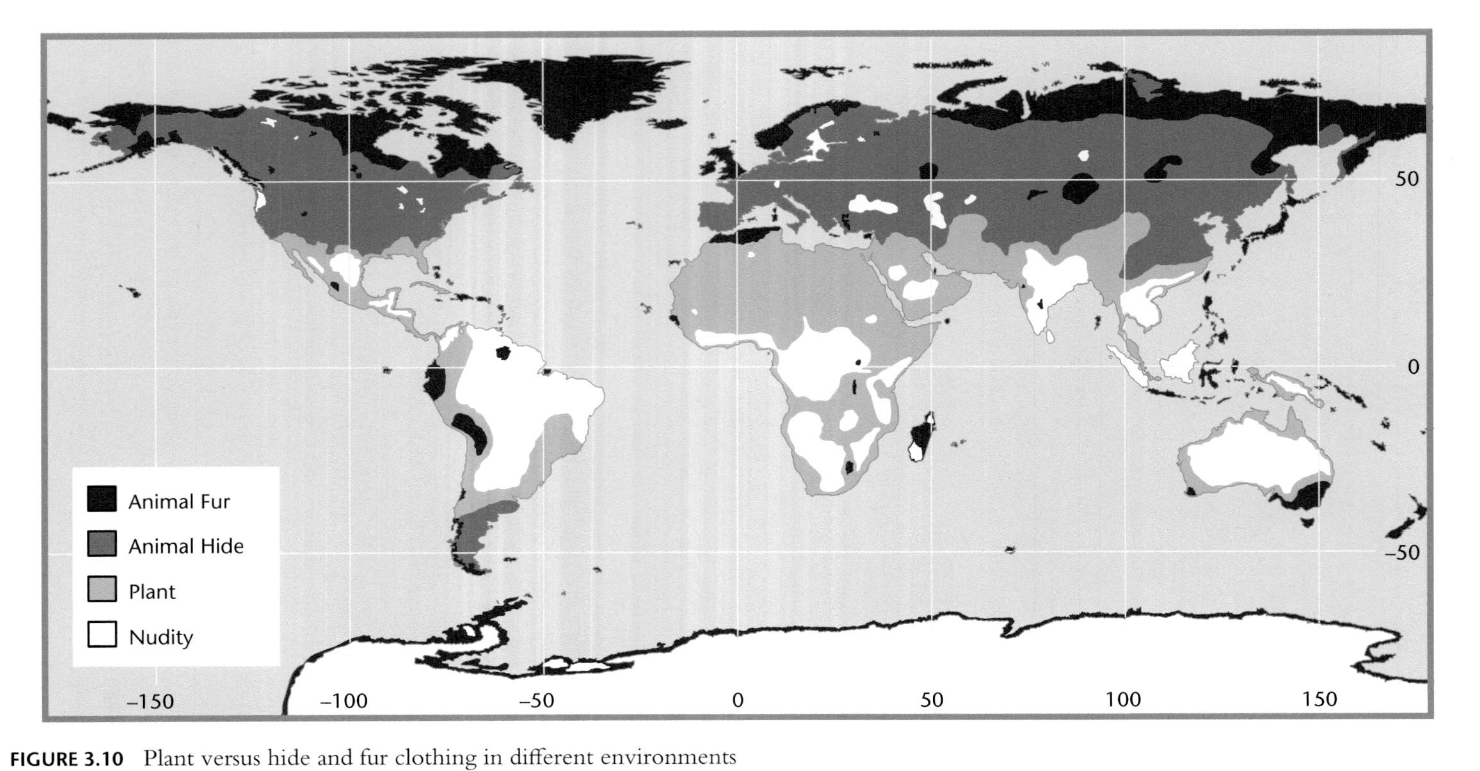

FIGURE 3.10 Plant versus hide and fur clothing in different environments

Note how the boundary between plant- and hide-based clothing closely matches the southern limit of the Palearctic Realm (see Figure 3.4).

Source: Davidson 2013, Fig. 3.

How did climatic change affect human groups?

In my previous book, I summarised the Pleistocene settlement of Asia as "a repeated theme of regional expansion and contraction, colonisation and abandonment, integration and isolation as rainfall increased or decreased" (Dennell 2009: 475). Neandertals, *H. erectus* in East and Southeast Asia, and *H. floresiensis* in Island Southeast Asia (ISEA) had each survived numerous climatic shifts from ice ages to interglacials (or their equivalent in tropical regions), although population levels would have fallen during climatic downturns. These all became extinct in the last glacial cycle which experienced a considerable degree of climatic and environmental change. It was also during the last glacial cycle that the main dispersal of *Homo sapiens* across Asia took place. The indigenous inhabitants of Asia thus had to contend with an unusually unstable and at times harsh climate but also the destabilising impact of an invasive competitor. A key factor here is how indigenous and colonising groups might have responded to habitat disruption, or specifically, loss, degradation and fragmentation that was caused by climate change.

Habitat loss, degradation and fragmentation

Although many researchers have suggested that periods of climatic instability had adverse consequences on human groups, and might even have been a prime factor behind Neandertal extinction (see e.g. Mellars 1998), there have been relatively few attempts to show how climatic change impacted on human groups at a regional level. We can consider this issue at two different scales, one regional and short-term, and the other at a sub-continental scale over a longer timescale.

Habitat loss, degradation and fragmentation: regional and short-term

Ecologists have paid considerable attention to the consequences of climatic and environmental change on plant and animal populations. Given current concerns over global warming, most of these studies consider the effects of increased temperatures: Pleistocene researchers of course have to consider also the consequences of climatic downturns involving lower temperatures. Climatic changes involve habitat fragmentation and loss that is most keenly felt at the edges of the inhabited range. Regarding habitat fragmentation, "Habitat fragmentation is an inherent consequence of habitat loss: the progressive disappearance of a given habitat entails changes in both the area of remnant fragments and their spatial configuration, with consequences on the structural connectivity among remnants" (Baguette et al. 2013: 381). They point out that when suitable habitat has fallen to <40% of its original extent, the distances between fragments suddenly increase, with significant consequences: "By increasing the distance among such patches, landscape fragmentation is expected to increase dispersal costs, including loss of time and energy and the risk of getting lost" (Baguette et al. 2013: 385). Dytham and Travis (2013: 400) point out that:

> It is not the loss of habitat per se that is the most important element of fragmentation. ... it is loss of "connectivity" that can be crucial for the survival of a species. This effect will be non-linear as there will be a threshold of habitat loss that will break the connectivity of a region.

Figure 3.11 shows how habitat loss and fragmentation might have impacted on a regional metapopulation. There are clear implications here for the importance for

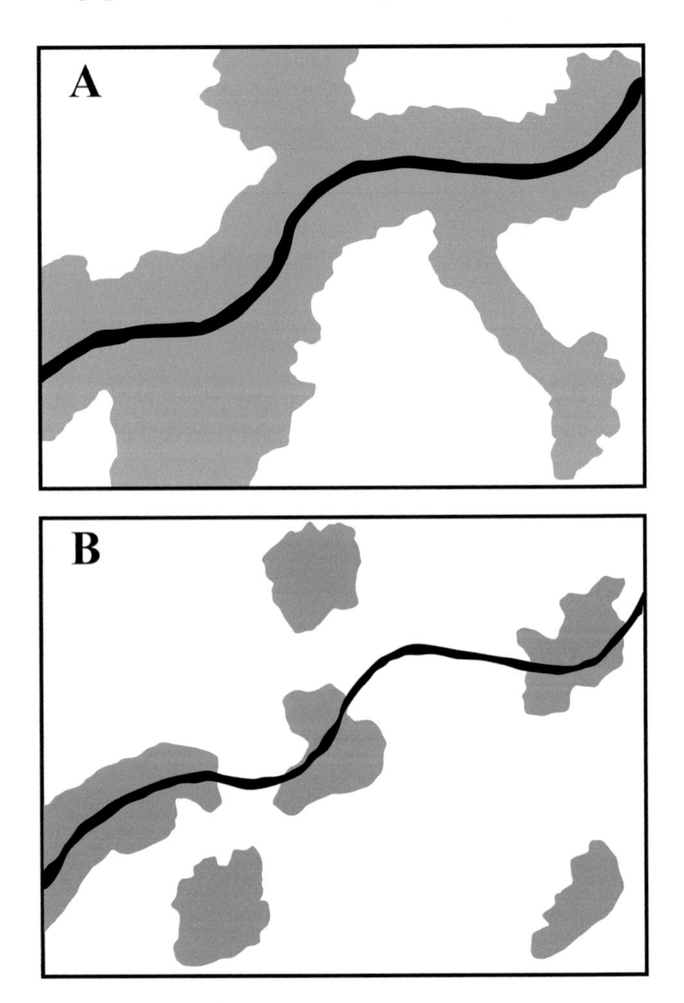

FIGURE 3.11 Habitat loss and fragmentation

Top: a hypothetical landscape containing a river valley and a continuous belt of highly productive resources.

Bottom: following a climatic downturn, the area of productive resources has been considerably reduced, but also fragmented into six patches. Animal (including human) populations within these patches have lost, or greatly reduced, their connectivity with neighbouring populations.

Source: The author.

widely dispersed groups in a metapopulation to maintain the connectedness of their social networks when habitats become fragmented, and particularly when confronted with an invasive competitor such as our species.

The most graphic example of habitat loss and fragmentation that affected Asia was the drowning of the Sunda Shelf of Southeast Asia in the last interglacial when at times of high sea level, when the enormous landmass of Sunda was transformed into the 17,000 islands of modern Island Southeast Asia (ISEA) (see above). As discussed in Chapter 6, changes in sea level in Southeast Asia and the consequent loss and fragmentation of habitats had profound consequences on animal (including human) populations.

Demography and climate change: source and sink populations

One way of thinking about the impact of climate change on metapopulations is to consider how they reacted to the type of major long-term shifts in climate that were a dominant feature of Pleistocene Asia. The dynamics of human metapopulations in response to major climatic shifts can be modelled as "sources and sinks". These models were first developed by geneticists (e.g. Eller et al. 2004; Hawks 2009; Pulliam 1988), and have been applied to early Palaeolithic Europe by myself and colleagues (Dennell et al. 2011), China by Martinón-Torres et al. (2016) and Southeast Asia by Louys and Turner (2012). Their utility lies in the way they can be used at a continental or sub-continental scale over time-scales of several tens of millennia to model how human populations responded to major climatic shifts between cold and usually dry glaciations, and warmer and moister interglacials such as the one we inhabit today. Palaeolithic archaeologists and Pleistocene scientists now have detailed records for some parts of the world showing how human and other populations responded to the repeated expansion and contraction of inhabitable environments. At the depths of an ice age, when conditions were at their coldest and driest, the hominin range across Eurasia contracted southwards. Populations also became more fragmented and would have been restricted to a small number of glacial refugia where resources were more abundant, predictable and sufficient for survival.

Populations – or rather, the metapopulations – within refugia would have formed the basis for expansion when conditions improved, as groups were able to recolonise areas previously vacated. During the process of re-colonisation, groups may have overlapped and mixed with others (see Figure 3.12). Those populations at the northern edge of the human range, or at the altitudinal limits of occupation in mountain areas, or in areas where resources are less abundant and predictable, can be called sink populations. To quote Hawks (2009), "a population sink is a region where the average rate of reproduction is below replacement levels. This region can remain populated only if individuals migrate in from other places. The places that reproduce above replacement are called population sources". Demographic expansion from source populations thus depends first upon the extinction rates in sink populations at the edge of the

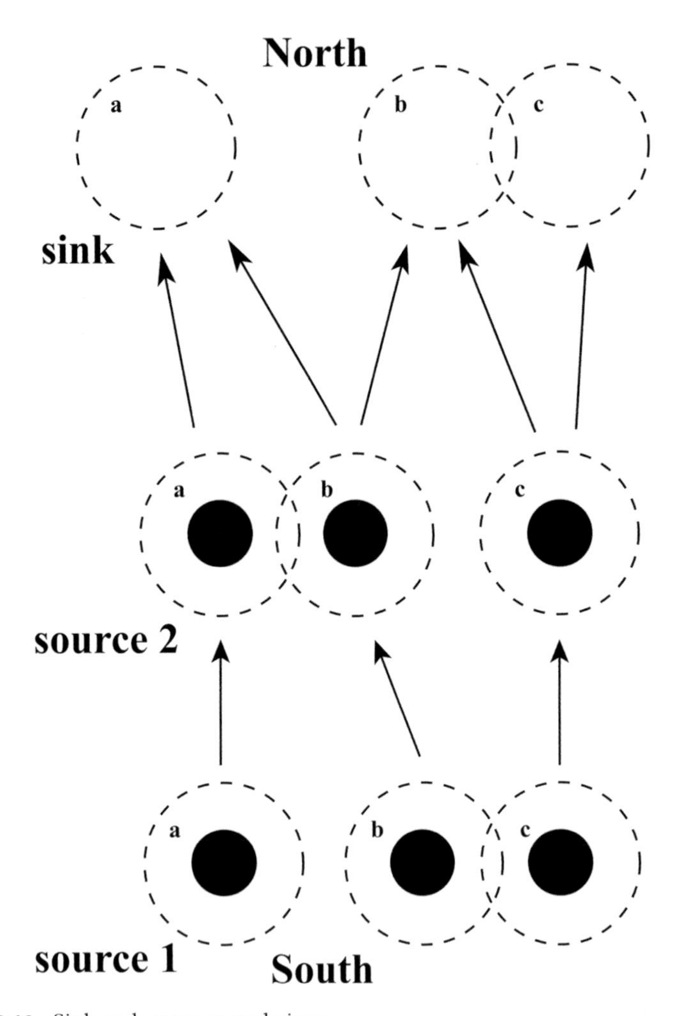

FIGURE 3.12 Sink and source populations

Source and sink populations: here, the bottom row indicates three metapopulations, or palaeodemes, in glacial refugia at the southern limit of the species' range. These are source populations that provide the basis for later expansion. The solid circles denote metapopulations during cold periods when populations contract into refugia; the dashed circles indicate interglacial or interstadial conditions when expansion from them is possible. Each is separated in glacial conditions, but in interglacial conditions, metapopulations b and c overlap. The middle row indicates how each expands in interglacial times and becomes a source population: here, metapopulations 2a and 2b overlap, but 2c (derived originally from demes b and c) remains isolated. The top row indicates the maximum expansion during an interglacial; here, deme 3a (derived from demes 2a and 2b) is isolated, but demes 3b and 3c overlap, although each has a different ancestry. At the northern edge of the species range, the metapopulations are sink populations in that they require recruitment from source populations to remain viable.

In this scenario, the process of expansion results in inter-demic mixing, and this schematic representation of population fragmentation, dispersal and recombination could lead to the type of demic variability exhibited by the Chinese fossil record during the Late Middle and Early Upper Pleistocene.

Source: The author.

inhabited range, and second, the ability of the main source populations to support sink populations, especially those at the edge of the range. This would become difficult when population densities were low and inter-group distances were high, thus making it difficult to maintain tight networks.

It is here that the contrast between short-term visitors and long-term residents (Dennell 2003) becomes important. In areas where resources are subject to fluctuations in availability, settlement is likely to have been intermittent and confined to moister and warmer intervals, punctuated by periods of settlement contraction or even abandonment during climatic downturns. In semi-arid and arid areas where rainfall is highly variable, settlement records are likely discontinuous, as also in high latitude regions where winter temperatures are the key variable. Two related points are that the first appearance of humans in these types of unstable environments should not be mistaken for the onset of colonisation: these might indicate, in Shea's (2008) words, a "failed dispersal"; and secondly, these areas are demographic "sinks" that could only be sustained by being replenished from an external source population (Dennell et al. 2011).

Summary

The dispersal of *H. sapiens* across Asia was shaped by biogeography, geography and climate. The deserts and mountain ranges from Iran through Central Asia to Mongolia and north China would have made it inevitable that humans dispersed northwards across the Palearctic Realm of continental Asia or southwards through the Oriental Realm of southern Asia to Sunda and ultimately Wallacea and Australia. Dispersal into the Arctic would have been facilitated by the rivers that flow northwards across Siberia once humans had learnt how to adapt to its long and brutally cold winters. The climate of Asia changed enormously between the last interglacial of MIS 5e and the Last Glacial Maximum of MIS 2. There were periods when climatic change favoured dispersal, particularly in arid and semi-arid regions of Arabia and inland continental Asia. Here, even modest increases in rainfall would have lessened the frequency and severity of drought, and opened up new areas that could be utilised. Climatic downturns that involved colder and drier conditions would obviously have had the opposite effect in increasing habitat loss, degradation and fragmentation, so human occupation of most of arid and semi-arid Asia was likely intermittent and characterised by frequent failures and abandonment. These semi-arid landscapes were generally the most vulnerable to changes in rainfall and temperature. Although high rainfall and warmer temperatures were generally favourable to human dispersal, they were disastrous in the lowland areas of Southeast Asia because the rise in sea level drowned a substantial part of the habitable area. Sunda was perhaps the one area in Asia where lower temperatures (and hence lower sea levels) were beneficial for colonisation.

At this point, we can start to trace the dispersal of our species across southern Asia.

Notes

1 The term "monsoon" is derived from the Arabic "mausin", meaning "a wind in South West Asia and the Indian Ocean blowing from the south-west from April to October, and from the north-east the rest of the year" (Baker 1949: 940).
2 Mawsynram, also in northeast India, is currently listed as the wettest place on earth, with comparatively modest average rainfall (over 38 years) of 11,873 mm, or ca. 39 feet (*Guiness World Records* 2004: 68).

References

Baguette, M., Legrand, D., Freville, H., Dyck, H. van and Ducatez, S. (2013) Evolutionary ecology of dispersal in fragmented landscape. In J. Clobert, M. Baguette, T.G. Benton and J.M. Bullock (eds) *Dispersal Ecology and Evolution*, Oxford: Oxford University Press, pp. 381–391.

Baker, E.A. (ed.) (1949) *Cassell's New English Dictionary*. London: The Reprint Society.

Bartov, Y., Goldstein, S.L., Stein, M. and Enzel, Y. (2003) Catastrophic arid episodes in the eastern Mediterranean linked with the North Atlantic Heinrich Events. *Geology*, 31: 439–442.

Bar-Yosef, O. (2017) Facing climatic hazards: Paleolithic foragers and Neolithic farmers. *Quaternary International*, 428: 64–72.

Beeton, T.A., Glantz, M.M., Trainer, A.K., Temirbekov, S.S. and Reich, R.M. (2013) The fundamental hominin niche in Late Pleistocene Central Asia: a preliminary refugium model. *Journal of Biogeography*, 41: 95–110.

Bennett, K.D. and Provan, J. (2008) What do we mean by "refugia"? *Quaternary Science Reviews*, 27: 2449–2455.

Blinkhorn, J., Achyuthan, H., Petraglia, M. and Ditchfield, P. (2013) Middle Palaeolithic occupation in the Thar Desert during the Upper Pleistocene: the signature of a modern human exit out of Africa? *Quaternary Science Reviews*, 77: 233–238.

Blunier, T. and Brook, E.J. (2001) Timing of millennial-scale climate change in Antarctica and Greenland during the Last Glacial period. *Science*, 291: 109–112.

Breeze, P.S., Drake, N.A., Groucutt, H.S., Parton, A., Jennings, R.P., White, T.S. and Clark-Balzan, L. et al. (2015) Remote sensing and GIS techniques for reconstructing Arabian palaeohydrology and identifying archaeological sites. *Quaternary International*, 382: 98–119.

Burke, A., Kageyama, M., Latombe, G., Fasel, M., Vrac, M., Ramstein, G. and James, P.M. A. (2017) Risky business: the impact of climate and climate variability on human population dynamics in western Europe during the Last Glacial Maximum. *Quaternary Science Reviews*, 164: 217–229.

Davidson, I. (2013) Peopling the last new worlds: the first colonisation of Sahul and the Americas. *Quaternary International*, 285: 1–29.

Dennell, R.W. (2009) *The Palaeolithic Settlement of Asia*. Cambridge: Cambridge University Press.

Dennell, R.W. (2013) Hominins, deserts, and the colonisation and settlement of continental Asia. *Quaternary International*, 300: 13–21.

Dennell, R.W., Martinón-Torres, M. and Bermudez de Castro, J.M. (2011) Hominin variability, climatic instability and population demography in Middle Pleistocene Europe. *Quaternary Science Reviews*, 30: 1511–1524.

Dytham, C. and Travis, J.M.J. (2013) Modelling the effects of habitat fragmentation. In J. Clobert, M. Baguette, T.G. Benton and J.M. Bullock (eds) *Dispersal Ecology and Evolution*, Oxford: Oxford University Press, pp. 392–404.

Eller, E., Hawks, J. and Relethford, R. (2004) Local extinction size and recolonization, species effective population size, and modern human origins. *Human Biology*, 76(5): 689–709.

Encyclopedia of World Geography (1974). London: Octopus.

Gibbard, P.L. and Lewin, J. (2016) Partitioning the Quaternary. *Quaternary Science Reviews*, 151: 127–139.

Gilligan, I. (2010) The prehistoric development of clothing: archaeological implications of a thermal model. *Journal of Archaeological Method and Theory*, 17: 15–80.

Goebel, T. (2015) The overland dispersal of modern humans to eastern Asia: an alternative, northern route from Africa. In Y. Kaifu, T. Goebel, H. Sato and A. Ono (eds) *Emergence and Diversity of Modern Human Behavior in Paleolithic Asia*, College Station: Texas A&M University Press, pp. 437–452.

Groucutt, H.S., Petraglia, M.D., Bailey, G., Scerri, E.M., Parton, A., Clark-Balzan, L., Jennings, R.P. et al. (2015) Rethinking the dispersal of *Homo sapiens* out of Africa. *Evolutionary Anthropology*, 24: 149–164.

Guiducci, D. and Burke, A. (2016) Reading the landscape: legible environments and hominin dispersals. *Evolutionary Anthropology*, 25: 133–141.

Guinness World Records 2004 (2004) Guinness World Records Ltd.

Hawks, J. (2009) Local adaptation in population sinks. http://johnhawks.net/taxonomy/term/260.

Hemming, S.R. (2004) Heinrich Events: massive Late Pleistocene detritus layers of the North Atlantic and their global climate imprint. *Reviews of Geophysics*, 42, RG1005, doi:10.1029/2003RG000128.

Hewitt, G.M. (1999) Post-glacial re-colonization of European biota. *Molecular Genetics in Animal Ecology*, 68: 87–112.

Holt, B.G., Lessard, J.-P., Borregaard, M.K., Fritz, S.A., Araújo, M.B. et al. (2013) An update of Wallace's zoogeographic regions of the world. *Science*, 339: 74–78.

Hong, Y.T., Wang, Z.G., Jiang, H.B., Lin, Q.H., Lin, B. et al. (2001) A 6000-year record of changes in drought and precipitation in northeastern China based on a $\delta^{13}C$ time series from peat cellulose. *Earth and Planetary Science Letters*, 185: 111–119.

Hosfield, R. (2016) Walking in a winter wonderland? Strategies for Early and Middle Pleistocene survival in midlatitude Europe. *Current Anthropology*, 57(5): 653–682.

Hurcombe, L.M. (2014) *Perishable Material Culture in Prehistory: Investigating the Missing Majority*. London and New York: Routledge.

Izuho, M. (2014). Human technological and behavioural adaptation to landscape changes around the Last Glacial Maximum in Japan: a focus on Hokkaido. In K.E. Graf, C. V. Ketron and M. Waters (eds) *Paleoamerican Odyssey*, Center for the Study of the first Americans: Texas A&M Press, pp. 45–64.

Kohfield, K. E. and Chase, Z. (2017) Temporal evolution of mechanisms controlling ocean carbon uptake during the last glacial cycle. *Earth and Planetary Science Letters*, 472: 206–215.

Kuhn, S.L. (2010) Was Anatolia a bridge or a barrier to early hominin dispersals? *Quaternary International*, 223–234: 434–435.

Liu, T. and Ding, Z. (1998) Chinese loess and the paleomonsoon. *Annual Review of Earth and Planetary Sciences*, 26: 111–145.

Louys, J. and Turner, A. (2012) Environment, preferred habitats and potential refugia for Pleistocene *Homo* in Southeast Asia. *Compte Rendues Palévolution*, 11: 203–211.

Martinón-Torres, M., Song Xing, Wu Liu and Bermúdez de Castro, J.M. (2016) A "source and sink" model for East Asia? Preliminary approach through the dental evidence. *Comptes Rendues Palévolution*, 17(1–2): 33–43.

Mellars, P. (1998) The impact of climatic changes on the demography of late Neandertal and early anatomically modern populations in Europe. In T. Akazawa, K. Aoki and O. Bar-Yosef (eds) *Neandertals and Modern Humans in Western Asia*, New York: Plenum Press, pp. 493–507.

Meltzer, D. (2009) *First Peoples in a New World*. Berkeley: University of California Press, pp. 221–224.

Mulligan, C.J. and Kitchen, A. (2014) Three-stage colonization model for the peopling of the Americas. In K.E. Graf, C.V. Ketron and M. Waters (eds) *Paleoamerican Odyssey*, Center for the Study of the first Americans: Texas A&M Press, pp. 171–181.

Porter, S.C. and An, Z. (1995) Correlation between climate events in the North Atlantic and China during the last glaciation. *Nature*, 375: 305–308.

Pulliam, H.R. (1988) Sources, sinks, and population regulation. *American Naturalist*, 132: 652–661.

Rabett, R.J. (2012) *Human Adaptation in the Asian Palaeolithic: Hominin Dispersal and Behaviour during the Late Quaternary*. Cambridge: Cambridge University Press.

Roberts, P. and Petraglia, M.D. (2015) Pleistocene rainforests: barriers or attractive environments for early human foragers? *World Archaeology*, 47(5): 718–739.

Rosenberg, T.M., Preusser, F., Fleitman, D., Schwalb, A., Penkman, K., Schmid, T.W., Al-Shanti, M.A. et al. (2011) Humid periods in southern Arabia: windows of opportunity for modern human dispersal. *Geology*, 39(12): 1115–1118.

Schulz, H., Rad, U. von and Erlenkeuser, H. (1998) Correlation between Arabian Sea and Greenland climate oscillations of the past 110,000 years. *Nature*, 393: 54–57.

Shea, J.J. (2008) Transitions or turnovers? Climatically-forced extinctions of *Homo sapiens* and Neanderthals in the east Mediterranean Levant. *Quaternary Science Reviews*, 27: 2253–2270.

Smith, M. (2013) *The Archaeology of Australia's Deserts*. Cambridge: Cambridge University Press.

Stewart, J.R., Lister, A.M., Barnes, I. and Dalén, L. (2010) Refugia revisited: individualistic responses of species in space and time. *Proceedings of the Royal Society Series B*, 277: 661–671.

Tumel, N. (2002) Permafrost. In M. Shahgedanova (ed) *The Physical Geography of Northern Eurasia*. Oxford: Oxford University Press, pp. 149–168.

Voris, H. (2000) Maps of Pleistocene sea levels in Southeast Asia: shorelines, river systems and time durations. *Journal of Biogeography*, 27: 1153–1167.

Wallace, A.R. (1876) *The Geographical Distribution of Animals, with a Study of The Relations of Living and Extinct Faunas as Elucidating the Past Changes of The Earth's Surface*. London: Macmillan.

Wang, Y.J., Cheng, H., Edwards, R.L., An, Z.S., Wu, J.Y., Shen, C.-C. and Dorale, J.A. (2001) A high-resolution absolute-dated Late Pleistocene monsoon record from Hulu Cave, China. *Science*, 294: 2345–2348.

PART 1

Prologue

The southern dispersal across Asia

The southern dispersal across Asia

When and how often humans dispersed across southern Asia has been intensely debated for several years, with some researchers proposing a single dispersal event around or even after 60,000 years ago (e.g. Mellars 2006; Klein 2008) and others (including myself) arguing that there were probably numerous dispersals, some of which could have occurred during or before the last interglacial (e.g. Boivin et al. 2013; Dennell and Petraglia 2012; Lahr and Foley 1994; Rabett 2015; Reyes-Centeno et al. 2014). Recent skeletal discoveries of *H. sapiens* in the Levant at 177–190 ka (Hershkovitz et al. 2018; Chapter 8) and Arabia ca. 90 ka (Groucutt et al. 2018; Chapter 4) strengthen considerably the argument that our species was out of Africa long before 60,000 years ago. How far east these early immigrants may have gone is much less clear, but recent skeletal evidence from Sumatra at ca. 63–73 ka (Chapter 6) and south China at 80 ka (Chapter 10) indicates that humans were already in Southeast Asia before 60,000 years ago.

Two stages were involved in the southern dispersal(s) across southern Asia (Groucutt and Blinkhorn 2013). The first is when our species entered the Arabian Peninsula and dispersed as far east as the Thar Desert of northwest India. In biogeographical terms, humans still remained within the Afro-Arabian Realm, and did not leave Africa; instead, Africa came with them. I suggest in Chapter 4 that because the earliest evidence for *Homo sapiens* in northeast Africa now dates back to ca. 300 ka BP, our species might have dispersed several times into Arabia and possibly the southern Levant at times of high rainfall when Arabia was "green" and essentially an eastern extension of the "green Sahara". Skeletal evidence indicates that our species was in the Levant in the last interglacial (MIS 5e) and possibly MIS 7 (Chapter 8); humans might even have spread into Arabia in MIS 9.

They might also have dispersed eastwards through southern Iran and southern Pakistan to the Thar Desert during MIS 5, and perhaps in earlier phases of high rainfall. These dispersals did not require any innovations in behaviour or technology and can be treated as examples of range extension. On this scenario, most of the Middle Palaeolithic in these regions is attributed to *H. sapiens*, although the occasional presence of Neandertals is not excluded. Southern Iran and southern Pakistan are two areas where our species might have overlapped with Neandertals at the southern margins of their range.

The second stage occurred when our species left the Saharan-Arabian Realm and entered the Oriental Realm, beginning with peninsular India and then moving eastwards into mainland and island Southeast Asia, south China and Wallacea. The importance of the human colonisation of this realm cannot be overestimated. In the Oriental Realm, plant foods are available year-round, whether as leaves, fruits, nuts, roots or tubers. Herbivores could usually find sufficient browsing or grazing year-round so did not need to migrate long distances between winter and summer pastures. Human populations could therefore remain localised within small annual territories. Because winter temperatures remained above freezing, there was no need for the warm, insulated clothing made from hide or fur that was essential for the northern dispersal across continental Asia. Unlike life in northern latitudes, the provision of clothing did not require substantial amounts of time and labour. The Oriental Realm was probably the main refugium in Asia for human populations during the Last Glacial Maximum, and may have contained most of the total population of Asia in the Late Pleistocene (see e.g. Atkinson et al. 2008).

A sub-tropical origin of gendered subsistence?

Adapting to life in the Oriental Realm required learning about a substantially different type of landscape, with largely unfamiliar plant and animal resources. The pathway to success would have been through gendered subsistence strategies. Kuhn and Stiner (2006) argue that there were three reasons why gendered subsistence strategies first developed in tropical and sub-tropical regions such as (but not necessarily) Africa. First, regions with biodiversity have a greater variety of potential foods that are available for a greater part of the year than in high latitude, colder and strongly seasonal environments. These foods vary from large and small animals to birds, reptiles, insects, molluscs and plant foods that can be from leaf, stem, root, tuber, fruit, nut, berry, seed, fungi and even lichen.[1] Second, plant foods such as tubers are commoner in tropical and sub-tropical regions and although these need cooking (and sometimes detoxifying), they often provide more nutrition than the equivalent amount of seeds and nuts. Niah Cave in Borneo provides some evidence that humans were detoxifying otherwise poisonous plants ca. 40,000 years ago (Chapter 6). Thirdly, many tropical foods can be obtained by children, and they and their mothers or carers would therefore have had greater control over their food, and could have provided more food than their counterparts in high latitude regions. Africa need not have been the only place where gendered subsistence originated; the densest human populations in Asia were probably in the tropical and sub-tropical regions of South and Southeast Asia that had the greatest potential for dietary diversification.

Chapter 5 looks at what we know about the human colonisation of peninsular India. When this occurred is unclear, and India is currently the most serious gap in our evidence for when our species dispersed across southern Asia because of the lack of human skeletal evidence, the small number of detailed analyses of Middle Palaeolithic assemblages, and the almost total lack of faunal evidence. The earliest skeletal evidence from South Asia for *H. sapiens* is from Sri Lanka and dates to only ca. 36 ka – long after Australia had been colonised. One important debate has been over whether humans were in India before the Toba super-eruption of 74 ka that deposited significant amounts of ash across India. Some genetic analyses of modern South Asian populations suggest that their ancestors may have been in South Asia around that time, but the evidence is inconclusive. Analyses of Middle Palaeolithic stone tool assemblages from above and below the Toba ash suggest that there was no major technological change, and this might therefore indicate continuity in population. Some researchers suggest that these assemblages are similar to African Middle Stone Age ones, and these similarities might further imply that the Indian Middle Palaeolithic was made by our species. As an illustration of how little we know for certain, recent research has also indicated that the Indian Middle Palaeolithic began as early as ca. 380 ka ago and might be part of the same processes by the same type of hominin that led to the emergence of the African Middle Stone Age. There has been a major debate over the appearance of microlithic assemblages in South India and Sri Lanka ca. 48–40 ka. Some researchers have attributed these to immigrant populations of *H. sapiens* from Africa because of claimed similarities of these assemblages with those known as Howieson's Poort from South Africa, dated to ca. 55–60 ka. Subsequent analyses of these South Asian and South African microlithic assemblages indicate that the South Asian microliths are now better explained as a local development, perhaps in response to population pressure. One significant development in South Asia is that humans were living in rainforests in Sri Lanka after 46 ka and inhabited these through the glacial maximum into the Holocene. Besides providing an example of a very early adaptation of humans to rainforests, this evidence also provides a clear example of how our species was able to colonise what had previously been a barrier.

Chapter 6 examines the colonisation of mainland Southeast Asia and Sunda, the conjoined landmass formed by the joining of the Malay Peninsula, Sumatra, Java and Borneo at times of low sea level. Although much of this region would have been covered in rainforest, there would also have been a corridor of savannah grassland and open woodland through Sunda and in mainland Southeast

Asia towards south China that would have provided easy access for humans when they colonised this region. A small amount of human skeletal evidence from Laos and Sumatra places humans in this region in MIS 4, ca. 60–74 ka. By far the best record for our species in this region in MIS 3 comes from the great cave of Niah, Borneo, where humans were present after 45 ka, and probably as early as 50 ka. One remarkable recent discovery in Borneo is that the local cave art dates back ca. 40 ka and is thus as old as that in western Europe.

Chapter 7 looks at the earliest evidence for human in Wallacea – the islands between Sunda and Sahul, the conjoined landmass of Australia, New Guinea and Tasmania. This is a transitional zone between Sunda and Sahul: apart from Sulawesi, these islands contain a depauperate fauna, but one that also include marsupials such as cuscus and mammals such as dwarf buffalo, babirusa (deer pig) and rats. The colonisation of the islands of Wallacea could only have been accomplished by the use of skilful navigation and watercraft that could be steered, and by the ability to harvest marine resources because most of the island faunas are deficient in animal resources.

Wallacea has produced some remarkable surprises in recent years. As examples, Flores, Sulawesi and Luzon in the Philippines all had indigenous populations before the arrival of *Homo sapiens*. A new palaeo-species *Homo luzonensis* has recently been identified at Callao Cave, Luzon island, the Philippines and is dated to ca. 50–70 ka BP. Flores was inhabited by *H. floresiensis*, also known as the "hobbit". A recent re-examination of the stratigraphy of the cave of Liang Bua, in which the skeletal remains of the hobbit were found, has shown that it became extinct ca. 50 ka, which might coincide with the arrival of our species. The neighbouring island of Timor was colonised ca. 44 ka, and its inhabitants were catching pelagic (deep-water) fish 42,000 years ago. Recent dating of cave art on Sulawesi has shown that it is as old as that on Borneo, and therefore shares the same artistic traditions.

This chapter ends with a brief look at the colonisation of Sahul. A strong case can be made that the Australian Realm was colonised ca. 50–55 ka. However, recent excavations at the rock shelter of Madjedbebe in northern Australia indicate that the earliest stone artefacts may date from ca. 60–65 ka. This site is thus an outlier, and may indicate an earlier, and possibly unsuccessful attempt at colonisation. An alternative possibility is that the earliest artefacts have been displaced down the cave profile through agencies such as termites. At present, further excavation is required to confirm the presence of humans in Sahul before 55 ka.

The earliest evidence for our species in China deferred to Chapter 10. This is because China is best treated as a single system that encompasses both north China in the Palearctic Realm and south China in the Oriental Realm. We therefore need to review the evidence from Siberia and Mongolia (Chapter 9) before examining developments in north and south China and the interactions between these two regions.

Note

1 I can recommend (almost) all of these from my times in southern China.

References

Atkinson, Q.D., Gray, R.D. and Drummond, A.J. (2008) mtDNA variation predicts population size in humans and reveals a major southern Asian chapter in human prehistory. *Molecular Biology and Evolution*, 25(2): 468–474.

Boivin, N., Fuller, D.Q., Dennell, R.W., Allaby, R. and Petraglia, M. (2013) Human dispersal across diverse environments of Asia during the Upper Pleistocene. *Quaternary International*, 300: 32–47.

Dennell, R.W. and Petraglia, M.D. (2012) The dispersal of *Homo sapiens* across southern Asia: how early, how often, how complex? *Quaternary Sciences Reviews*, 47: 15–22.

Groucutt, H.S. and Blinkhorn, J. (2013) The Middle Palaeolithic in the desert and its implications for understanding human adaptation and dispersal. *Quaternary International*, 300: 1–12.

Groucutt, H., Grün, R., Zalmout, I.S.A., Drake, N.A., Armitage, S.J. et al. (2018) *Homo sapiens* in Arabia by 85,000 years ago. *Nature Ecology and Evolution*, 2: 800–809, doi:10.1038/s41559-018-0518-2.

Hershkovitz, I., Weber, G.W., Quam, R., Duval, M. and Grün, R. (2018) The earliest modern humans outside Africa. *Science*, 359: 456–459.

Klein, R.G. (2008) Out of Africa and the evolution of human behavior. *Evolutionary Anthropology*, 17: 267–281.

Kuhn, S.L. and Stiner, M.C. (2006) What's a mother to do? The division of labor among Neandertals and modern humans in Eurasia. *Current Anthropology*, 47(6): 953–980.

Lahr, M.M. and Foley, R. (1994) Multiple dispersals and modern human origins. *Evolutionary Anthropology*, 3(2): 48–60.

Mellars, P. (2006) Why did modern human populations disperse from Africa ca. 60,000 years ago? A new model. *Proceedings of the National Academy of Sciences USA*, 103(25): 9381–9386.

Rabett, R.J. (2018) The success of failed *Homo sapiens* dispersals out of Africa and into Asia. *Nature Ecology and Evolution*, 2: 212–219.

Reyes-Centeno, H., Ghirotto, S., Détroit, F., Grimaud-Hervé, D., Barbujani, G. and Harvati, K. 2014. Genomic and cranial phenotype data support multiple modern human dispersals from Africa and a southern route into Asia. *Proceedings of the National Academy of Sciences USA*, 111(20): 7248–7253.

4

ARABIA TO THE THAR DESERT

Introduction

Until a few years ago, the Middle Palaeolithic of Arabia could have been summarised on a postcard. It was also devoid of a single date. Unsurprisingly, it was ignored in almost every synthesis about the dispersal of our species out of Africa, with the Levant always dominating discussion about Southwest Asia and the earliest evidence for our presence outside Africa. Thanks to an extraordinary series of surveys and excavations in the last decade, there is now a large and still-growing literature that can no longer be ignored. A clear storyline is now emerging: Arabia was colonised by hominins – probably *Homo sapiens* – by MIS 7 (ca. 200–250 ka), and perhaps earlier; the environmental and archaeological evidence indicates repeated episodes of colonisation as far north as the Levant in moist periods, followed by local developments in lithic traditions, and local extinction and abandonment when aridity returned. One issue that is unresolved is whether the whole of Arabia was abandoned in arid episodes such as MIS 4 (ca. 60–80 ka), or whether some populations survived in refugia. Looking further east, humans may have reached the Thar Desert of northwest India in MIS 5 (ca. 80–125 ka BP) via a corridor along southern Iran and Pakistan. Our species probably encountered Neandertals in northern Arabia, and perhaps also in southwest and south Iran, southern Pakistan and the Indus Valley, and it may have been through these contacts that non-Africans first acquired some Neandertal DNA. These are all new narratives about the earliest history of our species outside Africa. The key factor that underpins all these themes is rainfall, but first, some background information about Arabia is required.

The Arabian Peninsula

The Arabian Peninsula is vast: if defined arbitrarily at its northern end by the border of Saudi Arabia with Jordan, Iraq and Kuwait, it covers 1,200,000 square miles (ca. 3,200,000 sq km) (an area 12 times larger than the United Kingdom) but it extends further northwards into the Jordanian and Syrian Deserts. Its western edge along the Red Sea is a giant escarpment that is cut by numerous wadis and slopes downwards towards the Arabian/Persian Gulf. In moist periods, this escarpment received higher rainfall from the African monsoon that activated (or re-activated) rivers that flowed eastwards from the peninsula and provided resource zones, drinkable perennial water as well as corridors for movement. The key to "green Arabia" is therefore these active river systems rather than the whole area being green.

Today, the Arabian/Persian Gulf is a shallow sea not more than 90 m deep but during much of the Pleistocene, it would have been a broad river valley, with lakes and marshes (Lambeck 1996). Mountain ranges along the western, southern and southeast coasts of Arabia rise to over 3,000 m in Yemen and Oman. Entry into Arabia from Africa during the Pleistocene would have been either via the land bridge of the Sinai Peninsula at the northern end of the Red Sea, or across the Bab el Mandab Strait at its southern end. At times of low sea level, this would have narrowed from ca. 30 km to ca. 18 km; with island hopping, it would not have been necessary to cross >4 km of open water (Lambeck et al. 2011). By considering sea level changes and regional climate records, Rohling and colleagues (2013: 183) identified several "windows of opportunity" at 458–448, 345–340, 272–265, 145–140, and 70–65 ka as the optimal times for crossing.

Today, much of the Arabian Peninsula is desert: this is dominated by the Nefud Desert in the north, and in the south, the enormous sand sea of the Empty Quarter, or Rub'al Khali, which covers 600,000 sq km (an area larger than France) and was immortalised by Bertram Thomas's (1932) *Arabia Felix*, and William Thesiger in his book *Arabian Sands* (1960). Most of Arabia is arid, with typically <100 mm rain per year, but the Yemeni/Asir Highlands along the coast with the Indian Ocean receive up to 400 mm of rainfall from the African and Indian monsoon systems (see Figure 4.1). As in other arid regions, annual rainfall is highly variable, and often 100–200 mm of rain will fall in one event; variability, not annual average, is the key factor in this type of high-risk environment. Winter rainfall over northern Arabia is derived from westerly Mediterranean winds, but the most important contribution comes from the summer rains of the African and Indian monsoon systems. Here, the critical variable in Arabian prehistory is the location of the Inter-Tropical Convergence Zone (ITCZ). When this moved north, Arabia became green as lakes refilled and rivers flowed (Groucutt and Petraglia 2012; Jennings et al. 2015).

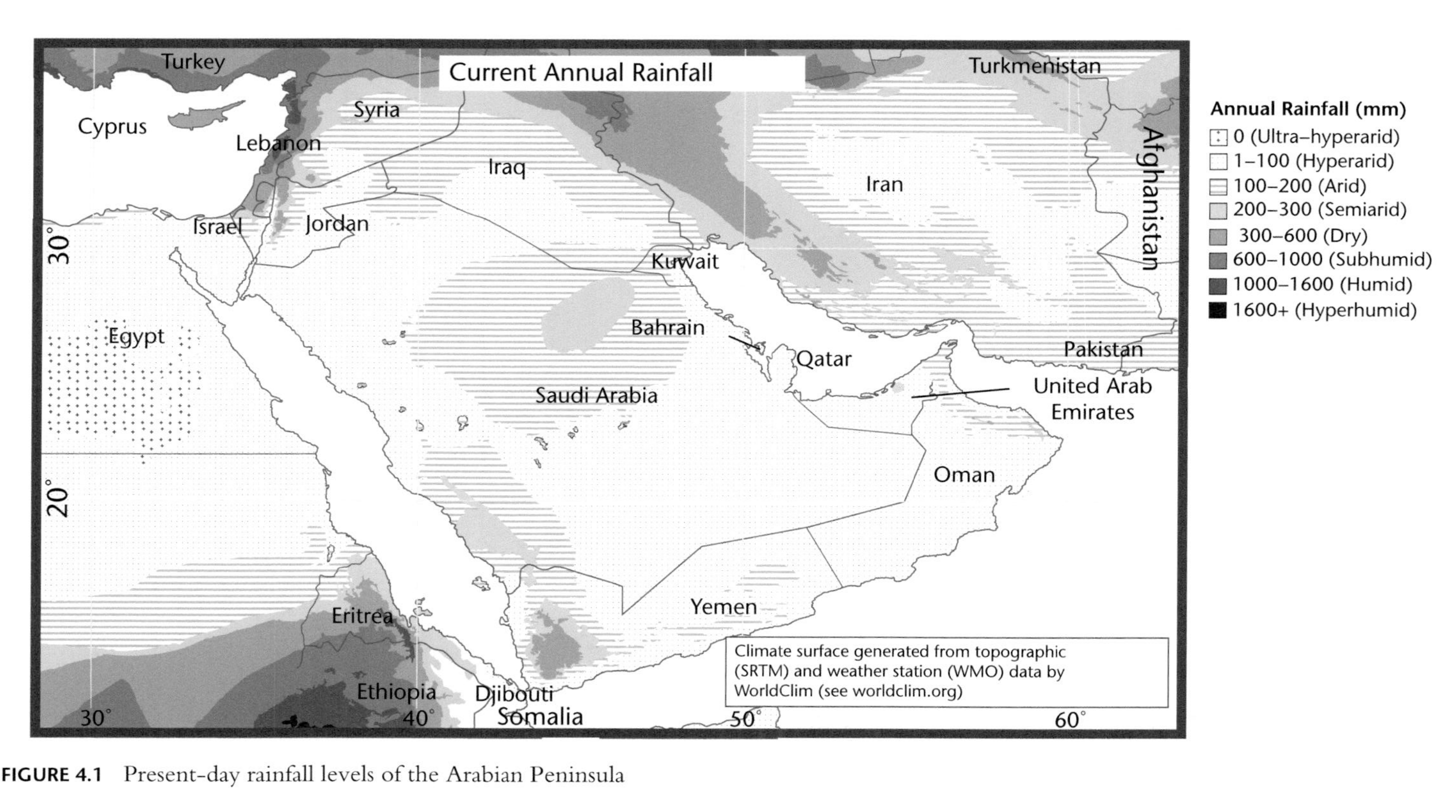

FIGURE 4.1 Present-day rainfall levels of the Arabian Peninsula

Arid to hyper-arid levels prevail across the Peninsula, with exceptions in Yemen, Oman, and parts of Jordan, where levels exceed 300 mm per annum.

Source: Jennings et al. 2015, Fig. 1 (the original is in colour).

Green Arabia

As with the Sahara, Arabia was not always the arid desert that is seen today (see Drake et al. 2013; Chapter 2). In the last interglacial, and probably during earlier interglacials, Arabia had hundreds of lakes and was traversed by numerous river systems (see Figure 4.2). Astonishingly, the palaeolake at Mundafan in the now hyper-arid Rub'al Khali was up to 30 m metres deep and covered ca. 300 sq km during MIS 5e (ca. 125 ka), MIS 5c (ca. 100 ka) and MIS 5a (ca. 80 ka) (Rosenberg et al. 2011; Groucutt et al. 2015a), and was home to hippopotamus (*H. amphibius*) as well as groups with Middle Palaeolithic tool-kits (Crassard et al. 2013). Even larger was palaeolake Saiwan that covered ca. 1,400 sq km in the interior of Oman between 132 and 104 ka (Rosenberg et al. 2012), but the largest so far recorded is a palaeolake dated to MIS 5 in the Mudawwara depression of southern Jordan that covered 2,000 sq km and was over 40m deep (Petit-Maire et al. 2010). Similarly, in the hyper-arid Nefud Desert in northern Saudi Arabia, there was a succession of palaeolakes in MIS 11 (ca. 420 ka), MIS 9 (ca. 320 ka), MIS 7 (ca. 200 ka), MIS 5e (ca. 125 ka) and MIS 5c (ca. 100 ka) (Groucutt et al. 2015b; Petraglia et al. 2012; Rosenberg et al. 2013). Other lakes in southeast Saudi Arabia date to the early part of MIS 3 (58–61 ka) (Parton et al. 2013). The large river systems that flowed eastwards from the Red Sea escarpment into the Arabian/Persian Gulf (see Breeze et al. 2015; Figure 4.2) are particularly important beyond their obvious importance as a source of perennial water because they are obvious corridors for any incoming populations from Africa.

The climatic record of Arabia is now known in considerable detail from offshore marine sediment cores (Almogi-Labin et al. 2000; Clemens and Prell 2003; Emeis et al. 1995), speleothem records from the Negev Desert, Israel (Vaks et al. 2007), Hoti Cave, Oman (Fleitmann et al. 2003), Mukalla Cave, Yemen (Fleitmann et al. 2011), studies of sand dune formation (Farrant et al. 2015), alluvial fans (Parton et al. 2015a, 2015b), plant phytoliths (see Groucutt et al. 2015a; Jennings et al. 2016) as well as the palaeolakes mentioned above. These show a succession of moist and arid episodes; unsurprisingly, the record of occupation is mostly restricted to the moist episodes (see Figures 4.3 and 4.4). Vegetational modelling shows a "green Arabia", with semi-arid grasslands replacing deserts in many parts; deserts now receiving <100 mm rain each year became semi-arid grasslands with an annual rainfall of 300–600 mm. (Figure 4.4; see Jennings et al. 2015). Speleothem growth at Mukalla Cave in southern Yemen between 123 and 130 ka indicates that more than 300 mm of rainfall occurred in a location that today receives only 150 mm per year (Fleitmann et al. 2011). In Oman and other parts of southeast Arabia, evidence from speleothems, palaeolakes and fluvial deposits all indicate increased rainfall and re-activated drainage systems in the last interglacial (Jennings et al. 2015) and during the Holocene (Matter et al. 2016).

(One should note an ongoing debate over whether there were palaeolakes in Arabia. Enzel et al. [2015, 2017] argue that these were in fact shallow marsh environments that required only a modest increase in rainfall, a claim that is

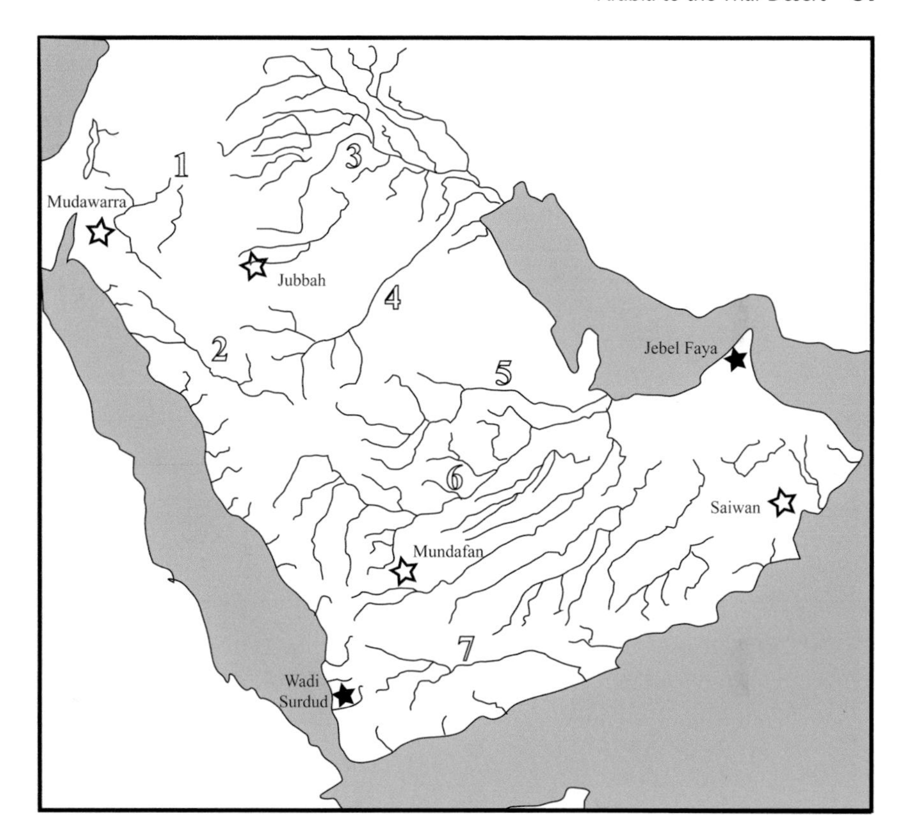

FIGURE 4.2 Palaeodrainage of Arabia

The main drainage systems are: 1 Wadi as Sirhan; 2 Wadi al Hamd; 3 The Palaeo-Euphrates and associated Widyan; 4 Wadi Al Batin; 5 Wadi Sahba; 6 Wadi ad Dawasir; and 7 Wadi Hadramawt. Open stars indicate large palaeolakes. Filled stars indicate archaeological sites.

Not all stream channels would have been active at the same time, although most were probably active in MIS 5. Some channels are covered by subsequent dune formations.

Source: This figure has been simplified from Fig. 1 in Crassard et al. 2013 (the original version is in colour).

countered by Engel et al. [2017]. Whilst not denying that there may have been marshes in Arabia, multiple lines of evidence offer convincing evidence that there were numerous and sometimes extensive lakes in Arabia.)

Faunal data shows a grassland fauna in the last interglacial and during similar moist episodes, with wild ass (*Equus hemionus*), gazelle and oryx as the main ungulates, and even hippopotamus in some of the lakes (Stewart et al. 2019). Because hippos require water at least two metres deep and abundant grasses on the lake margins for night-time feeding, their presence at some Arabian lakes is a graphic illustration of how green Arabia was in the last interglacial and similar moist episodes, compared to the present. At T'is Ghadah in the now hyper-arid Nefud Desert of northern Saudi Arabia during MIS 9, there was sufficient foliage to

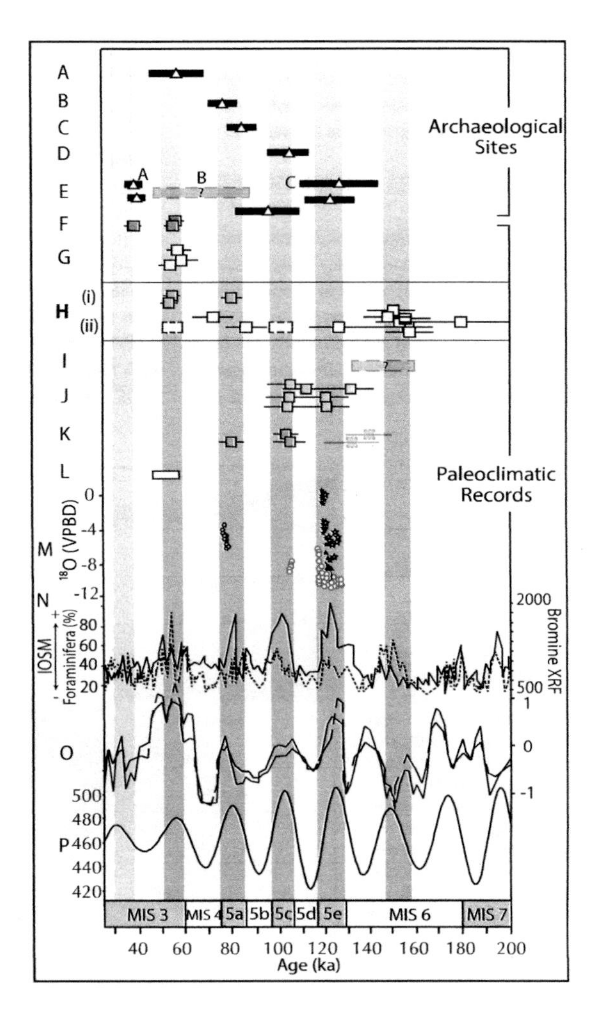

FIGURE 4.3 Comparison of palaeo-environmental records and dated Middle Palaeolithic sites in Arabia

Grey bars indicate humid periods. MIS = Marine isotope stage; IOSM = Indian Ocean summer monsoon. Archaeological sites: A = Shi'bat Dihya, Yemen; B = JQ1, northern Saudi Arabia; C = JKF, northern Saudi Arabia; D = Aybut Auwal, southern Oman; E = Jebel Faya assemblages A–C, United Arab Emirates (UAE). Palaeoclimatic records: F = fluvial deposits, central Saudi Arabia; G = Aqabah paleolake, UAE; H = Al Ain alluvial fan, UAE; I = palaeosols, Wahiba Sands, Oman; J = palaeolake Saiwan, Oman; K = palaeolakes Mundafan and Khujaimah, southern Saudi Arabia; L = Moomi cave speleothems, Socotra, Yemen; M = cave speleothems from Hoti, Oman, and Mukalla, Yemen; N = Arabian Sea productivity from foramiferal assemblages (solid line) and bromine counts (dashed line); O = summer monsoon; P = summer insolation at 30° N. Dashed boxes represent either age uncertainties or inferred ages.

Source: Parton et al. 2015, Fig. 3.

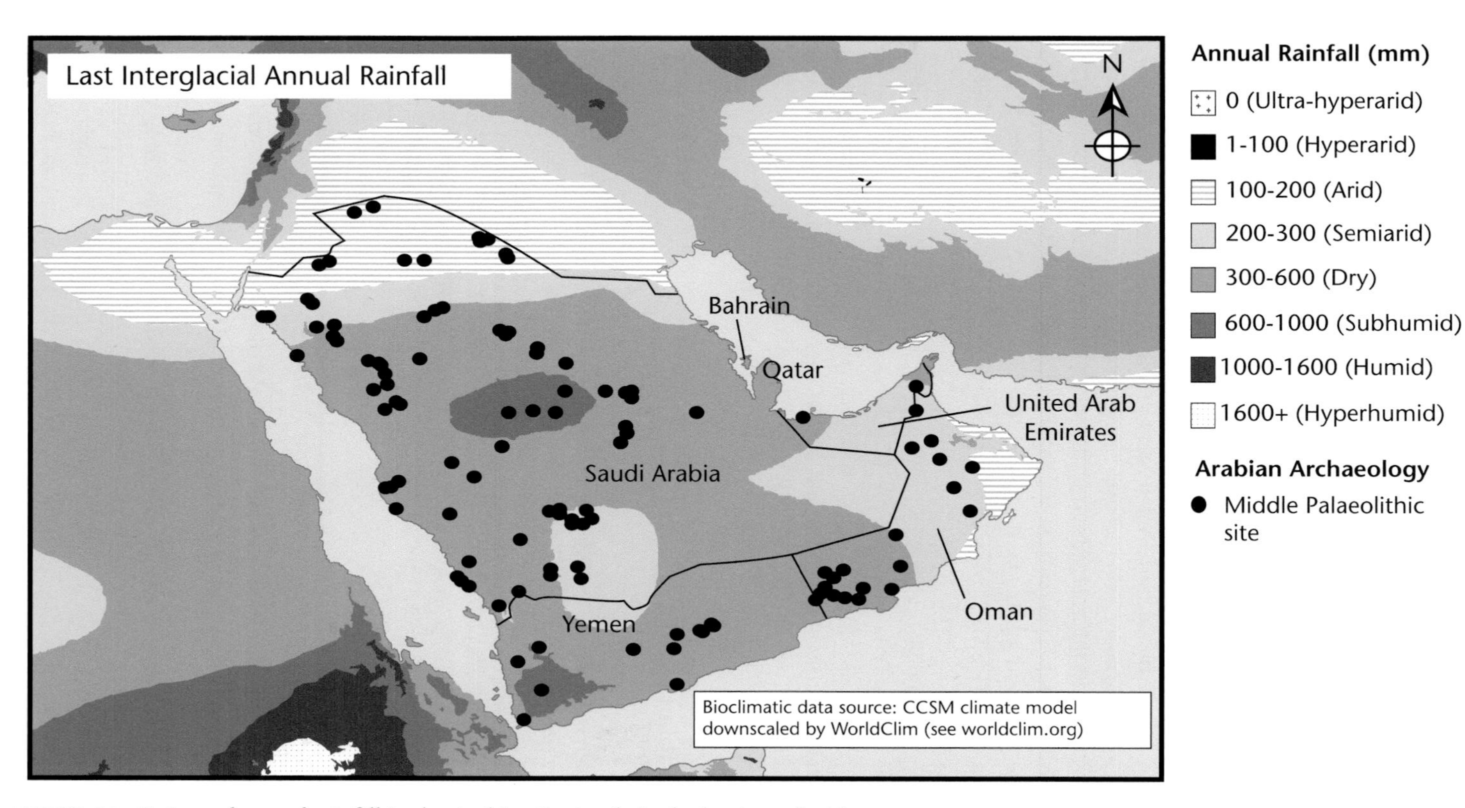

FIGURE 4.4 Estimated annual rainfall in the Arabian Peninsula in the last interglacial

When Arabian Middle Palaeolithic archaeological sites are plotted, the majority of sites are in areas of dry to sub-humid rainfall.

Source: Jennings et al. 2015, Fig. 10.

maintain elephants, and enough prey to support the large predator *Panthera cf. gombaszogensis* (Stimpson et al. 2015). There was also a diverse bird fauna at T'is Ghadah (Stimpson et al. 2016) that must have been an attractive food resource for humans, and there is indirect evidence for fish populations in the palaeolake at Mundafan in the Rub'al Khali (Matter et al. 2015); birds and fish were presumably also abundant in other Arabian lakes. According to genetic evidence, the hamadryas baboon (*Papio hamadryas*) may have dispersed from Africa into southern Arabia (but not the Levant) during moist episodes of the last 130,000 years (Kopp et al. 2014; see also Figure 4.3). Animal population densities and levels probably mapped onto these fluctuations in rainfall, with growth and expansion in moist times, and contraction and depopulation when arid. Refugia for some animals (including humans) may have been in the Yemeni/Asir and Omani Highlands in the south and east of Arabia (Delagnes et al. 2013), where rainfall would have remained sufficiently high for their survival, and along the rivers and marshes on the floor of the Gulf (Rose 2010). In the Last Glacial Maximum (LGM; = MIS 2), and probably during most of MIS 4 and MIS 6, most of Arabia was likely too arid to be inhabited.

It is thought unlikely that the greening of Arabia resulted from increased rainfall from the westerly winds of the Mediterranean. Instead, it was the shifting northwards of the Inter-Tropical Convergence Zone (ITCZ) that brought higher rainfall from the African and Indian monsoons (Groucutt and Petraglia 2012; Jennings et al. 2015). According to Rosenberg and colleagues (2013), the African monsoon was likely the more important.

The Arabian Middle Palaeolithic

When set against our previous ignorance, a phenomenal amount has been learnt in the last ten years about the Middle Palaeolithic of the Arabian Peninsula. Several sites have now been dated, and there are now detailed descriptions of some carefully collected lithic assemblages. Figure 4.5 shows a selection of Middle Palaeolithic artefacts from Arabia. Unfortunately, few of these have any associated faunal material. Two research questions are dominant: first, where did the inhabitants come from; second, were the makers of these assemblages *Homo sapiens* or Neandertals (or even another hominin species)?

Early Homo sapiens *in Levant and Arabia*

Two recent discoveries, and perhaps a third, considerably clarify when *Homo sapiens* entered Arabia. The first is from Misliya Cave, Israel, where skeletal remains attributed to *H. sapiens* are dated to between 177 ka and 194 ka (Hershkovitz et al. 2018; see Chapter 8), and the second is a finger bone from Wusta, Saudi Arabia, which is also classified as *H. sapiens* and is dated to ca. 85 ka (Groucutt et al. 2018). These discoveries make it probable that our species was in the Arabian Peninsula after 200 ka. (In the case of Wusta, however, we don't yet know if it entered Arabia from

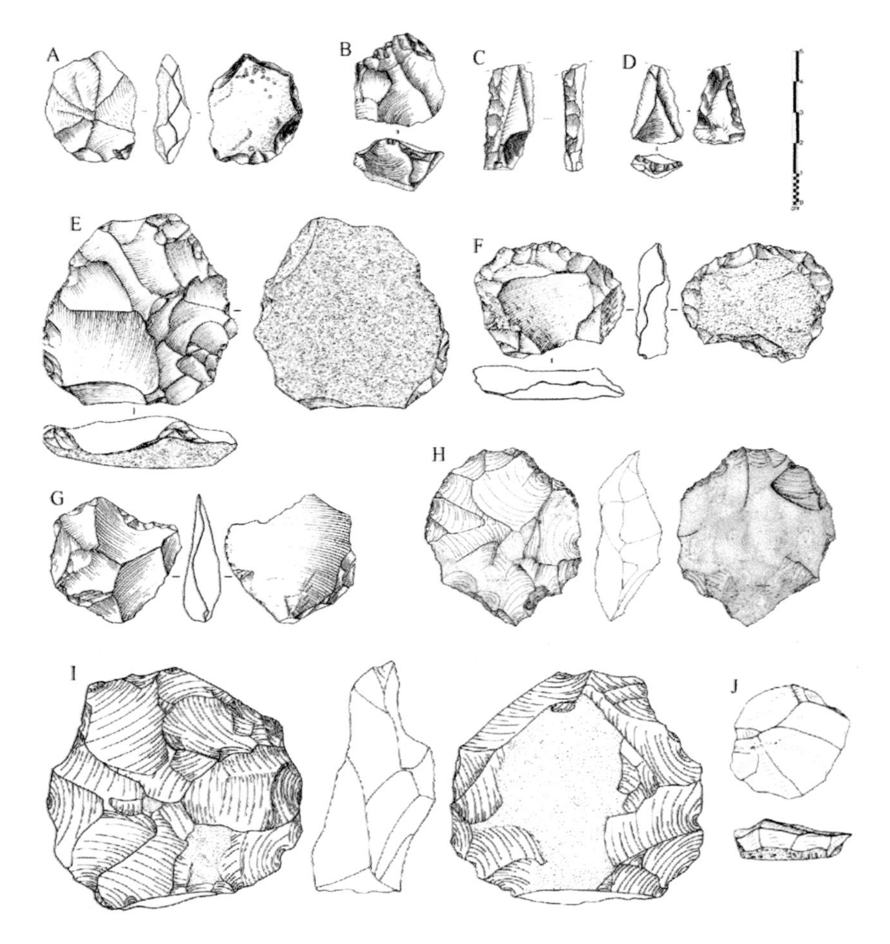

FIGURE 4.5 A selection of Middle Palaeolithic artefacts from Arabia

A–D: Jubbah, northern Saudi Arabia. E–G: Mundafan, southwest Saudi Arabia. H, I: Jebel Barakah, United Arab Emirates, J: Locality specimen 212-46, central Saudi Arabia. A, B, E, F, H, I, J: Levallois cores; C: side retouched flake; D: retouched point; E: Levallois flake. Note that specimen 212-46 is not to scale as this was not included in the original publication.

Source: Groucutt et al. 2015c, Fig. 12.

Africa or the Levant.) The third is the recent publication of a cranium identified as *H. sapiens* from Apidima, Greece, and dated to 210 ka (Harvati et al. 2019; see Chapter 8). Because *Homo sapiens* had emerged in Northeast Africa by at least 320 ka and was likely distributed thereafter across much of Africa (see Chapter 2), and because the earliest Arabian Middle Palaeolithic assemblages now date from MIS 9 (260–360 ka) or MIS 11 (340–440 ka) (Stimpson et al. 2015), it might even have been in Arabia before 200 ka. Additionally, the environments of the "green Sahara" and "green Arabia" were very similar, and populations would not have had any problems moving into the eastern part of the Saharan-Arabian Realm (see Chapter 3).

What's the earliest evidence for the Middle Palaeolithic in Arabia?

At present, the earliest Middle Palaeolithic assemblage in Arabia may come from the fossil locality of T'is al Ghadah in the western Nefud Desert of northern Saudi Arabia where, as noted above, a wide variety of fossil species were found in a stratified context in the lakeshore deposits that have been dated to MIS 9 or MIS 11 (Stimpson et al. 2015). Low density but discrete scatters of lithic were found on the basin surface, but unfortunately not in association with the stratified fauna (Scerri et al. 2015). Future work may confirm the association.

The earliest securely dated Middle Palaeolithic artefacts in Arabia are those from the lowest exposed layers at Jebel Qattar 1 in the western Nefud Desert, where 28 pieces of debitage were found in a deposit dated to 211 ± 16 ka that can be placed in MIS 7 (Petraglia et al. 2012). There are also assemblages D and E from Jebel Faya, Oman (Bretzke et al. 2014). These were found below assemblage C, which is dated to MIS 5e (Armitage et al. 2011) (see below).

Studies of the Arabian Palaeolithic are still in their infancy, and we might confidently expect more examples of Middle Palaeolithic artefacts in contexts pre-dating MIS 5e. It is perhaps relevant that at Misliya Cave, Israel, the transition from the Lower to the Middle Palaeolithic has now been dated to ca. 250 ka (Valladas et al. 2013; see Chapter 8). As this transition is assumed to indicate the arrival of *Homo sapiens* (as in Africa), similarly early evidence might be expected from Arabia.

The Arabian Middle Palaeolithic of the last interglacial (MIS 5)

Most of what we have learnt about the Arabian Middle Palaeolithic relates to the "green Arabia" of the last interglacial sensu lato, i.e. MIS 5e–MIS 5a. At this time, Arabia was open to colonisation from at least three directions: from Africa across the Sinai Peninsula, across the Bab-el-Mandeb strait at the entrance to the Red Sea; from the Levant to the north, and perhaps from southwest Iran to the east. Recent GIS modelling suggests that the Levant and north Arabia were linked by the "Tabuk corridor" during moist parts of the last interglacial (MIS 5a and MIS 5e) (Breeze et al. 2017). When Arabia became green, as in MIS 9, MIS 7 and MIS 5, populations had only to follow rivers and lake systems to be able to disperse over the entire peninsula. In theory, analysis of stone tool assemblages should indicate their affinities to the assemblages of their source populations in any of these regions. In practice, this is extremely difficult because we lack a reliable chronological framework that shows the relative sequencing of assemblages, confidence over the stratigraphic integrity of each assemblage, and confidence that different assemblages have been analysed in a comparable manner. Additionally, and most importantly, we need to be able to distinguish between cultural transmission and technological convergence.

Cultural transmission and technological convergence

Similarities in lithic assemblages from different regions may indicate the movement of groups from one region to another providing that the chronological gap between them is sufficiently small that we can assume continuity in population. (We need, of course, secure dates to demonstrate this, and at present we lack that degree for resolution for the region between Arabia and India.) These similarities could also indicate technological convergence: in other words, people in different regions changed some aspects of their technology in the same way without any contact. Examples might be that people independently developed a microlithic technology or used a blade technology more frequently. There is also the issue of drift: over time and distance, people change their habits, and so their lithic technology would change from what it had originally been. As we will see in this and the next chapter, these problems are central to discussions of the dispersal of humans from Africa to Arabia and India.

As so little is known about the Middle Palaeolithic of southern Iran, most discussions about the origin of the Arabian Middle Palaeolithic have concerned Northeast Africa and the Levant. Experienced lithic analysts can sometimes distinguish between the two, even though both regions relied heavily on the Levallois technique of core preparation.

The Nubian connection

Nubian assemblages are a distinct part of the African Middle Stone Age and are distributed over Egypt and Sudan and other parts of Northeast Africa. They are characterised by a highly specific method of producing elongated points or pointed flakes from a distinctive method of Levallois core preparation (Crassard and Hilbert 2013; Rose et al. 2011; Usik et al. 2013; see Figure 4.6), and by an equally specific output, the Nubian point. At Taramsa Hill, Egypt in the lower Nile Valley, a late Nubian assemblage was associated with the burial of a *H. sapiens* child, dated at 55.5 ± 3.7 ka BP[1] (Vermeersch et al. 1998), so it seems safe to attribute Nubian assemblages with our species. Nubian assemblages are also found at several places in Arabia, in central Yemen, and in Dhofar Province, southern Oman. At Aybut Al Auwal in Dhofar, a Nubian assemblage was dated to 106 ± 9 ka (Rose et al. 2011), which corresponds to the moist episode MIS 5c (c.110–100 ka). In Dhofar, Nubian assemblages are followed by local ones known as Mudayyan (Usik et al. 2013). Nubian-like assemblages have also been reported from the Nejd in central Saudi Arabia (Crassard and Hilbert, 2013), Al-Jawf province in northern Saudi Arabia (Hilbert et al. 2017) and the Sinai Peninsula (Rose et al. 2011). Although these are not dated, they are likely related to moist episodes of MIS 5.

At present, Nubian assemblages in Arabia (and perhaps the Thar Desert; see below) provide the best lithic proxy signal of *Homo sapiens*. However, as Usik and colleagues (2013) point out, it would be overly simplistic to interpret these as a single dispersal event; rather, there may have been recurring bi-directional movements between Arabia and the African side of the Red Sea.

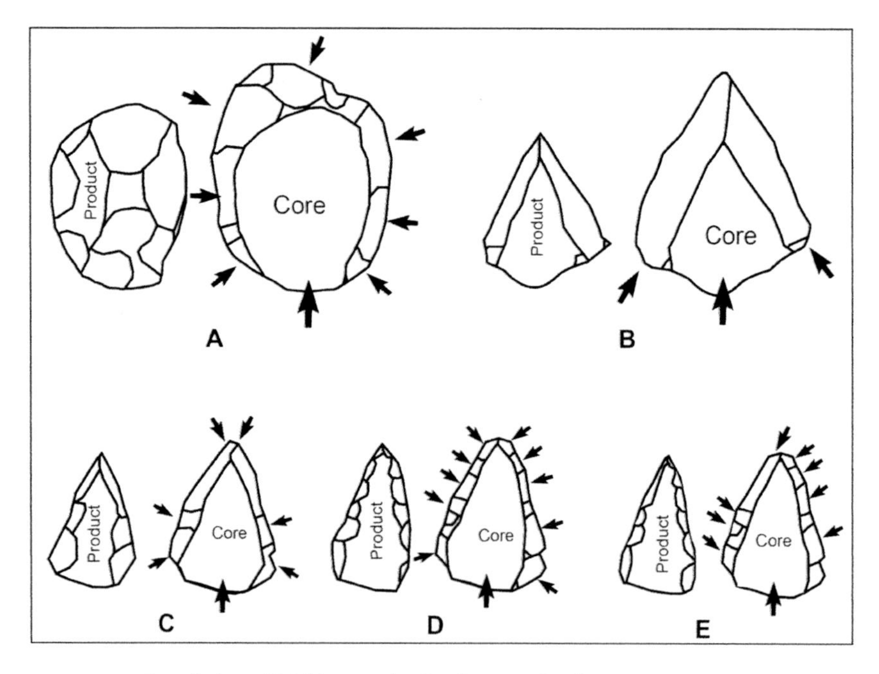

FIGURE 4.6 Levallois and Nubian methods of core reduction

A: Preferential Levallois flake production with centripetal preparation; B: Preferential Levallois point production with unidirectional convergent preparation; C: Nubian Levallois type 1 with distal divergent preparation; D: Nubian Levallois type 2 with double lateral preparation; E: Nubian Levallois type 1/2 with mixed type 1 and type 2 preparation.

Source: Crassard and Hilbert 2013, Fig. 5.

Jebel Faya assemblage C

Jebel Faya is a low mountain up to 350 m high in the UAE (United Arab Emirates) that is currently 55 km from the Gulf of Oman and the Arabian/Persian Gulf; the site of Jebel Faya, named FAY-NE1, is a rock shelter 180 m a.s.l. (above sea level) at its northern end. Excavations in 2003 and 2004 cut a section 24 m long and up to 5 m deep. Six stratified assemblages (A–E) were found, with E the earliest (Bretzke et al. 2014).

Three OSL dates associated with the lithic material of assemblage C provided age estimates of 127 ± 16 ka, 123 ± 10 ka, and a minimum age of 95 ± 13 ka (at one SE [standard error] of uncertainty). The earliest two dates allow correlation with MIS 5e. Little information is available on the lithics but they are said to show a variety of reduction strategies, with blank production by the Levallois method, volumetric blade, and simple parallel methods. Small hand axes, foliates, foliate preforms, end scrapers, sidescrapers and denticulates are present (see Figure 4.7). The small hand axes and foliates are features of MSA assemblages in Northeast Africa and are not found in Levantine Middle Palaeolithic that are

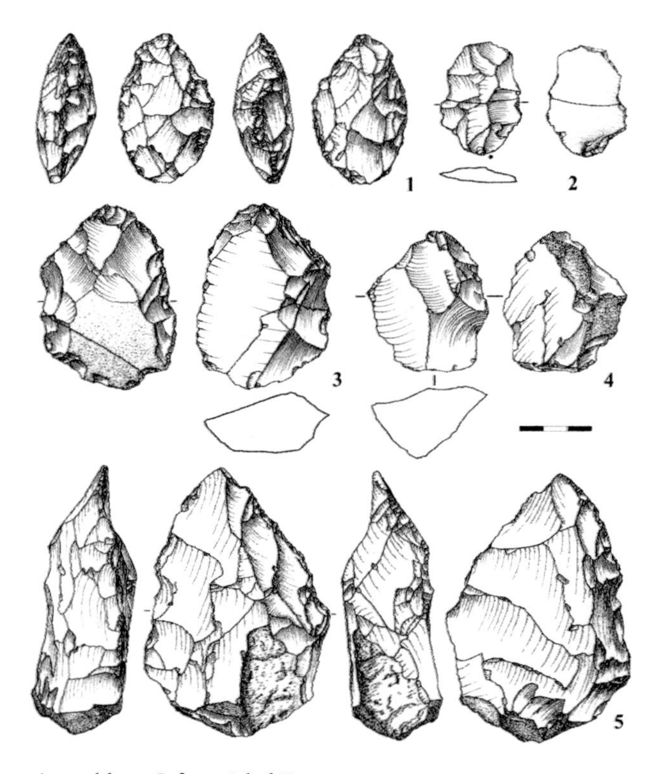

FIGURE 4.7 Assemblage C from Jebel Faya
1 bifacial foliate; 2 Levallois flake; 3 bifacial preform; 4 radial core; 5 hand axe preform.
Source: Armitage et al. 2011, Fig. 1.

contemporary with assemblage C. Marks (2009: 305) points out an absence of Levantine features such as elongated blanks, Levallois points and unidirectional-convergent reduction strategies. Because of the features that are shared by assemblage C with the African MSA, and because the African MSA was made by *Homo sapiens*, the investigators proposed that Jebel Faya was first colonised by groups of our species dispersing out of Africa at the beginning of MIS 5e, when rainfall was higher but sea levels still lower across the Bab al Mandab Strait at the southern end of the Red Sea (Armitage et al. 2011). In this scenario, our species was already across Arabia at the beginning of the last interglacial. Yet the claimed similarities between Jebel Faya C and the African MSA are very general, and we lack comparable data from neighbouring southwest Iran. (Paul Mellars emphatically rejected any comparisons of the Jebel Faya C assemblage with East Africa [see Lawler 2011], although Bretzke and Conard [2017] side with Armitage and his team; we shall return later to this criticism in Chapter 5 after examining the evidence from southern Iran, the Thar Desert and peninsular India.)

Rub'al Khali, the Mundafan palaeolake of MIS 5

The Mundafan palaeolake in the Rub'al Khali was first investigated in the 1970s by McClure (1976), who reported a rich fauna that included wild goat or sheep (*Capra* sp./*Ovis* sp.), oryx (*Oryx* sp.), gazelle (*Gazella* sp.), horse and ass (*Equus* sp.), camel (*Camelus* sp.), wild cattle (*Bos primigenius, Bubalus* sp.), *Hippopotamus amphibius*, and birds such as ostrich (*Struthio* sp.). Most of these species live in grassland and open woodlands and have large foraging ranges, but the presence of hippopotamus suggested that palaeolake Mundafan was at times a deep permanent water body (Crassard et al. 2013). The presence of mussels from the genus *Unio* indicates the presence of freshwater fish as these are intermediate hosts for their larvae (Matter et al. 2015). Radiocarbon dates initially (and incorrectly) suggested that the palaeolake was formed in MIS 3 but recent investigations, using OSL dating techniques that were not available to McClure, now indicate that the Mundafan palaeolake dates to the moist episodes MIS 5a and MIS 5c ca. 80 and 100 ka (Rosenberg et al. 2011). A major breakthrough in investigating the Palaeolithic evidence around the palaeolake was the discovery and excavation of a stratified assemblage at Mundafan Al-Buhayrah (MDF-61). The site was occupied during MIS 5a when the area was arid but still contained a large lake and was capable of supporting a diverse ungulate fauna. The lithics (N=935) from MDF-61 are similar to those with fossils of *Homo sapiens* in Africa and the Levant, as well as assemblages in India that lack them (see Chapter 5). The main shared characteristics of these assemblages are the combination of high frequencies of recurrent centripetal and centripetally prepared Levallois methods of core reduction, and retouched tools that are focused on side-retouched flakes and retouched points. The MDF-61 assemblage is therefore important in providing the first stratified and dated human occupation in the Empty Quarter, and also a reference point for lithic comparisons by being the first to demonstrate distinct similarities between contemporaneous African, Levantine and Arabian assemblages, as well as late MIS 5 with material from the site of Jwalapuram 22 in southern India (Groucutt et al. 2015a, 2015b).

The Nefud Desert, Jubbah

Numerous lakes and streams once watered the now hyper-arid Nefud Desert of northern Saudi Arabia. These could have been reached from several directions – north and west from the Levant and Sinai Peninsula, north and east from the Palaeo-Euphrates Basin, or east from the Gulf. Whilst newcomers might inadvertently show where they had come from by their stone tool kits, over time these groups might have developed their own traditions. The diversity of lithic assemblages found on or near the shores of these extinct lakes shows very clearly the problems of following these "stone breadcrumbs" (Rose et al. 2011: 18) back to their area of origin, and the difficulties of inferring the routes of dispersals from stone tool assemblages.

Several assemblages have been found recently that can be related to palaeo-lakes that flourished, even if briefly, during the moist episodes MIS 5e, 5c and 5a between ca. 125 and 80 ka in the area near the modern town of Jubbah (Petraglia et al. 2012). We have already noted Jebel Qattar 1, in which a palaeosol containing 28 pieces of debitage has been dated to ca. 210 ka in MIS 7 (see above). A larger assemblage of 160 artefacts was recovered at Jebel Qattar from an upper palaeosol dated to 75 ± 5 ka and attributed to MIS 5a. Of these artefacts, 123 pieces were debitage, and there are 22 retouched tools (Petraglia et al. 2012). A selection is shown in Figure 4.8. At the nearby site of Jebel Katefeh (JFK-1) 1,222 artefacts (300 of which were excavated) were recovered from a sand deposit. The proportions are similar to those at JQ-1: 91% (N=1113) debitage, 8% (N=67) cores and only 1% (20) retouched tools. Seven artefacts were tested for residues, which showed usage on both animal and plant materials (Petraglia et al. 2012: 10). Overall, the assemblage is interpreted as the outcome of a short-term occupation on a dune overlooking an area that may have had

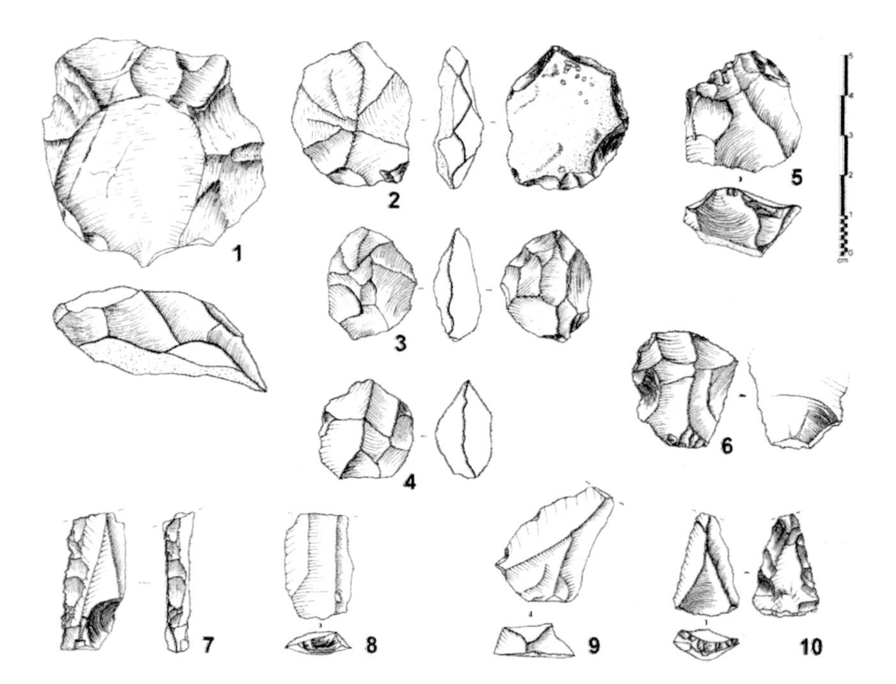

FIGURE 4.8 Middle Palaeolithic artefacts from Jebel Qattar 1, Nefud Desert

1 Preferential Levallois core with centripetal preparation and facetted striking platform. 2 Flat centripetal Levallois core. 3 Discoidal core with centripetal debitage. 4 Discoidal core with centripetal debitage. 5 Discoidal core with bidirectional debitage. 6 Flake struck from centripetal Levallois core. 7 Side retouched blade. 8 Facetted blade (proximal fragment). 9 Pseudo-Levallois point with dihedral platform. 10 Thick, retouched unifacial Levallois point with facetted platform and ventral stepped retouch along two lateral edges.

Source: Petraglia et al. 2011, Fig. 3.

standing water. Dating has proved difficult because the grains used in OSL dating indicate equally burial ages of 90–85 ka (MIS 5a–b) and 50 ka (MIS 3). It is possible that the younger grains are intrusive, in which case the site dates to MIS 5a (Groucutt et al. 2015b).

A third assemblage was found at the nearby site of Jebel Umm Sanman (JSM-1) in a narrow, yellow quartzite seam at the base of the local Jebel, or mountain. The 11 buried artefacts and 77 ones on the surface were homogenous, made from the local quartzite, and flaked by the Levallois technique. As with JFK-1, dating by OSL has proved difficult, with burial ages indicating a likely age between 100 and 60 ka and an occupation in a humid period in MIS 5.

All three sites – JQ-1 upper, JKF-1 and JSM-1 – are variations on a Levallois theme. The main take-home message is that they are not obviously North African or Levantine in origin, but share features with both, and also have their own characteristics. It is likely that groups migrating into this area and attracted by the resources available around the palaeolakes came from a different source population, may have interbred, and developed their own local traditions. It is even possible that both Neandertals and *Homo sapiens* inhabited this part of northern Saudi Arabia. As remarked by Scerri and colleagues (2014: 125), "MIS 5 hominin demography at the interface between Africa and Asia was complex".

Local developments in MIS 3: Wadi Surdud, Yemen, and Jebel Faya assemblages A and B

Most of Arabia was likely depopulated by the aridity of MIS 4 between 80 and 60 ka, with perhaps a few small populations isolated in refugia in the hills of southwest and southeast Arabia. Two possible examples are sites in the Wadi Surdud, western Yemen, and the upper assemblages A and B from the Jebel Faya (see above). In both cases, the assemblages have no obvious counterparts elsewhere in Arabia or in neighbouring regions.

The Wadi Surdud

The Wadi Surdud is a small sedimentary basin in the foothills of the western highlands of Yemen. It is considered a favourable habitat, with year-round water, and rainfall that occurs throughout the year. The local vegetation is described as rich and varied, with several types of palm trees, acacias, succulents and fruit trees, and the local fauna includes baboons, gazelle and several types of birds. If not a garden of Eden, it was at least a viable habitat. As it is not on a natural corridor leading inland, people probably settled here because it was dependable and good-quality flakable stone was locally available.

The basin contains two main stratigraphic units. The lower one is the Al-Sharj Member; this has been dated to ca. 85 ka and contains only a few artefacts. Although these are not diagnostic, they are useful in indicating the presence of hominins in Yemen at this time. The upper unit is the Shi'bat Dihya Member. This is a 30 m sequence of alluvial terrace deposits that was deposited early in MIS 3, and dated to between 42 and 63 ka. It contains several dense Middle Paleolithic archaeological concentrations, two of which – Shi'bat Dihya 1 (SD1) and Shi'bat Dihya 2 (SD2) – were partially excavated between 2006 and 2008 by a French-Yemeni team (Delagnes et al. 2013). There are also a few find-spots (AS1 and AS2) in the Al Sharj wadi parallel to and 150 m west of Shi'bat Dihya.

The SD1 assemblage lies 7 m below SD2, which is 6 m below AS1 and AS2. All four are estimated as being ca. 55 ka in age. The main features of SD1 are that over 90% of artefacts are made from local rhyolite; pointed and elongated flakes and blades are common; these were made with a hard hammer, and were probably used as cutting tools, but not as projectile points; very few pieces were retouched; and artefacts were made where they were found during an occupation phase at which many activities were carried out. The slightly younger SD2 assemblage also has very few retouched tools and many blades and pointed flakes (although not as many as in SD1), but the flaking methods are different, and result in a prevalence of unprepared discoidal-like cores.

These assemblages from the Wadi Surdud are unique in Arabia, and have no obvious parallels elsewhere in Arabia, or the Levant, or North Africa. As Scerri and colleagues (2014) point out, "While the cultural and biological background of the MIS 3 hominins occupying Wadi Surdud are currently unclear, the key point is that their material culture represents a distinctive localised technology". Delagnes and colleagues (2013) suggest that the Wadi Surdud (and that part of southwest Arabia) may have served as a refugium at the beginning of MIS 3 because of its stable water supply and animal resources (Delagnes et al. 2013). In this scenario, most of Arabia would have been depopulated during the harshest parts of MIS 4, with only a few refugia such as the highlands of southwest Arabia providing source populations that could expand back into the interior when conditions improved. As seen next, the same might be true of Jebel Faya during MIS 3.

Jebel Faya, assemblages A and B

At Jebel Faya, there were two archaeological horizons (A and B) above the lowest (assemblage C) that was dated to the last interglacial (see above). Similar assemblages were recovered from the parallel excavations of Jebel Faya Shelter (Bretzke et al. 2014) in archaeological horizons (AH) IV and V respectively. Two OSL samples from within assemblage A yielded ages of 38.6 ± 3.1 and 40.2 ± 3.0 ka, and two samples from the overlying sterile layer yielded ages of 38.6 ± 3.2 and 34.1 ± 2.8 ka (Armitage et al. 2011). Assemblage A is thus ca. 40,000 years old. The age of assemblage B is not established, but somewhere between 125 ka (the age of assemblage C) and 40 ka.

In assemblage B, blanks, mainly flakes and a few blades, were largely produced from flat flaking surfaces with parallel, converging and crossed removals. Tools include sidescrapers, end scrapers, denticulates, retouched pieces, burins and perforators. Unlike assemblage C, there was no evidence for the Levallois technique. The number and range of tools suggests that the occupation was multipurpose.

The uppermost assemblage A contained mainly flakes struck from multiple platform cores with parallel removals on each face. Blades were rare. Tools included burins, retouched pieces, end scrapers, sidescrapers and denticulates. Both assemblages A and B lack Levallois flaking and bifacial reduction, and, typologically, the use of backing was unknown.

As with the Wadi Surdud, the key point here is that assemblages A and B have no affinities with assemblages from East Africa, the Levant, the Zagros or elsewhere in Arabia – they appear to be entirely home-grown entities (Armitage et al. 2011; Marks 2009; Rose 2010). One explanation is increased aridity in MIS 4 isolated the inhabitants of Jebel Faya from neighbouring populations. Rose (2010) suggests that this part of southeast Arabia might have been recolonised in MIS 3 by groups living in a refugium during MIS 4 on the exposed floor of the Gulf; when temperatures and rainfall increased at the start of MIS 3, sea levels would have risen, and groups living in a refugium on the floor of the Gulf might have been displaced inland to neighbouring regions such as the Jebel Faya (Bretzke et al. 2013). Other alternatives for potential source populations are the nearby Hajar Mountains of northern Oman (Bretzke et al. 2013) or southern Zagros (Bretzke et al. 2014), both of which might have served as refugia during arid periods such as MIS 4.

Assessment

A brief summary is appropriate before we turn to the corridor between Arabia and the Thar Desert. We have learnt a phenomenal amount since Petraglia and Alsharekh (2003) published their tentative assessment of the Arabia Middle Palaeolithic 15 years ago. Six highly impactive features emerge. First, it certainly dates back to MIS 7 (ca. 200–250 ka), and probably/perhaps to MIS 9 (ca. 280–340 ka). Second, the occupation record of Arabia is limited overwhelmingly to those fairly brief episodes when Arabia was "green" and replete with lakes and rivers: MIS 7, MIS 5e, 5c, 5a and MIS 3. There may have been refugia in the southwest and southeast in the intervening arid periods of MIS 6, 5d, 5b and MIS 4, but the interior was likely depopulated in those periods. Consequently, it is unsurprising that the Middle Palaeolithic inhabitants prior to MIS 3 do not appear to have left a genetic signature in modern populations. Third, it is probable that most of the Arabian Middle Palaeolithic was the product of *Homo sapiens*. Fourth, those green episodes mark simply an eastward extension of the Saharan Realm (see Chapter 3): it was not so much the case that humans left Africa as that Africa went with them. Their presence in Arabia was simply an example of range extension (Chapter 1) brought about by climate change, for which no specific adaptation or technological breakthrough was needed for humans to colonise Arabia. Nevertheless, it would have been a challenging and harsh environment, particularly in dry seasons, and the main adaptation may have been the maintenance of social structures that would support high mobility throughout the year, and from year to year. Fifth, Arabia was not just a corridor that led eastwards but a large region with its own complex record of occupation in its own right. Finally, the complex lithic record of Arabia indicates that its occupation probably involved several discrete populations that derived from several source populations that could have entered from several directions.

Southern Iran and Pakistan

Let's stay with the possibility that *Homo sapiens* may have dispersed from Arabia as far as the Thar Desert during the last interglacial, and now look at the evidence from the intervening 1,600 km (1,000 miles) of southern Iran and Pakistan. Little is known about their Palaeolithic past, so these regions are the weakest links in attempts to trace connections between Arabia and India. In southern Iran, the valley systems of the Zagros mountains swing west to east towards Iranian Baluchistan (see Figure 4.10). Today, the area is arid, and extremely hot in summer. Total rainfall is ca. 180 mm, with March and April the wettest, and virtually no rain in October. Just as there has been a "green Sahara", a "green Arabia" and a "green Thar Desert" (see below), so there may have been a "green Iranian Plateau" during moist episodes of the last glacial cycle (see Chapter 8). As with Arabia, we can expect that the main period of increased rainfall would have been during MIS 5e, and then MIS 5c and MIS 5a, and parts of MIS 3. As in Arabia, the Iranian Plateau would have been watered by several lakes (now salt playas) and rivers.

Some Palaeolithic surveys have been carried out (see Chapter 8), but we lack dates and the type of detailed assemblage analyses that we now have for Arabia. A few Levallois-type Middle Palaeolithic flakes were found in the Makran by Vita-Finzi and Copeland (1980), and a number of surveys in southwest and southern Iran in Khuzistan, the Makran and Baluchistan have reported Middle Palaeolithic open-air scatters; Figure 4.9 shows the location of the main ones (Bahramiyan and Shouhani 2016; Darabi et al. 2012; Dashtizadeh 2009; Nasab et al. 2013; Piperno 1972). Typologically these show a mixture of features, some of which are shared by sites in the Thar, and others by sites on the Iranian Plateau. Piperno (1972) noted that Levallois flakes were rare at Jahrom, Fars Province, but Darabi et al. (2012) reported that Levallois flakes were common at Amar Merdeg in southwest Iran; Biagi and Starnini (2011, 2018) noted the same in Sindh at Ongar, as did Blinkhorn (2014) in the Thar Desert (see below). Heydari-Guran and colleagues (2015) also reported that the Levallois technique was widely employed at open-air sites on the Central Iranian Plateau, which they suggested probably date to the last interglacial on circumstantial grounds. Levallois flakes were also very common at Mirak in northern Iran (Rezvani and Nasab 2010), so the presence of the Levallois technique in southern Iran is not perhaps very significant. At the same time, the assemblage from

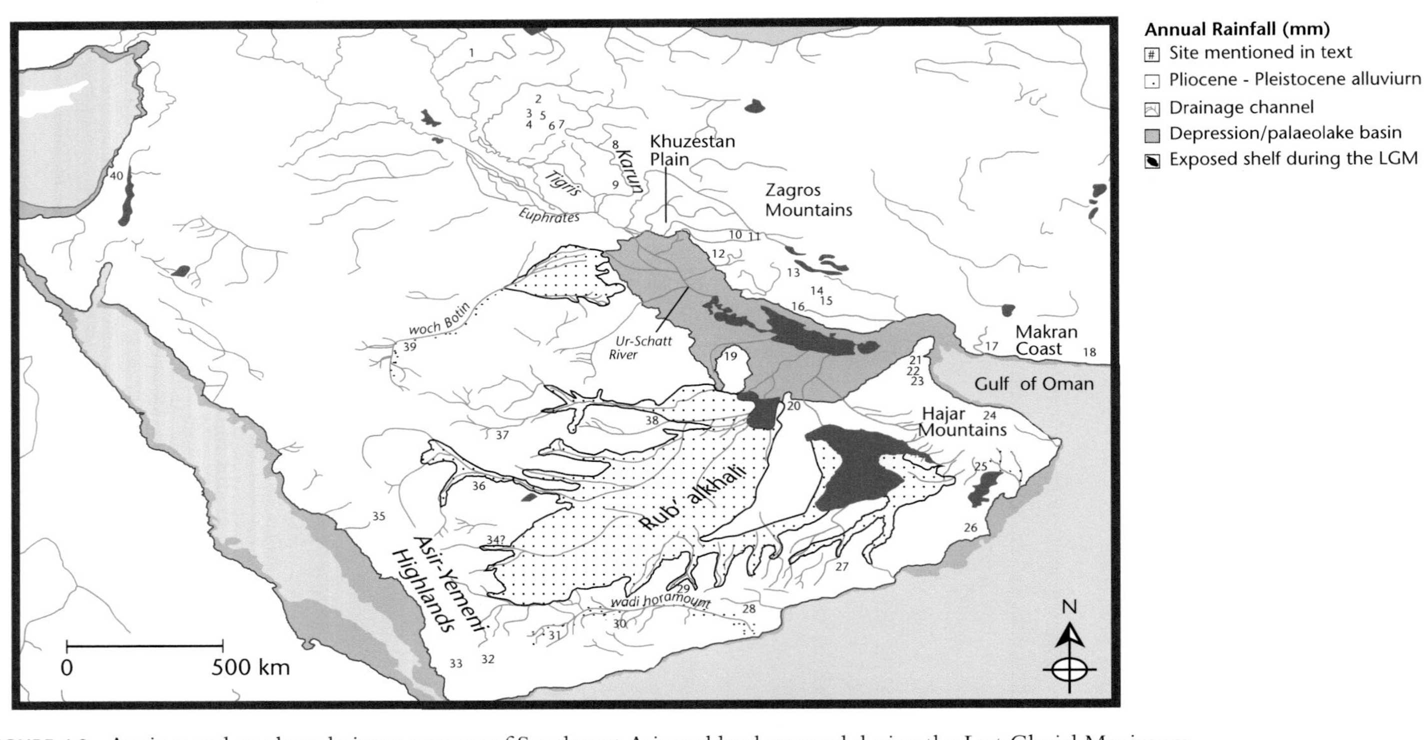

FIGURE 4.9 Ancient and modern drainage systems of Southwest Asia and land exposed during the Last Glacial Maximum

Sites: Iraq: 1 Shanidar; Iran: 2 Bisitun; 3 Warwasi; 4 Kobeh; 5 Ghar-e-Khar; 6 Yafteh; 7 Gar Arjeneh; 8 Kunji; 9 Izeh Plain; 10 Sarab Syah; 11 Qaleh Bozi; 12 Ghar-i-Boof; 13 Eshkaft-i-Gavi; 14 Bab Anar; 15 Jahrom; 16 Jam-o-Riz Plain; 17 Kuhestak; 18 Konarak; Qatar: 19 Ras 'Ushayriq; UAE: 20 Jebel Barakah; 21 Fili; 22 Jebel Faya; 23 Nad al-Thaman; Oman: 24 Wadi Wutayya; 25 Haushi-Huqf; 26 Ad Duqm; 27 Nejd Plateau; Yemen: 28 Mahra; 29 Wadi Wa'shah; 30 Hadramaut; 31 Shawba; 32 'Asir-Yemeni highlands; 33 Shi'bat Sihya; Saudi Arabia: 34 Faw Fell; 35 Western Province, Saudi Arabia; 36 Wadis Dawasir and Tathlith; 37 Saffaqah and Riyadh; 38 Yabrin Oasis and Wasdi Sahba; 39 Northern Province; Israel: 40 Skuhl and Qafzeh caves.

Source: Rose 2010, Fig. 2.

the site of Qaleh Bozi 2 (Biglari et al. 2009) in Central Iran is more like assemblages from the northern Zagros than sites on the Iranian Plateau, so its Middle Palaeolithic record may not be uniform and may be as complex as in Arabia (see Chapter 8). What may be more significant is the presence of pieces with a *chapeau de gendarme* platform in the Thar (as at Bap, Blinkhorn et al. 2015: 239) as well as in the Mehran Plain (Darabi et al. 2012) and Qaleh Gusheh and Zavyeh in the Central Plateau region (Heydari-Guran et al. 2015).

Although the Middle Palaeolithic evidence is very scant from both southern Iran and Pakistan, it is potentially extremely important. The key question is whether it represents the detritus left in a corridor by colonising populations of *Homo sapiens* moving from Arabia to the Thar, or the southernmost limit of Neandertals – or both. At first sight (see Figure 4.10), the topography of Iran suggests that the former possibility is more likely because the Zagros mountains trending west to east along the

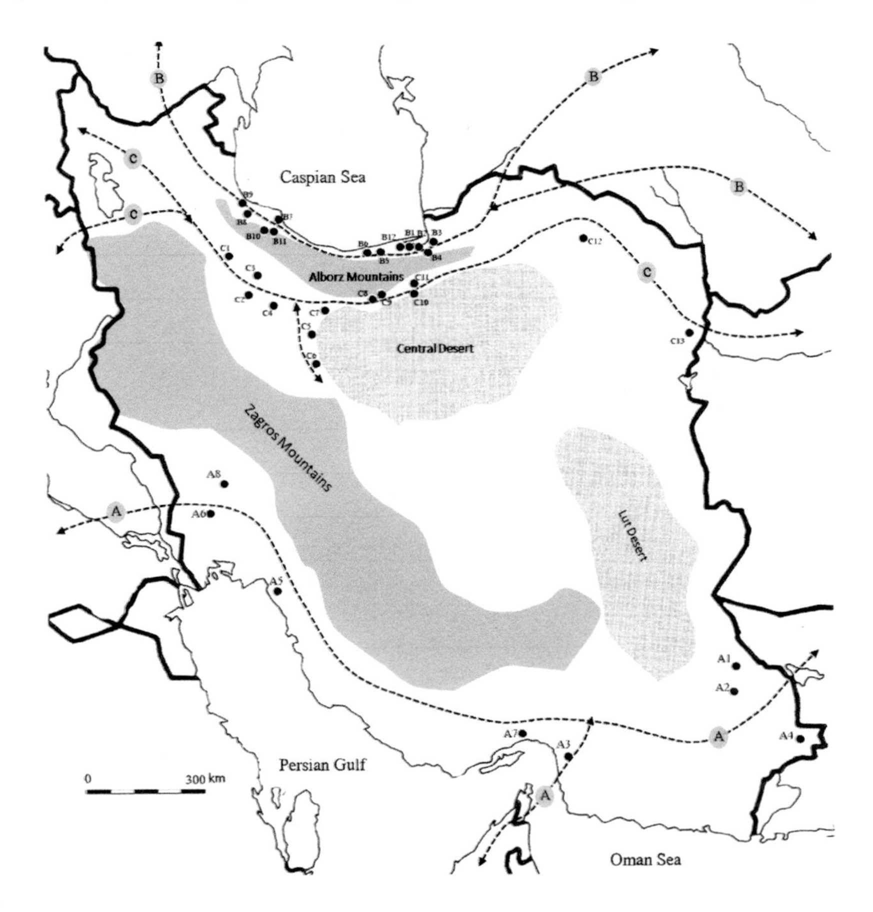

FIGURE 4.10 Potential corridors for dispersal across Iran

Note that the southern route along southern Iran lies within the Afro-Arabian Faunal Realm. (The northern routes are discussed in Chapter 8.)

Source: Nasab et al. 2013, Fig. 10.

Makran coast, and the coastal plain provide a corridor south of the barriers of the great deserts of the Dasht-i-Lut (Salt Desert) and Dasht-i-Kavir (Sand Desert) to the north. However, there are also corridors from southwest Iran into the Central Zagros, which we know from the evidence at Shanidar (see Chapter 8) was occupied by Neandertals. Roustaei (2010: 1) describes Mousterian artefacts from survey collections high in the Central Zagros at altitudes between 2,450 and 2,800 m a.s.l., and notes that "even today the population decreases sharply during Autumn and Winter, when all Bakhtiari nomads make their temporary camps in lowland Khuzistan, a few hundred kilometres to the west". The same pattern of migration between upland summer and lowland winter grazing is repeated by other nomadic pastoral groups along the Zagros, from the Kurds and Lurs to the north, and the Qashqai in southern Baluchistan. In other words, southwest Iran provides an east–west corridor for a colonising population, and northeast to southwest corridors for a different, indigenous population: it is an example of a convergent corridor used by two populations (see Chapter 3). The same scenario applies to the lower Indus Valley in Sindh province, Pakistan, where a west–east corridor intersects with a north–south one. The occurrence of typical Levallois cores, flakes, blades, points and scrapers at Ongar and other sites west of the Indus in Sindh may indicate the southeasternmost limit of Mousterian assemblages (Biagi and Starnini 2011, 2018), and perhaps, therefore, Neandertals. There are therefore two potential contact zones where incoming *Homo sapiens* populations could have encountered Neandertals who moved south in winter from the Zagros, or were resident in Sindh. Obviously, we will not know until we have some skeletal data.

The Thar Desert

The Thar Desert of northwest India covers ca. 300,000 sq km and is one of the largest sand bodies in the world. Today the area is semi-arid with an average rainfall (mainly in the summer monsoon) of ca. 370 mm (Achyuthan et al. 2007). In arid periods such as MIS 6 and MIS4, it would have been larger and effectively a barrier into peninsular India. In the last interglacial, it shrank during moist episodes of MIS 5e, 5c and 5a, and might therefore have become a corridor that allowed immigration from southern Iran and Pakistan. There is good evidence that it was less arid in the last glacial cycle. Blinkhorn et al. (2013) and Blinkhorn (2014) report evidence that there was intermittent occupation of the 16R dune (one of the major geological sections in the Thar Desert) between 130 and 90 ka, which indicates that the area was intermittently less arid, and there was increased fluvial action in the Orsang, Sabarmati and Mahi Valleys as well as braided stream channels in the Luni Valley.

In the absence of human skeletal data, we have to turn to any indication of immigration or population change. As we shall see in Chapter 5, Lower Palaeolithic Acheulean assemblages ceased to be made after ca. 130 ka, i.e. at the end of MIS 6. This does not necessarily imply a turnover in population, but it equally hints that there may have been. More definite evidence comes from the site of Katoati in the Thar Desert. Here, Blinkhorn and colleagues (2013) report that the earliest occupation in the Thar Desert occurred ca. 96 ka (=MIS 5c), with subsequent occupation dated to ca. 77 ka, ca. 60 ka and down to 45 ka, indicating continuity of these industries in MIS 5 and into MIS 4 and MIS 3. The assemblages are similar to each other – which may indicate population continuity. At this site and others in the Thar Desert, there appears to be a focus on the production of points that is evident by the presence of prepared point cores analogous to Nubian ones, and the retouching of some to produce a shoulder or tang suitable for hafting (Blinkhorn et al. 2015; Blinkhorn and Petraglia 2017). One of these points is shown in Figure 4.10, with a similar one from Jwallapurram 22, India (see Chapter 5); both are dated to ca. 77 ka (Blinkhorn et al. 2015). This might suggest that groups using a Nubian technology had dispersed as far east as the Thar Desert during MIS 5 and the end of MIS 4 and even into peninsular India; as pointed out above, however, it could equally indicate technical convergence.

One interesting point arises from a recent survey in the coastal region of Kachchh district of Gujrat (Blinkhorn et al. 2017). Although Middle Palaeolithic artefacts were found in contexts spanning the likely dates of the time frame for human expansion into India, there were no indications of microliths except in Holocene contexts. This finding does not support the hypothesis that India was colonised by immigrants using a microlithic technology and microliths ca. 60,000 years ago (see Chapter 5).

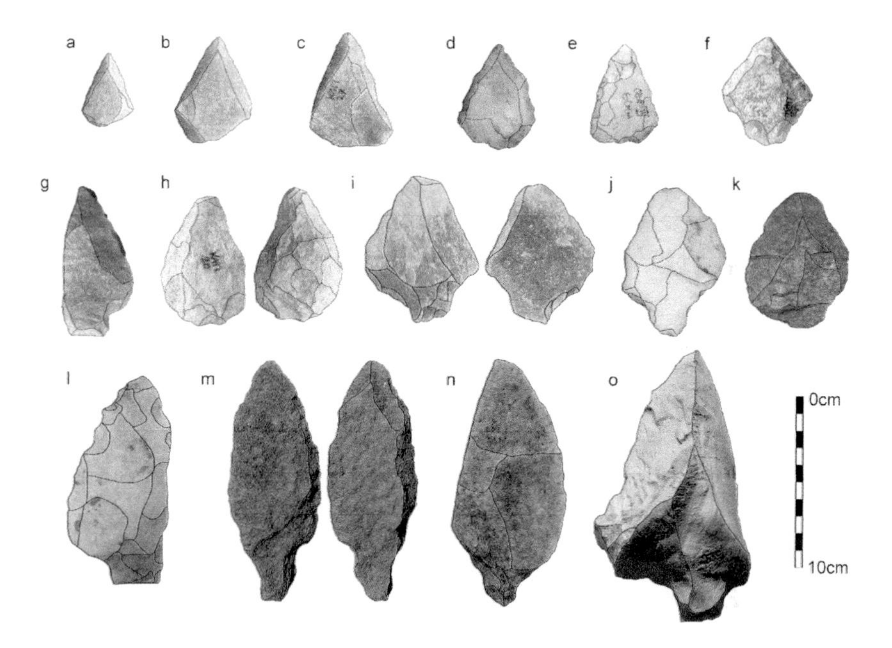

FIGURE 4.11 New evidence for point technologies from the Thar Desert

Key: (a–c) Levallois points from Katoati; (d) pseudo-Levallois point from Jogpura; (e) unifacial point from Jogpura; (f) retouched flake from Katoati with basal modification opposite convergent edge; (g) shouldered point from Katoati; (h) bifacial point from Katoati; (i) tanged point from Katoati; (j) tanged point from Jogpura; (k) shouldered point from Jogpura; (l) shouldered point from Sambhar; (m) bifacial point from Chamu; (n) tanged point from Sambhar; (o) tanged point from Damod.

Source: Blinkhorn et al. 2015, Fig. 6.

Overall assessment

The first of two most far-reaching changes in what we think we know about the earliest history of our species outside Africa is that the Arabian Peninsula was transformed from arid desert into a landscape of lakes, rivers and semi-arid grasslands in moist (=interglacial) episodes before and during the last interglacial. The second is that during these "green" episodes, Arabia was most likely colonised by *Homo sapiens*. On current evidence, this had occurred by MIS 5e, as at Jebel Faya, but there may have been earlier influxes of populations of our species as far back as MIS 9. This probability should require a major re-orientation in our perspectives: instead of seeing the skeletal evidence for *Homo sapiens* in the Levant in the last interglacial as indicative of a "failed dispersal", it now seems more likely that when Arabia was occupied, the Levant was the northern limit of human populations that had repeatedly inhabited the enormous territory of the Arabian Peninsula. Moreover, these inhabitants of Arabia were likely living in structured populations, having entered from several directions and maintaining and developing their own lithic traditions.

How far east these early colonists might have dispersed is at present uncertain. The corridor between Arabia and the Thar Desert through southern Iran and Pakistan is currently the weakest part of our evidence. This is doubly regrettable because it was also probably one of the areas where our species may have overlapped, and perhaps interbred, with Neandertals living to the north. Recent evidence from the Thar Desert offers a hint that groups were using a Nubian technology by at least the end of MIS 4. If so, our species may have been at the threshold of peninsular India by at least 60 ka. But as noted earlier, any such assessment is hampered by the difficulties of distinguishing in lithic analyses between cultural transmission and technological convergence.

We can turn now to the colonisation of the Oriental Realm of South Asia.

Note

1 This was a weighted average of nine OSL dates from ranging 49.8 ka BP to 80.4 ka BP (see Vermeersch et al. 1998, Table 1).

References

Arabia

Almogi-Labin, A., Schiedl, G., Hemleben, C., Siman-Tov, R., Segl, M. and Meischner, D. (2000) The influence of the NE winter monsoon on productivity changes in the Gulf of Aden, NW Arabian Sea, during the last 350 ka as recorded by foraminfera. *Marine Micropaleontology*, 40: 295–319.

Armitage, S.J., Jasim, S.A., Marks, A.E., Parker, A.G. and Uerpmann, H.-P. (2011) The Southern Route "Out of Africa": evidence for an early expansion of Modern Humans into Arabia. *Science*, 331: 453–456.

Breeze, P.S., Drake, N.A., Groucutt, H.S., Parton, A., Jennings, R.P. et al. (2015) Remote sensing and GIS techniques for reconstructing Arabian palaeohydrology and identifying archaeological sites. *Quaternary International*, 382: 98–119.

Breeze, P.S., Groucutt, H.S., Drake, N.A., White, T.S., Jennings, R.P. et al. (2017) Palaeohydrological corridors for hominin dispersals in the Middle East ~250–70,000 years ago. *Quaternary Science Reviews*, 144: 155–185.

Bretzke, K., Armitage, S.J., Parker, A.G., Walkington, H. and Uerpmann, H.-P. (2013) The environmental context of Paleolithic settlement at Jebel Faya, Emirate Sharjah, UAE. *Quaternary International*, 300: 83–93.

Bretzke, K. and Conard, N.J. (2017) Not just a crossroad. Population dynamics and changing material culture in Southwestern Asia during the Late Pleistocene. *Current Anthropology*, 58(Supplement 17): S449–S462.

Bretzke, K., Conard, N.J. and Uerpmann, H.-P. (2014) Excavations at Jebel Faya – The FAY-NE1 shelter sequence. *Proceedings of the Seminar for Arabian Studies*, 44: 69–81.

Clemens, S.C. and Prell, W.L. (2003) A 350,000 year summer-monsoon multi-proxy stack from the Owen ridge, northern Arabian Sea. *Marine Geology*, 201: 35–51.

Crassard, R. and Hilbert, Y.H. (2013) A Nubian Complex site from Central Arabia: implications for Levallois taxonomy and human dispersals during the Upper Pleistocene. *PLoS ONE*, 8(7): e69221, doi:10.1371/journal.pone.0069221.

Crassard, R., Petraglia, M.D., Drake, N.A., Breeze, P., Gratuze, B. et al. (2013) Middle Palaeolithic and Neolithic occupations around Mundafan Palaeolake, Saudi Arabia: implications for climate change and human dispersals. *PLoS ONE*, 8(7): e69665, doi:10.1371/journal.pone.0069665.

Delagnes, A., Crassard, R., Bertran, P. and Sitzia, L. (2013) Cultural and human dynamics in southern Arabia at the end of the Middle Paleolithic. *Quaternary International*, 300: 234–243.

Drake, N.A., Breeze, P. and Parker, A. (2013) Palaeoclimate in the Saharan and Arabian Deserts during the Middle Palaeolithic and the potential for hominin dispersals. *Quaternary International*, 300: 48–61.

Emeis, K.C., Anderson, D.M., Doose, H., Kroon, D. and Schulz-Bull, D. (1995) Sea-surface temperatures and the history of monsoon upwelling in the northwestern Arabian Sea during the last 500,000 years. *Quaternary Research*, 43: 355–361.

Engel, M., Matter, A., Parker, A.G., Petraglia, M.D., Preston, G.W. et al. (2017) Lakes or wetlands? A comment on "The middle Holocene climatic records from Arabia: reassessing lacustrine environments, shift of ITCZ in Arabian Sea, and impacts of the southwest Indian and African monsoons" by Enzel et al. 2015. *Global and Planetary Change*, 148: 258–267.

Enzel, Y., Kushnir, Y. and Quade, J. (2015) The middle Holocene climatic records from Arabia: reassessing lacustrine environments, shift of ITCZ in Arabian Sea, and impacts of the southwest Indian and African monsoons. *Global and Planetary Change*, 129: 69–91.

Enzel, Y., Quade, J. and Kushnir, Y. (2017) Response to Engel et al. (in press): lakes or wetlands? A comment on "The middle Holocene climatic records from Arabia: reassessing lacustrine environments, shift of ITCZ in Arabian Sea, and impacts of the southwest Indian and African monsoons" by Engel et al. (2015). *Global and Planetary Change*, 148: 268–271.

Farrant, A.R., Duller, G.A.T., Parker, A.G., Roberts, H.M., Parton, A. et al. (2015) Developing a framework of Quaternary dune accumulation in the northern Rub' al-Khali, Arabia. *Quaternary International*, 382: 132–144.

Fleitmann, D., Burns, S.J., Neff, U., Mangini, A. and Matter, A. (2003) Changing moisture sources over the last 330,000 years in Northern Oman from fluid-inclusion evidence in speleothems. *Quaternary Research*, 60: 223–232.

Fleitmann, D., Burns, S.J., Pekala, M., Mangini, A., Al-Subbary, A. et al. (2011) Holocene and Pleistocene pluvial periods in Yemen, southern Arabia. *Quaternary Science Reviews*, 30: 783–787.

Groucutt, H., Grün, R., Zalmout, I.S.A., Drake, N.A., Armitage, S.J. et al. (2018) *Homo sapiens* in Arabia by 85,000 years ago. *Nature Ecology and Evolution*, 2: 800–809, doi:10.1038/s41559-018-0518-2.

Groucutt, H.S. and Petraglia, M.D. (2012) The prehistory of the Arabian Peninsula: deserts, dispersals, and demography. *Evolutionary Anthropology*, 21: 113–125.

Groucutt, H.S., Scerri, E.M.L., Lewis, L., Clark-Balzan, L., Blinkhorn, J., Jennings, R.P., Parton, A. and Petraglia, M.D. (2015c) Stone tool assemblages and models for the dispersal of *Homo sapiens* out of Africa. *Quaternary International*, 382: 8–30.

Groucutt, H.S., Shipton, C., Alsharekh, A., Jennings, R., Scerri, E.M.L. and Petraglia, M.D. (2015b) Late Pleistocene lakeshore settlement in northern Arabia: Middle Palaeolithic technology from Jebel Katefeh, Jubbah. *Quaternary International*, 382: 215–236.

Groucutt, H.S., White, T.S., Clark-Balzan, L., Parton, A., Crassard. R. et al. (2015a) Human occupation of the Arabian Empty Quarter during MIS 5: evidence from Mundafan Al-Buhayrah, Saudi Arabia. *Quaternary Science Reviews*, 119: 116–135.

Harvati, K., Röding, C., Bosman, A.M., Karakostis, F.A., Grün, R. et al. (2019) Apidima Cave fossils provide earliest evidence of *Homo sapiens* in Eurasia. *Nature*, 571: 500–504, doi:10.1038/s41586-019-1376-z.

Hershkovitz, I., Weber, G.W., Quam, R., Duval, M. and Grün, R. (2018) The earliest modern humans outside Africa. *Science*, 359: 456–459.

Hilbert, Y.H., Crassard, R., Charloux, G. and Loretto, R. (2017) Nubian technology in northern Arabia: impact on interregional variability of Middle Paleolithic industries. *Quaternary International*, 435A: 77–93.

Jennings, R.P., Parton, A., Clark-Balzan, L., White, T.S., Groucutt, H.S. et al. (2016) Human occupation of the northern Arabian interior during early Marine Isotope Stage 3. *Journal of Quaternary Science*, 31(8): 953–966.

Jennings, R.P., Singarayer, J., Stone, E.J., Krebs-Kanzow, U., Khon, V. et al. (2015) The greening of Arabia: multiple opportunities for human occupation of the Arabian Peninsula during the Late Pleistocene inferred from an ensemble of climate model simulations. *Quaternary International*, 382: 181–199.

Kopp, G.H., Roos, C., Butynski, T.M., Wildman, D., Alagaili, A.N., Groeneveld, L.F. and Zinner, D. (2014) Out of Africa, but how and when? The case of hamadryas baboons (*Papio hamadryas*). *Journal of Human Evolution*, 76: 154–164.

Lambeck, K. (1996) Shoreline reconstructions for the Persian Gulf since the Last Glacial Maximum. *Earth and Planetary Science Letters*, 142: 43–57.

Lambeck, K., Purcell, A., Flemming, N.C., Vita-Finzi, C., Alsharekh, A.M. and Bailey, G. N. (2011) Sea level and shoreline reconstructions for the Red Sea: isostatic and tectonic considerations and implications for hominin migration out of Africa. *Quaternary Science Reviews*, 30: 3542–3574.

Lawler, A. (2011) Did Modern Humans travel out of Africa via Arabia? *Science*, 331: 387.

McClure, H.A. (1976) Radiocarbon chronology of late Quaternary lakes in the Arabian Desert. *Nature*, 263: 755–756.

Marks, A. (2009) The Paleolithic of Arabia in an inter-regional context. In M. Petraglia and J. Rose (eds) *Evolution of Human Populations in Arabia: Paleoenvironments, Prehistory and Genetics*, Dordrecht: Springer, pp. 293–309.

Matter, A., Mahjoub, A., Neubert, E., Preusse, F., Schwalb, A. et al. (2016) Reactivation of the Pleistocene trans-Arabian Wadi ad Dawasir fluvial system (Saudi Arabia) during the Holocene humid phase. *Geomorphology*, 270: 88–101.

Matter, A., Neubert, E., Preusser, F., Rosenberg, T. and Al-Wagdani, K. (2015) Palaeo-environmental implications derived from lake and sabkha deposits of the southern Rub' al-Khali, Saudi Arabia and Oman. *Quaternary International*, 382: 120–131.

Parton, A., Farrant, A.R., Leng, M.J., Schwenninger, J.-L., Rose, J.I. et al. (2013) An early MIS 3 pluvial phase in southeast Arabia: climatic and archaeological implications. *Quaternary International*, 300: 62–74.

Parton, A., Farrant, A.R., Leng, M.J., Telfer, M.W., Groucutt, H.S. et al. (2015a) Alluvial fan records from southeast Arabia reveal multiple windows for human dispersal. *Geology*, 43: 295–298.

Parton, A., White, T.S., Parker, A.G., Breeze, P., Jennings, R. et al. (2015b) Orbital-scale climate variability in Arabia as a potential motor for human dispersals. *Quaternary International*, 382: 82–97.

Petit-Maire, N., Carbonel, P., Reyss, J.L., Sanlaville, P., Abed, A. et al. (2010) A vast Eemian palaeolake in southern Jordan (29°N). *Global and Planetary Change*, 72: 368–373.

Petraglia, M.D. and Alsharekh, A. (2003) The Middle Palaeolithic of Arabia: implications for modern human origins, behaviour and dispersals. *Antiquity*, 77: 671–684.

Petraglia, M.D., Alsharekh, A.M., Crassard, R., Drake, N.A., Groucutt, H. et al. (2011) Middle Paleolithic occupation on a Marine Isotope Stage 5 lakeshore in the Nefud Desert, Saudi Arabia. *Quaternary Science Reviews*, 30: 1555–1559.

Petraglia, M.D., Alsharekh, A., Breeze, P., Clarkson, C., Crassard, R. et al. (2012) Hominin dispersal into the Nefud Desert and Middle Palaeolithic settlement along the Jubbah Palaeolake, Northern Arabia. *PLoS ONE*, 7(11): 1–21, e49840, doi:10.1371/journal.pone.0049840.

Rohling, E.J., Grant, K.M., Roberts, A.P. and Larrasoaña, J.-C. (2013) Paleoclimate variability in the Mediterranean and Red Sea regions during the last 500,000 years: implications for hominin migrations. *Current Anthropology*, 54(Supplement 8): S183–S201.

Rose, J.I. (2010) New light on human prehistory in the Arabo-Persian Gulf Oasis. *Current Anthropology*, 51(6): 849–883.

Rose, J.I., Usik, V.I., Marks, A.E., Hilbert, Y.H., Galletti, C.S. et al. (2011) The Nubian Complex of Dhofar, Oman: an African Middle Stone Age industry in southern Arabia. *PLoS ONE*, 6(11): e28239, doi:10.1371/journal.pone.0028239.

Rosenberg, T.M., Preusser, F., Blechschmidt, I., Fleitmann, D., Jagher, R. and Matter, A. (2012) Late Pleistocene palaeolake in the interior of Oman: a potential key area for the dispersal of anatomically modern humans out-of-Africa? *Journal of Quaternary Science*, 27 (1): 13–16.

Rosenberg, T.M., Preusser, F., Fleitmann, D., Schwalb, A., Penkman, K. et al. (2011) Humid periods in southern Arabia: windows of opportunity for modern human dispersal. *Geology*, 39: 1115–1118.

Rosenberg, T.M., Preusser, F., Risberg, J., Plikk, A., Kadi, K.A. et al. (2013) Middle and Late Pleistocene humid periods recorded in palaeolake deposits of the Nafud desert, Saudi Arabia. *Quaternary Science Reviews*, 70: 109–123.

Scerri, E.M.L., Breeze, P.S., Parton, A., Groucutt, H.S., White, T.S. et al. (2015) Middle to Late Pleistocene human habitation in the western Nefud Desert, Saudi Arabia. *Quaternary International*, 382: 200–214.

Scerri, E.M.L., Groucutt, H.S., Jennings, R.P. and Petraglia, M.D. (2014) Unexpected technological heterogeneity in northern Arabia indicates complex Late Pleistocene demography at the gateway to Asia. *Journal of Human Evolution*, 75: 125–142.

Stewart, M., Louys, J., Price, G.J., Drake, N.A., Groucutt, H.S. and Petraglia, M.D. (2019) Middle and Late Pleistocene mammal fossils of Arabia and surrounding regions: implications for biogeography and hominin dispersals. *Quaternary International*, 515: 12–29.

Stimpson, C.M., Breeze, P.S., Clark-Balzan, L., Groucutt, H.S., Jennings, R. et al. (2015) Stratified Pleistocene vertebrates with a new record of a jaguar-sized pantherine (*Panthera cf. gombaszogensis*) from northern Saudi Arabia. *Quaternary International*, 382: 168–180.

Stimpson, C.M., Lister, A., Parton, A., Clark-Balzan, L., Breeze, P.S. et al. (2016) Middle Pleistocene vertebrate fossils from the Nefud Desert, Saudi Arabia: implications for biogeography and palaeoecology. *Quaternary Science Reviews*, 143: 13–36.

Thesiger, W. (1960) *Arabian Sands*. London: Longmans, Green and Co Ltd.

Thomas, B. (1932) *Arabia Felix*. New York: Charles Scribner's Sons.

Usik, V.I., Rose, J.I., Hilbert, Y.H., Peer, P. van and Marks, A.E. (2013) Nubian Complex reduction strategies in Dhofar, southern Oman. *Quaternary International*, 300: 244–266.

Vaks, A., Bar-Matthews, M., Ayalon, A., Matthews, A., Halicz, L. and Frumkin, A. (2007) Desert speleothems reveal climatic window for African exodus of early modern humans. *Geology*, 35: 831–834, doi:10.1130/G23794A.

Valladas, H., Mercier, N., Hershkovitz, I., Zaidner, Y., Tsatskin, A. et al. (2013) Dating the Lower to Middle Paleolithic transition in the Levant: a view from Misliya Cave, Mount Carmel, Israel. *Journal of Human Evolution*, 65: 585–593.

Vermeersch, P.M., Paulissen, E., Stokes, S., Charlier, C., Peer, P. van et al. (1998) A Middle Palaeolithic burial of a modern human at Taramsa Hill, Egypt. *Antiquity*, 72: 475–484.

Southern Iran and Sindh

Bahramiyan, S. and Shouhani, L.A. (2016) Between mountain and plain: new evidence for the Middle Palaeolithic in the northern Susiana Plain, Khuzestan, Iran. *Antiquity*, 90: 354, e1: 1–6, doi:10.15184/aqy.2016.190.

Biagi, P. and Starnini, E. (2011) Neanderthals at the south-easternmost edge: the spread of Levalloisian Mousterian in the Indian Subcontinent. In K.T. Biró and A. Markó (eds) *Tanulmányok T. Dobosi Viola tiszteletére (Papers in honour of Viola T. Dobosi)*, Budapest: Magyar Nemzeti Múzeum, pp. 5–14.

Biagi, P. and Starnini, E. (2018) Neanderthals and Modern Humans in the Indus Valley? The Middle and Late (Upper) Palaeolithic Settlement of Sindh, a forgotten region of the Indian Subcontinent. In Y. Nishiaki and T. Akazawa (eds) *The Middle and Upper Paleolithic Archeology of the Levant and Beyond*. Replacement of Neanderthals by Modern Humans Series. Singapore: Springer.

Biglari, F., Javeri, M., Mashkour, M., Yazdi, M., Shidrang, S., Tengberg, M., Taheri, K. and Darvish, J. (2009) Test excavations at the Middle Paleolithic sites of Qaleh Bozi, southwest of central Iran, a preliminary report. In M. Otte, F. Biglari and J. Jaubert (eds) *Iran Palaeolithic/Le Paléolithique d'Iran*, BAR International Series 1968, pp. 29–38.

Darabi, H., Javanmardzadeh, A., Beshkani, A. and Jami-Alahmadi, M. (2012) Palaeolithic occupation of the Mehran Plain in southwestern Iran. *Documenta Praehistorica*, 39: 443–451.

Dashtizadeh, A. (2009) Palaeolithic remains from the north coast of the Persian Gulf: preliminary results from the Jam-o-Riz Plain, Bushehr Province, Iran. *Antiquity*, 83 (319): Project Gallery.

Heydari-Guran, S., Ghasidian, E. and Conard, N.J. (2015) Middle Paleolithic Settlement on the Iranian Central Plateau. In N. Conard and A. Delagnes (eds) *Settlement Dynamics of the Middle Palaeolithic and Middle Stone Age, Vol. 4*, Tübingen: Tübingen Publications in Prehistory. Tübingen: Kerns Verlag, pp. 171–204.

Nasab, H.V., Clark, G.A. and Torkamandi, S. (2013) Late Pleistocene dispersal corridors across the Iranian Plateau: A case study from Mirak, a Middle Paleolithic site on the northern edge of the Iranian Central desert (Dasht-e Kavir). *Quaternary International*, 300: 267–281.

Piperno, M. (1972) Jahrom, a Middle Palaeolithic Site in Fars, Iran. *East and West*, 22: 183–197.

Rezvani, H. and Nasab, H.V. (2010) A major Middle Palaeolithic open-air site at Mirak, Semnan Province, Iran. *Antiquity*, 84 (323):Project Gallery.

Roustaei, K. (2010) Discovery of Middle Palaeolithic occupation at high altitude in the Zagros Mountains, Iran. *Antiquity*, 84 (325): Project Gallery.

Vita-Finzi, C. and Copeland, L. (1980) Surface finds from Iranian Makran. *Iran*, 18: 149–155.

Thar Desert

Achyuthan, H., Quade, J., Roe, L. and Placzek, C. (2007) Stable isotopic composition of pedogenic carbonates from the eastern margin of the Thar Desert, Rajasthan, India. *Quaternary International*, 162–163:50–60.

Blinkhorn, J. (2014) Late Middle Palaeolithic surface sites occurring on dated sediment formations in the Thar Desert. *Quaternary International*, 350: 94–104.

Blinkhorn, J. and Petraglia, M.D. (2017) Environments and cultural change in the Indian Subcontinent. Implications for the dispersal of *Homo sapiens* in the Late Pleistocene. *Current Anthropology*, 58(Supplement 17): S463–S479.

Blinkhorn, J., Achyuthan, H. and Ajithprasad, P. (2015) Middle Palaeolithic point technologies in the Thar Desert, India. *Quaternary International*, 382: 237–249.

Blinkhorn, J., Achyuthan, H., Petraglia, M. and Ditchfield, P. (2013) Middle Palaeolithic occupation in the Thar Desert during the Upper Pleistocene: the signature of a modern human exit out of Africa? *Quaternary Science Reviews*, 77: 233–238.

Blinkhorn, J., Ajithprasad, P. and Mukherjee, A. (2017) Did modern dispersal take a coastal route into India? New evidence from Palaeolithic surveys of Kachchh, Gujrat. *Journal of Field Archaeology*, 42(3): 198–213, doi:10.1080/009344690.2017.1323543.

5

THE ORIENTAL REALM OF SOUTH ASIA

Introduction

It was only when humans entered the Oriental Realm (see Figure 3.4) of South Asia that they can be said to have left Africa. As I suggested in Chapter 4, the humans who moved into Arabia when it was replete with streams and lakes were living in environments already familiar to them in North Africa and did not need to change their behaviour or technology. South Asia was substantially different in its climate, fauna and flora, and the humans who colonised it had to learn new skills in completely new landscapes. India, Bangladesh, Sri Lanka and Myanmar cover an enormous area as large as the European Union,[1] so its colonisation could not have been a quick and easy affair. The new landscapes that humans would have encountered included the great mountain chain of the Himalayas, the wide floodplains of the Ganges and Brahmaputra, the gently undulating river basins of the Deccan Peninsula, the heavily forested slopes of the Western Ghats that form a giant escarpment along the western coast of India (see Figure 5.1), and the rainforests of southern India, Sri Lanka and northeast India. Each of these presented different challenges and opportunities for immigrant groups of humans and their predecessors in finding the necessary local availability of plant and animal resources, year-round water and stone for making artefacts. Floodplains, for example, were rich in food resources yet often deficient in stone, and also subject to periodic mega-floods (Dennell 2007); rainforests were rich in plant resources but poor in ground-dwelling animals that could be easily hunted (see below). So far, most of what we know about this colonisation comes from the western side of modern India and the rainforests of Sri Lanka.

Biogeographically, the Oriental Realm encompasses modern-day India (except for the Thar Desert), Bangladesh, Myanmar, as well as the countries of mainland

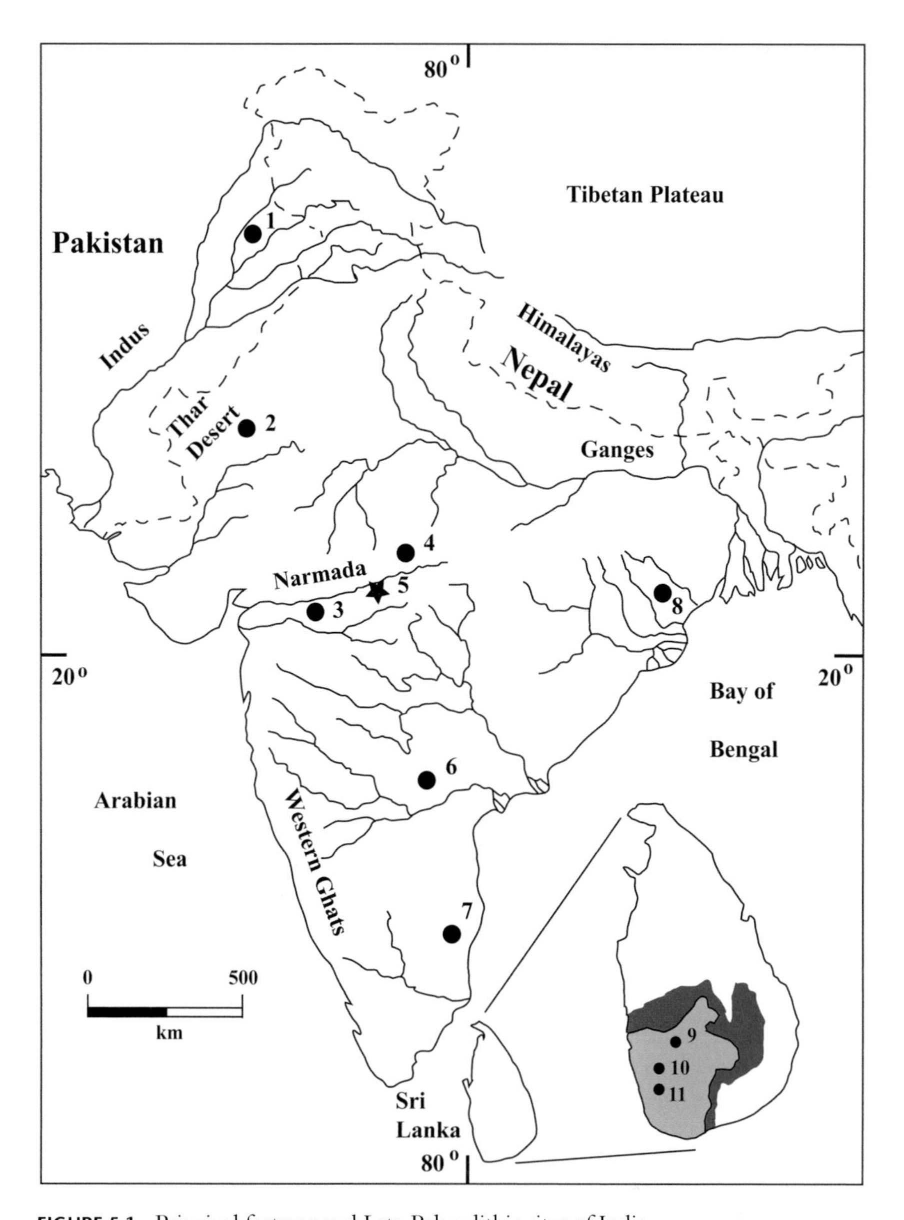

FIGURE 5.1 Principal features and Late Palaeolithic sites of India

Key: 1 Site 55, Riwat; 2 Katoati; 3 Mehtakheri; 4 Bhimbetka; 5 Hatnora; 6 Jwalapuram, Jerreru river; 7 Attirampakkam; 8 Mahadebbera/Kana; 9 Kitulgala lena; 10 Batadomba lena; 11 Fa Hien lena.

Sri Lanka inset: dark grey shows intermediate rainforest; light grey shows rainforest and grassland.

Source: Details of Sri Lanka based on Roberts 2015, Fig. 2c.

and island Southeast Asia (see Chapter 6) and south China (Chapter 10). The part comprising India, Bangladesh and Myanmar differs significantly from Southeast Asia in its faunal history. Myanmar shared with India and Pakistan the same Siwalik and post-Siwalik Early and Middle Pleistocene fauna (Nanda et al. 2018) that differs substantially from that of Southeast Asia. As examples, camel and horse never entered Pleistocene Southeast Asia, and tapir, sun bear and orangutan are not known to have inhabited South Asia. There are a few overlaps: for example, *Stegodon* originated in Southeast Asia (Saegusa et al. 2004) and expanded as far west as Israel in the Early Pleistocene (Tchernov et al. 1994), and panda was reported from caves in eastern Myanmar in excavations in 1938 (Colbert 1943). Although undated, the range extension of panda may have occurred in the last interglacial when rainfall was higher. Myanmar is thus a likely transitional zone but meanwhile remains a serious gap in our coverage of southern Asia. (Pakistan is another transitional zone as the western and northern parts have more in common with Iran and Afghanistan than with peninsular India.)

A unifying factor of the Oriental Realm is the monsoonal climate, with most of the rainfall falling during the summer months from June to September. There is a cold (but not sub-freezing) winter season from November to February, followed by a hot season from March to May, and then the summer monsoon that usually starts in June and tails off in October. Rainfall totals vary enormously. Over much of the Indian peninsula, annual rainfall is ca. 100–200 cm; over the great escarpment of the Western Ghats, 300–400 cm, and much higher in northeast India in the Himalayan foothills. Despite the monsoon, water scarcity can be serious in the dry season, or when the monsoon fails. Conversely, there can be catastrophic mega-floods in years when the monsoonal rainfall is unusually heavy.

The first – and successive – populations of humans that entered peninsular India from the west or north would have encountered a new range of animals, such as Indian lion and tiger, Indian elephant and rhinoceros, the gaur (*Bos gaurus*, the Indian buffalo), the four-horned antelope (*Tetracerus quadricornis*) and the nilgai (*Boselaphus tragocamelus*), the largest Asian antelope, and at least 14 types of non-human primates.[2] These included several species of macaques and langurs as well as lorises[3] and in northeast India, gibbon. Botanically, India, Bangladesh and Myanmar are incredibly diverse, with virtually all types of plant communities from desert to rainforest, and from coastal to high mountain. If humans entered India from the Thar Desert in MIS 5, the most familiar landscapes would have been a mosaic of tropical savannah and woodland which extends south down the entire peninsula (see Figure 5.2) and is bordered to both east and west by dry tropical woodland. The least familiar environments for initial colonisers would have been the tropical woodlands along the Western Ghats, and the rainforests of southernmost India, southern Sri Lanka and northeast India. As we will see, those of southern India and Sri Lanka were colonised remarkably early by at least 45,000 years ago (Wedage et al. 2019a, 2019b).

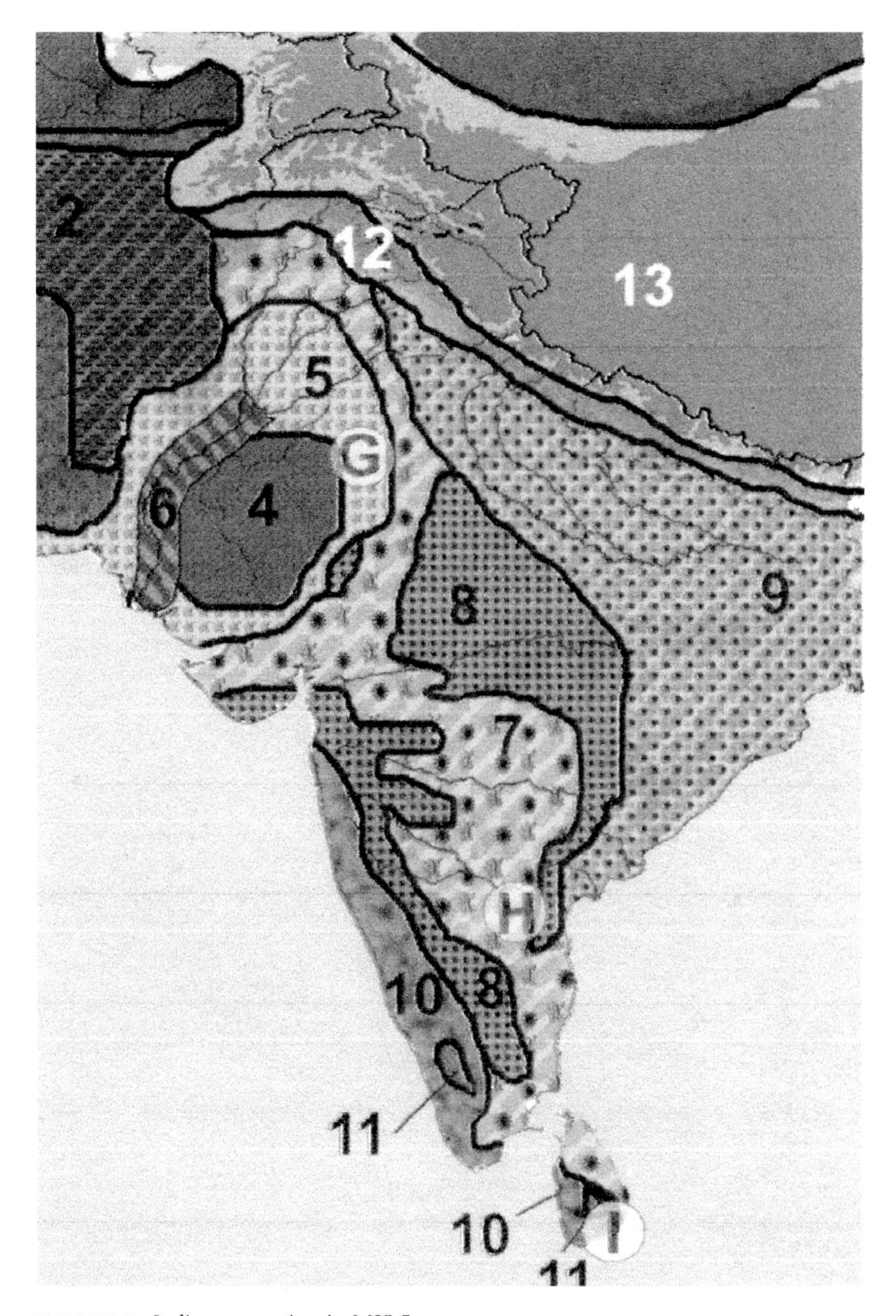

FIGURE 5.2 Indian vegetation in MIS 5

Key: 4 Desert; 5 Sub-desert; 6 Riverine corridors: marshes and gallery forests; 7 Tropical savannah/woodland grass mosaic, including the tropical evergreen zones of India; 8 Dry tropical woodland, including dry deciduous zones of India; 9 Moist tropical woodland and grassland mosaic; 10 Moist tropical woodlands, including both moist deciduous zones like those in India and true tropical rainforests; 11 Tropical montane vegetation; 12 Warm temperate hill and sub-montane vegetation; 13 High elevation coniferous forests and desert/the Tibetan plateau.

Source: Boivin et al. 2013, Fig. 2.

Climate history

Biogeographically, India shows a high degree of stability despite the climatic changes of the Late Pleistocene. Pollen from cores of marine sediments from the western coast of India that record the last 200 ka BP show a dominance of Chenopods, Amaranths and Poaceae,[4] with very few indications of extensive tree cover and very little coastal mangroves (see Figure 5.3). Perhaps unsurprisingly, MIS 1 (the Holocene), MIS 3 and MIS 5 (the last interglacial) were warm and wet, and MIS 2 (the glacial maximum) and MIS 4 were cooler and drier (Prabhu et al. 2004). An ice core in Tibet showed the same trends, but also similar short-term fluctuations of ca. 200 years during MIS 3 (Thompson et al. 1997). Petraglia and colleagues (2009) reconstruct the vegetation of peninsular India between 35 ka BP and 25 ka BP as "a semiglacial-period mosaic environment, consisting of deserts, savannahs, tropical deciduous woodlands, and limited tropical forests" (see Figure 5.4). In moister periods, such as MIS 3 and 5, the extent of deserts and savannahs would decline, and woodlands and forest would expand.

Faunal evidence indicates a considerable degree of continuity. As an example, the fauna from the cave of Billaspurgam in central India shows that 20 out of 21 taxa[5]

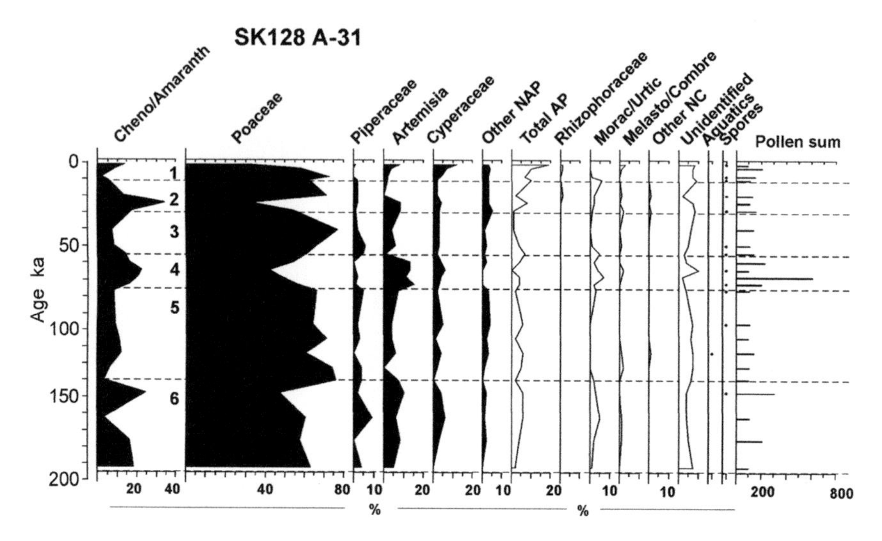

FIGURE 5.3 Pollen frequencies in core SK-128A-31 from the Indian Ocean west of India

The frequencies of Chenopodiaceae/Amaranthaceae and Poaceae pollen provide the best indicators of fluctuations between low and high rainfall. *Artemisia* indicates low rainfall. Bold numbers are marine isotope stages, and the dashed lines are the boundaries between them. AP and NAP = arboreal pollen and nonarboreal pollen. Rhizophoraceae are a mangrove family. Morac/Urtic are abbreviations of Moraceae and Urticaceae (the nettle and fig families). Melasto and Combre = Melastomataceae and Combretaceae; these are families that include rainforest shrubs and mangroves, respectively.

Source: Prabhu et al. 2004, Fig. 4.

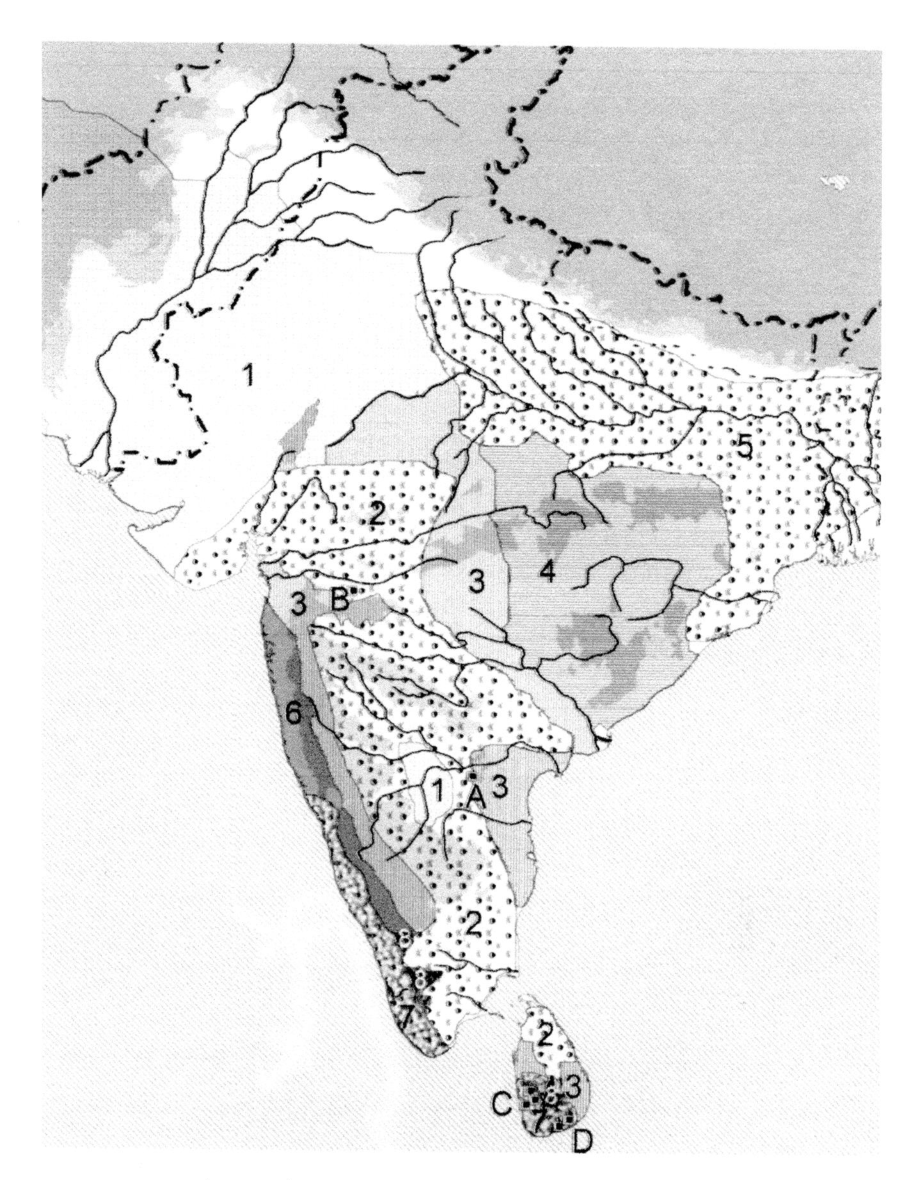

FIGURE 5.4 South Asia, showing reconstructed vegetation zones for ca. 30 ka ago and location of microlithic sites

Sites indicated by letters are as follows – A: Jurreru Valley; B: Patne; C: Sri Lanka caves, from north to south Beli-lena Kitulgala, Batadomba-lena and Fa Hien; and D: Sri Lanka coastal sites 49 and 50. Vegetation zones indicated by numbers are as follows – 1: desert and semidesert (*Caligonum-Salvadora-Prosopis-Acacia* and scattered grasses); 2: savannah and tropical dry deciduous woodland mosaic (*Acacia-Anogeissus-Terminalia, Hardwickia* in some localities, abundant gatherable grasses and legumes); 3: dry deciduous woodlands, including teak; 4: dry deciduous woodlands, including *Shorea-Hopea*; 5: deciduous *Shorea-Hopea* woodland and grassland/marsh mosaic; 6: moist deciduous and scattered evergreen taxa; 7: Tropical evergreen and semievergreen forest refugia; 8: tropical/subtropical mountain forests.

Source: Petraglia et al. 2009, Fig. 2.

have survived from 100,000 years ago into the present, and there have been no major changes over the last 200,000 years (Roberts et al. 2014; see also Chauhan 2008). This degree of continuity is attributed to interconnected mosaic habitats, and stability in topography and precipitation. Blinkhorn and Petraglia (2017) point out that geological "Purana" basins (Korisettar 2007) played an important part in maintaining environmental stability because aquifers would deliver reliable perennial water, and thus buffer areas against reductions in rainfall. Because of India's location in the Tropics, temperature changes are unlikely to have greatly impacted on faunal and botanical communities, and the summer monsoon probably ensured that most regions received sufficient rainfall to maintain stability. Even if vegetational boundaries changed in response to climate change, the same habitats would persist. Unlike Arabia, there would have been no switches as pronounced as hyper-arid desert to semi-arid grassland, and unlike Southeast Asia, changes in sea levels had little impact (Blinkhorn and Petraglia 2017).

These factors imply that India might have maintained a large population. Indeed, some geneticists have suggested that southern Asia contained at least half the world's population between 45 and 20 ka BP (Atkinson et al. 2008). Although I am sceptical of that claim,[6] it was probably one of the more densely populated parts of Palaeolithic Asia – just as today, half the world's population lives in the Oriental Realm. As Kuhn and Stiner (2006: 963) point out, biodiversity has demographic consequences:

> Tropical or subtropical latitudes have probably always supported the densest hominid populations, at the same time presenting the greatest potentials for dietary diversification because of their inherently higher biodiversity of all sorts. … Dietary diversification is simply more likely to emerge *repeatedly* in low-latitude habitats, and thus human populations in these areas are more likely to undergo repeated episodes of expansion.

Before turning to archaeological evidence for our species in India, we need first to assess the fossil skeletal record for when it might have arrived, and whom it might have replaced.

The Indian fossil skeletal record for Homo sapiens *and its predecessors*

The only fossil evidence for the hominin that was in South Asia before *Homo sapiens* is from Hatnora on the northern bank of the Narmada, which is the longest river of peninsular India and one of the few that flows from east to west. Here, a partial cranium was discovered in a conglomerate, along with two clavicles and part of a rib. The cranium is a frustrating find because it is difficult to date, and not particularly informative. Sankhyan (1997) suggested an age range of 0.2 to 0.7 Ma, with a probable range of 0.4–0.5 Ma, and U-series dating of a bovid scapula near where the cranium was found indicated a minimum age of 236,000 years

(Cameron et al. 2004: 419). Recent investigations suggest an age between 131 ± 5 ka BP and 236 ka BP if it is assumed that the cranium has not been reworked into younger deposits (Patnaik et al. 2009). On current evidence, the cranium is likely Late Middle Pleistocene and thus ca. 125–300 ka BP in age.

The cranium sadly lacks its face and dentition, both of which would sharpen up its identification. According to Sheela Athreya (2007), it is neither Neandertal nor *H. erectus*, so it seems distinct from its neighbours in West and East Asia. Nor does it seem to belong to *H. heidelbergensis*, which inhabited Middle Pleistocene Europe. She suggests it is best regarded as *Homo* sp. indet. (species indeterminate – or, in plain speak, we don't know), or perhaps as a specimen showing mixed African and Asian ancestry (Athreya 2015). I suspect that the indigenous inhabitants of South Asia were a type that we have not yet recognised. South Asia, particularly the peninsular part, has a highly endemic fauna and flora. It is also difficult to colonise from the west because of the Thar Desert, from the north because of the Himalayas, or the east because of the delta of the Ganges and the mountain ranges of Assam in northeast India. We might therefore expect its original inhabitants at the time of contact with our species to have been different from its contemporaries in West and East Asia.

Recent discoveries of fragments of six femurs and three humeri that are probably Late Middle Pleistocene are useful additions to our scant knowledge of the Indian human fossil record (Sankhyan 2017a), but we still lack a clear idea of what the original inhabitants of South Asia were like.

When did our species arrive in South Asia?

At present, the earliest unambiguous evidence of our species in South Asia is the skeletal remains from caves of Fa Hien and Batadomba-Lena in Sri Lanka that are dated to ca. 28–36,000 BP (see Kennedy 2003: 180–188 and below) – far younger than the Hatnora specimen (see Figure 5.5). There are two sacra from the Narmada Valley that are described as ca. 70–80 ka old and modern-looking as well as sexually dimorphic (Sankhyan 2017b). If – and it is a large "if" – these are from *H. sapiens*, they are the earliest skeletal evidence of our species in South Asia. The earliest definite evidence from mainland South Asia are four burned cranial vault fragments and a single tooth from a horizon dating to 12–20 ka BP at Jwalapuram 9 (Clarkson et al. 2009). There is also a skull of *H. sapiens* from deposits in the Orsang river, which is one of the tributaries of the Narmada. The deposits containing the skull are dated by IRSL (infrared stimulated luminescence) to 38 ± 5 ka BP and 45 ± 7 ka BP, and the sediments inside the skull were estimated as 30 ± 7 ka BP and 27 ± 4 ka old. However, the skull itself was dated to only 4,600 ± 200 ka BP by AMS C^{14} (Dambricourt-Malassé et al. 2013) which implies that the skull was incorporated into much older sediments and has little relevance to the earliest human populations in South Asia. Unfortunately, skeletal remains often end up in much older deposits when these are reworked. One example is a partial cranium from river gravels at Darra-i-Kur, Afghanistan, which

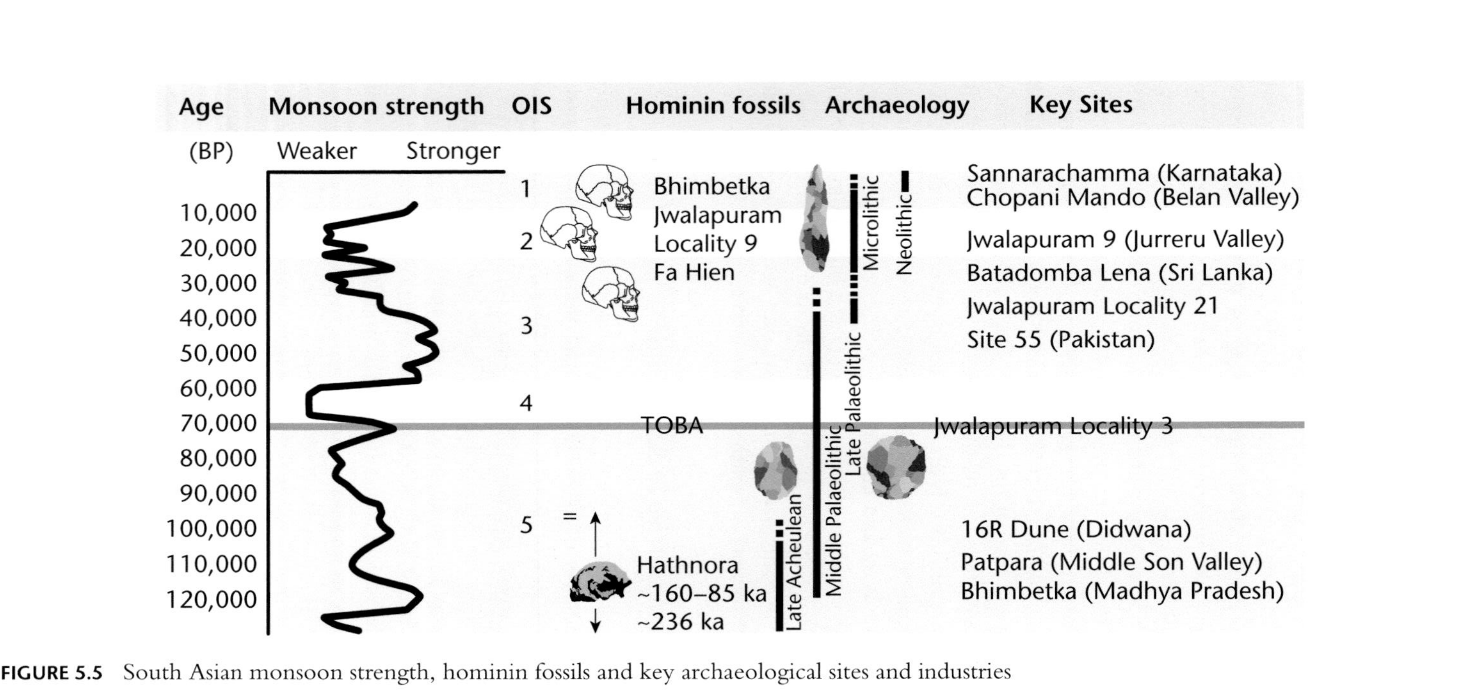

FIGURE 5.5 South Asian monsoon strength, hominin fossils and key archaeological sites and industries

Note the long gap in the fossil skeletal record between Hatnora and Fa Hien.

Source: Petraglia et al. 2010, Fig. 2.

was once thought to be ca. 33,000 years old and potentially the earliest *H. sapiens* specimen from that part of the world but now it seems is also Holocene (Douka et al. 2017; and see Chapter 9).

Because our species was in Southeast Asia by at least 63–73 ka BP (see Chapter 6), humans must have arrived in India before their first recorded skeletal presence of 28–36 ka BP in Sri Lanka (assuming of course that they arrived in South Asia before dispersing east into Southeast Asia). How much earlier than 63–73 ka BP is impossible to say because of the huge gap between the Hatnora specimen (125,000 to 300,000 years old) and the earliest Sri Lankan finds of *H. sapiens* that are ca. 36,000 years old.

There are currently two conflicting narratives about its colonisation by our species which can be summarised as a long or short chronology. Proponents of a "long chronology" propose that humans using a Middle Palaeolithic technology began to enter the Thar Desert in the last interglacial (see Chapter 4), and then gradually moved into inland peninsular India before the massive volcanic eruption of Toba ca. 74 ka ago (Blinkhorn and Petraglia 2014; James and Petraglia 2005; and see below). The opposing narrative rejects all of this and argues instead that there was a single, major episode of immigration by our species after 60 ka ago that introduced a microlithic technology that originated in southern Africa from a South African lithic tradition called the Howieson's Poort. Most supporters of this narrative also propose that these African colonists took a coastal route along the eastern coast of Africa, the Arabian Peninsula and western India, and ultimately continued onwards towards Southeast Asia and Australia. As I have expressed support for the "long chronology" (Dennell and Petraglia 2012; Boivin et al. 2013), I cannot claim to be a neutral observer, but here I hope to give an even-handed account of the relevant parts of both narratives.

Before discussing these, we should begin with a provocative suggestion that our species may already have been in India as early as 300 ka BP – around the same time as it is first evidenced in Africa.

A daring suggestion: Homo sapiens *in India at 250–300 ka BP?*

In a recent paper, a team led by Kumar Akhilesh and colleagues (2018) propose that the Middle Palaeolithic in India began 385 ± 64 ka BP – much earlier than previously proposed (see below). They argue that at Attirampakkam in the Kortallya Basin in southern India (see Figure 5.4), the gradual disuse of bifaces, the predominance of small tools, the appearance of distinctive and diverse Levallois flake and point strategies, and a blade component all indicate a shift away from the preceding Acheulian large-flake technologies. They also point out that these changes occur within the time range of the long interglacial period of MIS 11 340–405 ka BP (Loutre and Berger 2003) or 362–423 ka BP (Dennell 2009: 222). The warmer and wetter conditions of this interglacial would have been favourable to long-distance hominin dispersals across greener deserts of Africa and Southwest Asia (see Chapters 3 and 4). This seems plausible, and I made the same suggestion concerning dispersals between Central Asia and China during MIS 11 (Dennell 2016a: 16).

Akhilesh and colleagues also point out that these changes in technology occurred at roughly the same time in India as in Africa (see Chapter 2). To press this further, they suggest that the advent of a Middle Palaeolithic technology in India at around the same time as in Africa may be part of the same process. Although we cannot yet say whether this was entirely indigenous in India or the result of external influences, there is a possibility that it may have been introduced. Additionally, as the Middle Stone Age in Africa and Arabia is associated with *H. sapiens*, so too it might have been in South Asia with the Middle Palaeolithic. We may yet have to consider a dramatic transformation of our thinking about when, how early and how often our species left Africa.

Homo sapiens *not in India until after 120 ka BP?*

Not all researchers agree that the Middle Palaeolithic in India began as early as proposed above by Akhilesh and colleagues. Instead, several argue that the Lower Palaeolithic Acheulean assemblages persisted until ca. 120 ka BP. As examples, assemblages described as Late Acheulean from the sites of Bamburi 1 and Patpara in the middle Son Valley of north-central India were dated by optically stimulated luminescence (OSL) to ca. 120–140 ka BP, around the transition of MIS 5 to MIS 6 (Haslam et al. 2011). This dating places them among the youngest sites in Africa and Eurasia with Acheulean assemblages. In their view, the Middle Palaeolithic in India does not commence until after 120 ka BP.

These differences of opinion may not be as great as they appear. First, because of India's size and diversity, no single site or region will necessarily represent the entire sub-continent. Second, Late Acheulean technology is defined primarily by "an increase in smaller flake tools, a decrease in both size and relative abundance of bifaces, continued cleaver manufacture, and the introduction of prepared-core reduction techniques such as Levallois" (Haslam et al. 2011). Akhilesh and colleagues agree that in the Late Middle Pleistocene there was a decrease in the size and use of bifacial tools, an increase in smaller tools, and the use of prepared-core Levallois flaking. One way of resolving these differences is to propose that assemblages like those from Bamburi 1 and Patpara are Middle Palaeolithic with bifaces, so the change is not one from Lower to Middle Palaeolithic, but from Middle Palaeolithic with bifaces to a Middle Palaeolithic without them. At the rock shelter of Bhimbetka, for example, a large assemblage of ca. 18,000 items described as Acheulean was followed by a Middle Palaeolithic assemblage of ca. 8,000 items (Misra 1985). The differences between them are statistically insignificant apart from the discontinuation of cleavers (1.15%) and hand axes (0.9%), and a small increase in flake tools from 28% to 34%. The significance of the discontinuation of bifaces in favour of smaller flake tools may be primarily related to the development of hafting technologies (Shipton et al. 2013).

As argued in Chapter 4, Middle Palaeolithic assemblages from the last interglacial in the Thar Desert were likely made by our species. Without the necessary fossil skeletal evidence, we simply do not know whether these groups were the first populations of *Homo sapiens* to inhabit peninsular India or to use a Middle Palaeolithic technology in India.

We can next consider whether our species was in India before the mega-eruption of Mount Toba 74 ka ago.

The mega-eruption of Mount Toba

Toba was a volcano in the north of Sumatra, Indonesia, before it blew itself into the atmosphere in a massive mega-eruption ca. 74,000 ka ago (see Chapter 6). It was the largest, and probably loudest, explosion in the last two million years,[7] and produced ca. 2,800 km^3 of magma (Rose and Chesner 1990). All that remains today of Mount Toba is its caldera, or crater, which now forms a peaceful-looking lake about 100 kilometres long, 30 kilometres wide and up to 500 metres deep. For comparison, the greatest known eruption in historical times was that of Tambora on the island of Sumbawa, Indonesia, in 1815 which expelled ca. 50 km^3 of dense rock and caused widespread crop failure and famine in Europe in 1816 (Oppenheimer 2003); the eruption of Krakatoa in 1883 produced only 3 km^3 (Winchester 2003).

The ash fall-out from the Toba mega-eruption covered around four million sq km of South and Southeast Asia, including large parts of India (Rose and Chesner 1990; Blinkhorn et al. 2012, 2014). Because Toba erupted many times before blowing itself into extinction, the ash from the mega-eruption of 74 ka BP is known as the Young or Youngest Toba Tuff, or YTT. This is easily recognised as a white volcanic ash that has been accurately dated to 73.88 ± 0.32 ka BP (Storey et al. 2012) and forms a useful marker horizon in various geological sections across India. Because the 74 ka BP eruption was so massive, there has been speculation that it changed the earth's climate by blocking out sunlight, thus reducing global temperatures by 3–5° C. for a few years in a "volcanic winter" (Rampino and Self 1992). Ambrose (1998) further suggested that the eruption decimated human populations outside a few tropical refugia (of which the largest were in tropical Africa) and thus caused a major genetic bottleneck from which small populations began to colonise southern Asia after 70 ka BP. When viewed in those terms, Toba has immense significance for the history of our species outside Africa.

A recent study has shown that a tiny amount of Toba ash fell over southern Africa, where a few grains have been recovered from the cave of Pinnacle Point, South Africa (Smith et al. 2018). There is no indication from the long archaeological sequence in that cave that daily life was interrupted in any way. This is not surprising, as Pinnacle Point is 9,000 km from Toba and its main fall-out. The areas we need to know about most are those that lay under the main fall-out. For this reason, considerable attention has been given to the effects of the Toba mega-eruption in India. One study by Williams and colleagues (2009) of pollen in a marine core from the Bay of Bengal and soil carbonates above and below the YTT at four sites in Central India showed that the eruption was followed by a considerably cooler and drier climate for 1,000–2,000 years, in which forest was replaced by open woodland in Central India. The difficulty here lies in determining whether these changes occurred because of Toba or were part of the climatic deterioration into MIS 4. Another study, this time simulating the eruption, suggested that humans experienced "thermal discomfort, but not a real challenge for survival"; in other words, chilly but not cold; less rainfall, lower temperatures and more open vegetation, but not "for a period long enough to have dramatic consequences for their survival" (Timmreck et al. 2012). As noted above, the Indian fauna appears to have survived the Toba fall-out and no extinctions can be attributed to it. Likewise, the effects of Toba in Southeast Asia appear to have been short-lived (see Chapter 6).

There is some evidence from south and central India that human populations survived the fall-out from the Toba eruption. Fieldwork in the Jurreru and Son Valleys of south and northern India confirm drier conditions after Toba, but continuity in the type of stone toolkits (Petraglia et al. 2007, 2012). In the Jurreru Valley, cores were not only flaked in the same way before and after the eruption but were technologically similar to the way cores were flaked in sub-Saharan Africa, Southeast Asia and Australia, "suggesting modern humans may have entered India before the Toba eruption as part of an early eastward dispersal from Africa" (Clarkson et al. 2012: 165; see also Haslam et al. 2010, 2012). This trail of "stone breadcrumbs" from Africa to India is at odds with the proposal of some geneticists that humans did not arrive in India until after the Toba eruption and established genetic lineages that are still present in modern Indian populations.

M and N lineages

Modern Indian populations have as genetic signatures mtDNA haplogroups M and N, which are regarded as daughters, or descendants, of an African haplogroup L3. Most genetic studies show that *H. sapiens* groups with the M and N haplogroups dispersed into India ca. 60–70 ka BP, after the Toba eruption (e.g. Quintana-Murci et al. 1999; Palanichamy et al. 2004; Macaulay et al. 2005). Some geneticists and palaeoanthropologists propose that these groups were the first humans to colonise South Asia.

Sacha Jones (2012) offered an ingenious way of accepting both the archaeological suggestion that humans were present in India before Toba erupted, and the geneticists' proposal that they did not arrive until afterwards. She proposed that humans could have entered India at the end of MIS 6 when sea levels were low, or at the beginning of the last interglacial (MIS 5e) ca. 125 ka ago. During the cool and arid MIS 6, the indigenous inhabitants of peninsular India would likely have been restricted to refugia, with some local extinctions. (If one accepts Akhilesh et al.'s [2018] suggestion, the indigenous populations of India before the last interglacial may have been earlier populations of *Homo sapiens*.) When humans entered India during the last interglacial, they were able to expand into areas previously vacated by the local populations, out-compete them, and expand over most of India. By the end of MIS 5 (ca. 80 ka BP), she argued, the original inhabitants of India were likely extinct. When Toba erupted, many of the *H. sapiens* populations contracted, some became isolated and some became locally extinct. At some time after the eruption, new groups of humans arrived in India with haplogroups M and N, and gradually colonised the sub-continent. These new immigrant groups used a Middle Palaeolithic toolkit that cannot at present be confidently distinguished from that used by those human groups in India before Toba.

This model provides a convincing way of explaining what we know about the Indian Middle Palaeolithic, the Toba eruption, and the genetic make-up of modern Indians. Its main weakness – as with any synthesis of human evolution in India – is the lack of any supporting human skeletal evidence. In addition, there is currently too little archaeological evidence to indicate whether there was a fresh wave of emigration across Arabia and southern Iran towards India, or into northern Pakistan from Central Asia. As seen below, some researchers have argued that humans did disperse into India around 60,000 years ago, with a tool-kit rooted in sub-Sahara Africa.

Indian microliths and the African connection?

Microliths are small flakes or blades that could be slotted into a wooden or bone handle or shaft and used as part of a composite tool. An example might be a sickle, a knife or an arrow. In northwest Europe, where they were first recognised in the 19th century, they are one of the defining features of the Mesolithic, but in South Asia, they have a much greater antiquity, and were being used at least 45,000 years ago (Blinkhorn 2019). Microliths might be small, but the composite tools in which they were used were one of the greatest inventions of the palaeolithic. The concept itself is simple: a slotted handle or shaft, and a set of microliths that could be inset as barbs. If one of them broke or fell out, it could readily be replaced by another – this was the first invention of a spare part. The hard part of this technology is in producing an effective adhesive that could be melted when a microlith needed replacing but would be effective when set. The ingredient(s) to make an effective glue would vary from region to region, and humans had to relearn how to make one every time they shifted to a new environment. Ideally, the adhesive should either be from a source that is common and available year-round, or in a form that could be stored and carried in anticipation of future use. As noted in Chapter 2, Neandertals had the necessary chemical skills to produce a tar from birch bark (Boëda et al. 1996; Grünberg 2002), and the users of the Howieson's Poort assemblages clearly had also mastered this 60–65 ka BP (Wadley et al. 2009) by creating an adhesive that was effective but also flexible. We don't know what type of glue was used in South Asia.

Composite tools that use a handle, microliths and glue have several advantages over single-element tools. They are repairable (providing the handle does not break) and lightweight; if an animal is hit, the spear or arrow is less likely to fall out;[8] if used for processing plants, they can be made into highly effective knives, scrapers or shredders.

Microliths have also cast a long shadow over discussions of when humans first left Africa. Indeed, they are at the centre of one of the hottest debates in recent years over the timing of when our species left Africa. First, we need to see when and where microliths are first documented in South Asia.

The first microliths in South Asia

The oldest microliths in India are from the site of Mehtakheri (Mishra et al. 2013), which has been dated to ca. 44 ± 2 ka BP based on the weighted average of four dates and ~ 48 ka BP if the oldest date is accepted as the most accurate. At the rock shelter of Jwalapuram 9 in southern India, microliths were found in layer E that are dated to >34 ka BP (Clarkson et al. 2009) (Figure 5.6). Microliths of similar age have also been found in Sri Lanka in the caves of Kitulgala and Batadomba-lena, where they are dated to ca. 38 and 30 ka BP (Roberts and Petraglia 2015; Roberts et al. 2015a). At the cave of Fa Hien, the earliest microliths are now dated to ca. 48 ka BP (Wedage et al. 2019b). At Mahadebbera and Kana, in West Bengal, microliths have been dated to ca. 25 to 42 ka BP (Basak and Srivastava 2017).

Microliths and the South African connection: the "impossible coincidence"?

How might these microliths be explained? Paul Mellars (2006a, 2006b; Mellars et al. 2013) proposed a simple but daring hypothesis. The earliest microlithic

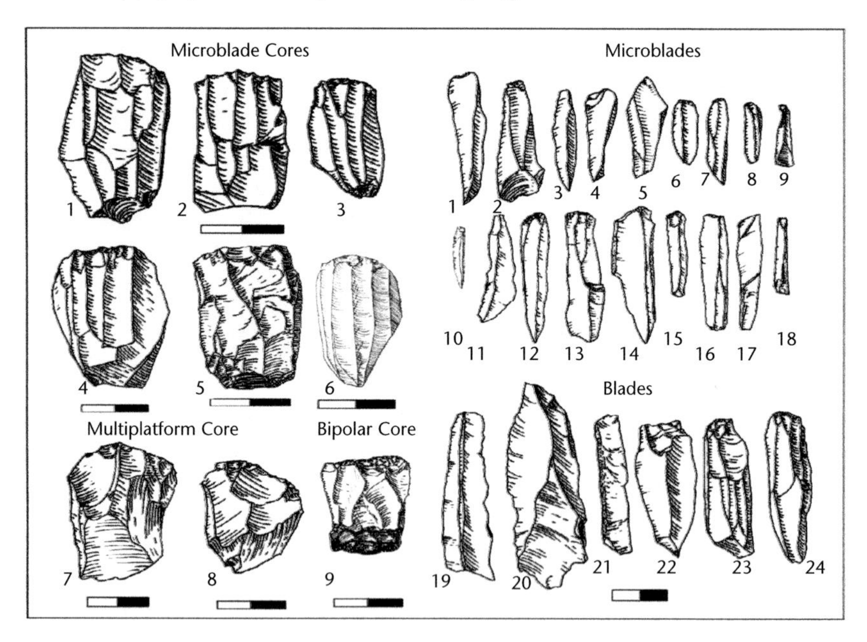

FIGURE 5.6 Blades, microblades and cores from Jwalapuram 9, Jerreru Valley
Source: Clarkson et al. 2009, Fig. 10.

assemblages we know about are those from southern Africa that are known as Howieson's Poort. These are dated to ca. 55–60 ka BP and are thus considerably older than those from South Asia. Because some of the geometric microliths that were illustrated in publications of the Howieson's Poort and South Asian microliths looked very similar, Mellars proposed that those in South Asia were made by people who originated in South Africa and had used a Howieson's Poort technology: "both the African and the Indian industries are dominated by a range of carefully shaped microlithic or larger 'backed-segment' forms of precisely the same range of shapes as those documented in the South Asian industries" (Mellars et al. 2013: 10702). The similarities were so close, he argued, that it was an "impossible coincidence" if these arose independently.

Mellars furthermore linked the microlithic evidence to genetic data on the origin of modern South Asians. As noted already, all Indians today carry haplotypes M and N, which are thought to be derived from haplotype L, which is thought to originate in Africa. According to some geneticists, the M and N lineages diverged from their African progenitor around 60 ka BP – or around the time of the Howieson's Poort assemblages. To Mellars, this was the "impossible coincidence" – the original carriers of the M and N haplotypes dispersed around 60,000 years ago from Africa, and with a microlithic tool-kit.

One potential problem with his argument is that there was no sign of any microlithic assemblages in Arabia, southern Iran or the Thar Desert. There was, however, a way of explaining this: if people using microliths dispersed out of Africa and into southern Asia, they would have done so at a time when sea levels were lower than today, and perhaps up to 60 m lower than today. If therefore they had followed a coastal route, one would not expect any indication of them until they moved inland. It was thus to be expected that there was no sign of them in Arabia or southern Iran. Regarding India, Mellars argued, the discovery of microlithic assemblages in cave deposits such as Jwalapuram 9 marked the time when their makers had moved inland, but not when they first arrived. As seen in Chapter 4, there is no indication of microliths in Pleistocene contexts in Kachchh, northwest India, which is one of the few coastal areas of India that has been surveyed.

Mellars presented a bold narrative for the origin of our species in India but it has not been confirmed by subsequent research. Laura Lewis (2015; Lewis et al. 2014) undertook a detailed comparative study of the Howieson's Poort and South Asian microlithic assemblages from India and Sri India. She found considerable variation within all three; additionally, the Howieson's Poort and South Asian assemblages indicated different technological systems and did not in fact have much in common. "Not only have geometric shapes such as segments shown to be in the minority at all of the sites analysed here, but proportions of microlith morphologies vary considerably between sites" (Lewis et al. 2014: 23). What they had in common was a small percentage of geometric microliths – exactly the artefacts that led Mellars to think that the two traditions were fundamentally similar – but these were a very minor component of these assemblages.

In many ways, it is a pity that the South Asian microlith connection with South Africa proved to be illusory because it leaves us without an adequate explanation for the M and N haplotypes in South Asia. If we assume that the age of these lineages has been estimated correctly, there is no archaeological trace of their arrival. Instead, the advent of microlithic assemblages in India and Sri Lanka is best seen as an indigenous development, perhaps in response to population pressure or environmental deterioration (Petraglia et al. 2009). (Alternatively, the adoption of microliths and a hafting technology may have been driven by the development of effective mastics for securing microliths in a handle; see Blinkhorn 2019).

The earliest blade assemblage in South Asia: site 55, Pakistan

The earliest blade assemblage comes from site 55, Riwat, Pakistan (Dennell et al. 1991; Rendell et al. 1989). Riwat is in the northern part of the Punjab, and although within the monsoon region, this region is part of the Palearctic Realm in terms of its modern and Pleistocene fauna. Up to 10 m of loess thick cover this part of Pakistan (Rendell and Townsend 1988), and it is thus one of the few areas in South Asia where there are extensive, undisturbed deposits from the last glacial cycle in which Palaeolithic artefacts could be found.

Excavation of site 55 produced an assemblage that included freshly flaked blades (Figure 5.7); these were found associated with what may have the stone footing of a structure, a slab-lined pit and a circular niche, function unknown (Figure 5.8). Several pieces of flaked stone conjoined, which indicated that stone was being flaked and used on the site. The archaeological horizon lay on top of a conglomerate surface and thermoluminescence (TL) dating of the loess directly over the site indicates that it is ca. 45,000 years old (Rendell and Dennell 1987). The site is difficult to interpret: it does not appear to have been a residential, butchery, ritual or transit site, or an area where refuse was dumped from a neighbouring activity zone. My own interpretation of site 55 was that the conglomerate surface may have been used for degreasing and drying animal skins in the autumn, as described by Binford (1978) for the Nunamiut of Alaska.

Pioneers of the South Asian rainforests

One of the most exciting recent reappraisals of how we regard human history in the Tropics has been about rainforests. For many years, the prevalent view was that hunter-gatherer-foragers were unable to live in tropical rainforests unless they could trade or exchange with neighbouring agricultural communities. The stark implication of this viewpoint was that pre-agricultural, Pleistocene inhabitation of rainforests was highly unlikely. This view dates to the 1980s, when (among others) Bailey and colleagues (1989) and Headland (Headland and Reid, 1989) independently and then jointly (Bailey and Headland, 1991) argued that there was no evidence that human foragers pre-dated agriculturalists in rainforests. They also made the entirely legitimate point that the fact that an archaeological

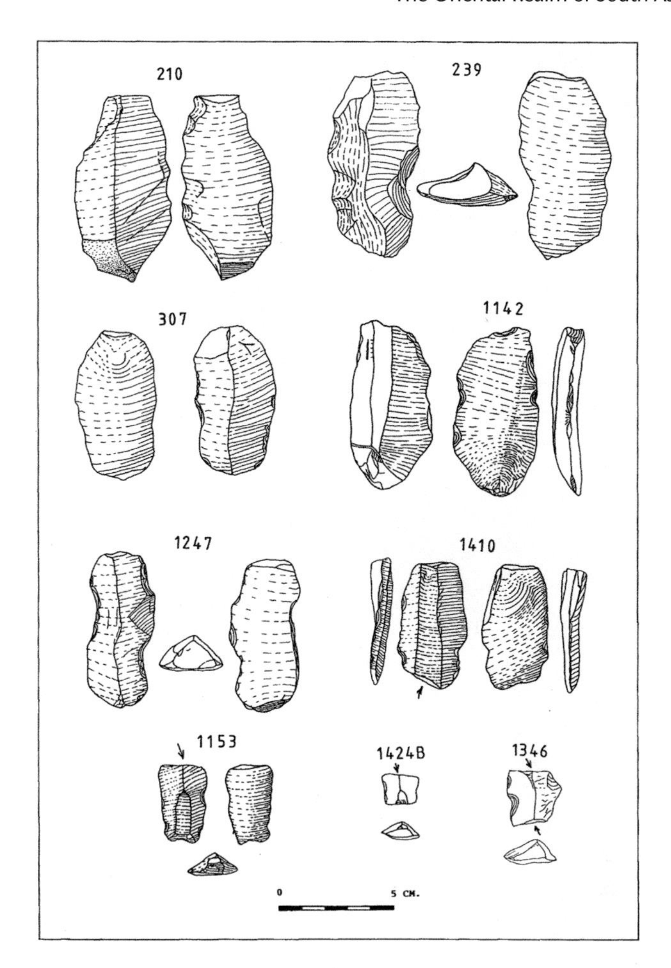

FIGURE 5.7 Blades from site 55, Riwat, Pakistan
Source: Dennell et al. 1991, Fig. 9.

site was located today in rainforest did not mean that the site lay in rainforest when it was originally occupied: rainforests would have expanded and contracted as much as any environment.

There is no denying the fact that pre- or non-agricultural communities of humans faced an enormous set of challenges in obtaining a regular and adequate diet in a tropical rainforest, even when living in an environment with such an amazing amount of biodiversity. Animals that were large enough to be worth hunting tended to be largely solitary or lived in small groups; those living on the ground were often difficult to see and track, and pursuing them was impeded by dense vegetation, and those (such as monkeys) that lived in the high canopy were even more difficult to hunt. With plants, it is necessary to know which can be eaten, and at what time of year, and whether it is the roots (as with tubers such as yams), stems or fruits that

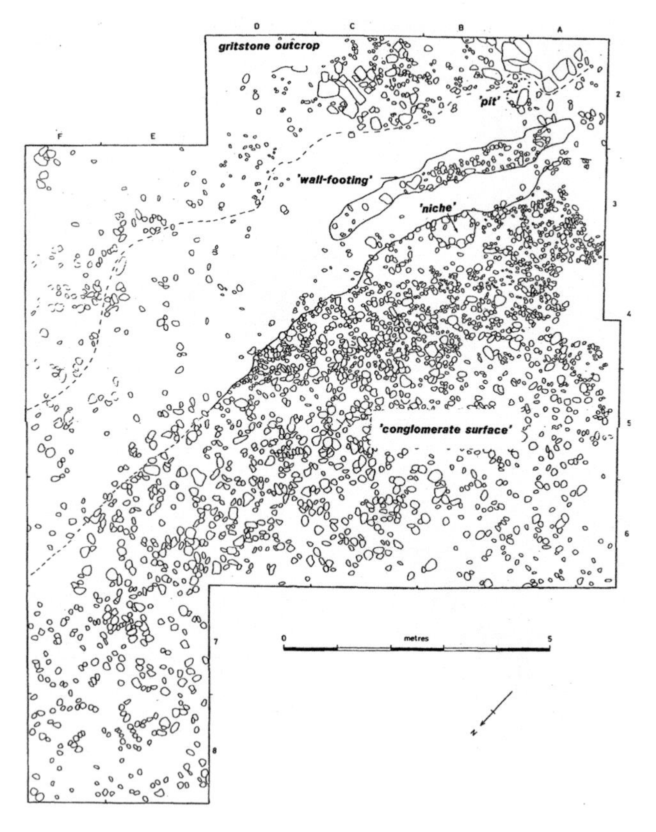

FIGURE 5.8 Plan of site 55, Riwat, Pakistan
Source: Dennell et al. 1991, Fig. 3.

are edible; some are poisonous (such as cassava) when picked and require processing by washing or boiling before they can be eaten. Rainforests can also be unhealthy places to live because of water- or insect-borne diseases since open wounds can easily fester and because much of the smaller fauna can be venomous. Although rainforests have been described as the world's largest natural pharmacy, great skill and knowledge is required in knowing what plants (and which parts of them) can serve as medicines, and how these can be used. For humans to adapt to living in rainforests was thus a major achievement, particularly for a creature that originated and long flourished in grasslands and open woodlands. Rainforests were thus unsurprisingly seen as a barrier to human settlement before farming. It is only in the last few years that we now realise that people have been living in rainforests for over 45,000 years (Wedage et al. 2019a). The earliest definite evidence comes from South and Southeast Asia. (People may have inhabited the rainforests of West and Central Africa for a similar length of time and perhaps even longer but the evidence is inconclusive because of the problems of associating a site with a forested environment; see Roberts and Petraglia 2015).

Recent evidence shows that humans were living wholly in rainforests in Sri Lanka before 40 ka BP. Excavations of the caves of Batadomba lena, Fa-Hien lena and Kitulgala lena have produced human remains associated with a microlithic assemblage (Lewis et al. 2014; Roberts et al. 2015a; and see Figure 5.9) and an extensive bone industry that included points that were probably used as projectiles or in snares as early as 36 to 38 ka BP (Perera et al. 2011). Skeletal evidence indicates that the inhabitants were *H. sapiens* (Kennedy 2003) and this is currently the earliest direct evidence of our species in South Asia. Faunal data indicates that semi-arboreal and arboreal primates comprise most of the mammalian assemblages; other resources exploited at Batadomba-lena include mouse deer, giant squirrel, mongoose, jungle cat and civet (Figure 5.10), as well as *Canarium* sp. nuts and starchy rainforest plants (Perera et al. 2011), all of which imply dedicated rainforest subsistence. In a groundbreaking investigation, Roberts and colleagues (2015b) showed from stable carbon and oxygen isotopic analysis of human and other animal bones from these caves that human diet was overwhelmingly from rainforest foods. This lifestyle was maintained from 38 ka BP through the rest of the Pleistocene (including the Last Glacial Maximum) and into the Holocene up to when agriculture was introduced ca. 3,000 years ago. Far from being a barrier to human settlement before the advent of agriculture, the rainforests of Sri Lanka have been a resource zone and were also a refugium during the Last Glacial Maximum. We will see in the next chapter even earlier evidence that humans were inhabiting the rain forests of Southeast Asia.

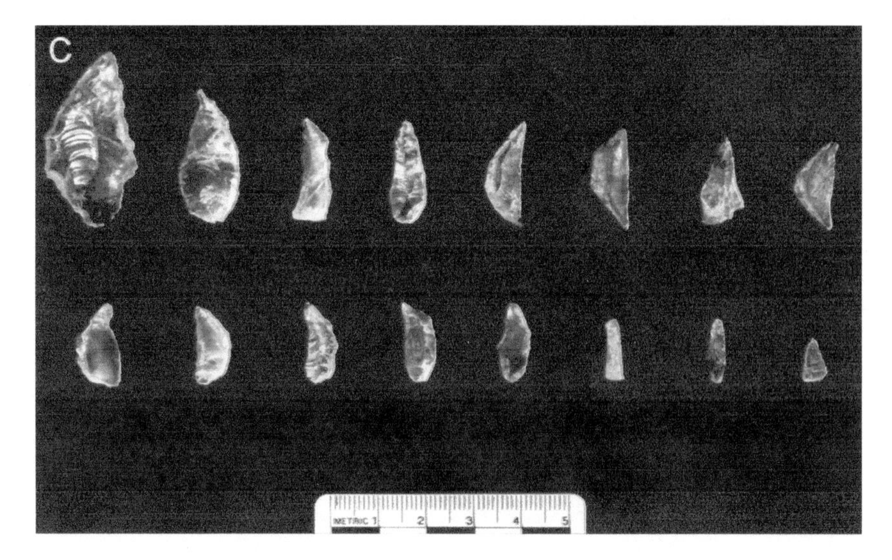

FIGURE 5.9 Microliths from Batadomba lena, Sri Lanka

Source: Lewis 2014, Fig. 6.

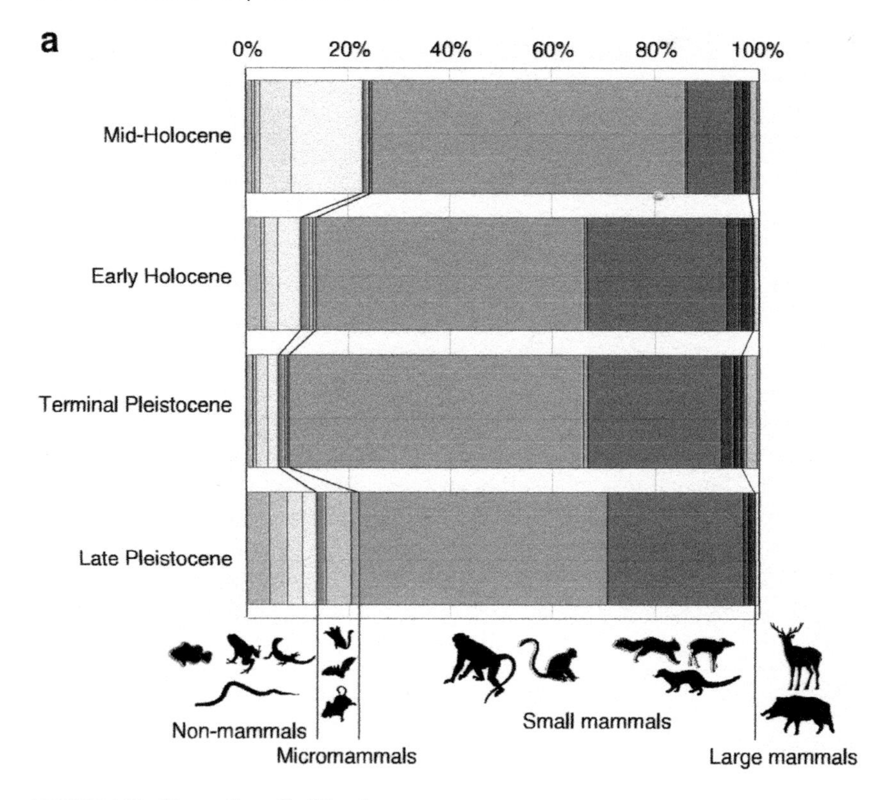

FIGURE 5.10 Fauna from Fa Hien lena

The abundance of each taxon is based on NISP (numbers of identified specimens). These proportions are not changed greatly when based on MNI (minimum numbers of individuals).

Source: Wedage et al. 2019a, Fig. 3.

Recent evidence from South and Southeast Asia (Chapter 6) has important implications on how rainforests are envisaged and managed today. Perhaps the most important point to emerge from recent archaeological research in rainforests is that these are not "not wild, pristine expanses of raw nature, but rather are – to varying degrees – anthropogenic landscapes" (Moore 2014: 153) that have been modified by humans for millennia, whether South and Southeast Asia, Africa or in Amazonia and Central America. This realisation has important consequences for conservation policies: first, it is unrealistic to think about restoring rainforests to their "natural" state; and second, the indigenous inhabitants are not necessarily a threat to biodiversity but often an active agent in promoting it. This is one area where Palaeolithic archaeology can make a positive contribution to modern conservation measures (see e.g. Roberts 2019; Dennell 2016b).

Evidence here strongly implies that *Homo sapiens*, and not its predecessors, were the first to inhabit rainforests. In what we have seen so far about the dispersal of our species outside Africa, the colonisation of the rainforests of southern Asia is a very eloquent example of how good we are as a colonising species. The adaptation to living in rainforests is as significant as that of living in the Arctic and provides an outstanding example of human adaptability as a colonising species after 50 ka BP; as Roberts and colleagues (2016) point out, *Homo sapiens* seems to be the only member of our genus that has been able to adapt to rainforests.

Putting it together

The main area of disagreement is over whether the Middle Palaeolithic of India was a local development from the final Early Palaeolithic of the indigenous inhabitants, or was intrusive and made by incoming groups of *H. sapiens*. Until the all-important skeletal evidence is obtained, there is no clear indication that our species was or was not in India before the Toba mega-eruption of 74 ka BP. As indicated above, I suggest we do not reject out of hand the proposal by Akhilesh et al. (2018) that the Indian Middle Palaeolithic may be as early as 385 ka BP and may have been introduced from Africa, perhaps by early populations of our species. My own suspicion is that Early Middle Palaeolithic assemblages before 80,000 years ago were made by both the indigenous population and the incoming population of our species, but after that data – i.e. post-Toba – it was mostly made by *H. sapiens*. My main reason for arguing that our species was already in South Asia before the Toba eruption is that there is (in my view) convincing evidence that it was in Southeast Asia and south China before 60 ka BP (see Chapters 6 and 10), and it is therefore reasonable to assume that it was in South Asia before that date. Interestingly, Sheila Mishra – one of the strongest proponents of the view that our species did not arrive in India until the Late Palaeolithic – accepts that our species is evidenced in Southeast Asia and South China before it is clearly evidenced in India. Her explanation for this is that our species could not compete with the indigenous inhabitants of India, and therefore bypassed it by taking a route to the north across Central Asia and then entering south China (Mishra et al. 2013). Although this is a daring suggestion, I see no support for it in the current evidence from Central Asia (Chapter 9), so I incline to the view that the Indian Middle Palaeolithic is the product of our species.

We need also recognise that we have a very limited view of the Middle and Late Palaeolithic of India. Almost all the current relevant evidence comes from the savannah and open woodland of western India, and the rainforests of Sri Lanka. Almost nothing is known about east and northeast India, apart from a few sites such as Mahadebbera and Kana, in West Bengal (Basak and Srivastava 2017). Very few South Asian Palaeolithic sites (apart from the cave sites in Sri Lanka) have any faunal data, so we know almost nothing about hunting and foraging. Meanwhile, our main certainties are that *H. sapiens* was producing composite artefacts that used microliths ca. 48,000 years ago and in Sri Lanka had already adapted to living in rain forests by 45,000 years ago. Although no skeletal remains were found at site 55, it is probable that it was our species that was making blade assemblages in north Pakistan ca. 45,000 years ago. Prior to that time, we are dealing with probabilities, possibilities and uncertainties, and clearly there is still much to learn about when and how our species colonised the Oriental Realm of India, Bangladesh and Myanmar.

Notes

1 The EU minus the UK and India, Bangladesh, Sri Lanka and Myanmar each cover ca. 1.6 million sq miles (ca. 4 million sq km).
2 In terms of landscape learning, I wonder if incoming humans learnt by observing the non-human primates which plants were edible or medicinal; see Huffman (1997).
3 A loris is a small nocturnal primate, largely insectivorous, that lives in the tree canopy in tropical forests.
4 Chenopods and Ameranths: these are informal names for Chenopodiaceae and Amaranthaceae. They are large families of herbs and shrubs, whereas the Poaceae are grasses. In simple terms, increases in the frequencies of grass pollen indicate cooler and drier conditions.
5 The exception is a single fossil specimen of *Theropithecus*, a giant gelada (Roberts et al. 2014: 5849).
6 Surely Africa, Europe, Southwest Asia and China must have had substantial populations by this time?
7 The eruption of Krakatoa in 1883 was heard 3,000 miles away on Rodriguez Island (Winchester 2003: 274) but was puny compared to Toba.
8 Many years ago I spent a week in the New Guinea highlands and stayed in a local village. The local young men who used bow and arrows told me that they used barbed arrows when hunting birds because the arrow would not fall out, and that made it easier for them to catch the bird. Needle-like arrows were used in war parties because these caused only a flesh wound and could easily be removed with only a little bleeding – mutual restraint on both sides.

References

Akhilesh, K., Pappu, S., Rajapara, H.M., Gunnell, Y., Shukla, A.D. and Singhvi, A.K. (2018) Early Middle Palaeolithic culture in India around 385–172 ka reframes Out of Africa models. *Nature*, 554: 97–101.

Ambrose, S.H. (1998) Late Pleistocene human populations bottlenecks, volcanic winter, and differentiation of modern humans. *Journal of Human Evolution*, 34: 623–651.

Athreya, S. (2007) Was *Homo heidelbergensis* in South Asia? A test using the Narmada fossil from central India. In M.D. Petraglia and B. Allchin (eds) *The Evolution and History of Human Populations in South Asia: Inter-disciplinary Studies in Archaeology, Biological Anthropology, Linguistics and Genetics*, Dordrecht: Springer, pp. 137–170.

Athreya, S. (2015) Modern human emergence in South Asia. In Y. Kaifu, M. Izuho, T. Goebel, H. Sato and A. Ono (eds) *Emergence and Diversity of Modern Human Behavior in Paleolithic Asia*. College Station: Texas A&M University Press, pp. 61–79.

Atkinson, Q.D., Gray, R.D. and Drummond, A.J. (2008) mtDNA variation predicts population size in humans and reveals a major southern Asian chapter in human prehistory. *Molecular Biology and Evolution*, 25(2): 468–474.

Bailey, R., Head, G., Jenike, M., Owen, B., Rechtman, R. and Zechenter, E. (1989) Hunting and gathering in tropical rain forest: Is it possible? *American Anthropologist*, 91: 59–82.

Bailey, R.C. and Headland, T.N. (1991) The tropical rain forest: is it a productive environment for human foragers? *Human Ecology*, 19(20): 261–285.

Basak, B. and Srivastava, P. (2017) Earliest dates of microlithic industries (42–25 ka) from West Bengal, eastern India: new light on modern human occupation in the Indian Subcontinent. *Asian Perspectives*, 56: 237–259.

Binford, L.R. (1978) *Nunamiut Ethnoarchaeology*. London: Academic Press.

Blinkhorn, J. (2019) Examining the origins of hafting in South Asia. *Journal of Paleolithic Archaeology*, 2: 466–481, doi:10.1007/s41982-019-00034-4.

Blinkhorn, J., Parker, A.G., Ditchfield, P., Haslam, M. and Petraglia, M. (2012) Uncovering a landscape buried by the super-eruption of Toba, 74,000 years ago: a multi-proxy environmental reconstruction of landscape heterogeneity in the Jurreru Valley, south India. *Quaternary International*, 258: 135–147.

Blinkhorn, J. and Petraglia, M.D. (2014) Assessing models for the dispersal of modern humans to South Asia. In R. Dennell and M. Porr (eds) *Southern Asia, Australia and the Search for Human Origins*, Cambridge: Cambridge University Press, pp. 64–75.

Blinkhorn, J. and Petraglia, M.D. (2017) Environments and cultural change in the Indian Subcontinent implications for the dispersal of *Homo sapiens* in the Late Pleistocene. *Current Anthropology*, 58(Supplement 27): S463–S479.

Blinkhorn, J., Smith, V.C., Achyuthan, H., Shipton, C., Jones, S.C., Ditchfield, P.D. and Petraglia, M.D. (2014) Discovery of Youngest Toba Tuff localities in the Sagileru Valley, south India, in association with Palaeolithic industries. *Quaternary Science Reviews*, 105: 239–243.

Boëda, E., Connan, J., Dessort, D., Muhesen, S., Mercier, N., Valladas, H. and Tisnérat, N. (1996) Bitumen as a hafting material on Middle Palaeolithic artefacts. *Nature*, 380: 336–338.

Boivin, N., Fuller, D.Q., Dennell, R.W., Allaby, R. and Petraglia, M. (2013) Human dispersal across diverse environments of Asia during the Upper Pleistocene. *Quaternary International*, 300: 32–47.

Cameron, D., Patnaik, R. and Sahni, A. (2004) The phylogenetic significance of the Middle Pleistocene Narmada cranium from Central India. *International Journal of Osteoarchaeology*, 14: 419–447.

Chauhan, P. (2008) Large mammal fossil occurrences and associated archaeological evidence in Pleistocene contexts of peninsular India and Sri Lanka. *Quaternary International*, 192: 20–42.

Clarkson, C., Jones, S. and Harris, C. (2012) Continuity and change in the lithic industries of the Jurreru Valley, India, before and after the Toba eruption. *Quaternary International*, 258: 165–179.

Clarkson, C., Petraglia, M., Korisettar, R., Haslam, M., Boivin, N., Crowther A., Ditchfield, P., Fuller, D., Miracle, P., Harris, C., Connell, K., James, H. and Koshy, J. (2009) The oldest and longest enduring microlithic sequence in India: 35,000 years of modern human occupation and change at the Jwalapuram Locality 9 rockshelter. *Antiquity*, 83: 326–348.

Colbert, E.H. (1943) Pleistocene vertebrates collected in Burma by the American Southeast Asiatic Expedition. *Transactions of the American Philosophical Society*, 32: 395–429.

Dambricourt-Malassé, A., Raj, R. and Shah, S. (2013) Orsang Man: a robust *Homo sapiens* in Central India with Asian *Homo erectus* features. *Evolving Humanity, Emerging Worlds*, 17th World Congress of the International Union of Anthropological and Ethnological Sciences, August 2013, Manchester, United Kingdom. <halshs-00873939>.

Dennell, R.W. (2007) "Resource-rich, stone-poor": early hominin land use in large river systems of northern India and Pakistan. In M. Petraglia and B. Allchin (eds) *The Evolution and Diversity of Humans in South Asia*, Dordrecht: Springer, pp. 41–68.

Dennell, R.W. (2009) *The Palaeolithic Settlement of Asia*. Cambridge: Cambridge University Press.

Dennell, R.W. (2016a) Life without the Movius Line. *Quaternary International*, 400: 14–22.

Dennell, R.W. (2016b) Tropical rainforests as long-established cultural landscapes. In N. Sanz (ed.) *Exploring Rainforests as Long-Established Cultural Landscapes*. Paris: UNESCO, pp. 15–26.

Dennell, R.W. and Petraglia, M.D. (2012) The dispersal of *Homo sapiens* across southern Asia: how early, how often, how complex? *Quaternary Sciences Reviews*, 47: 15–22.

Dennell, R.W., Rendell, H., Halim, M. and Moth, E. (1991) Site 55, Riwat: a 42,000 yr.-bp. open-air Palaeolithic site from northern Pakistan. *Journal of Field Archaeology*, 19: 17–33.

Douka, K., Slon, V., Stringer, C., Potts, R., Hübner, R., Meyer, M., Spoor, F., Pääbo, S. and Higham, T. (2017) Direct radiocarbon dating and DNA analysis of the Darra-i-Kur (Afghanistan) human temporal bone. *Journal of Human Evolution*, 107: 86–93.

Grünberg, J.M. (2002) Middle Palaeolithic birch-bark pitch. *Antiquity*, 76: 15–16.

Haslam, M., Clarkson, C., Petraglia, M., Korisettar, R., Jones, S. et al. (2010) The 74 ka Toba super-eruption and southern Indian hominins: archaeology, lithic technology and environments at Jwalapuram Locality 3. *Journal of Archaeological Science*, 37: 3370–3384.

Haslam, M., Clarkson, C., Roberts, R.G., Borad, J., Korisettar, R. et al. (2012) A southern Indian Middle Palaeolithic occupation surface sealed by the 74 ka Toba eruption: further evidence from Jwalapuram Locality 22. *Quaternary International*, 258: 148–164.

Haslam, M., Roberts, R.G., Shipton, C., Pal, J.N., Fenwick, J.L., Ditchfield, P., Boivin, N., Dubey, A.K., Gupta, M.C. and Petraglia, M. (2011) Late Acheulean hominins at the Marine Isotope Stage 6/5e transition in north-central India. *Quaternary Research*, 75: 670–682.

Headland, T.N. and Reid, L.A. (1989) Hunter-gatherers and their neighbors from prehistory to the present. *Current Anthropology*, 3: 43–66.

Huffman, M. (1997) Current evidence for self-medication in primates: a multidisciplinary perspective. *Yearbook of Physical Anthropology*, 40: 171–200.

James, H.A.V. and Petraglia, M. (2005) Origins and the evolution of behaviour in the Later Pleistocene record of South Asia. *Current Anthropology*, 46: S3–S27.

Jones, S. (2012) Local- and regional-scale impacts of the ~74 ka Toba supervolcanic eruption on hominin populations and habitats in India. *Quaternary International*, 258: 100–118.

Kennedy, K.A.R. (2003) *God-Apes and Fossil Men: Palaeoanthropology in South Asia*. Ann Arbor: University of Michigan Press.

Korisettar, R. (2007) Toward developing a basin model for Paleolithic settlement of the Indian subcontinent: geodynamics, monsoon dynamics, habitat diversity and dispersal routes. In M. D. Petraglia and B. Allchin (eds) *The Evolution and History of Human Populations in South Asia*, Dordrecht: Springer, pp. 69–96.

Kuhn, S.L. and Stiner, M.C. (2006) What's a mother to do? The division of labor among Neandertals and Modern Humans in Eurasia. *Current Anthropology*, 47(6): 953–980.

Lewis, L. (2015) *Early Microlithic Technologies and Behavioural Variability in Southern Africa and South Asia*. Oxford: D. Phil thesis.

Lewis, L., Perera, N. and Petraglia, M. (2014) First technological comparison of southern African Howiesons Poort and South Asian microlithic industries: an exploration of inter-regional variability in microlithic assemblages. *Quaternary International*, 350: 7–25, doi:10.1016/j.quaint.2014.09.013.

Loutre, M.F. and Berger, A. (2003) Marine Isotope Stage 11 as an analogue for the present interglacial. *Global and Planetary Change*, 36: 209–217.

Macaulay, V., Hill, C., Achilli, A., Rengo, C., Clarke, D. et al. (2005) Single, rapid coastal settlement of Asia revealed by analysis of complete mitochondrial genomes. *Science*, 308: 1034–1036.

Mellars, P. (2006a) Going East: new genetic and archaeological perspectives on the modern human colonization of Eurasia. *Science*, 313: 796–800.

Mellars, P. (2006b). Why did modern human populations disperse from Africa ca. 60,000 years ago? A new model. *Proceedings of the National Academy of Sciences USA*, 103(25): 9381–9386.

Mellars, P., Gorí, K.C., Carr, M., Soares, P.A. and Richards, M.B. (2013) Genetic and archaeological perspectives on the initial modern human colonization of southern Asia. *Proceedings of the National Academy of Sciences USA*, 110(26): 10699–10704.

Mishra, S., Chauhan, N. and Singhvi, A.K. (2013) Continuity of microblade technology in the Indian Subcontinent since 45 ka: implications for the dispersal of modern humans. *PLoS ONE*, 8(7): e69280, doi:10.1371/journal.pone.0069280.

Misra, V.M. (1985) The Acheulean succession at Bhimbetka, Central India. In V.N. Misra and P. Bellwood (eds) *Recent Advances in Indo-Pacific Prehistory*, New Delhi: Oxford and IBH Publishing Co., pp. 35–47.

Moore, J.D. (2014) *A Prehistory of South America: Ancient Cultural Diversity on the Least Known Continent*. Boulder, Colorado: University Press of Colorado.

Nanda, A.C., Sehgal, R.K. and Chauhan, P.R. (2018) Siwalik-age faunas from the Himalayan Foreland Basin of South Asia. *Journal of Asian Earth Sciences*, 162: 54–68.

Oppenheimer, C. (2003) Climatic, environmental and human consequences of the largest known historic eruption: Tambora volcano (Indonesia) 1815. *Progress in Physical Geography*, 2(2): 230–259.

Palanichamy, M.G., Sun, C., Agrawal, S., Bandelt, H.-J., Kong, Q.-P. et al. (2004) Phylogeny of mitochondrial DNA macrohaplogroup N in India, based on complete sequencing: implications for the peopling of South Asia. *American Journal of Human Genetics*, 75: 966–978.

Patnaik, R., Chauhan, P.R., Rao, M.R., Blackwell, B.A.B., Skinner, A.R., Sahni, A., Chauhan, M.S. and Khan, H.S. (2009) New geochronological, palaeoclimatological and Palaeolithic data from the Narmada Valley hominin locality, central India. *Journal of Human Evolution*, 5: 114–133.

Perera, N., Kourampas, N., Simpson, I.A., Deraniyagala, S.U., Bulbeck, D., Kamminga, J., Perera, J., Fuller, D.Q., Szabo, K. and Oliviera, N.V. (2011) People of the ancient rainforest: Late Pleistocene foragers at the Batadomba-lena rockshelter, Sri Lanka. *Journal of Human Evolution*, 61: 254–269.

Petraglia, M., Clarkson, C., Boivin, N., Haslam, M., Korisettar, R., Chaubey, G., Ditchfield, P., Fuller, D., James, H., Jones, S., Kisivild, T., Koshy, J., Lahr, M.M., Metspalu, M., Roberts, R. and Arnold, L. (2009) Population increase and environmental deterioration correspond with microlithic innovations in South Asia ca. 35,000 years ago. *Proceedings of the National Academy of Sciences USA*, 106: 12261–12267.

Petraglia, M.D., Ditchfield, P., Jones, S., Korisettar, R. and Pal, J.N. (2012) The Toba volcanic super-eruption, environmental change, and hominin occupation history in India over the last 140,000 years. *Quaternary International*, 258: 119–134.

Petraglia, M.D., Haslam, M., Fuller, D.Q., Boivin, N. and Clarkson, C. (2010) Out of Africa: new hypotheses and evidence for the dispersal of *Homo sapiens* along the Indian Ocean rim. *Annals of Human Biology*, 37(3): 288–311.

Petraglia, M., Korisettar, R., Boivin, N., Clarkson, C., Ditchfield, P. et al. (2007) Middle Paleolithic assemblages from the Indian Subcontinent before and after the Toba Super-Eruption. *Science*, 317: 114–116.

Prabhu, C.N., Shankar, R., Anupama, K., Taieb, M., Bonnefille, R., Vidal, L., and Prasad, S. (2004) A 200-ka pollen and oxygen-isotope record from two sediment cores from the eastern Arabian Sea. *Palaeogeography, Palaeoclimatology, Palaeoecology*, 214: 309–321.

Quintana-Murci, L., Semino, O., Bandelt, H.-J., Passarino, G., McElreavey, K. et al. (1999) Genetic evidence of an early exit of *Homo sapiens sapiens* from Africa through eastern Africa. *Nature Genetics*, 23: 437–441.

Rampino, M.R. and Self, S. (1992) Volcanic winter and accelerated glaciation following the Toba super-eruption. *Nature*, 359: 50–52.

Rendell, H. and Dennell, R.W. (1987) The dating of an Upper Pleistocene archaeological site at Riwat, northern Pakistan. *Geoarchaeology*, 1: 6–12.

Rendell, H.R., Dennell, R.W. and Halim, M. (1989) *Pleistocene and Palaeolithic Investigations in the Soan Valley, Northern Pakistan.* British Archaeological Reports International Series 544, 1–364.

Rendell, H.M. and Townsend, P.D. (1988) Thermoluminescence dating of a 10 m loess profile in Pakistan. *Quaternary Science Reviews*, 7: 251–255.

Roberts, P. (2019) *Tropical Forests in Prehistory, History, and Modernity.* Oxford: Oxford University Press.

Roberts, P., Boivin, N., Lee-Thorp, J., Petraglia, M. and Stock, J. (2016) Tropical forests and the genus *Homo. Evolutionary Anthropology*, 25(6): 306–317.

Roberts, P., Boivin, N. and Petraglia, M.D. (2015a) The Sri Lankan "microlithic" tradition c. 38,000 to 3000 years ago: tropical technologies and adaptations of *Homo sapiens* at the southern edge of Asia. *Journal of World Prehistory*, 28(2): 69–112, doi:10.1007/s10963-015-9085-5.

Roberts, P., Delson, E., Miracle, P., Ditchfield, P., Roberts, R.G. et al. (2014) Continuity of mammalian fauna over the last 200,000 yr in the Indian subcontinent. *Proceedings of the National Academy of Sciences USA*, 113: 5848–5853.

Roberts, P., Perera, N., Wedage, O., Deraniyagala, S., Perera, J., Eregama, S., Gledhill, A., Petraglia, M.D. and Lee-Thorpe, J. (2015b) Direct evidence for human reliance on rainforest resources in Late Pleistocene Sri Lanka. *Science*, 347: 1246–1249.

Roberts, P. and Petraglia, P. (2015) Pleistocene rainforests: barriers or attractive environments for early human foragers? *World Archaeology*, 47(5): 718–739.

Rose, W.I. and Chesner, C.A. (1990) Worldwide dispersal of ash and gases from Earth's largest known eruption: Toba, Sumatra, 75 ka. *Palaeogeography, Palaeoclimatology, Palaeoecology*, 89: 269–275.

Saegusa, H., Thasod, Y. and Ratanasthien, B. (2004) Notes on Asian stegodontids. *Quaternary International*, 126–128: 31–48.

Sankhyan, A.R. (1997) Fossil clavicle of a Middle Pleistocene hominid from the Central Narmada Valley, India. *Journal of Human Evolution*, 32: 3–16.

Sankhyan, A.R. (2017a) Pleistocene hominin fossil femora and humeri. *International Journal of Anatomy and Research*, 5(4.1): 4510–4518.

Sankhyan, A.R. (2017b) First record and study of prehistoric sacra from central Narmada Valley (m.p.). *International Journal of Anatomy and Research*, 5(3.1): 4144–4151.

Shipton, C., Clarkson, C., Pal, J.N., Jones, S.C., Roberts, R.G., Harris, C., Gupta, M.C., Ditchfield, P.W. and Petraglia, M.D. (2013) Generativity, hierarchical action and recursion in the technology of the Acheulean to Middle Palaeolithic transition: a perspective from Patpara, the Son Valley, India. *Journal of Human Evolution*, 65: 93–108.

Smith, E.I., Jacobs, Z., Johnsen, R., Ren, M., Fisher, E.C., Oestmo, S., Wilkins, J., Harris, J.A., Karkanas, P., Fitch, S., Ciravolo, A., Keenan, D., Cleghorn, N., Lane, C.S., Matthews, T. and Marean, C.W. (2018) Humans thrived in South Africa through the Toba eruption about 74,000 years ago. *Nature Communications*, 555: 511–515, doi:10.1038/nature25967.

Storey, M., Roberts, R.G. and Saidin, M. (2012) Astronomically calibrated $^{40}Ar/^{39}Ar$ age for the Toba supereruption and global synchronization of late Quaternary records. *Proceedings of the National Academy of Sciences USA*, 109: 18684–18688.

Tchernov, E., Horwitz, K., Ronen, A. and Lister, A.I. (1994) The faunal remains from Evron Quarry in relation to other lower paleolithic hominid sites in the southern Levant. *Quaternary Research*, 42: 328–339.

Thompson, L., Yao, G., Davis, M.E., Henderson, K.A., Mosley-Thompson, E. et al. (1997) Tropical climate instability: the last glacial cycle from a Qinghai-Tibetan ice core. *Science*, 276: 1821–1825.

Timmreck, C., Graf, H.-F., Zanchettin, D., Hagemann, S., Kleinen, T. and Krüger, K. (2012) Climate response to the Toba super-eruption: regional changes. *Quaternary International*, 258: 30–44.

Wadley, L., Hodgskiss, T. and Grant, M. (2009) Implications for complex cognition from the hafting of tools with compound adhesives in the Middle Stone Age, South Africa. *Proceedings of the National Academy of Sciences USA*, 106: 9590–9594.

Wedage, O., Amano, N., Langley, M.C., Douka, K., Blinkhorn, J. et al. (2019a) Specialized rainforest hunting by *Homo sapiens* ~45,000 years ago. *Nature Communications*, 10: 739, doi:10.1038/s41467-019-08623-1.

Wedage, O., Picin, A., Blinkhorn, J., Douka, K., Deraniyagala, S., Kourampas, N. et al. (2019b) Microliths in the South Asian rainforest ~45-4 ka: New insights from Fa-Hien Lena Cave, Sri Lanka. *PLoS ONE*, 14(10): e0222606, doi:10.1371/journal.pone.0222606.

Williams, M.A.J., Ambrose, S.H., Kaars, S. van der, Ruehlmann, C., Chattopadhya, U. et al. (2009) Environmental impact of the 73 ka Toba super-eruption in South Asia. *Palaeogeography, Palaeoclimatology, Palaeoecology*, 284: 295–314.

Winchester, S. (2003) *Krakatoa: The Day the World Exploded 27th August 1883*. London: Penguin Group.

6

SUNDA AND MAINLAND SOUTHEAST ASIA

Introduction

One of the great ironies of the Pleistocene is that many of the people who were most affected by the advances and retreats of ice sheets never saw a snowflake or a lump of ice. This point applies particularly to Southeast Asia, where the advances and retreats of ice sheets in faraway Europe and North America were matched by the exposure and flooding of massive areas of land. In a post-glacial world in which agriculture and city life were developed, we tend to think of interglacials as more benign than glacial periods. Maybe so in Europe, but in Southeast Asia, the last and the present interglacial were catastrophic for those humans (and other animals) whose world was swallowed by the sea. According to Hanebuth and colleagues (2000), sea levels between 14.6 ka and 14.3 ka rose by 16 m at an accelerated rate of 5.33 m per century – well within the perception of local inhabitants. Even more noticeable would have been the rate of marine transgression: to quote an extreme example, the shoreline of southern Vietnam retreated by 200 km in 800 years between 9.0 and 8.2 ka BP (Tjallingii et al. 2014) – an average rate of 250 m per year.[1]

We know a great deal about the environmental history of Sunda, and especially the way its geography has been repeatedly transformed by changes in sea level, but far less about its human history before the Last Glacial Maximum. Coverage is very uneven, especially on mainland Southeast Asia, but its early human record has improved greatly in recent years through the discovery of new sites or the re-examination of sites that were first investigated many decades ago. The region is also well covered by some major syntheses. For those wishing to explore this region's Palaeolithic history in depth, there are excellent syntheses by Peter Bellwood (2007, 2017), Ryan Rabett (2012), Charles Higham (2013) and Jon de Vos (2014). Before proceeding any further, we must first look at Southeast Asia in terms of three variables of sea level, vegetation and faunal communities. With these in place, we can then see how humans might have mapped onto those landscapes.

Sea levels, vegetation and climate faunal communities

The overall picture is simple: in the last interglacial in the Tropics, sea surface temperatures were much the same as today (Hoffman et al. 2017) but sea levels were ca. 5.5–9 m higher (Dutton and Lambeck 2012). During the last glacial cycle, sea temperatures in the tropical Pacific fell by 5° C. during the LGM (Gagan et al. 1998). Sea levels fell by up to 120 m, thus exposing the enormous area of land covering c. 1.5 million sq km (or three quarters the combined size of Myanmar, Thailand, Laos, Cambodia, Vietnam, the Malay Peninsula and Singapore) that is known as Sunda (Voris 2000: 1155). For half of the past 150,000 years, sea levels have been c. 30–40 m below present levels. A fall of 40 m would have been sufficient to close the Singapore Strait (Bird et al. 2006) and join the Malay Peninsula with Sumatra and Borneo. When sea levels were 50 m below the present, Java was joined to Sunda (see Figure 6.1) and there was a land bridge perhaps 150 km wide between Sumatra and Borneo (Voris 2000: 1164). At this point, the area of land doubled, but the length of coastline halved (Woodruff and Turner 2009). However, even when sea levels fell by their maximum of 120 m below current levels, Palawan (between Borneo and the Philippines), Sulawesi and the other islands of Wallacea would still have remained as islands (see Chapter 7).

Sunda was drained by numerous rivers, some of which were very large when sea levels fell by 75 m (see Figure 6.2). The largest systems were the North Sunda between Sumatra and Borneo, the East Sunda that drained the area between Java and Borneo, the Malacca Straits system between the Malay Peninsula and Sumatra, and the Siam Gulf river system (Voris 2000: 1164–1165). These may have been corridors for humans (and may also have provided the first impetus for watercraft) but would have been barriers for animals such as orangutans and gibbons that are not capable swimmers. When combined with the reduction and fragmentation of rainforest at times of low sea level, these rivers would have contributed to the isolation and fragmentation of forest species (Harrison et al. 2006).

Most accounts of Sunda describe only the long-term fall in sea level and overlook the short-term regional consequences on climate, vegetation and fauna. Falling sea levels would have created a "dead zone" in coastal regions until coastal plant and animals became established and marine communities adapted to new conditions. Rapid rises in sea levels would have been locally catastrophic in coastal regions. The flooding of coastal areas would have killed salt-sensitive vegetation and forced people and their prey further inland. Estuaries would gradually have widened and extended further inland to become channels, and

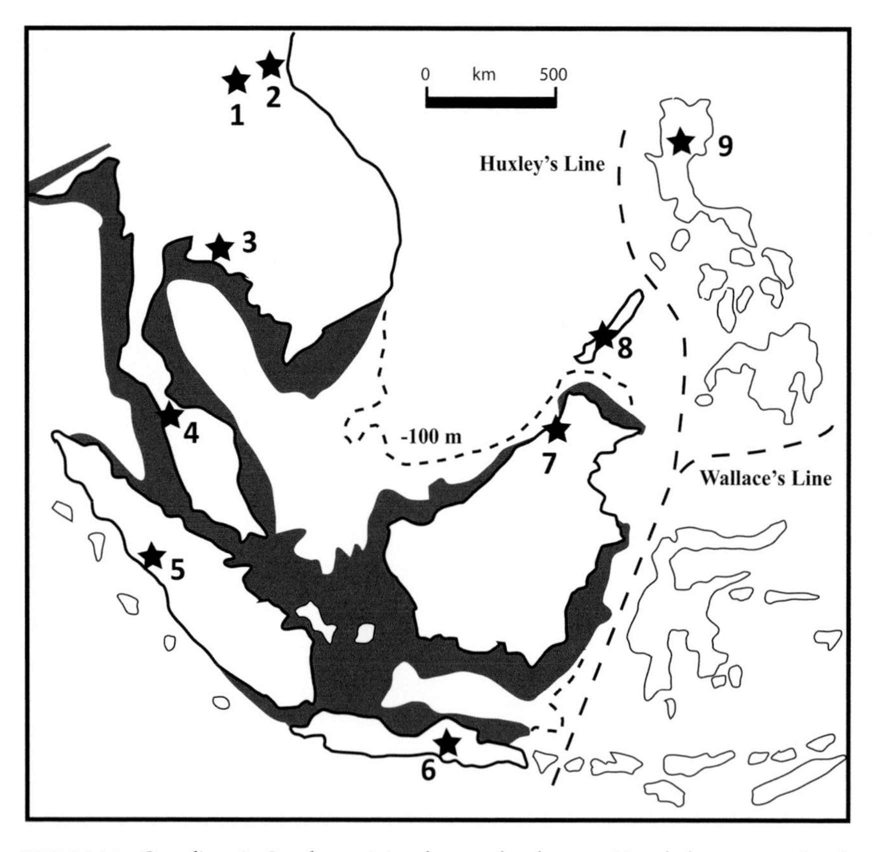

FIGURE 6.1 Coastlines in Southeast Asia when sea levels were 40 m below present levels At this level, mainland Southeast Asia is conjoined with Sumatra, Java and Borneo. Sites mentioned in the text: 1 Tam Pa Ling, Nam Lot (Laos); 2 Duoi U'Oi (Vietnam); 3 Boh Dambang (Cambodia); 4 Moh Khiew, Lang Rongrien (Thailand); 5 Lida Ajer (Sumatra); 6 Braholo Cave, Punung, Ngandong, Song Gupuh, Wajak, (Java); 7 Niah Cave (Borneo); 8 Tabon Cave Complex (Palawan); 9 Callao Cave (Luzon, Philippines).

Source: Redrawn and modified from Tougard, 2001, Fig. 2.

severed land connections. These problems would have been compounded by tidal surges during cyclones and by tsunamis after earthquakes (see Hanebuth et al. 2011: 107; Hinton 2000; Mei and Shang-Ping 2016).

The consequences of long-term falls in sea level were summarised by Heaney (1991). A key point he made was that the fall in sea level was a global phenomenon but in Sunda had regional climatic and environmental consequences. He noted that in some regions the increased land area would have reduced the surface area of adjacent shallow seas, thereby decreasing evaporation and the moisture content of the monsoonal winds. When the huge continental shelf of Australia/New Guinea (Sahul) was exposed as dry land, less moisture would have been carried by the summer monsoons to the Lesser Sunda Islands, Java,

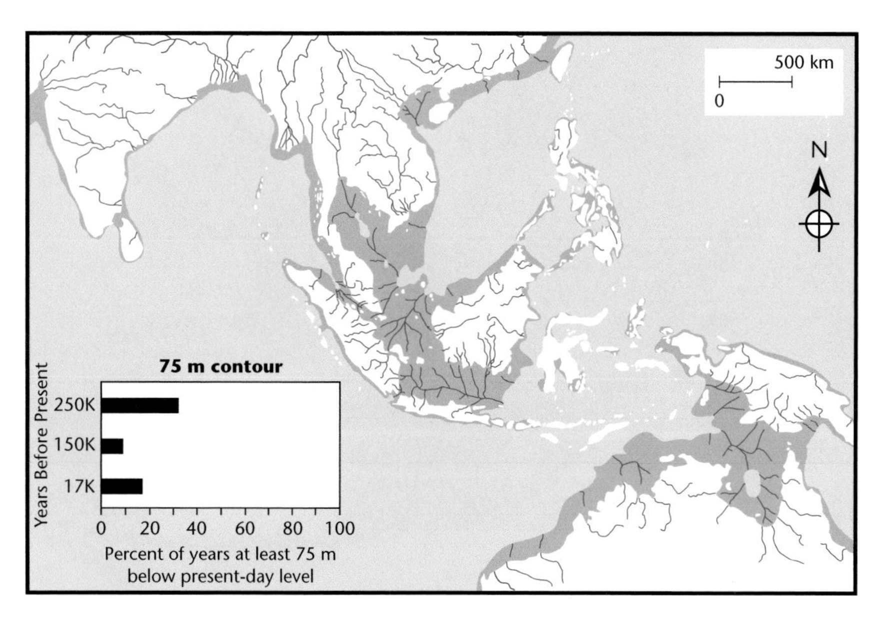

FIGURE 6.2 The principal rivers on Sunda and adjacent mainland Southeast Asia when sea levels were 75 m below the present
Source: Voris 2000, Fig. 1c.

and southeast Borneo, causing an extension of the current area of seasonal forest and savannah. The South China Sea and Java Sea would have been much smaller, and summer monsoons would have carried less moisture to southern Borneo and the land areas to the west of Borneo (now beneath the South China Sea) and to southern Indochina. Winter monsoons would have carried less moisture to the Malay Peninsula, eastern Sumatra, western Borneo, and intervening land, as well as the southwestern Philippines. Additionally, lower temperatures would have reduced the rate of evaporation over water and the moisture-carrying capacity of air, and hence, less rainfall. However, in those areas that continued to receive significant rainfall, lower temperatures would have increased effective rainfall by decreasing rates of evaporation. From a human point of view, Heaney's most important conclusion was that there would probably have been a corridor of low rainfall passing through the centre of the Sunda Shelf, extending in an arc from southern Thailand to eastern Java (Figure 6.3). Vegetation in this corridor was likely seasonal forest and savannah although the large rivers that flowed across these relatively dry regions probably supported gallery forest, so that high-canopy forest habitat may have persisted in a mosaic of forest and savannah in even the driest regions. Palynological evidence shows that during the warm and wet conditions of the last interglacial, rainforest expanded; when the climate became cooler and drier in MIS 4 to MIS 2, grassland and open woodland expanded, and when seasonal forest expanded in some lowland areas,

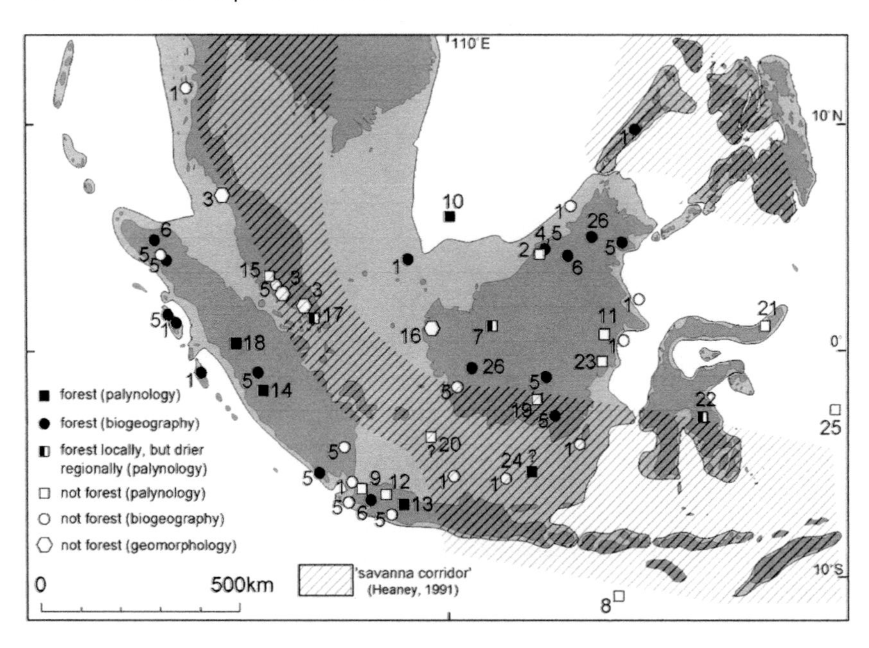

FIGURE 6.3 Vegetation zones and major rivers on Sunda during the Last Glacial Maximum (LGM)

Note the "savannah corridor" in areas of low rainfall through Sunda that separated the forested areas of present-day Sumatra and Borneo.

Source: Bird et al. 2005, Fig. 2.

montane forest shifted to lower elevations, perhaps covering regions many times larger than now (see e.g. Dam et al. 2001; Kaars and Dam 1995; Kaars et al. 2000; Kaars et al. 2010; Sémah and Sémah 2012). The overall effect of the conditions associated with glacial maxima may well have been to reduce Southeast Asian lowland rainforest to its smallest area since the Miocene, prior to the advent of human disruption. Many subsequent researchers (e.g. Urushibara-Yoshino and Yoshino, 1997; Meijaard, 2003; Bird et al. 2005; Harrison et al. 2006; Louys and Meijaard 2010; Lohman et al. 2011; Wurster and Bird 2014; Wurster et al. 2010) have accepted Heaney's proposed "savannah corridor" from the Malay Peninsula to Java; we will see shortly that it could also have extended through Cambodia and Laos towards southern China.

A corridor of savannah and seasonal forest 50–150 km wide when sea levels fell more than 50 metres below present levels would have greatly facilitated human dispersal from the modern Malay Peninsula to Java but would have separated the rainforest regions of Sumatra from those of modern Borneo and restricted gene flow between them (Harrison et al. 2006) This separation, no doubt repeated in previous glacial cycles, resulted in the orangutans in Sumatra and Borneo becoming different species, *Pongo abelii* and *P. pygmaeus* respectively.[2]

Molecular evidence from apes and elephants indicates deep temporal separation between the two rainforest areas (Cannon et al. 2009).

According to Louys and Meijgaard (2010), Southeast Asia had significant areas of mixed habitats, but rainforests were less extensive than now in MIS 4. The dominant vegetation type for mainland Southeast Asia during much of the Pleistocene appears to have been lowland evergreen, semi-evergreen and coniferous forests, with varying amounts of grasses and other herbaceous vegetation (see Figures 6.4 and 6.5). As shown in Figure 6.4, the vegetation of mainland Southeast Asia and south China did not change significantly between MIS 5 and MIS 4. In MIS 4, the lowering of sea levels in MIS 4 and consequent exposure of Sunda was accompanied by a major increase in tropical savannah and open woodland that also separated the rainforests of Sumatra and Borneo (see Figure 6.5). Glacial periods are associated with drier and more mixed habitats, and

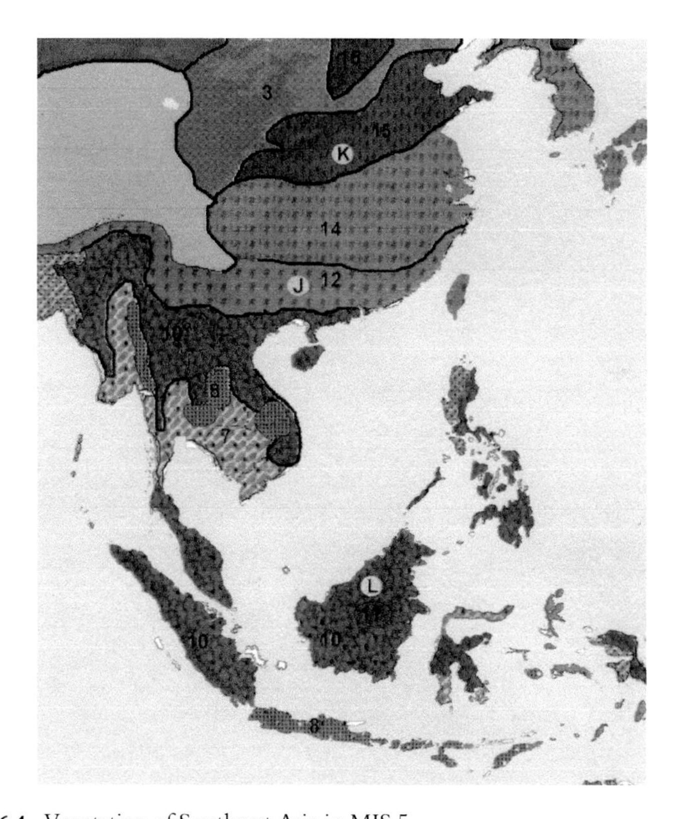

FIGURE 6.4 Vegetation of Southeast Asia in MIS 5

Key: 3 Steppe (north China); 7 Tropical savannah/woodland-grass mosaic; 8 Dry tropical woodland; 10 Moist tropical woodlands and rainforest; 12 Warm temperate hill and submontane vegetation, including the evergreen broadleaf forests of southern China; 14 Mixed evergreen-deciduous forests; 15 Mixed conifer and temperate deciduous.

Source: Boivin et al. 2013, Fig. 4.

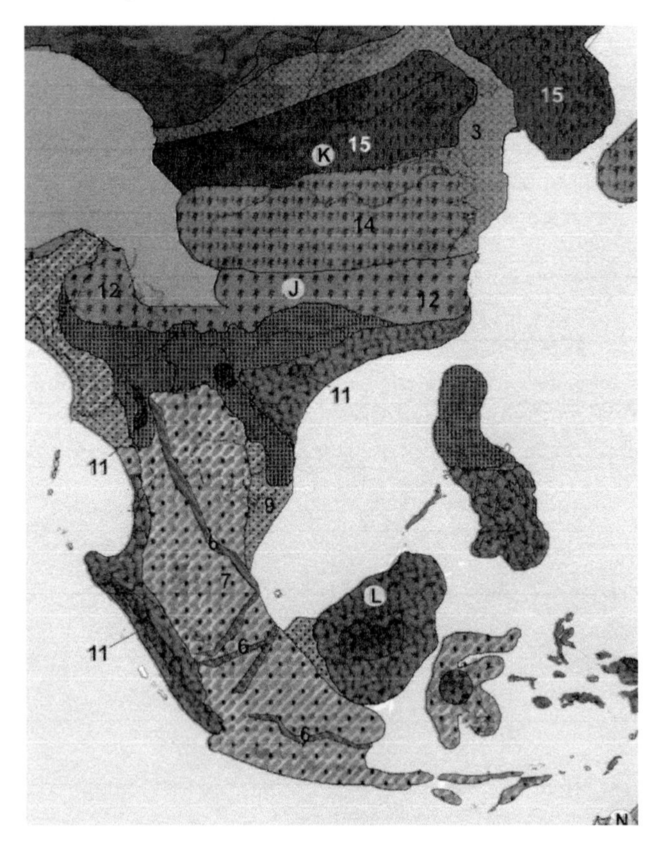

FIGURE 6.5 Vegetation of Southeast Asia in MIS 4

Key: 6 Riverine corridors: marshes and gallery forests; the rest as in Figure 6.4.

Source: Boivin et al. 2013, Fig. 5.

interglacials with more humid, closed habitats. Evidence for environmental changes in the Indochinese sub-region is still limited but indicates the dominance of mixed habitats of open and closed habitats through most of the Pleistocene. One long-term trend was that the rainforests that supported panda and orangutan contracted; panda retreated northwards into the hill forests of Sichuan in south China (Jablonski et al. 2000; Tougard et al. 1996), and orangutan either became locally extinct or shifted its range southwards in the Late Pleistocene from southern China, northern Vietnam (Kahlke 1972; Bacon and Long 2001; Harrison et al. 2014; Tshen 2015) and the Malaysian Peninsula (Ibrahim et al. 2013) into a glacial refugium in Sunda. (Sadly for them, and because they cannot swim, they were unable to recolonise mainland Southeast Asia because of rising sea levels, which left them trapped in Sumatra, Borneo and [until historical times] Java.) Contraction of rainforest in Southeast Asia was matched by the expansion of open and mixed habitats. Bacon and colleagues (2018a) argue that the presence of spotted hyaena and other carnivores and their prey species at Boh Dambang in southern Cambodia and other sites in Laos and northern Vietnam indicates open tropical savannah at latitudes of 10° N. The 72–86-ka-old fauna from Nam Lot, Laos includes several large herbivores, notably rhinoceros, *Elephas, Stegodon*, tapir and two large bovids as well as dhole (*Cuon alpinus*) and spotted hyaena (*Crocuta crocuta ultima*) (Bacon et al. 2018b). These researchers propose that:

> In the plausible scenario that modern humans entered Southeast Asia during MIS 5, a Nam Lot-type habitat could have constituted a relatively attractive environment for people with small to large game available and exploitable on the ground. The presence of intermediate zones in rainforest environments, open forests, forest-edges, and grassland zones rich in prey species, could have procured many tuberous and leafy species for food. (Bacon et al. 2018b: 141)

Additionally, the 60–70-ka-old fauna from Duoi U'Oi, Vietnam (which included fossils of tiger and leopard and two teeth of *Homo*) also shows that "during MIS 4, the locality constituted a relatively attractive environment for predators (hominins and non-hominins)" (Bacon et al. 2018b: 141). Immigration into south China from mainland Southeast Asia along corridors of open vegetation should not therefore have been difficult for most of the last glacial cycle (see Chapter 10). The regional extinction of orangutans in mainland Southeast Asia probably reflects the repeated contraction and fragmentation of rainforest, combined with their limited dispersal abilities and human predation: the distribution in southern China and northern Vietnam is unlikely to indicate continuous closed vegetation across this region. Slightly later, the hunting groups that occasionally used the rock shelter of Lang Rongrien, near Moh Khiew in Thailand, between 43 and 27 ka were hunting deer and bovids and collecting turtles and tortoises in a savannah-like and open forest (Anderson 2005; Mudar and Anderson 2007).

Faunal consequences of sea level changes

Sea level changes had profound consequences for animal communities. Figure 6.6 shows the impact of rises in sea level on coastal topography and a hypothetical source population. In this scenario, a large area of land that could support a single source population when sea levels were 40 m below the present is reduced by rises in sea level to several islands, many of which are too small to support a viable population (especially, if like orangutan, they cannot swim). (See also the discussion of habitat fragmentation in Chapter 3.) Large conjoined islands at times of low sea level therefore have greater potential for source populations than small isolated ones at times of high sea level (see Heaney [1986] for discussion of Southeast Asian island faunas). Island populations, of course, also have their own idiosyncratic histories of gigantism (as with a giant stork on Flores [Meijer and Due 2010]) or dwarfing

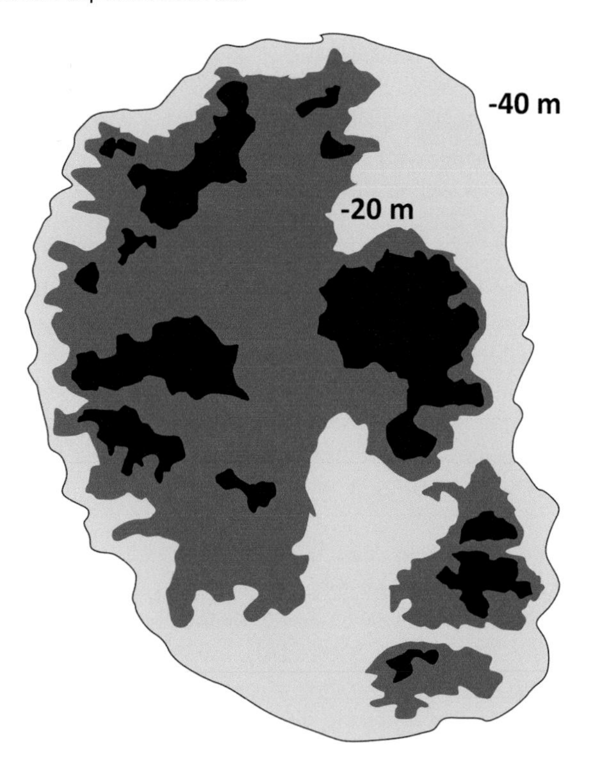

FIGURE 6.6 The consequences of rises in sea level on animal populations

In this hypothetical situation, there is a large area for animal populations on a single large island when sea levels are 40 m below present levels. When sea levels rise to 20 m below present levels, this area is split into three smaller islands and animal populations are fragmented and confined to smaller areas. Those on the smallest might face local extinction. Present-day populations are fragmented and dispersed among several small islands (shown in black), some of which may be too small to support a viable population.

(as with dwarf stegodons on Flores [Bergh et al. 2008], deer on Jersey [Lister 1995] and mammoths on Wrangel Island [Vartanyan et al. 1993]). In Southeast Asia, therefore, high sea levels have an adverse effect on animal and plant species by fragmenting and isolating them, and sometimes reducing the area of land beyond the point that a population is viable: Palawan (see below) provides a dramatic example. For these reasons, Southeast Asia is much more likely to have acted as a source population for humans and other animals at times of low sea levels, such as MIS 5d, MIS 5b and MIS 4 (see Figure 3.2) than during MIS 5e. (However, we need to bear in mind that human populations were capable of purposeful sea travel in MIS 3, and perhaps earlier.) For an incoming immigrant population of *H. sapiens*, the combination of low sea levels (and therefore more available land) and open vegetation (and therefore easier movement) in MIS 4 (or MIS 5d or MIS 5b) would have offered better prospects for colonisation than a combination of high sea levels and extensive dense forest in MIS 5e.

Sea levels and coastal mangroves

Mangroves form an important part of coastal vegetation in many areas of Southeast Asia and are rich in fish and other resources. Erlandson and Braje (2014) have suggested that there may have been a "mangrove highway" that provided an easy route for coastal colonisation, in much the same way as there may have been a "kelp highway" along the Pacific coast (see Chapter 11). Unfortunately, there is no way of testing this idea as the relevant evidence would now be below sea level. One cautionary point is that mangroves take about 80–100 years to become established, and thus depend upon a stable coastline (Hanebuth et al. 2011). In periods when sea levels are either steadily rising or falling, they are unlikely to form, and we need better information than we have about when coasts were stable enough for these to develop. According to Cannon et al. (2009), there was little coastal mangrove during the LGM.

Did Toba affect human dispersals in SE Asia?

As we saw in Chapter 5, the Toba super-eruption on Sumatra ca. 74 ka was the largest volcanic eruption in the last two million years, and its ash fall-out covered around four million sq km of South and Southeast Asia (Rose and Chesner 1987). Although one might expect its regional consequences on plant and animal life to have been catastrophic, its impact in Southeast Asia on large mammals appears to have been negligible when compared with the consequences of long-term climate change and sea level changes (Gathorne-Hardy and Harcourt-Smith 2003; Louys 2012; Louys et al. 2007). Kaars et al. (2011) noted severe short-term reduction in pine forest, but regeneration appears to have been rapid. Orangutans survived on Sumatra despite their proximity to Toba and their poor resilience to environmental disruption[3] (Louys 2012). Although Wilting et al. (2012) observed that the fauna of the Mentawi Islands west of Sumatra has more in common with the fauna of Borneo than with Sumatra, they also concluded that this reflects long-term changes of climate and vegetation during the Pleistocene and not necessarily the effect of the Toba eruption. Overall, if – and it is currently a big "if" – humans were in Southeast Asia and Sunda when Toba erupted, they are unlikely to have been impacted significantly by the super-eruption.

Demography, Sunda and mainland Southeast Asia

Louys and Turner (2012) proposed that during the last glacial cycle, much of the exposed land surface of Sunda and adjacent Southeast Asia would have been covered by grassland and open woodland, and was therefore a suitable habitat for medium and large mammals such as cervids, bovids and perhaps suids as well as other large herbivores. Because of its location in the Tropics, its climate would also have been suitable for hominins without the need for adaptation to cold conditions, and the region probably served as a refugium for humans. It would thus have provided a source population that could expand northwards into areas such as northern China when conditions improved. Their model was derived from a "source and sink" model that I and colleagues developed for Middle Pleistocene Europe, in which areas such as Iberia, Italy and southwest Europe served as glacial refugia (Dennell et al. 2010). These refugia would have been continuously occupied throughout glacial downturns and provided the source populations for when hominins recolonised northern Europe. At the northern limit of the hominin range in Europe, there would have been "sink" populations that needed recruitment and replenishment from these source populations to remain viable. In Louys and Turner's (2012) model, Sunda and adjacent mainland Southeast Asia would have served the same function as a refugium and source population for regions further north where continuous occupation was difficult or impossible. As we will see in Chapter 10, animals such as *Hystrix* (porcupine), *Bubalus* (water buffalo) and *Elephas* (elephants) show repeated incursions into and out of northern China during the last glacial cycle similar to those of their counterparts into and out of northern Europe, and humans probably did the same. A further indication of Sunda's importance as a refugium is that it would also have provided the source population for the colonisation of Wallacea (see Chapter 7).

The skeletal evidence for early *Homo sapiens* in Southeast Asia and Sunda

The location of sites with human remains is shown in Figure 6.1 and their details summarised in Table 6.1. In the following discussion, we need to bear in mind the problems of recognising the first recorded, and first actual presence of an invasive species in a new region (see Figure 1.10). Most of the key sites are caves or rock shelters, and these have complex sedimentary histories that have only recently been recognised (see e.g. Anderson 1997; Morley 2017). This point applies also to the caves and rock shelters in Wallacea (Chapter 7) and also south China (Chapter 10.)

Mainland Southeast Asia

One very important site that has been investigated recently is the cave of Tam Pa Ling (TPL), Laos. Here, the remains of five individuals have been found (see Figure 6.7 and Table 1) (Demeter et al. 2012, 2015, 2017; Shackelford et al. 2018). A small partial cranium (TPL1), two mandibles (TPL2, TPL3), a partial rib (TPL4) and a proximal pedal phalange (TP5) represent the earliest well-dated, anatomically modern humans in Southeast Asia. The TPL1 skull and TPL2 mandible are dated to 43 ± 7 ka and 46 ± 6 ka, respectively (Shackelford et al. 2018: 103), and the *maximal* depositional age of the TPL3 mandible is 70 ka (Demeter et al. 2017), which implies the presence of our species in Southeast Asia in or by MIS 4 (Bacon et al. 2018b). (Work in progress should clarify

TABLE 6.1 Human skeletal remains from Southeast Asia and Sunda

Country	*Site*	*Skeletal remains*	*Context*	*Age*	*Identification*	*Reference*
Thailand	Moh Khiew	Partial skeleton, adult, probably female	Layer 2, cave	25,000 ± 600 BP (University of Tokyo: TK-933Pr), or 30,989 ± 707 cal BP (Rabett 2012: 110)	*H. sapiens*	Matsumura and Pookajorn (2005)
Vietnam	Lang Trang	Five teeth	Cave fissure	146 ± 2 ka to 480 ± 40 ka, but on faunal grounds, ca. 80 ka	*H. erectus* or *H. sapiens*, depending on age estimate; only one tooth definitely *H. sapiens*	Schwartz et al. (1995)
Vietnam	Duoi U'Oi,	Two teeth	Cave fissure		*H. sapiens*	Bacon et al. (2018b)
Laos	Tam Pa Ling	TP1: partial cranium TP2: mandible TP3: mandible TP4: postcranial TP5: postcranial		44–63 ka 44–63 ka 70 ka maximum		Demeter et al. (2012, 2015, 2017); Shackelford et al. (2018)
Indonesia	Ngandong	12 crania, two tibiae	River terrace	28–54 ka or 143–546 ka; older dates more likely than 28–54 ka	*H. erectus*	Indriati et al. (2011); Swisher et al. (1996)
	Lida Ajer	One right upper central incisor and left M^2	Cave	63–73 ka	*H. sapiens*	Westaway et al. (2017)
	Punung I or II	Left P^3 (PU-198)	Cave locality	Possibly 115–128 ka; more likely 80 ka	*H. sapiens?* *H. erectus?*	Storm et al. (2005) Polanski et al. (2016)
	Wajak	Cranium and femur	Open locality	37.4 and 28.5 ka-old	*H. sapiens*	Storm et al. (2013)

(*Continued*)

TABLE 6.1 (Cont).

Country	*Site*	*Skeletal remains*	*Context*	*Age*	*Identification*	*Reference*
Malaysia (Borneo)	Niah	Cranium	Cave	40,984 ± 305 cal BP 40,344 ± 415 cal BP $37^{+2}/_{-4.7}$ KBP on mandible $34.5^{+3}{}_{-3}$ KBP on mandible Combined age: 35–42 KBP	*H. sapiens*	Barker et al. (2007); Rabett (2012: 117)
Philippines	Tabon	Frontal fragment (P-XIII-T-288)	Cave	16,500 ± 2000 BP		Dizon et al. (2002)
		Right mandibular fragment (PXIII-T436 Sg 19)		31 + 8/−7 ka (U-series)		Détroit et al. (2004: 710)
		Right temporal bone (IV-2000-T-188)				
		Occipital bone (IV-2000-T-372)				
		Tibia fragment (IV-2000 T-197)		47 + 11/−10 ka (U-series)		

the age of the TPL 3 specimen.) These remains are therefore older than the 45-ka-old cranium from Niah Cave (see below), and are currently the oldest from mainland Southeast Asia and east of the Levant. Additionally, the inland location of Tam Pa Ling "suggests that Pleistocene modern humans may have followed inland migration routes or used multiple migratory paths" (Demeter et al. 2012: 14379) rather than following coast lines, as some have suggested (e.g. Stringer

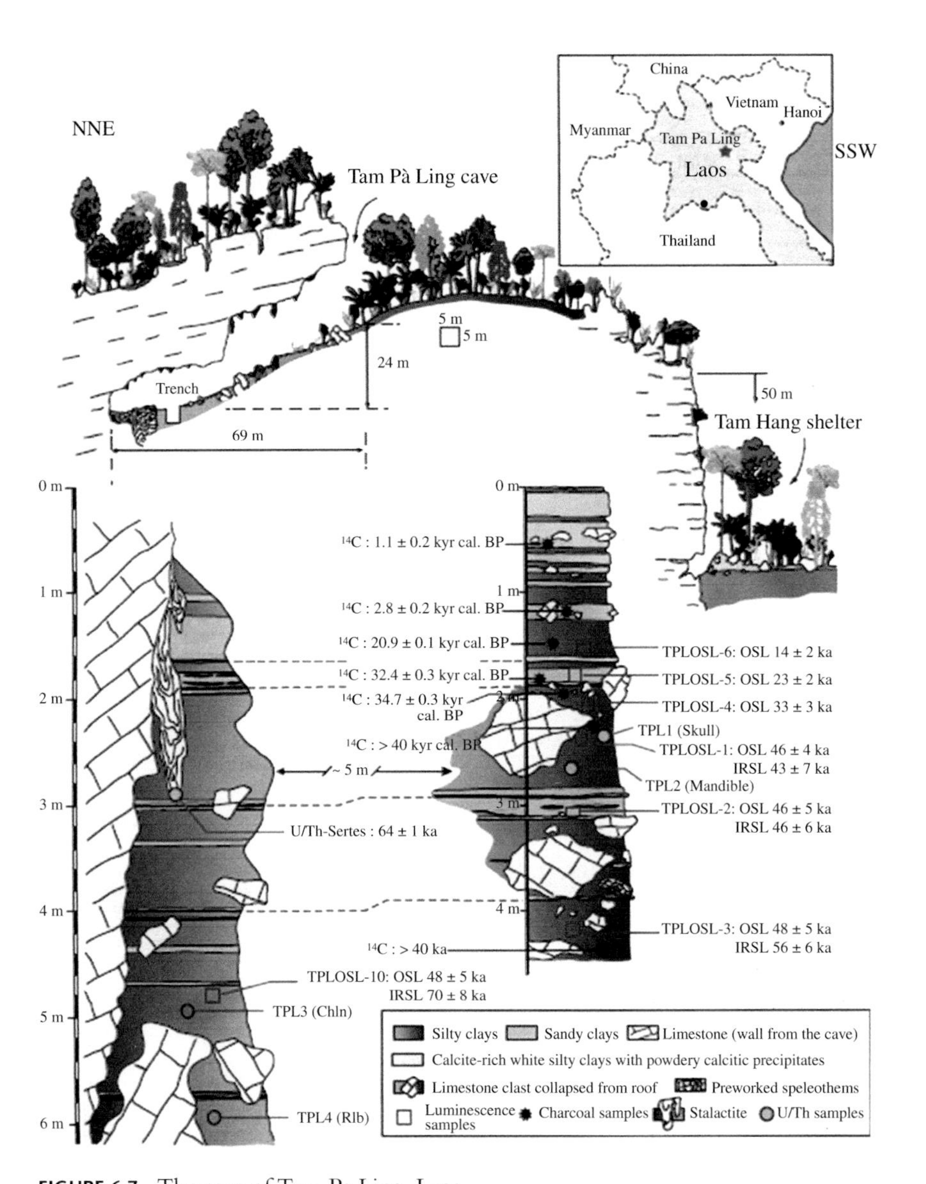

FIGURE 6.7 The cave of Tam Pa Ling, Laos

The 5.0 m stratigraphic section showing the accumulation of sandy and silty clay layers from Tam Pa Ling. Provenance of the charcoals sampled for 14C dating, soils sampled for OSL and TL dating and stalactite sampled for U-series dating are identified on the stratigraphy. TPL1 was recovered at a depth of 2.35 m; TPL2 was recovered at a depth of 2.65 m; TPL3 was recovered at a depth of 5.0 m. The 5.0 m between the two stratigraphic sections represents the distance between where the TPL human fossils were found; there is lateral continuity between the sections.

Source: Shackelford et al. 2018, Fig. 1.

2000). The latest assessment of the local environment during MIS 4 based on stable isotope analysis of gastropods is that it largely comprised woodlands with a minor component of open habitats although "the degree of exploitation of closed forests versus open landscapes is still an open question" (Milano et al. 2018: 361).

Moh Khiew, southern Thailand

This site is a rock shelter, with a sequence of five cultural layers that extends from ca. 30 ka to the Holocene. The burial was found at the boundary of cultural layers 1 and 2. Cultural layer 1 contained unifacial pebble tools, and cultural layer 2 flake tools and waste flakes. An AMS radiocarbon date from charcoal from the burial gave a date of 25,800 ± 600 BP, or 30,989 ± 170 cal. BP. The skeleton is said to resemble Late Pleistocene ones from Coobool Creek, Australia, and is considered to represent a member of a Sunda population that shared its ancestry with present-day Australian Aborigines and Melanesians (Matsumura and Pookajorn 2005).

Island Southeast Asia

Punung, Java

The earliest claimed evidence for our species in Southeast Asia is a left upper third premolar (P^3) (PU-198) in the Punung faunal collections held at the Senckenburg Institute, Frankfurt (Storm et al. 2005). These fossils had been excavated by Ralph von Koenigswald in the 1930s from two caves (Punung I and II) in East Java. Because they appeared to have had similar faunal assemblages, they were unfortunately thereafter mixed together. Even more unfortunately, five human teeth from Punung were lost sometime between 1959 and 2001 (Storm and Vos 2006). In the study by Storm and colleagues (2005), the premolar fell within the size range of modern *H. sapiens* from Australasia (N=46), and outside that of *H. erectus* (N=7). At a new and nearby cave site (Punung III), the Punung Fauna (including orangutan, siamang and sun bear) was dated by TL and OSL to between 128 ± 15 and 118 ± 3 ka, i.e. MIS 5e), which "would imply that *H. sapiens* arrived in Southeast Asia during the Last Interglacial" (Westaway et al. 2007: 715). However, human remains have not yet been found at Punung III, and Barker et al. (2007) and Bacon et al. (2008) were sceptical of both the dating and identification of the Punung I/II premolar at that of *H. sapiens*. Polanski and colleagues (2016) recently suggested that the tooth could indicate *H. erectus*. If so, *H. erectus* would be outside its normal habitat, which is predominantly one of open woodland (Bergh et al. 2001). As Sémah and Sémah (2012: 124) point out, there is no clear evidence from Africa and Asia that *H. erectus* inhabited rainforest, and therefore "the presence of *H. sapiens* in a rainforest environment is more likely than occupation of this habitat by *H. erectus*". So far, *H. sapiens* is the only hominin that is definitely known to have inhabited rainforest (see Roberts et al. 2016).

Wajak

Whilst on Java, one should mention the crania and a partial skeleton from Wajak,[4] Java. The first cranium (Wajak 1) was discovered in 1888 by geologist B.D. van Rietschoten, who sent it to Eugene Dubois, who visited the site and excavated the second skull (Wajak 2) and a partial skeleton in 1890 during his early search for *Pithecanthropus*. The associated fauna is consistent with an open woodland setting (Storm 1995). The Wajak crania have had several identities since their discovery (Storm and Nelson, 1992). Initially, Dubois (1921) regarded the Wajak skulls as a proto-Australian, as did Weidenreich (1945), who incorporated them into his scheme of multiregional evolution. Radiocarbon dating of a human femur from Wadjak II that was dated to 7670–7210 cal. BP (Shuttler et al., 2004) later indicated that they were Holocene in age, and Wajak 1 and 2 thus became indigenous post-glacial Javans. They have since been re-dated by U-series dating and are now thought to be between 37.4 and 28.5 ka old (Storm et al. 2013), which now makes them inhabitants of ice-age Sunda.

Lida Ajer, Sumatra

Lida Ajer is a cave in western Sumatra that was excavated by Eugene Dubois between 1887 and 1890. Most of the fossil remains were of orangutan but two human teeth were also found that were identified as *H. sapiens* by Hooijer in 1948. Unfortunately, no archaeological traces were present. Recent field investigations succeeded in relocating the site from Dubois's field notes (Westaway et al., 2017). As a result of an intensive re-examination of the cave deposits, the age of the breccia from which the teeth likely derived has been established by uranium-series, luminescence and coupled uranium-series/ESR dating methods to 63–73 ka, which places them in MIS 4 (58–74 ka). Hooijer's (1948) identification of the human teeth as *H. sapiens* has also been unequivocally confirmed after thorough re-examination. The significance of the evidence from Lida Ajer is that it definitely places humans in a rainforest setting by 63–73 ka and is thus the earliest example not only of humans in Sunda, but also the earliest unambiguous indication of humans living in rainforest (Westaway et al. 2017).

Niah Cave, Borneo

The best-known and least ambiguous specimen of *H. sapiens* from Island Southeast Asia is from the cave of Niah (Sarawak, Malaysia; see Figure 6.8), where a partial cranium was found in "Hell Trench" in 1958 by a team led by Tom and Barbara Harrisson.[5] It was later radiocarbon dated to ca. 40 ka and was for many years the oldest example of our species worldwide. The skull is a partial cranium of an individual (probably female) in their late teens to mid-twenties (Barker et al. 2007: 251). An almost complete left femur and a right proximal tibia fragment were found near the cranium and are probably from the same individual, as well as a human talus. The small size of the femur suggests an individual ca. 145 cm in height and a body mass of ca. 35 kg (Curnoe et al. 2019).

Following a lengthy and thorough re-investigation of the deposits and materials from Niah, the cranium has now been dated by an exhaustive series of AMS procedures and uranium-series dating to 41–34 ^{14}C ka BP (ca. 45–39 ka cal. BP), some five millennia after the cave was first occupied, and a uranium-thorium date of 35,200 ± 2,600 BP was obtained on part of the mandible. For those who rely solely upon cranial evidence (excluding isolated teeth) to record the first appearance of our species, the Niah Cave cranium provides a vital anchor point of ca. 40 ka for its appearance in Southeast Asia.

In a clever piece of detective work, Hunt and Barker (2014) analysed sediments from inside the Deep Skull and from samples immediately around it and found differences between the two that should not have been there if the two were of the same age. They concluded that the Deep Skull was in fact a secondary burial that had been placed in a pit from a slightly higher level. This is consistent with the uranium-thorium date on part of the mandible of 35,200 ± 2,600 BP. In addition, the sample from the skull contained numerous angular quartz crystals that are not found anywhere else in the cave. Their nearest source is in granite mountains 200–500 km distant. The clear implication is that these were deliberately added to the burial, perhaps as part of a ritual.

FIGURE 6.8 Niah Cave, Borneo

The hut at the entrance to the cave gives an indication of the size of this cave, which *Homo sapiens* occupied for ca. 50,000 years.

Source: Reproduced with kind permission of Graeme Barker.

There were other skeletal remains from Niah. A faunal sample from the same level as and 10 m south of the cranium contained 27 cranial fragments, three of which had traces of a red pigment, possibly from a tree resin (Pyatt et al. 2010). These are currently undated, but a tortoise plastron (shell) fragment with pigment was found in a nearby layer that was of comparable age to the Deep Skull. As shown below, this evidence complements the recent dramatic discovery that figurative cave art in Borneo is among the oldest in the world. There was also a partial mandible dated at ca. 28–30 ka and two others that are terminal Pleistocene in age (Curnoe et al. 2018).

Homo sapiens teeth from Lida Ajer, Sumatra, are associated with a rich rainforest fauna, and definitely place our species in a rainforest by 63 ka to 73 ka (Westaway et al. 2017). A rainforest (or at least closed lowland forest) setting is also indicated at Niah Cave, Borneo by 50 ka (Barker et al. 2007; Reynolds et al. 2013), and at Sri Lankan caves by 40 ka (Roberts et al. 2015, 2016; see Chapter 5). It is significant that both Lida Ajer and Niah Cave are in areas where rainforest would persist, even though these were separated by the intervening "savannah corridor" (see Meijaard 2003). Current evidence thus strongly supports the proposition that our species was the first hominin to inhabit rainforest, and thus the Punung premolar could well belong to *H. sapiens* rather than *H. erectus*. The timing of arrival of *H. sapiens* in Java remains unclear but has to predate the earliest Australian evidence of ca. 50–55 ka (O'Connell et al. 2018) or ca. 60–65 ka (Clarkson et al. 2017) (see Chapter 7). If the archaeological sequence at the cave of Song Gupuh, Java, is entirely that of *H. sapiens*, our species may have been present on Java at ca. 70 ka (Morwood et al. 2008).

When did H. erectus *become extinct?*

A key issue in Sunda and Southeast Asia is the last appearance of *H. erectus*. A key site here is Ngandong in Java. The site was excavated by a Dutch team in the 1930s (Oppenoorth 1932; Koenigswald 1933) and produced a remarkable assemblage of 12 crania and two tibiae, more than 25,000 other mammalian fossils that define the Ngandong Fauna, but unfortunately no archaeological evidence. Parietal and pelvic fragments were later excavated by an Indonesian team in 1976 and 1978. Most researchers regard the Ngandong crania as *H. erectus*, although Schwartz and Tattersall (2000: 20) suggested they probably belong to a sister taxon, and Hawks and colleagues (2000) suggested they should be reclassified as *H. sapiens* because of their similarities to WLH-50 from the Willandra Lakes, Australia. In the 1990s, Carl Swisher and colleagues (Swisher et al. 1996) caused a sensation when they dated bovid teeth from Ngandong and proposed that these crania were only 28–54 ka old. On those grounds, *H. erectus* may have co-existed for a time with *Homo sapiens* although their dating was questioned on technical grounds by Grün and Thorne (1997). A similar range of dates was obtained by Yokoyama et al. (2008), who attempted to date two crania from Ngandong (Ng-1 and Ng-7) and Sambungmacan 1 by uranium-thorium (^{234}U/^{230}Th), uranium-protactinium (^{235}U/^{231}Pa) and thorium-thorium (^{227}Th/^{230}Th) series dating. These produced dates as young as 39.6 ± 9.5 ka for Ng-1 by thorium-thorium, and a maximum age of 88.1 ± 10.8 ka by uranium-thorium (early uptake model). They thus suggested that their likely age was between 40 and 60–70 ka, in which case *H. erectus* and *H. sapiens* may have overlapped in time. Nevertheless, uranium leaching was observed in the dated specimens, and despite the ingenuity and thoroughness of the investigators, their dates are not problem-free.

It now seems likely that the Ngandong *H. erectus* crania and associated fauna are considerably older than indicated by Swisher's team in the 1990s. Recent re-examination of the stratigraphic context of the finds, and greatly improved dating techniques by a team that included Swisher indicates that the Ngandong crania are probably between 146 and 546 ka in age, with a preference for an earlier date (Indriati et al. 2011).

On faunal grounds, there are sound reasons for placing the Ngandong Fauna (and the associated hominins) in the Late Middle Pleistocene. The Ngandong Fauna contains several extinct species such as *Hexaprotodon sivalensis* (pygmy hippopotamus), *Bubalus palaeokerabau* (a type of water buffalo), *Elephas hysudrindicus* and *Stegodon trigonocephalus*, and no indications of rainforest. In contrast, the succeeding Upper Pleistocene Punung Fauna consists entirely of extant species, and the first indications in East Java of dense, humid forest[6] (as at Punung itself), with the first appearance of ten taxa, including the orangutan (*Pongo pygmaeus*), gibbon (*Hylobates syndactylus*) and sun bear (*Helarctos malayanus*).

At present, we do not know when *H. erectus* became extinct in Java or elsewhere in Sunda. Arguments hang on the assessments of one tooth: if the Punung premolar (see above) is accepted as *H. erectus*, its last recorded presence is ca. 125 ka and it may have persisted into and even beyond the last interglacial. If it is regarded as *H. sapiens*, then *H. erectus* might have been extinct in Java when rainforest expanded over it in the last interglacial. The first recorded presence of *H. sapiens* in Southeast Asia is currently at Lida Ajer, Sumatra, ca. 63–73 ka. The last and first actual appearance (see Figure 1.10) of these species lies somewhere between 125 ka and 73 ka. We clearly need more evidence to narrow this gap.

Palawan

The island of Palawan, between northern Borneo and southern Mindoro, provides a graphic example of the devastating consequences for some animals of the post-glacial rise in sea levels and associated vegetational changes. At the height of the Last Glacial Maximum when sea levels were 120 metres lower than now, it was still an island, but more than six times larger, at 78,000 sq km, with extensive coastal plains (see Figure 6.9). It was then sufficiently large to maintain a viable tiger population of perhaps 800–1,000 until the early

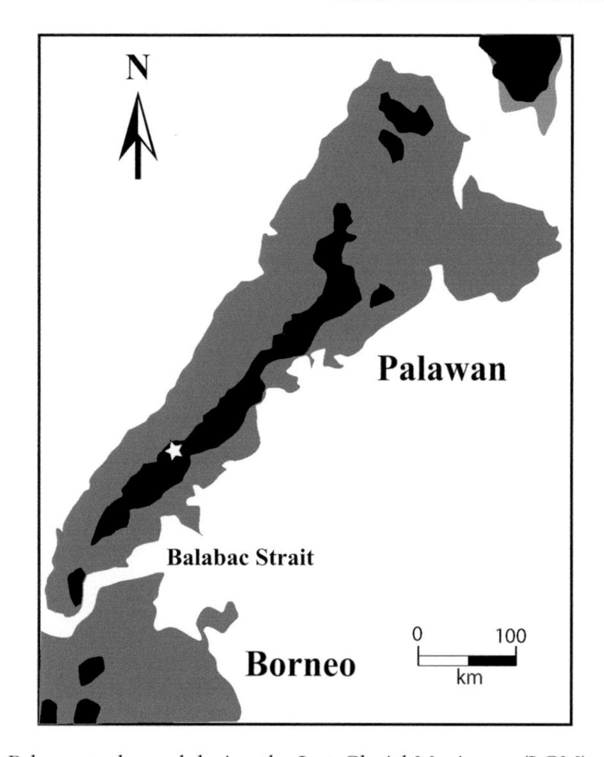

FIGURE 6.9 Palawan today and during the Last Glacial Maximum (LGM)

During the LGM, sea levels were 120 metres lower than today and Palawan was seven times larger. The asterisk marks the Tabon cave complex.

Source: Redrawn and simplified from Piper et al. 2011, Fig. 1.

Holocene (Piper et al. 2008). The post-glacial rise in sea level reduced the island's area to its present size of ca. 12,000 sq km, and this, combined with the replacement of savannah and open woodland by closed rainforest (Piper et al. 2011; Choa et al. 2016) and the extinction of the deer that were its main prey caused the tiger's extinction here during the early Holocene. Palawan's history makes the point that changes in vegetation can be just as important as changes in sea level when considering animal extinctions. As Piper et al. (2011) point out, tiger survived on the much smaller island of Bali until the last century, so the extinction of tiger on Palawan was not entirely a consequence of a rise in sea level.

Palawan today has a very limited fauna of only 58 native and four non-native species compared with neighbouring Borneo's 285 species of mammals (including 103 bat species). In particular, it lacks rainforest animals such orangutan (*Pongo pygmaeus*), gibbons (*Hylobates* spp.), leaf monkeys (*Presbytis* spp.) and sun bear (*Helarctos malayanus*), which are all present on Borneo. Because the most recent land bridges between Sunda and Palawan were probably during the Middle Pleistocene, ca. 420 ka and 620 ka (Piper et al. 2011), their absence from Palawan implies that these animals (and also tiger) colonised Sunda in the later Middle Pleistocene, and there was no land bridge between the two islands when sea levels dropped 120 m during the Last Glacial Maximum. According to Voris (2000), the Balabac Straits between Palawan and Borneo were reduced to a width of only about 12 km, which is too far for an orangutan but would still be feasible for a tiger.

Humans managed to cross the Balabac Straits and have inhabited Palawan for at least 40,000 years. Exploration of the Tabon Cave Complex of more than 200 caves (Dizon et al. 2002; Reis and Garong, 2001) has produced ca. 480 skeletal fragments (Corny et al. 2016) of at least three individuals (see Table 1). The oldest specimen, a tibia fragment, was dated by U-series to a maximum age of 58 ka (Détroit et al. 2004: 710). This age is compatible with the age of Tam Pa Ling 3 (see above), so is no longer as anomalous as first thought. Further ongoing research should result in further discoveries at Tabon, and hopefully some associated archaeological evidence that is so far lacking.

Archaeological evidence

Niah Cave

Most of what we think we know about the subsistence, technology and adaptations of the Palaeolithic inhabitants of Sunda and Southeast Asia comes from one site, namely Niah Cave in Borneo. A smaller amount of information comes from Lang Rongrien, Thailand (see above) and the lower levels of the Javanese caves of Song Gupuh, Java (Morwood et al. 2008) and Braholo (Amano et al. 2016), in which deer, bovid and pig were hunted.

Niah is an exceptional cave because of its sheer size, its length of occupation, the quality of its evidence, the high standard of its recent re-investigation, and probably also its fortunate location in or near several types of resource zone. Nevertheless, we need to resist the temptation of regarding it as "typical" or representative of Pleistocene life on Sunda. Most sites would have been in the open air, probably near lakes and rivers, and caves generally in Southeast Asia were used mainly as places for burial, or shelter, or temporary refuge (Anderson 1997).

This enormous cave was first tentatively explored in the 1870s (Sherratt 2002) and in the 1950s and 1960s by Tom and Barbara Harrisson (see above). From 2000 onwards, the Harrissons' archives and excavations were re-investigated by the Niah Cave Project, led by Graeme Barker. This project has generated an impressive number of specialist publications and two excellent monographs (Barker 2013; Barker and Farr 2016) that make Niah by far the most important Palaeolithic site in Southeast Asia. There are excellent syntheses of the new investigations by Barker (2012; Barker et al. 2007), and by Rabett (2012: 116–122), who was one of the Niah Cave Project team.

Although Niah is best known for its human skull, people had been visiting the cave for several millennia before the cranium was buried, as the earliest calibrated radiocarbon dates extend back to ca. 48,000 cal. BP. The stone artefacts are mainly flakes, some of which can be classified as scrapers and burins. Some of these show signs of polish. Use-wear analysis shows that 25% of 80 scrapers had been used on hard materials such as palm wood, and 12% on softer materials. Most of the stone was a local metamorphic sandstone, with some limestone, quartz and jasper, the last of which came from ca. 48 km away. Bone tools were also found. The earliest of these date from ca. 44 ka cal. BP and were possibly used for piercing. There was also a spatula and a worked pig tusk, but no evidence of armatures. Hunting arboreal species such as orangutan must therefore have relied on other materials such as bamboo or nets (Barker et al. 2007; Rabett 2012: 116–122), and perhaps the use of blow-pipes.

There was also a large faunal assemblage of ca. 10,000 bones and teeth. Many of these show signs of burning, and some have cut-marks indicating butchery events. In terms of NISPs (number of individual specimens), terrestrial as well as arboreal and aquatic species were hunted. In a sample of 1,213 specimens from the Hell Trench that documents the early part of the Niah sequence, the main terrestrial species were bearded pig (29%), followed by monitor lizard (11%) and pangolins[7] (3%); the main arboreal species was orangutan (6%); and cercopiths (leaf monkeys, macaques) (16%) were the main species that are both terrestrial and arboreal. In addition, ca. 20% of specimens were of tortoise and turtle (Barker et al. 2007). The hunting of pig appears to have been non-selective, and one possibility is that they hunted by using traps, and possibly at times of fruit abundance, when they form large groups. There was no indication that marine resources were used; all the fish are freshwater, and the coast was probably tens of kilometres from the cave when it was occupied between 50–30,000 BP. (The coast may also have been unproductive if sea levels were changing and coastal resources had not yet stabilised.)

Niah also gives us a rare and detailed glimpse of the use of plant foods. There is evidence from starch grains of deep-rooted yam (*Dioscorea hispida* and *Dioscorea* spp.) and sago palm (*Caryota mitis* or *Eugeissona* spp.) at 48 ka cal. BP. These are particularly interesting because they need detoxifying before they are edible. According to Barker and colleagues (2007), a piece of *Dioscorea hispida* the size of an apple can be fatal if eaten raw. Seeds of *Pangium edule* were also found; in ethnographic contexts, the hydrocyanic acid in these nuts could be removed by burying the ripe fruits or boiled seeds in a pit for 10–14 days and then boiling them, or by burying the seeds with ash for up to 40 days[8] (Barker et al. 2007). In the cave, there are pits containing nut fragments dated at 39 ka and 35 ka that may have been used for detoxifying nuts.

There is also evidence of deliberate forest burning to create or enlarge open spaces that would provide edible tubers and other plant foods, and for trapping those animals attracted to such patches. High frequencies of pollen of *Justicia* – a coloniser of burned areas – consistently occur in forest phases. This evidence is consistent with high frequencies of charcoal fragments in pollen cores from the

Sulu Sea north of Borneo and dating back 50,000 years, and possibly an indicator of when our species arrived in the region and began to modify their surroundings. According to Hunt and colleagues' (2007, 2012) study of the local vegetation at Niah, the vegetation was cyclic, and varied between lowland and montane forest, and more open savannah-like environments indicating lower temperatures and rainfall compared with the present day. The interesting point that they make is that people may have reached the site ca. 50 ka during a climatic phase when habitats were fairly open, and biomass burning may have taken place during more forest-rich phases as a way of creating or enlarging open spaces; people were not therefore necessarily foraging under closed canopy high forest (Hunt et al. 2007).

As discussed in Chapter 5, life in tropical forests was not impossible for Palaeolithic hunter-foragers, and Niah provides a fascinating and detailed account of how small groups with a very simple stone and bone technology, and probably a versatile and effective one based on plants, were able to utilise a complex environment. In some respects, it shows the reverse of the European Upper Palaeolithic, in which a complex technology of stone, bone and ivory was used in a simple environment with a small range of resources; in situations such as Niah, a simple tool-kit was used in a complex landscape. What emerges from the recent work at Niah is that its inhabitants were highly mobile, and probably moved camp several times throughout the year because food resources are widely dispersed, and often found in low frequencies. It was not, however, a random "catch and catch can" existence. Survival in a complex and changing mosaic of different types of habitats – lowland dipterocarp forest, swamp forest, open woodland and scrub – required an impressive degree of forward planning, knowledge and understanding of potential resources, and ingenuity in, for example, catching arboreal species or detoxifying plant resources. It was indeed, as Barker and colleagues (2007) point out, a "human revolution" that allowed a site such as Niah to be used repeatedly over 50,000 years.

Cave art in Kalimantan by 52 ka?

The discovery of cave art on the neighbouring island of Sulawesi that was at least 40,000 years old (see Chapter 7) was one of the most dramatic Palaeolithic discoveries in recent years. Comparable and even older evidence has recently been reported from the cave of Lubang Jeriji Saléh in eastern Kalimantan, Borneo (Aubert et al. 2018). Here, uranium-series dating of the calcium carbonate crust over one hand stencil showed a minimum age of 51.8 ka, and two others were dated by the same method to ca. 37.2 ka. A large reddish-orange figurative painting of an animal was dated to a minimum of ca. 40 ka, which makes it the earliest example worldwide. What is now emerging is an artistic tradition of rock art in Sunda and Wallacea (see Chapter 7) that is at least as old as that in western Europe. This raises several major questions, such as whether there were two traditions of rock art that developed independently on opposite sides of Eurasia after 40 ka (or perhaps even earlier in Sunda), or whether we have so far detected the eastern and western limits of a once continuous distribution; and whether figurative and symbolic rock art developed only after our species left Africa, or whether it was part of their behavioural capability prior to that event. We are on the verge of re-writing a significant part of our history as an artistic species.

Summary and discussion

Several points can be made about the colonisation of Sunda and mainland Southeast Asia by our species. First, *H. sapiens* was present at Lida Ajer, Sumatra, in Sunda by 63–74 ka and may have been present at Tam Pa Ling in mainland southeast Asia by 70 ka. Because we are unlikely ever to pinpoint the exact date when a species first appeared in a new region (see Chapter 1), we don't know how much earlier it might have been present. Current data from Southeast Asia do not at present confirm or refute the idea that our species was in this region before Toba's dramatic eruption. We have little idea of when *H. erectus* became extinct in Sunda and Southeast Asia. It was present at Ngandong between 146 and 546 ka in age (with a preference for the earlier date) but presumably persisted for much longer. If the Punung premolar represents *H. erectus*, it may have persisted in Java into the last interglacial, but not if it represents *H. sapiens*.

Second, there probably was a savannah corridor through the Malaysian Peninsula to Java when sea levels fell in the last glacial cycle, and this possibly extended into southwest China (see Chapter 10). Colonisation would thus have been easy through areas of open vegetation. Cool, drier periods were more favourable to colonisation because vegetation would have been more open, and more land was available because of lower sea levels; conversely, warm, moist periods were generally detrimental to the inhabitants of Sunda because vegetation was less open, and rising sea levels flooded large areas.

Third, recent evidence from Lida Ajer and Niah show remarkable adaptations to living in rainforest before 40,000 years ago. Fourth, the recent discovery of cave art on Borneo reveals a previously unsuspected artistic tradition extending back 40,000 years – as old as in western Europe. Next, each of the main islands in Sunda – Sumatra, Borneo, Java and Palawan – has its own fauna, and likely human history. We should therefore be wary of generalising from one island to the entire region. Finally, most of the key evidence is now under water. We know nothing about the use of coastal resources, and for that, we now need to look at Wallacea.

Notes

1 On a much smaller scale, the same type of catastrophic flooding occurred in northwest Europe with the drowning of Doggerland and formation of the present North Sea between England and Denmark (see Coles 1998).

2 This situation is complicated by the discovery of a new species of *Pongo* on Sumatra with genetic connections to both Sumatran and Bornean orangutans (Nater et al. 2017).

3 Although one suspects that any animals in the vicinity of Toba would have been stone deaf after the initial blast, which raises some interesting issues on their immediate survival.
4 In older publications, Wajak is spelt Wadjak.
5 Tom Harrisson was variously an anthropologist, archaeologist, sociologist, conservationist and guerrilla leader. He was also described as the "most offending soul alive" by one of his many detractors. He may well have been like Bryon, "mad, bad, and dangerous to know", and was loathed and loved in equal measure. For a thoroughly entertaining and detailed account of this highly gifted but deeply flawed man, read Judith Heimann's (1998) book *The Most Offending Soul Alive.*
6 Because rainforest was widespread in west and south Sumatra, there may have been rainforest in neighbouring West Java when the open woodland Ngandong Fauna dominated East Java.
7 Pangolins, sometimes known as scaly anteaters, are solitary, nocturnal insectivores. Those in Southeast Asia are up to ca. 3–5 kg in weight. Today pangolins are the most heavily poached animal worldwide for their meat and alleged medicinal properties.
8 How many died or nearly died before they worked out how to do this? Who thought it a good idea to process a plant that was so obviously toxic?

References

Amano, N., Moigne, A.M., Ingicco, T., Sémah, F., Awe, R.D. and Simanjuntak, T. (2016) Subsistence strategies and environment in Late Pleistocene–Early Holocene Eastern Java: evidence from Braholo Cave. *Quaternary International*, 416: 46–63.

Anderson, D.D. (1997) Cave archaeology in Southeast Asia. *Geoarchaeology*, 12(6): 607–638.

Anderson, D. (2005) The use of caves in Peninsular Thailand in the Late Pleistocene and Early and Middle Holocene. *Asian Perspectives*, 44(10): 137–153.

Aubert, M., Setiawan, P., Oktaviana, A.A., Brumm, A., Sulistyarto, P.H., Saptomo, E.W., Istiawan, B., Ma'rifat, T.A., Wahyuono, V.N., Zhao, J.-X., Huntley, J., Taçon, P.S.C., Howard, D.L. and Brand, H.E.A. (2018) Palaeolithic cave art in Borneo. *Nature*, 564: 254–257.

Bacon, A.-M., Bourgon, N., Dufour, E., Zanolli, C., Duringer, P., Ponche, J.-L., Antoine, P.-O., Shackelford, L., Nguyen Thi Mai Huong, Thongsa Sayavonkhamdy, Elise Patole-Edoumba, E. and Demeter, F. (2018b) Nam Lot (MIS 5) and Duoi U'Oi (MIS 4) Southeast Asian sites revisited: zooarchaeological and isotopic evidences. *Palaeogeography, Palaeoclimatology, Palaeoecology*, 512: 132–144.

Bacon, A.-M., Demeter, F., Duringer, P., Helm, C., Bano, M., Long, Vu The, Nguyen Thi Kim Thuy, Antoine, P.-O., Bui Thi Mai, Nguyen Thi Mai Huong, Dodo, Y., Chabaux, F. and Rihs S. (2008) The Late Pleistocene Duoi U'Oi cave in northern Vietnam: palaeontology, sedimentology, taphonomy and palaeoenvironments. *Quaternary Science Reviews*, 27: 1627–1654.

Bacon, A.-M., Duringer, P., Westaway, K., Joannes-Boyau, R., Zhao, J.-X. et al. (2018a) Testing the savannah corridor hypothesis during MIS2: the Boh Dambang hyena site in southern Cambodia. *Quaternary International*, 464: 417–439.

Bacon, A.-M. and Long, Vu The (2001) The first discovery of a complete skeleton of a fossil orang-utan in a cave of the Hoa Binh Province, Vietnam. *Journal of Human Evolution*, 41: 227–241.

Barker, G.W. (2012) The Niah cave in Malaysia: potential for World Heritage nomination, justification of Outstanding Universal Value, conditions of integrity and authenticity, and guidelines for optimal protection and management of the site. In N. Sanz (ed.) *Human Origin Sites and the World Heritage Convention in Asia*, Paris: UNESCO, pp. 220–234.

Barker, G. (ed.) (2013) *Rainforest Foraging and Farming in Island Southeast Asia: The Archaeology of the Niah Caves, Sarawak* (Volume I). Cambridge: McDonald Institute for Archaeological Research.

Barker, G.W.W., Barton, H., Bird, M., Daly, P., Datan, I., Dykes, A., Farr, L., Gilbertson, D., Harrisson, B., Hunt, C., Higham, T., Kealhofer, L., Krigbaum, J., Lewis, H., McLaren, S., Paz, V., Pike, A., Piper, P., Pyatt, B., Rabett, R., Reynolds, T., Rose, J., Rushworth, G., Stephens, G., Stephens, M. and Stringer, C. (2007) The "human revolution" in lowland tropical Southeast Asia: the antiquity and behavior of anatomically modern humans at Niah Cave (Sarawak, Borneo). *Journal of Human Evolution*, 52: 243–261.

Barker, G. and Farr, L. (eds) (2016) *The Archaeology of the Niah Caves, Sarawak* (Volume II). Cambridge: McDonald Institute for Archaeological Research.

Bellwood, P. (2007) *Prehistory of the Indo-Malaysian Peninsula* (3rd edition). Canberra: ANU Press.

Bellwood, P. (2017) *First Islanders: Prehistory and Human Migration in Island Southeast Asia*. Oxford: Wiley Blackwell.

Bergh, G.D. van den, Due Awe, Rokhus, Morwood, M.J., Sutikna, T., Jatmiko and Wahyu Saptomo, E. (2008) The youngest *Stegodon* remains in Southeast Asia from the Late Pleistocene archaeological site Liang Bua, Flores, Indonesia. *Quaternary International*, 182: 16–48.

Bergh, G.D. van den, Vos, J. de and Sondaar, P.Y. (2001) The Late Quaternary palaeogeography of mammal evolution in the Indonesian Archipelago. *Palaeogeography, Palaeoclimatology, Palaeoecology*, 171: 385–408.

Bird, M.I., Pang, W.C. and Lambeck, K. (2006) The age and origin of the Straits of Singapore. *Palaeogeography, Palaeoclimatology, Palaeoecology*, 241: 531–538.

Bird, M.I., Taylor, D. and Hunt, C. (2005) Palaeoenvironments of insular Southeast Asia during the Last Glacial Period: a savanna corridor in Sundaland? *Quaternary Science Reviews*, 24: 2228–2242.

Boivin, N., Fuller, D.Q., Dennell, R.W., Allaby, R. and Petraglia, M. (2013) Human dispersal across diverse environments of Asia during the Upper Pleistocene. *Quaternary International*, 300: 32–47.

Cannon, C.H., Morley, R.J. and Bush, A.B.G. (2009) The current refugial rainforests of Sundaland are unrepresentative of their biogeographic past and highly vulnerable to disturbance. *Proceedings of the National Academy of Sciences USA*, 106(27): 11188–11193.

Choa, O., Lebon, M., Gallet, X., Dizon, E., Ronquillo, W. et al. (2016) Stable isotopes in guano: potential contributions towards palaeoenvironmental reconstruction in Tabon Cave, Palawan, Philippines. *Quaternary International*, 416: 27–37.

Clarkson, C., Jacobs, Z., Marwick, B., Fullagar, R., Wallis, L. et al. (2017) New evidence for human occupation of northern Australia by 65,000 years ago. *Nature*, 547: 306–310.

Coles, B.J. (1998) Doggerland: a speculative survey. *Proceedings of the Prehistoric Society*, 64: 45–81.

Corny, J., Garong, A.M., Sémah, F., Dizon, E.Z., Bolunia, M.J.L.A., Bautista, R. and Détroit, F. (2016) Paleoanthropological significance and morphological variability of the human bones and teeth from Tabon Cave. *Quaternary International*, 416: 210–218.

Curnoe, D., Datan, I., Zhao, J.-X., Leh Moi Ung, C., Aubert, M., Sauffi, M.S. et al. (2018) Rare Late Pleistocene-early Holocene human mandibles from the Niah Caves (Sarawak, Borneo). *PLoS ONE*, 13: 1–13: e0196633.

Curnoe, D., Ipoi Datan, Hsiao Mei Goh and Sauffi, M.S. (2019) Femur associated with the Deep Skull from the West Mouth of the Niah Caves (Sarawak, Malaysia). *Journal of Human Evolution*, 127: 133–148.

Dam, R.A.C., Kaars, S. van der and Kershaw, A.P. (2001) Quaternary environmental change in the Indonesian region. *Palaeogeography, Palaeoclimatology, Palaeoecology*, 171: 91–95.

Demeter, F., Shackelford, L.L., Bacon, A.-M., Duringer, P., Westaway, K., Sayavongkhamdy, T., Braga, J. et al. (2012) Anatomically modern human in Southeast Asia (Laos) by 46 ka. *Proceedings of the National Academy of Sciences of the USA*, 109 (36):14375–14380.

Demeter, F., Shackelford, L.L., Braga, J., Westaway, K., Duringer, P., Bacon, A.-M., Ponche, J.-L. et al. (2015) Early modern humans and morphological variation in Southeast Asia: fossil evidence from Tam Pa Ling, Laos. *PLoS ONE*, 10(4): e0121193.

Demeter, F., Shackelford, L., Westaway, K., Barnes, L., Duringer, P. et al. (2017) Early modern humans from Tam Pà Ling, Laos: fossil review and perspectives. *Current Anthropology* 58(Supplement 17): S527–S538.

Dennell, R.W., Martinón-Torres, M. and Bermudéz de Castro, J.M. (2010) Hominin variability, climatic instability and population demography in Middle Pleistocene Europe. *Quaternary Science Reviews*, 30(11–12): 1511–1524, doi:10.1016/j.quascirev.2009.11.027.

Détroit, F., Dizon, E., Falguères, C., Hameau, S., Ronquillo, W. and Sémah, F. (2004) Upper Pleistocene *Homo sapiens* from the Tabon Cave (Palawan, the Philippines): description and dating of new discoveries. *Comptes Rendues Palévolution*, 3: 705–712.

Dizon, E., Détroit, F., Sémah, F., Falguères, C., Hameau, S., Ronquillo, W. and Cabanis, E. (2002) Notes on the morphology and age of the Tabon Cave *Homo sapiens*. *Current Anthropology*, 43: 660–666.

Dubois, E. (1921) The proto-Australian fossil man of Wadjak, Java. *Koninklijke Nederlandsche Akademie van Wetenschappen Proceedings*, 23(2): 1013–1051.

Dutton, A. and Lambeck, K. (2012) Ice volume and sea level during the last interglacial. *Science*, 337: 216–219.

Erlandson, J. and Braje, T.J. (2014) Coasting out of Africa: the potential of mangrove forests and marine habitats to facilitate human coastal expansion via the Southern Dispersal Route. *Quaternary International*, 382: 31–41.

Gagan, M.K., Ayliffe, L.K., Hopley, D., Cali, J.A., Mortimer, G.E., Chappell, J., McCulloch, M.T. and Head, M.J. (1998) Temperature and surface-ocean water balance of the Mid-Holocene Tropical Western Pacific. *Science*, 279: 1014–1018.

Gathorne-Hardy, F.J. and Harcourt-Smith, W.E.H. (2003) The super-eruption of Toba, did it cause a human bottleneck? *Journal of Human Evolution*, 45: 227–230.

Grün, R. and Thorne, A. (1997) Dating the Ngandong humans. *Science*, 276: 1575.

Hanebuth, T., Stattegger, K. and Grootes, P.M. (2000) Rapid flooding of the Sunda Shelf: a Late-Glacial sea-level record. *Science*, 288: 1033–1035.

Hanebuth, T.J.J., Voris, H.K., Yokoyama, Y., Saito, Y. and Okuno, J. (2011) Formation and fate of sedimentary depocentres on Southeast Asia's Sunda Shelf over the past sea-level cycle and biogeographic implications. *Earth-Science Reviews*, 104: 92–110.

Harrison, T., Jin, C., Zhang, Y., Wang, Y. and Zhu, M. (2014) Fossil *Pongo* from the Early Pleistocene *Gigantopithecus* fauna of Chongzuo, Guangxi, southern China. *Quaternary International*, 354: 59–67.

Harrison, T., Krigbaum, J. and Manser, J. (2006) Primate biogeography and ecology on the Sunda Shelf Islands: A paleontological and zooarchaeological perspective. In S.M. Lehman and J.G. Fleagle (eds) *Primate Biogeography*, New York: Springer, pp. 331–372.

Hawks, J., Oh, S., Hunley, K., Dobson, S., Cabana, G., Dayalu, P. and Wolpoff, M.H. (2000) An Australasian test of the recent African origin theory using the WLH-50 calvarium. *Journal of Human Evolution*, 39: 1–22.

Heaney, L.R. (1986) Biogeography of mammals in SE Asia: estimates of rates of colonization, extinction and speciation. *Biological Journal of the Linnean Society*, 28: 127–165.

Heaney, L.R. (1991) A synopsis of climatic and vegetational change in Southeast Asia. *Climatic Change*, 19: 53–61.

Heimann, J.M. (1998) *The Most Offending Soul Alive: Tom Harrisson and His Remarkable Life*. Honolulu: University of Hawai'i Press.

Higham, C. (2013) Hunter-gatherers in Southeast Asia: from prehistory to the present. *Human Biology*, 85 (1–3): 21–43.

Hinton, A.C. (2000) Tidal changes and coastal hazards: past, present and future. *Natural Hazards*, 21: 173–184.

Hoffman, J.S., Clark, P.U., Parnell, A.C. and Feng He (2017) Regional and global sea-surface temperatures during the last interglaciation. *Science*, 355: 276–279.

Hooijer, D.A. (1948) Prehistoric teeth of man and of the orang-utan from central Sumatra, with notes on the fossil orang-utan from Java and southern China. *Zoologische Mededelingen*, 29: 175–301.

Hunt, C. and Barker, G. (2014) Missing links, cultural modernity and the dead: anatomically modern humans in the Great Cave of Niah (Sarawak, Borneo). In R. Dennell and M. Porr (eds) *Southern Asia, Australia and the Search for Human Origins*. Cambridge: Cambridge University Press, pp. 90–107.

Hunt, C.O., Gilbertson, D.D. and Rushworth, G. (2007) Modern humans in Sarawak, Malaysian Borneo, during Oxygen Isotope Stage 3: palaeoenvironmental evidence from the Great Cave of Niah. *Journal of Archaeological Science*, 34: 1953–1969.

Hunt, C.O., Gilbertson, D.D. and Rushworth, G. (2012) A 50,000-year record of late Pleistocene tropical vegetation and human impact in lowland Borneo. *Quaternary Science Reviews*, 37: 61–80.

Ibrahim, Y.K., Tshen, L.T., Westaway, K.E., Cranbrook, Earl of, Humphrey, L., Muhammad, R.F., Zhao, J.-X. and Peng, L.C. (2013) First discovery of Pleistocene orangutan (*Pongo* sp.) fossils in Peninsular Malaysia: biogeographic and paleoenvironmental implications. *Journal of Human Evolution*, 65(6): 770–797.

Indriati, E., Swisher, C.C., Lepre, C., Quinn, R.L., Suriyanto, R.A., Hascaryo, A.T., Grün, R., Feibel, C.S., Pobiner, B.L., Aubert, M., Lees, W. and Antón, S.C. (2011) The age of the 20 meter Solo River Terrace, Java, Indonesia and the survival of *Homo erectus* in Asia. *PLoS*, 6: 1–10.

Jablonski, N., Whitfort, M.J., Roberts-Smith, N. and Xu, Q. (2000) The influence of life history and diet on the distribution of catarrhine primates during the Pleistocene in eastern Asia. *Journal of Human Evolution*, 39(2): 131–157.

Kaars, W.A. van der and Dam, M.A.C. (1995) A 135,000-year record of vegetational and climatic change from the Bandung area, West-Java, Indonesia. *Palaeogeography, Palaeoclimatology, Palaeoecology*, 117(1–2): 55–72.

Kaars, S. van der, Wang, X. Kershaw, P., Guichard, F. and Arifin Setiabudi (2000) A Late Quaternary palaeoecological record from the Banda Sea, Indonesia: patterns of vegetation, climate and biomass burning in Indonesia and northern Australia. *Palaeogeography, Palaeoclimatology, Palaeoecology*, 155 (1–2): 135–153.

Kaars, S. van der, Bassinot, F., Deckker, P. de and Guichard, F. (2010) Changes in monsoon and ocean circulation and the vegetation cover of southwest Sumatra through the last 83,000 years: the record from marine core BAR94-42. *Palaeogeography, Palaeoclimatology, Palaeoecology*, 296: 52–78.

Kaars, S. van der, Williams, M.A.J., Bassinot, F., Guichard, F., Moreno, E. et al. (2011) The influence of the c. 73 ka Toba super-eruption on the ecosystems of northern Sumatra as recorded in marine core BAR94-25. *Quaternary International*, 258: 45–53.

Kahlke, H.D. (1972) A review of the Pleistocene history of the orang-utan (*Pongo* Lacépède 1799). *Asian Perspectives*, 15: 5–14.

Koenigswald, G.H.R. von (1933) Ein neuer Urmensch aus dem Diluvium Javas. *Zeitblatt Miner Geol. A. und B. Palaontologie*, 29–42.

Lister, A.M. (1995) Sea-levels and the evolution of island endemics: the dwarf red deer of Jersey. In R.C. Preece (ed.) *Island Britain: A Quaternary Perspective*. Geological Society Special Publication 96: 151–172.

Lohman, D.J., Bruyn, M. de, Page, T., Rintelen, K. von, Hall, R., Ng, P.K.L., Shih, Hsi-Te, Carvalho, G.R. and Rintelen, T. von (2011) Biogeography of the Indo-Australian Archipelago. *Annual Review of Ecology, Evolution, and Systematics*, 42: 205–226.

Louys, J. (2012) Mammal community structure of Sundanese fossil assemblages from the Late Pleistocene, and a discussion on the ecological effects of the Toba eruption. *Quaternary International*, 258: 80–87, doi:10.1016/j.quaint.2011.07.027.

Louys, J., Curnoe, D. and Tong, Haowen (2007) Characteristics of Pleistocene megafauna extinctions in Southeast Asia. *Palaeogeography, Palaeoclimatology, Palaeoecology*, 243: 152–173.

Louys, J. and Meijaard, E. (2010) Palaeoecology of Southeast Asian megafauna-bearing sites from the Pleistocene and a review of environmental changes in the region. *Journal of Biogeography*, 37: 1432–1449.

Louys, J. and Turner, A. (2012) Environment, preferred habitats and potential refugia for Pleistocene *Homo* in Southeast Asia. *Comptes Rendues Palévolution*, 11: 203–211.

Matsumura, H. and Pookajorn, S. (2005) A morphometric analysis of the Late Pleistocene Human Skeleton from the Moh Khiew Cave in Thailand. *HOMO—Journal of Comparative Human Biology*, 56: 93–118.

Mei, W. and Xie, S.-P. (2016) Intensification of landfalling typhoons over the northwest Pacific since the late 1970s. *Nature Geoscience*, 9: 753–757.

Meijaard, E. (2003) Mammals of south-east Asian islands and their Late Pleistocene environments. *Journal of Biogeography*, 30: 1245–1257.

Meijer, H.J.M. and Due, R.A. (2010) A new species of giant marabou stork (Aves: Ciconiiformes) from the Pleistocene of Liang Bua, Flores (Indonesia). *Zoological Journal of the Linnean Society*, 160: 707–724.

Milano, S., Demeter, F., Hublin, J.-J., Duringer, P., Patole-Edoumba, E. et al. (2018) Environmental conditions framing the first evidence of modern humans at Tam Pà Ling, Laos: a stable isotope record from terrestrial gastropod Carbonates. *Palaeogeography, Palaeoclimatology, Palaeoecology*, 511: 352–363.

Morley, M.W. (2017) The geoarchaeology of hominin dispersals to and from tropical Southeast Asia: a review and prognosis. *Journal of Archaeological Science*, 77: 78–93.

Morwood, M.J., Sutikna, T., Saptomo, E.W., Westaway, K.E., Due, R.A., Moore, M.W., Yuniawati, D.Y., Hadi, P., Zhao, J.X., Turney, C.S.M. and Fifield, K. (2008) Climate, people and faunal succession on Java, Indonesia: evidence from Song Gupuh. *Journal of Archaeological Science*, 35(7): 1776–1789.

Mudar, K. and Anderson, D. (2007) New Evidence for Southeast Asian Pleistocene foraging economies: faunal remains from the early levels of Lang Rongrien rockshelter, Krabi, Thailand. *Asian Perspectives*, 46: 298–334.

Nater, A., Mattle-Greminger, M.P., Nurcahyo, A., Nowak, M.G., De Manuel, M., Desai, T., Groves, C., Pybus, M., Sonay, T.B., Roos, C. and Lameira, A.R. (2017) Morphometric, behavioral, and genomic evidence for a new orangutan species. *Current Biology*, 27(22): 3487–3498.

Oppenoorth, W.E.F. (1932) *Homo (Javanthropus) soloensis*: een Pleistocene Mensch van Java. *Scientific Proceedings of the Mining Company of the Dutch East Indies*, 20: 49–75.

Piper, P.J., Ochoa, J., Lewis, H., Paz, V. and Ronquillo, W.P. (2008) The first evidence for the past presence of the tiger *Panthera tigris* (L.) on the island of Palawan, Philippines: extinction in an island population. *Palaeogeography, Palaeoclimatology, Palaeoecology*, 264: 123–127.

Piper, P.J., Ochoa, J., Robles, E.C., Lewis, H. and Paz, V. (2011) Palawan palaeozoology of Palawan Island, Philippines. *Quaternary International*, 233: 142–158.

Polanski, J., Marsh, H.E. and Maddux, S.D. (2016) Dental size reduction in Indonesian *Homo erectus*: implications for the PU-198 premolar and the appearance of *Homo sapiens* on Java. *Journal of Human Evolution*, 90: 49–54.

Pyatt, F.B., Barker, G.W., Rabett, R.J., Szabo, K. and Wilson, B. (2010) Analytical examination of animal remains from Borneo: the painting of bone and shell. *Journal of Archaeological Science*, 37: 2102–2105.

Rabett, R.J. (2012) *Human Adaptation in the Asian Palaeolithic: Hominin Dispersal and Behaviour during the Late Quaternary*. Cambridge: Cambridge University Press.

Reis, K.R. and Garong, A.M. (2001) Late Quaternary terrestrial vertebrates from Palawan Island, Philippines. *Palaeogeography, Palaeoclimatology, Palaeoecology*, 171: 409–421.

Reynolds, T., Barker, G., Barton, H., Cranbrook, G., Farr, L., Hunt, C., Kealhofer, L., Paz, V., Pike, A., Piper, P.J. and Rabett, R.J. (2013) The first modern humans at Niah, c. 50,000–35,000 years ago. In G. Barker (ed.) *Rainforest Foraging and Farming in Island Southeast Asia*, Cambridge: McDonald Institute for Archaeological Research, pp.135–172.

Roberts, P., Boivin, N., Lee-Thorp, J., Petraglia, M. and Stock, J. (2016) Tropical forests and the genus *Homo*. *Evolutionary Anthropology*, 25(6): 306–317.

Roberts, P., Perera, N., Wedage, O., Deraniyagala, S., Perera, J., Eregama, S., Gledhill, A., Petraglia, M.D. and Lee-Thorpe, J. (2015) Direct evidence for human reliance on rainforest resources in Late Pleistocene Sri Lanka. *Science*, 347: 1246–1249.

Rose, W.I. and Chesner, C.A. (1987) Dispersal of ash in the great Toba eruption. *Geology*, 15: 913–917.

Schwartz, J.H. and Tattersall, I. (2000) What constitutes *Homo erectus*? *Acta Anthropologica Sinica*, Supplement 19: 18–22.

Schwartz, J.H., Long, V.T., Cuong, N.L., Kha, L.T. and Tattersall, I. (1995) A review of the Pleistocene hominoid fauna of the Socialist Republic of Vietnam (excluding Hylobatidae). *Anthropological Papers of the American Museum of Natural History*, 76: 2–24.

Sémah, A.-M. and Sémah, F. (2012) The rain forest in Java through the Quaternary and its relationships with humans (adaptation, exploitation and impact on the forest). *Quaternary International*, 249: 120–128.

Shackelford, L., Demeter, F., Westaway, K., Duringer, P., Ponche, J.-L., Sayavongkhamdy, T., Zhao, J.-X., Barnes, L., Boyon, M., Sichanthongtip, S., Sénégas, F., Patole-Edoumba, E., Coppens, Y., Dumoncel, J. and Bacon, A.-M. (2018) Additional evidence for early modern human morphological diversity in Southeast Asia at Tam Pa Ling, Laos. *Quaternary International*, 466: 93–106.

Sherratt, A.G. (2002) Darwin among the archaeologists: the John Evans nexus and the Borneo caves. *Antiquity*, 76: 151–157.

Shuttler, R., Head, J.M., Donahue, D.J., Jull, A.J.T., Barbetti, J.F., Matsu'ura, S., Vos, J. de and Storm, P. (2004) AMS radiocarbon dates on bone from cave sites in southeast Java, Indonesia, including Wadjak. *Modern Quaternary Research in SE Asia*, 18: 89–94.

Storm, P. (1995) The evolutionary significance of the Wajak skulls. *Scripta Geologia*, 110: 1–247.

Storm, P., Aziz, F., Vos, J. de, Kosasih, D., Baskoro, S., Ngaliman and Hoek Ostende, L.W. van den (2005) Late Pleistocene *Homo sapiens* in a tropical rainforest fauna in East Java. *Journal of Human Evolution*, 49: 536–545.

Storm, P. and Nelson, A. (1992) The many faces of Wadjak Man. *Archaeology in Oceania*, 27: 7–46.

Storm, P. and Vos, J. de (2006) Rediscovery of the Late Pleistocene Punung hominin sites and the discovery of a new site Gunung Dawung in East Java. *Senckenburgiana Lethaea*, 86: 271–281.

Storm, P., Wood, R., Stringer, C., Bartsiokas, A., Vos, J. de, Aubert, M., Kinsley, L. and Grün, R. (2013) U-series and radiocarbon analyses of human and faunal remains from Wajak, Indonesia. *Journal of Human Evolution*, 64: 356–365.

Stringer, C.B. (2000) Coasting out of Africa. *Nature*, 405: 24–25.

Swisher, C.C. III, Rink, W.J., Antón, S.C., Schwarcz, H.P., Curtis, G.H., Suprijo, A. and Widiasmoro (1996) Latest *Homo erectus* of Java: Potential contemporaneity with *Homo sapiens* in Southeast Asia. *Science*, 274: 1870–1874.

Tjallingii, R., Stattegger, K., Stocchi, P., Saito, Y. and Wetzel, A. (2014) Rapid flooding of the southern Vietnam shelf during the early to mid-Holocene. *Journal of Quaternary Science*, 29(6): 581–588.

Tougard, C. (2001) Biogeography and migration routes of a large mammal faunas in South-East Asia during the Late Middle Pleistocene: focus on the fossil and extant faunas from Thailand. *Palaeogeography, Palaeoclimatology, Palaeoecology*, 168: 337–358.

Tougard, C., Chaimanee, Y., Suteethorn, V., Triamwichanon, S. and Jaeger, J.-J. (1996) Extension of the geographic distribution of the giant panda (*Ailuropoda*) and search for the reasons for its progressive disappearance in Southeast Asia during the latest Middle Pleistocene. *Comptes Rendues de l'Academie des Sciences Paris IIa*, 323: 973–979.

Tshen, L.T. (2015) Biogeographic distribution and metric dental variation of fossil and living orangutans (*Pongo* spp.). *Primates*, doi:10.1007/s10329-015-0493-z.

Urushibara-Yoshino, K. and Yoshino, M. (1997) Palaeoenvironmental change in Java island and its surrounding areas. *Journal of Quaternary Science*, 12: 435–442.

Vartanyan, S.L., Garutt, V.E. and Sher, A.V. (1993) Holocene dwarf mammoths from Wrangel Island in the Siberian Arctic. *Nature*, 362: 337–340.

Voris, H. (2000) Maps of Pleistocene sea levels in Southeast Asia: Shorelines, river systems and time durations. *Journal of Biogeography*, 27: 1153–1167.

Vos, J. de (2014) The history of paleoanthropological research in Asia: reasons and priorities for future cooperation in research and preservation of sites and collections. In N. Sanz (ed.) *Human Origin Sites and the World Heritage Convention in Asia*, Paris: UNESCO, pp. 68–82.

Weidenreich, F. (1945) The Keilor skull: a Wadjak type from Southeast Australia. *American Journal of Physical Anthropology*, 3(1): 21–32.

Westaway, K.E., Louys, J., Dur Awe, R., Morwood, M.J., Prices, G.J. et al. (2017) An early modern human presence in Sumatra 73,000–63,000 years ago. *Nature*, 548: 322–325, doi:10.1038/nature23452.

Westaway, K.E., Morwood, M.J., Roberts, R.G., Rokus, A.D., Zhao, J.-X., Storm, P., Aziz, F., Bergh, G. van den, Hadi, P., Jatmiko and Vos, J. de (2007) Age and biostratigraphic significance of the Punung rainforest fauna, East Java, Indonesia, and implications for *Pongo* and *Homo*. *Journal of Human Evolution*, 53, 709–717.

Wilting, A., Sollman, R., Meijaard, E., Helgen, K.M. and Fickel, J. (2012) Mentawai's endemic, relictual fauna: is it evidence for Pleistocene extinctions on Sumatra? *Journal of Biogeography*, 39(9): 1608–1620.

Woodruff, D.S. and Turner, L.M. (2009) The Indochinese-Sundaic zoogeographic transition: a description and analysis of terrestrial mammal species distributions. *Journal of Biogeography*, 36: 803–821.

Wurster, C.M. and Bird, M.I. (2014) Barriers and bridges: early human dispersals in equatorial SE Asia. In J. Harff, G. N. Bailey and F. Lüth (eds) *Geology and Archaeology: Submerged Landscapes of the Continental Shelf: An Introduction. Geological Society, London, Special Publications*, 411: 235–250.

Wurster, C.M., Bird, M.I., Bull, I.D., Creed, F., Bryant, C., Dungait, J.A.J. and Paz, V. (2010) Forest contraction in north equatorial Southeast Asia during the Last Glacial Period. *Proceedings of the National Academy of Sciences USA*, 107: 15508–15511.

Yokoyama, Y., Falguères, C., Semah, F., Jacob, T., Grün, R. (2008) Gamma-ray spectrometric dating of Late *Homo erectus* skulls from Ngandong and Sambungmacan, Central Java, Indonesia. *Journal of Human Evolution*, 55: 274–277.

7

WALLACEA AND SAHUL

Introduction

If one travels east from Java to the well-known holiday island of Bali and then across the Lombok Strait to Lombok, one crosses the Wallace Line and enters the biogeographical province of Wallacea, named after Alfred Russel Wallace[1] who recognised it in 1859. Wallace's Line "demarcates the most abrupt faunal transition in the world" (Lohman et al. 2011: 208[2]), and denotes a transitional zone between the biological realms of Asia and Australia. Unlike Sahul, Wallacea (excluding the Philippines) contains mammals such as dwarf buffalo, babirusa (deer pig) and rats, but unlike Sunda, it also contains marsupials such as cuscus.[3] Additionally, because it has a highly endemic fauna and flora, the first colonists encountered significantly different plants and animals from those encountered on Sunda, so it provides another example of humans learning a new type of landscape. Wallacea takes in the 2,000 or so islands east of Java and Borneo (see Figure 7.1) and ends at the western edge of the continental shelf of Sahul, the great conjoined landmass of Australia, New Guinea and Tasmania. That eastern boundary is often defined by Lydekker's Line,[4] which supposedly demarcates the Australian from the Wallacean fauna. In fact, both Wallace's and Lydekker's Line[5] are better seen as porous boundaries (Lohman et al. 2011). Wallace excluded the Philippines from Wallacea, but Thomas Huxley[6] redrew his line to include them (apart from Palawan), and most researchers have followed suit. On heuristic grounds, I include them in Wallacea, as part of the offshore land mass of Southeast Asia.

Wallacea is often seen simply as part of Southeast Asia or as a corridor to Australia but it deserves to be seen as a major area of colonisation and settlement in its own right. Its size should not be underestimated: Lombok to Cape York (the northernmost tip of Australia) is 2,900 km; Lombok to Manila in the Philippines

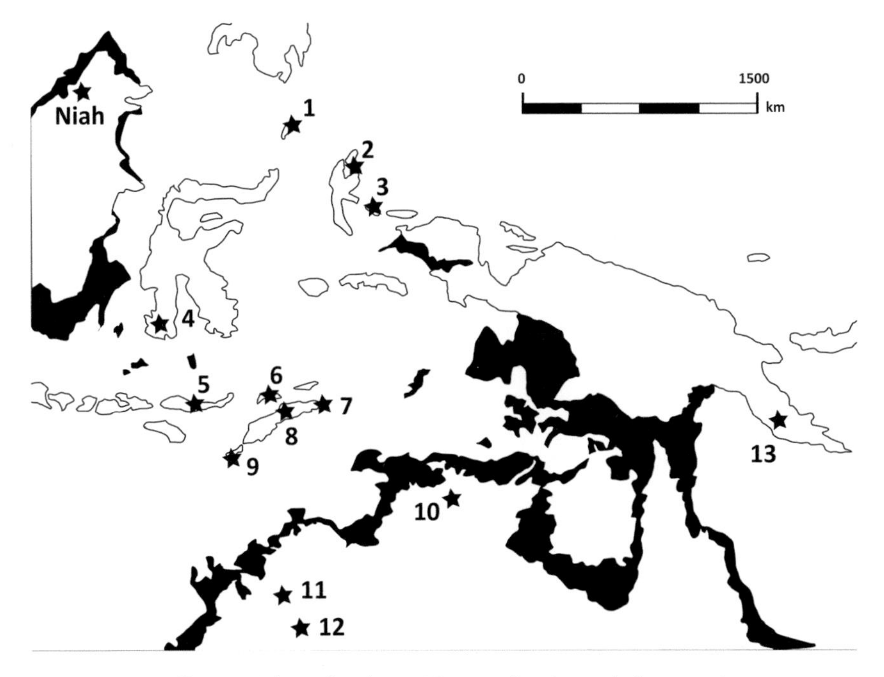

FIGURE 7.1 Wallacea, with sea levels at –50 m, and major early human sites

1 Leang Sarru (33.9–35 ka); 2 (Halmahera) Daeo 2 (15.9–16.8 ka); 3 Golo Cave (35–36.4 ka); 4 Sulawesi: Leang Timpuseng (42–40 ka); Leang Sakapao (28.4–29.8 ka); Leang Burung (31.3 ka); 5 Flores (46 ka onwards); 6 Alor (Nusa Tengarra); Tron Bon Lei (21–21.6 ka); 7 (Timor-Leste) Jerimalai (41.3–43 ka); Lene Hara (37–40.3 ka); Bui Ceri Uato (29.6–31 ka); Uai Bobo 2 (14.5–17.7 ka); 8 Laili (43,283–44,631 cal. BP); 9 Lua Meko (28–29 ka); Lua Manggetek (14.1–16.6 ka); Sahul: 10 Madjedbede (54–?65 ka); 11 Carpenters Gap (44–49 ka); 12 Riwi (45–46 ka); 13 Ivane Valley (43–49 ka).

Data from Kealy et al. (2016). Note reservations over the basal dates for the occupation of Madjedbede. Dates are calibrated (OxCal 4.2).

Source: Redrawn from Kealy et al. 2017, Fig. 1.

is 2,600 km and Manila to Cape York is 3,700 km. In recent years, it has produced three remarkable surprises. The first is that at least three islands were colonised before our species arrived; the second is that Wallacea contains the world's earliest maritime economies in which the sea became a resource zone instead of a barrier; and the third is that its earliest figurative cave art is as old as that in western Europe.

Wallacea before humans

The earliest evidence for hominins in Wallacea comes from the islands of Luzon in the Philippines, Flores and Sulawesi.

The Philippines

The Philippines comprise over 7,000 islands, of which the main ones are Luzon, Minoro, Panay and Mindanao (see Figure 7.2). Some islands are surrounded by deep ocean and never conjoined other islands, but most were conjoined with other islands when sea levels were 50 m below the present. When sea levels were at their lowest, ca. 120 m below the present, the total land area would have increased from ca. 285,000 sq km to 420,000 sq km (excluding Palawan) (Robles 2013). At that point, islands would have merged to form mega-islands or platforms known as Pleistocene Aggregate Island Complexes (PAICs) (Siler et al. 2010): the largest were Greater Palawan (see Chapter 6), Greater Luzon, Greater Mindanao and Greater Negros-Pana. As in Sundaland, sea level changes had profound faunal consequence in removing marine barriers between islands when sea levels fell, but fragmenting and isolating them when sea levels rose (Siler et al. 2010).

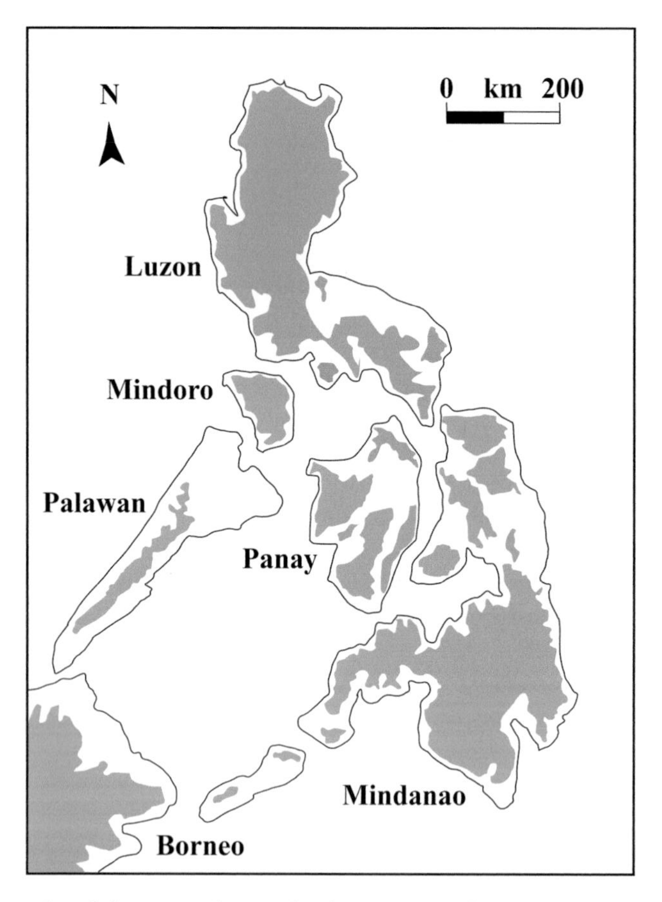

FIGURE 7.2 The Philippines, when sea levels were 125 m lower than now
Source: Coastlines from Esselstyn and Brown 2009, Fig. 1.

As with most isolated islands, the Philippines has an impoverished fauna that was basically limited to animals that were good swimmers – in this case, a giant tortoise (*Geochelone*), *Stegodon*, *Elephas*, *Rhinoceros*, and some bovines, cervids and suids. Like most of Southeast Asia, Sunda and Wallacea, the Philippines have an extraordinary range of plants that can be used as food, medicine, poison, clothing, ornaments and for making artefacts. In a three-month field season in the forests of Palawan (see Chapter 6) Xhauflair and colleagues (2017) recorded that the local inhabitants used 95 species from 34 families and noted that all parts of plants could be used at all life stages. This number pales into insignificance against the 1,100 species of usable plants that ethnobotanist Harold Conklin recorded among the Hanunóo of Mindoro in the southern Philippines (Xhauflair et al. 2017).

The Philippines has produced two of the most remarkable discoveries in the last few years: 700,000-year-old artefacts, and a new paleo-species, *Homo luzonensis*.

Kalinga, Cagayan Valley, northern Luzon

The publication of 700-ka-old artefacts from the Philippines (Ingicco et al. 2018) was one of the least expected discoveries of 2018. At the site of Kalinga in the Cagayan Valley, 57 stone artefacts (six cores, 49 flakes, two possible hammer stones, and a possible manuport) were found associated with the almost-complete, disarticulated skeleton of a rhinoceros (*R. philippenensis*). Thirteen ribs and metacarpals that in life were covered in soft tissue showed clear cut marks, and both humeri had percussion damage consistent with attempts to smash the bone open for its marrow. These were all found in the base of an erosional channel that was filled with over 3 m of a thick mudflow. The dating of this site was very thorough and was accomplished by ESR, ESR/U-Th, palaeomagnetism, and argon-argon. A tektite[7] was also present, which implies a maximum age of ca. 780–800 ka. The resulting age estimate of the artefacts and associated bone was 709 ka.

We have no idea who these first inhabitants of the Philippines were, or how they got there. The most likely candidates are *H. erectus*, known from Java, an ancestor of *H. floresiensis*, as known from Flores (see below), or an ancestor of *H. luzonensis* (see below). They probably arrived by accidental drifting on rafts of vegetation after, for example, a typhoon, as suggested by Smith (2001) for the colonisation of Flores (see below). Because the Philippines have been colonised from almost every direction (Lohman et al. 2011), they could have arrived from south China via Taiwan, or from Borneo/Kalimantan via Palawan, or from Sulawesi. If they arrived from Borneo or Sulawesi, there were at least two sea crossings. Because the direction of flow in this region is mainly from north to south, from the South China Sea into to Indian Ocean (see Figure 7.3), dispersal southwards from Taiwan is perhaps the most likely. We await further discoveries from this area.

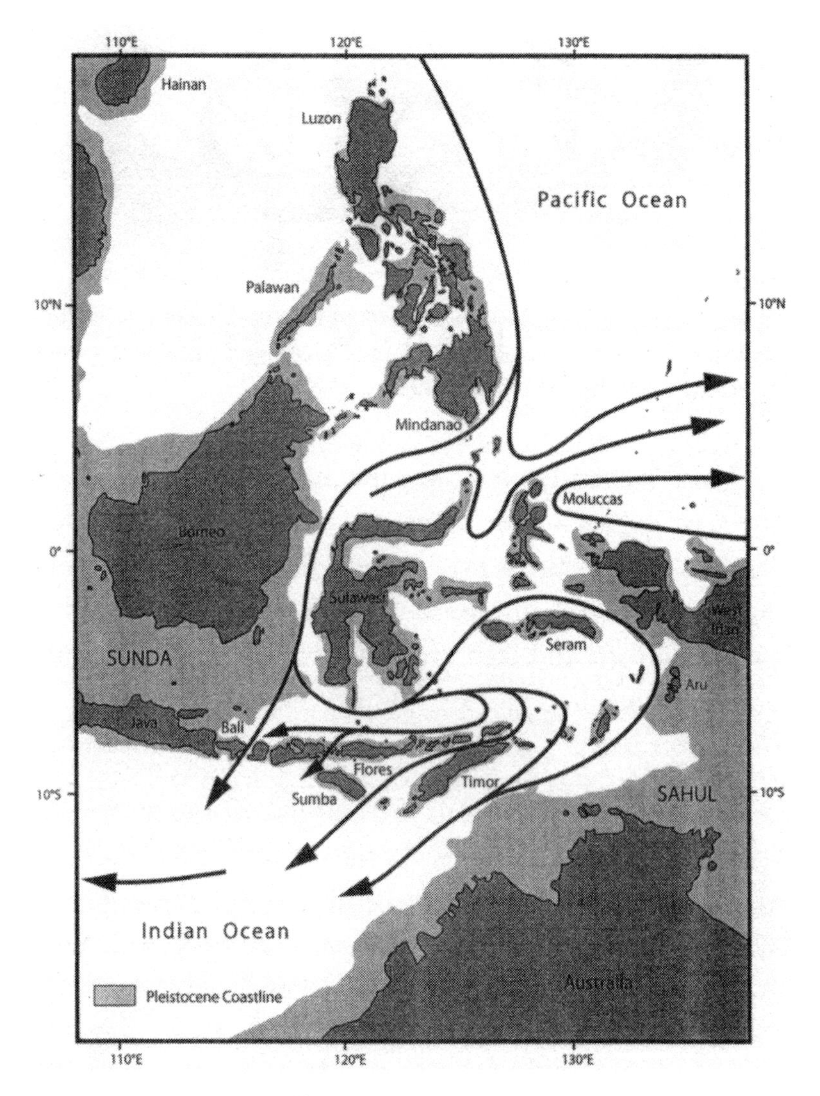

FIGURE 7.3 Sea currents in Wallacea

Note the strong north–south currents between the Pacific and Indian Oceans. These would make west–east crossings from southern Sunda to southern Wallacea less likely than those from further north.

Source: Morwood and Oosterzee 2007, p. 225.

Callao Cave and Homo luzonensis

Callao cave in northern Luzon (Mijares et al. 2010; Pawlik et al. 2014) contains over 3 m of deposit. Layer 14 contained a hominin right third metatarsal (foot bone) as well as bones (some with cut marks) and tooth fragments. Two deer teeth from this layer produced uranium-series and ESR minimum ages of 52 ± 1.4 ka and 54.3 ± 1.9 ka, and the metatarsal was dated by uranium-series to 66.7 ± 1 ka. At the time, it was tentatively identified as a small-bodied *H. sapiens*, but its size fell within the range of the hobbit *H. floresiensis*.

Further excavation of layer 14 produced seven teeth, two fingers and two toe phalanges, and a femur shaft from at least three individuals and has led to the identification of a new hominin species, *Homo luzonesis* (Détroit et al. 2019). Put simply, the teeth are small but *Homo*-like in their cusp morphology, but the hand and foot bones are like those of *Australopithecus afarensis* and *A. africanus* that are known only in Africa and are at least two million years earlier (Tocheri 2019). The finger and toe bones are also curved, suggesting that it may have been a climber, as with earlier forms of hominins. Overall, "our picture of hominin evolution in Asia during the Pleistocene just got even messier, more complicated and a whole lot more interesting" (Tocheri 2019: 178).

Sulawesi

As with Luzon, there has been a suspicion that surface finds of artefacts were associated with Middle Pleistocene fossils, but until now confirmation of this "Cabengan" industry has been lacking (Keates and Bartstra 2015). Recent excavations at Talepu in southwest Sulawesi, however, have discovered artefacts in stratified contexts between 100 and 200 ka old, and in association with fossils of *Bubalus* sp., *Stegodon* and *Celebochoerus* (an endemic type of large pig). Two trenches (T2 and T4) were excavated to a combined depth of 18 metres. Five units were recognised: in Trench 2 at the top, Unit A comprised 4 metres of gravel and sands, and contained 318 heavily rolled stone artefacts,[8] with an upper date of ca. 100 ka and a basal date of ca. 130 ka; Unit B was ca 1 metre of silt and sand, with dates of 130 and 140 ka; below this, Unit C comprised 3.5 metres of silt and clay, and finally Unit D was more gravel and sand, with a date of 156 ± 19 ka.

The excavation downslope of trench T4 cut through Units C and D into over 6m of silt and clay (Unit E), with a basal date of >190 ka. This unit contained three unretouched flakes with clear flake features; importantly, these are not water-rolled, and there is no evidence in the Unit of water action (unlike in Unit A). Hominins are therefore the most likely agent.

This discovery raises more questions than answers. Obviously, without skeletal evidence we have no idea about the makers of these artefacts, but they are too old to have been made by *Homo sapiens* (unless we are due for a really major surprise!). Potential regional candidates are *H. erectus* (as on Java) or *H. floresiensis* (as on Flores), in which case Sulawesi might have provided the source population of this species. We also have no idea how they reached Sulawesi, but the most direct route is across the Makassar Strait from Borneo, perhaps by drifting – or should we exclude the possibility of watercraft by this time?

One interesting question raised by these discoveries on Luzon, Sulawesi and Flores for settlement before our species arrived is how many other islands might have been reached by accidental drifting as it is unlikely that these three islands were the only ones to which earlier groups might have drifted. However, if they did, it is unlikely that they would have survived because most islands in Wallacea at the time of human contact had a limited terrestrial fauna – typically, bats, rodents and the small marsupial cuscus.[9] If hominins did manage to make land, they would most likely have starved unless they could fish. As seen below, marine resources were the key to survival in Wallacea.

Flores

The discovery of the hobbit – *Homo floresiensis* – in the cave deposits of Liang Bua (Brown et al. 2004; Morwood et al. 2004) is still one of the most unexpected surprises in palaeoanthropology this century, and no island in Southeast Asia has generated more interest and debate than Flores. No one expected an island in Southeast Asia – or anywhere else – to have been inhabited by a hominin that was only c. 1 metre tall, with a body weight of only ca. 12–16 kg, a cranial capacity of ca. 400 cc (scarcely larger than a chimpanzee) and the ability to make stone tools.

To summarise briefly what we know of the deep history of Flores: an unknown hominin was making stone tools on the island before 1 Ma (Brumm et al. 2010), and also at 800 ka (Morwood et al. 1998). A mandible fragment ca. 700 ka old may or may not be an ancestor of the hobbit because its identification is uncertain (Bergh et al. 2016; Brumm et al. 2016). The skeletal remains of *H. floresiensis* and numerous associated stone tools at Liang Bua were initially dated as between 95,000 and 12,000 years old, implying a long overlap between it and our species (Morwood et al. 2004). It was also suggested that its extinction – as well as that of the local dwarf *Stegodon* – may have been caused by the volcanic eruptions, as there are numerous active volcanos on the island (Westaway et al. 2007).

The evidence from Flores continues to throw up more questions than answers. It is not clear if Flores was occupied continuously for over a million years: do the gaps indicate an absence of hominins making artefacts, or an absence of artefacts in observed sequences? We do not know how they arrived, although accidental arrival on rafts of natural vegetation after a cyclone or tsunami is perhaps the most parsimonious mechanism: it is after all unlikely that *Homo erectus* was building boats or rafts a million years ago.[10] Nor do we know where they arrived from: the shortest route is from Java to the west, but the currents between Java and Flores are strongly north to south (Figure 7.3), so Sulawesi or Kalimantan are probably more likely starting points (Dennell et al. 2013). There has been a major debate over its origins, with some researchers arguing that the *H. floresiensis* represents a primitive lineage that extended back to early forms of *Homo* or even *Australopithecus* that dispersed from Africa in the earliest part of the Pleistocene (see e.g. Argue et al. 2006; Jungers et al. 2009), and others regarding it a local descendant of local *H. erectus* that dwarfed after its arrival on the island (Kaifu and Fujita 2012).[11]

The dating of the Liang Bua sequence is crucial, because it originally indicated a long overlap with *H. sapiens*. The reason why this is problematic is that *H. sapiens* was present on numerous islands in Wallacea and had almost certainly visited most by the Late Pleistocene, yet *H. floresiensis* supposedly continued to live on Flores for perhaps 30,000 years until it was snuffed out by volcanic eruptions. As John Shea (2011) pointed out, *H. sapiens* is not known for getting on well with his classmates, and such a long period of peaceful co-existence seems highly improbable.

For this reason, a recent paper that presents a new chronology for Liang Bua (Sutikna et al. 2016) is extremely important. Excavations between 2007 and 2014 allowed the opportunity to re-evaluate the observations made in the original excavations of 2001–2004. Fresh and larger excavations in different part of the cave floor showed that the stratigraphy was more complex than first realised. Most importantly, there was an erosional unconformity that had not been recognised. As a result of these fresh stratigraphic and dating analyses, all skeletal remains of *H. floresiensis* are now dated to approximately 100–60 ka ago, while stone artefacts attributed to this species range from about 190 ka to 50 ka in age (Sutikna et al. 2016: 368).

The latest assessment (Sutikna et al. 2018) is that *H. floresiensis, Stegodon floresensis insularis*), the giant marabou stork[12] (*Leptoptilos robustus*) and a vulture (*Trigonoceps* sp.) likely became extinct ca. 50 ka ago. There was also an important shift towards using red chert for making stone tools around 46 ka cal BP that may indicate the arrival of *H. sapiens*. Recent analysis of the cave sediments indicates the use of fire after 41 ka that is another probable indirect indicator that our species was present (Morley et al. 2017). It is not clear whether the hobbit was already extinct before our species arrived but on current evidence, it did not survive after it encountered *Homo sapiens*.

Humans on the High Seas

Wallacea deserves a special place in accounts of our success as a colonising species because it was the first region where the sea – a previous barrier to dispersal – became a corridor and a resource zone (see Chapter 1). Wallacea is where maritime economies were born: where people learnt not only to sail safely, navigate across open water and make return trips, but also to harvest the sea as a way of compensating for the lack of protein resources on many of its islands. The successful colonisation of Wallacea required four breakthroughs: the ability to construct sea-worthy, steerable sailing craft; the ability to navigate by day and night[13]; the ability to store fresh water when sailing between islands; and the ability to fish with lines, nets or harpoons. (Swimming was probably an optional extra.)

Emergent and nearly emergent islands

Sea level changes were not as dramatic across Wallacea as in Sunda, where huge areas of land were exposed, but were no less important. Falls in sea level affected each island differently. As seen in Chapter 6, Palawan was seven times larger than now when sea levels were at their lowest at –120m, but other islands with steep offshore gradients were affected much less. The site of Jermilai on Timor-Leste, for example, was still near enough to the coast 42,000 years ago for it to be used for offshore fishing, whereas coastal sites of that age on Palawan or between Borneo and Java are long submerged. Some islands merged with their neighbours and could therefore support larger populations of resources and people. Lower sea levels also meant more islands as some would emerge as sea levels fell, just as some would disappear when sea levels rose. With a fall in sea level of 45 metres, over 100 islands more than 5 sq km would have emerged, and these "emergent" islands would have provided stepping stones between other islands, thereby facilitating dispersal and colonisation (Kealy et al. 2016). (Additionally, there would have been "nearly emergent" islands that were near the surface: these would create different wave patterns, and perhaps affect the movements of fish, and attract predatory birds. For experienced sailors, these "nearly emergent islands" would provide additional navigational points.)

When considering the amount of land exposed when sea levels fell, it is necessary to consider tectonic factors in a region as active as Wallacea. Uplift rates vary considerably across the region, and unless these are considered, island size could be over-estimated. Timor and Sumba, for example, have been rising at an average rate of ca. 1.0–1.5 km/Ma (1 m–1.5 m/ka) (Lohman et al. 2011). As uplift rates cannot be calculated for all 2,000 islands of Wallacea, O'Connor et al. (2017b) applied an average rate of 0.5 m/ka across the region and produced a modified change in sea levels for the last 100,000 years (see Figure 7.4).

Island hopping and navigation

As Kealy and colleagues point out (2017), lower sea levels also mean higher islands, and that would increase visibility between them. As a rough guide, a person two metres tall at sea level can see ca. 5 km (3 miles); if sitting down, only 2.5 km. On the top of a hill 100 m high, the horizon is 36 km (22 miles) distant, but from a hill 150 m high, the horizon is 44 km (26 miles) away. As another example, a standing two-metre-tall person at sea level could see Alor Island, with a height of 1,717 m up to 150 km away.[14] To explore this issue further, Kealy and colleagues (2017) distinguished between relative and absolute visibility. Relative visibility was when two islands could be seen by

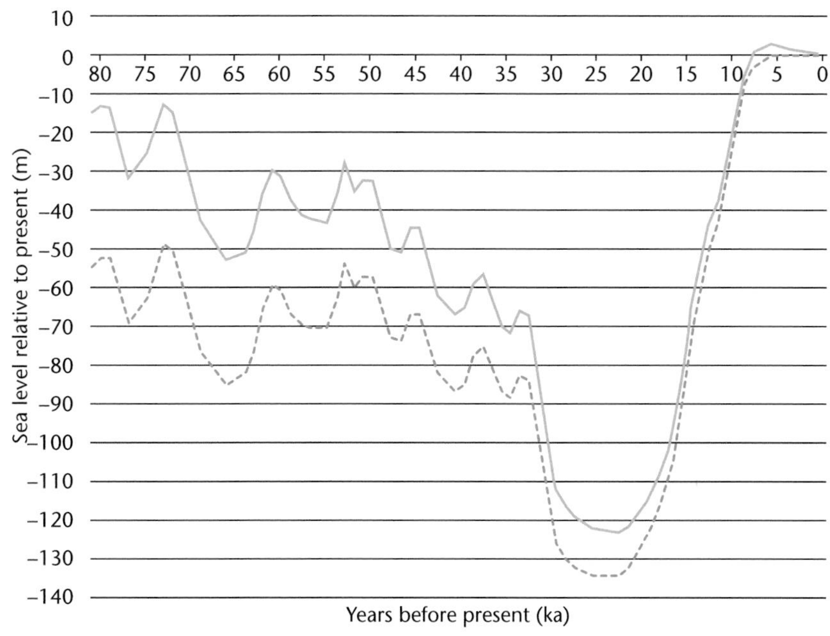

FIGURE 7.4 Sea levels and uplift rates in Wallacea

Because Wallacea is tectonically active, uplift rates need to be factored into reconstructions of Pleistocene coastlines at times of low sea level. Here, an average uplift rate of 0.5 m/kyr has been used.

Source: Kealy et al. 2018, Fig. 2.

someone in a boat; absolute visibility occurs when an island is visible from the shore of another. In an elegant paper, they calculated that most of Wallacea between 45 ka and 60 ka (including the emergent islands) would have been visible from land, and only a few parts would have been out of sight of land for people in a boat. Wallacea was thus an archipelago that was visually interconnected during the period of colonisation, and its colonisation was not therefore a matter of sailing into the unknown. Additionally, navigation would have been helped by indirect island indicators – such as "different cloud formations, smoke from bushfires, cloud reflections, wind and current directions, phosphorescence, altered wave patterns, and the presence of birds in addition to migratory movements of birds and marine species" (Kealy et al. 2017: 269). Risk would also have been reduced as settlers on an unsuitable island could readily move to neighbouring ones. This was particularly important because most of the smaller islands have a very impoverished fauna: for example, in modern times, the Talaud islands had only bats (including flying fox), rats and cuscus (Ono et al. 2009).

The colonisation of Wallacea

Several features of the early colonisation of Wallacea need consideration. The first is when this occurred. On most estimates, humans reached Sahul – the great conjoined landmass of Australia, New Guinea and Tasmania – around 50,000–55,000 years ago (see e.g. O'Connell et al. 2018), but others claim that the earliest evidence dates from ca. 60–65 ka (see Clarkson et al., 2017). In Wallacea, not a single site is as old as 47 ka, let alone 65 ka: as shown in Table 7.1, the oldest site so far investigated is Laili, Timor-Leste, at which the earliest dates are 43,283–44,631 cal. BP (Hawkins et al. 2017). (It should be noted, however, that this date came from spit 35 but artefacts continued to the base of the section in spit 40, so the oldest evidence will be earlier.) The discrepancy between Wallacea and Sahul regarding their initial colonisation is likely the outcome of two factors. The first is the small number of very small excavations and the very large number of islands in Wallacea, most of which are still unexplored archaeologically. The second factor is that radiocarbon (with an effective limit of around 50,000 years) is the primary method for dating Palaeolithic sites in Wallacea, whereas in Australia, the earliest dates have been obtained by methods such as uranium-series and OSL (see Hawkins et al. 2017). Likewise, as seen in Sunda and mainland Southeast Asia (Chapter 6), the dating of Lida Ajer, Sumatra (63–73 ka), and Tam Pa Ling was by OSL/TL, as was the dating of Callao Cave, Luzon (67 ka). It seems a reasonably safe prediction that the chronological crevasse between Wallacea and neighbouring regions (including northern Australia) will narrow when a wider range of dating techniques is used.

The second feature of Wallacea is that the occupation of several sites appears to have been highly intermittent (see Table 7.2). This pattern may simply be an artefact of inadequate sampling. Most of the excavations in Wallacea – and also Australia, see below – are tiny by European standards, and often just one or two test pits 1 metre square. As Langley et al. (2011: 197) point out, much of the research in Sahul (and Wallacea) is a matter of proceeding from "small holes to grand narratives". In one respect, the reliance on small excavations reflects a responsible attitude because cave deposits are a finite resource, and some should be left intact for future investigators. It also reflects the realities of fieldwork on remote parts of islands with a limited budget and short field season (O'Connor et al. 2010). Another consideration is the amount of biological and other data that is generated if – as at Jerimalai, Lene Hara and Tron Bon Lei – all cultural deposits are washed through a ≤2mm mesh: the post-excavation analysis of that residue will easily dominate research time between field seasons. What O'Connor and her team (2010) have shown is that the deposits at Lene Hara on Timor-Leste were stratigraphically complex and a complete account of the human occupation could only be obtained by integrating data from different stratigraphic sections (see Figure 7.5). An additional consideration is that some of the "missing" occupation units in cave sediments may be preserved as breccia along the walls of the cave or rock shelter, and these too need to be incorporated into site histories (O'Connor et al. 2017a).

TABLE 7.1 Palaeolithic sites in Wallacea and Sahul with C14 dates

No.	*Island (group)*	*Site name*	*C14/U-series age*	*Cal BP age 95% OxCal 4.2*	*Reference*
1	Salibabu (Talaud)	Leang Sarru	30,850 ± 340	33,860–35,030	Ono et al. (2009)
2	Morotai (Halmahera)	Daeo 2	13,930 ± 140	5,890–16,770	Bellwood et al. (1998)
3	Gebe (Halmahera)	Golo Cave	32,210 ± 320	35,000–36,350	Bellwood et al. (1998)
4	Sulawesi	Leang Timpuseng	Corrected U-Series 39,860–41,570		Aubert et al. (2014)
		Leang Burung 2	31,260 ± 330		Glover (1981)
		Leang Burung 2	31,2606 ± 320	35,2486 ± 420	Glover (1981); Aubert et al. (2014)
		Leang Sakapao	25,390 ± 310	28,380–29,810	Bulbeck et al. (2004)
5	Flores	Liang Bua	100–60 ka 190–50 ka 46 ka onwards		*H. floresiensis*; Sutikna et al. (2018) Stone tools associated with *H. floresiensis*; Sutikna et al. (2018) Presumed arrival of *Homo sapiens*; Sutikna et al. (2018)
6	Alor (Nusa Tengarra)	Tron Bon Lei	17,630 ± 70	21,030–21,580	Samper Carro et al. (2016)
7	Timor-Leste	Jerimalai		43,381–41,616	Test pit A, *Trochus* shell; Langley and O'Connor (2016)
		Jerimalai		42,475–41,125	Test Pit B spit 66; Langley and O'Connor (2016); Langley et al. (2016)
	Timor-Leste	Lene Hara	37,267 ± 453 BP	42,475–41,125	Langley and O'Connor (2016), test pit B
		Lene Hara	34,650 ± 630	36,970–40,270	O'Connor et al. (2002)
		Lene Hara		35,192–33,896	Langley and O'Connor (2016), test pit F; base not reached
		Matja Kuru 1		16,355–15,566	Langley and O'Connor (2016)
		Matja Kuru 2		36,268–34,649	Langley and O'Connor (2016); spit 41

(*Continued*)

TABLE 7.1 (Cont.)

No.	*Island (group)*	*Site name*	*C14/U-series age*	*Cal BP age 95% OxCal 4.2*	*Reference*
		Matja Kuru 2		35,882–34,575	Langley and O'Connor (2016); O'Connor et al. (2014); spit 44
		Matja Kuru 2		36,866–35,285	Langley and O'Connor (2016), spit 47
		Bui Ceri Uato	26,520 ± 340	29,560–30,970	Selimiotis (2006); Hawkins et al. (2017)
		Uai Bobo 2	13,400 ± 520	? 14,380–17,600?	Glover (1969); Hawkins et al. (2017)
8	Timor-Leste	Laili		43,283–44,631 cal BP	Hawkins et al. (2017); spit 35. Occupation continues to spit 40
9	Roti (Timor)	Lua Meko	24,420 ± 250	27,680–28,600	Mahirta (2003)
		Lua Manggetek	13,390 ± 430	14,090–16,800	Mahirta (2003)
	Sahul				
10	Madjedbebe			60–65 ka	Clarkson et al. (2017)
11	Carpenter's Gap			43.0–45.7 ka (44.3 ka)	O'Connell and Allen (2015)
12	Riwi			43.1–46.9 ka (45.0 ka)	O'Connell and Allen (2015)
13	Ivane Vilakauv			43.1–49.1 ka (46.1 ka)	O'Connell and Allen (2015)

The third feature is that the earliest dates for human settlement in Wallacea are in the south, from sites such as Laili, Jerimalai and Lene Hara on Timor-Leste. This pattern need not indicate that the southern part of Wallacea was occupied before the northern part. Because few sites have been investigated in Wallacea, it would be premature to conclude that current data provide an accurate indication of settlement history. It seems a safe prediction that sites as old as Laili or Jerimalai will be found in due course in northern Wallacea.

The fourth impression to counter is that the material culture of these islanders was "simple" compared with, for example, the European Upper Palaeolithic. At first sight, this is true: the lithics are generally very simple and unstandardised, and show little change over time (see e.g. Marwick et al. 2016 for Jerimalai). There is also little evidence in the way of jewellery, bone or shell. Two points can be raised: one is that much of the material culture was probably organic, from wood, plants and their fibres, and has not survived. The second is the size of the excavations. A complex material culture is most unlikely to emerge from a test pit one

TABLE 7.2 Intermittent occupation records from Wallacea

Island	*Site*	*Dates BP*	*Reference*
Talauds	Leang Sarru	Layer 3: 35–32,000 BP Layer 2 lower: 21–18,000 BP Layer 2 upper: 10–8,000 BP	Ono et al. (2009)
Gebe	Golo	36,194 ± 457 cal BP 21–19 ka 12–10 ka 7–3 ka	Szabó et al. (2007); O'Connor et al. (2010)
Timor-Leste	Jerimalai	40–38 ka 17–9 ka 6.5 ka – present	O'Connor et al. (2011)
Alor	Tron Bon Lei pit B shelter	21 – 18.9 ka 12.5–7.5 ka 4–3 ka	Samper Carro et al. (2017)
Timor-Leste	Matja Kuru 2	36–30 cal BP 13–9.5 ka cal BP 4,000 onwards, cal BP	Samper Carro et al. (2016) O'Connor and Aplin (2007)
Timor-Leste	Laili	27,000 and 44,631 cal BP 18,500 and 22,500 cal BP 11,161 and 17,468 cal BP	Hawkins et al. (2017)

metre square; unless one is very lucky, rare items (such as carved bone objects) are more likely to be found in a large excavation than from one the size of a small table. I suspect if excavations of European Upper Palaeolithic sites had been limited to a couple of one metre square test pits from a few sites, the results would not be very different from what we see in Wallacea. It is worth noting the rare exceptions. There is, for example, from the rock shelter of Matja Kuru 2, Timor-Leste (O'Connor et al. 2014: 111) what appears to be (see Figure 7.6):

> the broken butt of a formerly hafted projectile point, and it preserves evidence of a complex hafting mechanism including insertion into a shaped or split shaft, a complex pattern of binding including lateral stabilization of the cordage within a bilateral series of notches, and the application of mastic at several stages in the hafting process.

Radiocarbon samples taken above and below it produced dates of dates of 36,268–34,649 cal. BP above and 36,866–35,285 cal. BP below, so its age is tightly constrained at ca. 35 ka cal. BP. This piece is at least as complex as any comparable European Upper Palaeolithic bone point. Shellfish hooks are rare, but one from Jerimalai was dated to between ca. 23,000 and 16,000 cal. BP and another from

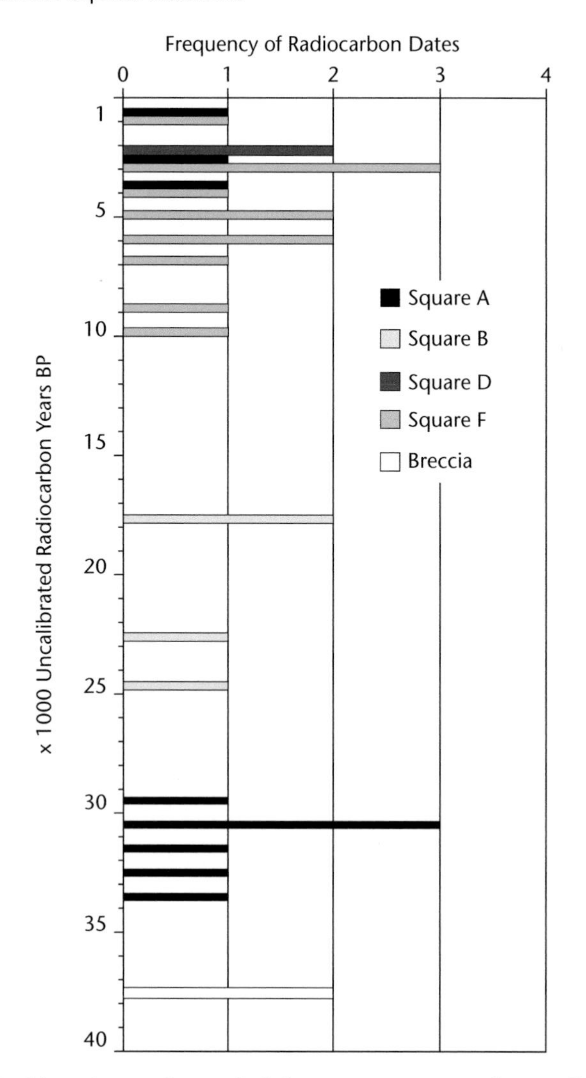

FIGURE 7.5 Problems in sampling rock shelter occupation records in Wallacea

A complete occupation record is unlikely to emerge from the excavation of a single test pit, and several may be needed to sample the entire history of occupation. The example shown here is from Lene Hara, East Timor.

Source: O'Connor et al. 2010, Fig. 9.

Lene Hara was been directly AMS dated to ca. 11,000 cal. BP (O'Connor et al. 2011). Excavation at the rock shelters of Jerimalai, Lene Hara, and Matja Kuru 1 and 2 produced almost 500 small (ca. 9 mm) shell beads of *Oliva* spp., of which the earliest from Jerimalai was directly dated to 38,246–36,136 cal. BP (ca. 37,000 cal. BP) (Langley and O'Connor 2016), making it the oldest in Southeast Asia, and

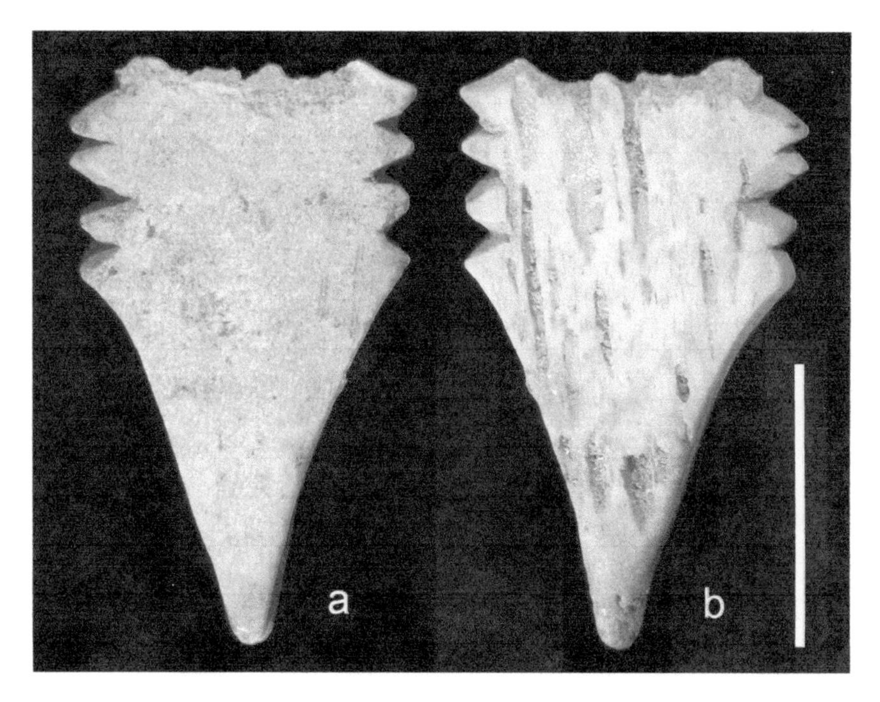

FIGURE 7.6 The broken, but elaborate, 35,000-year-old projectile point from Matja Kuru 2, Timor Leste
Source: O'Connor et al. 2014, Fig. 2.

comparable in age to European examples. It is worth noting that these beads were only recovered because all excavated deposits were washed through a 1.5 mm mesh (O'Connor 2007: 528) – doubtless many were missed in other less thorough excavations.

As shown below, shell was also incorporated into social life and ritual used for decoration and ornament.

Subsistence in Pleistocene Wallacea

Apart from Sulawesi, which is the largest Wallacean island and the only one with a large, diverse fauna that includes large animals such as pigs and monkeys (Dennell et al. 2013), the islands of Wallacea have a highly impoverished terrestrial fauna that is usually limited to different types of bats, rats and lizards. Some are moderately large, like flying foxes or giant rats (up to 6 kg, or the size of a moderate domestic cat; O'Connor 2007); a few are large and nasty, like the Komodo dragon, nowadays found on Flores and a few other islands. Otherwise, any human group on most Wallacean islands would probably starve if they tried to live off the local terrestrial fauna. It is perhaps not surprising that there were no animal bones from Leang Serru, or in the Pleistocene levels of Golo Cave.

At Jerimalai on Timor-Leste, rodents, bats, birds, lizards and snakes form a minute proportion of the faunal assemblages (O'Connor et al. 2011), and they are equally negligible at Tron Bon Lei on Alor Island (Samper Carro et al. 2017). On the other hand, a surprisingly diverse range of terrestrial resources were utilised at Laili, Timor-Leste (Hawkins et al. 2017).

The key to survival in Wallacea was to harvest the sea, and in particular, shellfish and fish. Shellfish are ubiquitous in coastal and near-coastal sites across Wallacea, whether in the Talaud islands, as at Leang Sarru on Salebabu (Ono et al. 2009), Golo Cave on Gebe Island (Szabó et al. 2007), at Jerimalai (O'Connor 2007), Lene Hara (O'Connor et al. 2002), or Tron Bon Lei (O'Connor et al. 2002). Strangely, neither Leang Sarru nor Golo had any evidence of fishing: this might indicate local soil conditions, that fishing sites have yet to be found, or it could be a genuine absence. The clearest evidence for fishing comes from Jerimalai and Tron Bon Lei, not least because all cultural deposits were washed through a mesh of only 1.5 mm. At Jerimalai, a large assemblage of ca. 39,000 fish bones shows that from 38–42 ka, 15 fish taxa were caught, and that half were coastal, and half pelagic (O'Connor et al. 2011). Pelagic fish are "those that inhabit both shallow inshore and oceanic offshore regions. They are not found exclusively in the deep ocean below 200m" (O'Connor and Ono 2013: 885). Many of those from Jerimalai are Scombrids, a class of fish that include tuna and mackerel. Yellowfin tuna grows up to 180 kg but the Jerimalai specimens are much smaller, and probably juvenile. Although these might have been caught by rod and line, it is perhaps more likely that they were caught in nets. Tuna can swim at speeds up to 80 km an hour, but they are also attracted to floating logs or even boats, which makes them much easier to catch. O'Connor and colleagues (2011) proposed that the Jerimalai evidence showed that the inhabitants were engaging in deep-sea fishing – or at least fishing for deep sea species that might also be found in inshore waters. Although this claim was challenged (see Anderson 2013), her argument is persuasive (see O'Connor and Ono 2013). At Tron Bon Lei, an assemblage of ca. 50,000 bones was found (Samper Carro et al. 2017). Of these, 57% were identified as fish, and only 1.85% as rats, bats, lizards, snakes, frogs, birds, and the remainder (41%) were unidentifiable, largely because of fragmentation and carbonate encrustation. Although only 3.5% of the fish bones could be identified to family, the proportion of offshore fish was higher in the earliest phases, as at Jerimalai.

What is currently missing from Wallacea is any indication of plants as food or materials, despite the intensive sieving at Jerimalai, Lene Hara, Tron Bon Lei and other sites. These must have been used, as at Niah Cave, but so far any traces of them have proved elusive.

Art, ritual and social life

There are several indications of a socially constructed world in Wallacea. Brumm and colleagues (2017) reported beads, perforated bone pendants, incised stone artefacts and ochre from the cave of Leang Bulu Bettue in Sulawesi dating from ca. 30,000 to 22,000 BP. Haematite nodules and ochre-smeared stone tools were found at the rock shelters of Leang Gurung 2 and Leang Sakapao (see Table 7.1), which were radiocarbon dated to ca. 35 ka and 28 ka respectively. The most dramatic and visually striking example of a socially constructed world is the cave art in the Maros karst region of southern Sulawesi that has recently been dated in excess of 30,000 years. By applying uranium-series dating to flowstones associated with paintings (Aubert et al. 2014, 2107), a hand stencil at Leang Timpuseng was dated to a minimum age of 39.9 ka, thus making it is the oldest example worldwide until even older examples were found on Borneo (Aubert et al. 2018; see Chapter 6). A nearby painting of a babirusa ("pig-deer") was dated to at least 35.4 ka, and a second animal painting (probably of another pig) at another site, Leang Barugayya 2, has a minimum age of 35.7 ka. These are among the oldest examples of figurative art worldwide.

There are two massive implications of these developments. The first is that the cave art on Sulawesi is part of the same tradition as recently demonstrated on Borneo (see Chapter 6), so the Southeast Asian cave art that is already known from Sunda and Southeast Asia (see Taçon et al. 2014) may prove to be far older and widespread than previously suspected. As noted in the previous chapter, there may be a regional tradition of art dating back to at least 40 ka and comparable to that in Europe. The second is that if figurative art was part of human culture 40,000 years ago, we might expect similar examples to be found between western Europe and Southeast Asia. Either way, Palaeolithic cave art is not unique to western Europe.

The cave art on Sulawesi is the most dramatic example of art from Palaeolithic Wallacea, but there are more modest indications of using pigments. Langley and colleagues (2016) report ochre-stained *Nautilus* shells from the earliest layers at Jerimalai. These show traces of drilling, flaking and grinding. By widening their study to include younger examples from Late Pleistocene and Holocene contexts, they have shown that people were incorporating coastal resources such as shellfish into their social systems. Another graphic example of the importance of marine resources to the ideology of these islanders comes from a burial at Tron Bon Lei on Alor (O'Connor et al. 2017c). Here, five fish hooks and a perforated sea shell were buried next to a burial ca. 12,000 years old. This association of fishing equipment and funerary rites illustrates "cosmological status of fishing in this island environment, probably because of the role that fishing played during daily life for the inhabitants of an island that largely lacked other sources of protein" (O'Connor et al. 2107b).

Sailing to Sahul

Irrespective of when humans first reached Sahul, they had to arrive by boat after navigating across perhaps 40–70 km of open sea and also be able to make return voyages to encourage other potential migrants. There have been long discussions over the route(s) they may have taken. The main two were proposed by the American biogeographer Joseph Birdsell (1977), and most discussions of this topic refer to his landmark paper. In this, he took into account distance between islands, their height (and thus visibility) and width (wide islands being less easily missed). He proposed two main routes: a southern one from Java and then eastwards via Flores and Timor to northern Australia, and a northern route from Kalimatan to Sulawesi and then to Halmahera or Ceram, with a landing in eastern New Guinea. Two other routes have been proposed: one by Sondaar (1989; cited by Kealy et al. 2017), a "giant rat route" from Taiwan through the Philippines to Halmahera and New Guinea, and another by Morwood and Osterzee (2007), from northern Borneo to the southern Philippines.

None of these can be either proven or falsified, and certainly the paucity of dated sites in Wallacea nullifies any attempt to infer which route may have been used. We can, however, discuss which may have been the easiest or safest. In practice, this means identifying routes with the shortest distances between islands, and between the last island and the nearest landing point. These distances vary according to upon when colonisation is thought to have taken place, because they depend upon the sea level at the time of sailing. As discussed earlier, Kealy and colleagues have shown that inter-visibility between islands would increase when sea levels fell – because islands became higher – and "emergent islands" would also reduce distances between present-day islands. Most researchers tend to favour Birdsell's northern route from Sulawesi to western New Guinea as involving the shortest distances between islands, especially if one includes "emergent islands" (see Kealy et al. 2017, 2018). The southern route has its advocates: Bird and colleagues (2018) have recently suggested that the islands of Timor and Roti were favourable starting points for reaching Australia. They take into account "emergent islands" off the north coast of Australia and point out that many of these would have been visible from the tops of the highest peaks on Timor and Roti. On Timor, the highest peak close to the coast has an elevation of 1,447 m, from which one could, on a fine clear day, see 135 km, and the nearest emergent island on the Sahul Banks was only 87 km distant. In their opinion, the emergent islands in the Sahul Banks could have been reached after 4–7 days – a bold undertaking, but not an impossible one. In all likelihood, Sahul was colonised by numerous groups using both the northern and southern routes (Norman et al. 2018).

How many, how few?

Another area of speculation concerns the numbers of people that might have colonised Sahul. One estimate is John Calaby's (1976) imaginative suggestion that it may have been accomplished by a young pregnant woman who was swept out to sea on a log and managed to land in Sahul. (Presumably, she had twins, one girl and a boy, and a very broad mind about incest.) Most researchers now favour founding populations of several boat loads totalling perhaps 1,000–2,000 people for mainland Australia but with no subsequent immigration (Williams 2013).[15] On current thinking, Sahul was not colonised by accident as an unplanned one-way trip, but a result of planned migration involving numerous groups and numerous return voyages (Bird et al. 2019). In a recent simulation study, Bradshaw et al. (2019) suggest a founding population of between 1,300 and 1,550 individuals who could have arrived after one voyage, or (perhaps more likely) through several voyages of ca. 130 people over 700–900 years.

A very brief look at Sahul

Although this book is primarily concerned with Asia, it is worth a brief look at some of the earliest evidence from Sahul because it provides a fascinating perspective on the colonisation of Eurasia.

Several sites across Sahul indicate the arrival of *H. sapiens* ca. 50–55 ka. These include Narwala Gabarnmang, Carpenter's Gap and Riwi in the north of Australia, Boodie Cave in the west, Devils Lair in the southwest, Menindee in the southeast, and Warratyi in the Central Australian Desert as well as the Ivane valley of New Guinea (see Figure 7.7; and O'Connell et al. 2018). As Boodie Cave, Riwi and Menindee were dated by both C14 and OSL, it is unlikely that these dates reflect the limits of C14 dating. There is a long-running debate over the dating of the earliest assemblages from the rock shelter of Madjedbebe (formerly Malakunanja II) in north Australia, which were recently dated to ca. 60–65 ka by Clarkson et al. (2017) and are thus considerably older than other known sites in Sahul. It is of course possible that Madjedbebe is an outlier and sites between 55 ka and 60 ka have not yet been found in Australia. O'Connell and colleagues (2018) contest the dating of Madjedbebe by pointing out the likely problems of post-depositional disturbance at this site, particularly from termites. These, they argue, form their mounds from sub-surface burrowing, and when their galleries collapse, stone objects (including artefacts) would be displaced downwards into much older strata,[16] as pointed out many years ago by McBrearty (1990) for African contexts. Minimally, they suggest, confirmation of this dating "would require evidence from another site in Sahul in a more secure geomorphic setting". Until these issues have been addressed, it seems prudent to place Madjedbebe in a suspense account.

Leaving Madjedbebe aside, humans colonised a truly impressive range of novel environments in Australia after their arrival. These included high-altitude tropical forest-grassland in New Guinea (Summerhayes et al. 2010; Summerhayes and Ford 2014), subtropical savannah at Narwala Gabarnmang, Carpenter's Gap and Riwi, semi-arid woodland and grassland at Boodie Cave (Veth et al. 2016), Warratyi (Hamm et al. 2016) and Menindee, and temperate forest at Devils Lair (O'Connell et al. 2018).

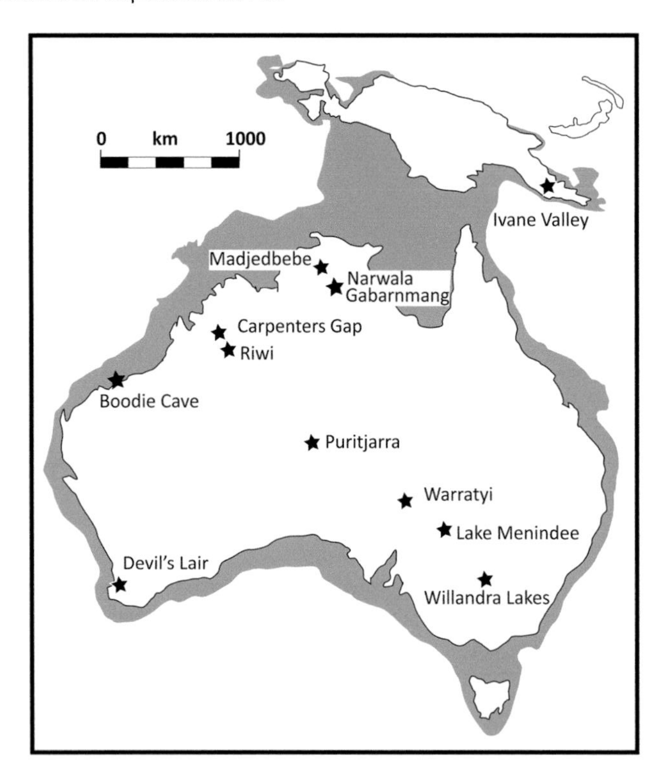

FIGURE 7.7 Early human sites on Pleistocene Sahul

The shaded area shows sea levels 125 m below present levels in the LGM. Based on data in Habgood and Franklin (2008) and O'Connell et al. (2018).

Ivane Valley (43–49 ka); Madjedbebe (54–?65 ka); Narwala Gabarnmang (44.2–49.6 (47.0) ka); Carpenter's Gap (44–49 ka); Riwi (45–46 ka); Boodie Cave (51.1–46.2 ka); Devil's Lair (42.9–48.1 (45.4) ka); Puritjarra(34.6 ± 1.6 ka [TL]); Warratyi (49.2–46.3 ka); Menindee (42.9–49.0 (45.9) ka); Willandra (40.7–44.5 (42.6) ka).

Sources: O'Connell and Allen 2015, except for Boodie Cave (Veth et al. 2016); Puritjarra (Smith et al. 1997); and Warratyi (Hamm et al. 2016).

All this was happening while, at the opposite end of the inhabited world, Neandertals were the sole occupants of western Europe. By 45,000 years ago, the first inhabitants of Sahul had colonised a previously alien continent, learnt how to utilise a completely new fauna and flora, and adapted to some of the harshest environments that had ever been encountered. All this was accomplished with a simple lithic tool-kit, but, as remarked by Balme and O'Connor (2014), they were using a simple tool-kit to do complex things. In short, they were "fully modern" in their cognitive abilities, their flexibility, ingenuity and adaptability. The earliest Australian evidence brings home the point that the Upper Palaeolithic of western Europe, which denotes the colonisation of Europe by our species, is but a footnote in the deep history of the colonisation of Eurasia and Australia.

Notes

1 Alfred Russell Wallace (1823–1913) is often regarded as the father of biogeography and was an outstanding naturalist, explorer in Amazonia and Southeast Asia, biologist and anthropologist. Independently of Darwin, he developed a theory of natural selection, which was presented jointly as an idea in London in 1859. His classic book *The Malay Archipelago*, first published in 1869, is still in print. He was a prolific author: 22 books, over 700 scientific papers (including 191 in *Nature*!).
2 Wallace (1860: 174) observed "South America and Africa, separated by the Atlantic, do not differ so widely as Asia and Australia" (cited by Lohman et al. 2011).
3 A cuscus is an arboreal marsupial most closely related to the Australian possum. Body weight is only 2–7 kg, so they are hardly a sustainable protein source for a human group.
4 Richard Lydekker (1849–1915): English geologist, palaeontologist and writer; member of the Geological Survey of India (1874–1882), where he studied Siwalik fossils. Lydekker's Line (1895) defined the easternmost extension of Oriental fauna into the Australian faunal zone, whereas Wallace's line denoted the westernmost extension of marsupials.
5 Seven lines have been proposed for Wallacea, but those by Müller, Murray, Sclater and Weber are now historical curios. For the historically inclined, see Simpson (1977).
6 Thomas Huxley (1825–1895): also known as "Darwin's bulldog" for championing Darwin's theory of natural selection; biologist and comparative anatomist; 1846, member of HMS Rattlesnake's scientific voyage to New Guinea and Australia. Huxley's Line (1868) includes the Philippines (excluding Palawan) as part of Wallacea because of its unique fauna.
7 Tektites are small glassy objects that resulted from the impact of a meteor. The one found here would have resulted from a massive impact that hit Southeast Asia around 780 ka ago; the fall-out covered one tenth of the earth's surface (see Fiske et al. 1999; Schnetzler and McHone 1996).
8 This type of context is problematic because naturally rolled stone can often look like stone artefacts, although most of the ones shown from Talepu look convincing. For that reason, the least ambiguous evidence are the three artefacts from the clay deposit of Unit E, as this was deposited under gentle conditions under which stone could not be flaked naturally.
9 There were stegodons on Timor ca. 130 ka, and also on Flores, Sangihe and Sulawesi, but these were likely extinct when humans arrived (Louys et al. 2016).
10 Bednarik (2003) challenged this notion by building a raft with the type of simple tools found in Early Palaeolithic assemblages. A crew of five later sailed it from Timor to Australia. However, the fact that *H. sapiens* today can build a raft with simple stone tools and sail it to a known destination does not necessarily imply that *H. erectus* had the cognitive and co-operative skills (or the impulse) to build a raft and sail into the unknown.
11 I remain unconvinced by suggestions that the hobbit is a pygmy population of *H. sapiens* (Jacob et al. 2006; Perry and Domini 2009), or a population afflicted by Laron's disease (Hershkovitz et al. 2007).
12 The giant marabou stork on Flores is appropriately named as it was up to 1.8 m tall and would have towered over the 1.0 m-high hobbit. As this stork weighted up to 16 kg, it was probably flightless, and fed off carrion, immature Komodo dragons and giant rats (see Meijer and Due Awe 2010).
13 Particularly important because of the abrupt transition between day and night in that part of the world.
14 The formula is $d = 3.57\sqrt{h}$, where d = distance (km), and h = height in metres of observer. This ignores refraction, but is a useful rule of thumb for estimating distance to horizon.

15 Williams (2013) notes that New Guinea and the flooded coastal shelf of northern Australia may have had different population histories.
16 The Supplementary Information of the paper by O'Connell et al. (2018) contains a detailed critique of the dating and probable site formation processes at Madjedbebe.

References

Anderson, A. (2013) The antiquity of sustained offshore fishing. *Antiquity*, 87: 879–895.

Argue, D., Donlon, D., Groves, C. and Wright, R. (2006) *Homo floresiensis*: Microcephalic, pygmoid, *Australopithecus*, or *Homo*? *Journal of Human Evolution*, 51: 360–374.

Aubert, M., Brumm, A., Ramli, M., Sutnika, T., Wahyu Saptomo, E., Hakim, B., Morwood, M.J., Bergh, G.D. van den, Kinsley, L. and Dosseto, A. (2014) Pleistocene cave art from Sulawesi, Indonesia. *Nature*, 514: 223–227.

Aubert, M., Brumm, A. and Taçon, P.S.C. (2017) The timing and nature of human colonization of Southeast Asia in the Late Pleistocene: a rock art perspective. *Current Anthropology*, 58(Supplement 17): S553–S566.

Balme, J. and O'Connor, S. (2014) Early modern humans in Island Southeast Asia and Sahul: adaptive and creative societies with simple lithic industries. In R. Dennell and M. Porr (eds) *Southern Asia, Australia and the Search for Modern Human Origins*, Cambridge: Cambridge University Press, pp. 164–174.

Bednarik, R.G. (2003) Seafaring in the Pleistocene. *Cambridge Archaeological Journal*, 13: 41–66.

Bellwood, P., Nitihaminoto, G., Irwin, G., Gunadi, A.W. and Tanudirjo, D. (1998) 35,000 years of prehistory in the northern Moluccas. In G.-J. Bartstra (ed.) *Bird's Head Approaches: Irian Jaya Studies – A Programme for Interdisciplinary Research*, Rotterdam, Netherlands: A. A. Balkema, pp. 233–275.

Bergh, G. van den, Li, B., Brumm, A., Grün, R., Yurnaldi, D., Moore, M.W., Kurniawan, I., Setiawan, R., Aziz, F., Roberts, R.G., Suyono, Storey, M., Setibudi, E. and Morwood, M. (2016) Earliest hominin occupation of Sulawesi, Indonesia. *Nature*, 529: 208–211.

Bird, M.I., Beaman, R.J., Condie, S.A., Cooper, A., Ulm, S. and Veth, P. (2018) Palaeogeography and voyage modelling indicates early human colonization of Australia was likely from Timor-Roti. *Quaternary Science Reviews*, 191: 431–439.

Bird, M.I., Condie, S.A., O'Connor, S., O'Grady, D., Reepmeyer, C., Ulm, S., Zega, M., Saltré, F. and Bradshaw, C.J.A. (2019) Early human settlement of Sahul was not an accident. *Scientific Reports*, 9: 8220, doi:10.1038/s41598-019-42946-9 1.

Birdsell, J.B. (1977) The recalibration of a paradigm for the first peopling of greater Australia. In J. Allen, J. Golson and R. Jones (eds) *Sunda and Sahul: Prehistoric Studies in Southeast Asia, Melanesia and Australia*, London: Academic Press, pp. 113–167.

Bradshaw, C.J.A., Ulm, S., Williams, A.N., Bird, M.I., Roberts, R.G., Jacobs, Z., Laviano, F., Weyrich, F., Friedrich, T., Norman, K. and Saltré, F. (2019) Minimum founding populations for the first peopling of Sahul. *Nature Ecology and Evolution*, doi:10.1038/s41559-019-0902-6.

Brown, P., Sutkina, T., Morwood, M.J., Soejono, R.P., Jatniko and Wahyu Saptomo, E. (2004) A new small-bodied hominin from the Late Pleistocene of Flores, Indonesia. *Nature*, 431: 1055–1068.

Brumm, A., Bergh, G.D. van den, Storey, M., Kurniawans, I., Alloway, B.V., Setiawan, R. et al. (2016) Age and context of the oldest known hominin fossils from Flores. *Nature*, 534: 249–253.

Brumm, A., Jensen, G.M., Bergh, G.D. van den, Morwood, M.J., Kurniawan, I., Aziz, F. and Storey, M. (2010) Hominins on Flores, Indonesia, by one million years ago. *Nature*, 464: 748–752.

Brumm, A., Langley, M.C., Moore, M.W., Hakim, B., Ramli, M. et al. (2017) Early human symbolic behavior in the Late Pleistocene of Wallacea. *Proceedings of the National Academy of Sciences USA*, 114: 4105–4110.

Bulbeck, D., Sumantri, I. and Hiscock, P. (2004) Leang Sakapao 1, a second dated Pleistocene site from South Sulawesi, Indonesia. *Modern Quaternary Research in Southeast Asia: Quaternary Research in Indonesia*, 18: 111–128. Leiden: A. A. Balkema Publishers.

Calaby, J.H. (1976) Some biogeographical factors relevant to the Pleistocene movement of man in Australasia. In R.L. Kirk and A.G. Thorne (eds) *The Origin of Australians*, Canberra: Australian Institute of Aboriginal Studies, pp. 23–28.

Clarkson, C., Jacobs, Z., Marwick, B., Fullagar, R., Wallis, L. et al. (2017) Human occupation of northern Australia by 65,000 years ago. *Nature*, 547: 306–310.

Dennell, R.W., Louys, J.L., O'Regan, H.J. and Wilkinson, D.M. (2013) The origins and persistence of *Homo floresiensis* on Flores: biogeographical and ecological perspectives. *Quaternary Science Reviews*, 96: 98–107.

Détroit, F., Mijares, A.S., Corny, J., Daver, G., Znolli, C., Dizon, E., Robles, E., Grün, R. and Piper, P.J. (2019) A new species of *Homo* from the Late Pleistocene of the Philippines. *Nature*, 568: 181–186.

Esselstyn, J.A. and Brown, R.M. (2009) The role of repeated sea-level fluctuations in the generation of shrew (Soricidae: *Crocidura*) diversity in the Philippine Archipelago. *Molecular Phylogenetics and Evolution*, 53: 171–181.

Fiske, P.S., Schnetzler, C.C., McHone, J., Chanthva Vaitchith, K.K., Homsombath, I., Phouthakayalat, T., Khenthavong, B., and Xuan, P.T. (1999) Layered tektites of Southeast Asia: field studies in central Laos and Vietnam. *Meteoritics and Planetary Science*, 34: 757–761.

Glover, I. (1969) Radiocarbon dates from Portuguese Timor. *Archaeology & Physical Anthropology in Oceania*, 4(2): 107–112.

Glover, I.C. (1981) Leang Burung 2: An Upper Palaeolithic rock shelter in South Sulawesi, Indonesia. *Modern Quaternary Research in Southeast Asia*, 6: 1–38.

Habgood, P.J. and Franklin, N.R. (2008) The revolution that didn't arrive: a review of Pleistocene Sahul. *Journal of Human Evolution*, 55: 187–222.

Hamm, G., Mitchell, P., Arnold, L.J., Prideaux, G.J. and Questiaux, D. (2016) Cultural innovation and megafauna interaction in the early settlement of arid Australia. *Nature*, 539: 280–283.

Hawkins, S., O'Connor, S., Maloney, T.R., Litster, M., Kealy, S., Fenner, J.N., Aplin, K., Boulanger, C., Brockwell, S., Willan, R., Piotto, E. and Louys, J. (2017) Oldest human occupation of Wallacea at Laili Cave, Timor-Leste, shows broad-spectrum foraging responses to Late Pleistocene environments. *Quaternary Science Reviews*, 171: 58–72.

Hershkovitz, I., Kornreich, L. and Laron, Z. (2007) Comparative skeletal features between *Homo floresiensis* and patients with primary growth hormone insensitivity (Laron Syndrome). *American Journal of Physical Anthropology*, 134: 198–208.

Ingicco, T., Bergh, G.D. van den, Jago-on, C., Bahain, J.-J., Chacón, M.G., Forestier, H., King, C. et al. (2018) Earliest known hominin activity in the Philippines by 709 thousand years ago. *Nature*, 557: 233–237.

Jacob, T., Indriati, E., Soejono, R.P., Hsu, K., Frayer, D.W., Eckhardt, R.B., Kuperavage, A. J., Thorne, A. and Henneberg, M. (2006) Pygmoid Australomelanesian

Homo sapiens skeletal remains from Liang Bua, Flores: population affinities and pathological abnormalities. *Proceedings of the National Academy of Sciences USA*, 103: 13421–13426.

Jungers, W.L., Larson, S.G., Harcourt-Smith, W., Morwood, M.J., Sutkina, T., Due Awe, Rokhus and Djubianto, T. (2009) Descriptions of the lower limb skeleton of *Homo floresiensis*. *Journal of Human Evolution*, 57: 548–554.

Kaifu, Y. and Fujita, M. (2012) Fossil record of early modern humans in East Asia. *Quaternary International*, 248: 2–11.

Kealy, S., Louys, J. and O'Connor, S. (2016) Islands under the sea: a review of early modern human dispersal routes and migration hypotheses through Wallacea. *Journal of Island and Coastal Archaeology*, 11: 364–384, doi:10.1080/15564894.2015.1119218.

Kealy, S., Louys, J. and O'Connor, S. (2017) Reconstructing palaeogeography and inter-island visibility in the Wallacean Archipelago during the likely period of Sahul colonization, 65–45000 years ago. *Archaeological Prospection*, 24: 259–272.

Kealy, S., Louys, J. and O'Connor, S. (2018) Least-cost pathway models indicate northern human dispersal from Sunda to Sahul. *Journal of Human Evolution*, 125: 59–70.

Keates, S. and Bartstra, J. (2015) Island migration of early modern *Homo sapiens* in Southeast Asia: the artifacts from the Walanae depression, Sulawesi, Indonesia. *Palaeohistoria*, 15: 19–30.

Langley, M.C. and O'Connor, S. (2016) An enduring shell artefact tradition from Timor-Leste: *Oliva* bead production from the Pleistocene to Late Holocene at Jerimalai, Lene Hara, and Matja Kuru 1 and 2. *PLoS ONE*, 11(8): e0161071, doi:10.1371/journal.pone.0161071.

Langley, M.C., Clarkson, C. and Ulm, S. (2011) From small holes to grand narratives: the impact of taphonomy and sample size on the modernity debate in Australia and New Guinea. *Journal of Human Evolution*, 61: 197–208.

Langley, M.C., O'Connor, S. and Piotto, E. (2016) 42,000-year-old worked and pigment-stained *Nautilus* shell from Jerimalai (Timor-Leste): evidence for an early coastal adaptation in ISEA. *Journal of Human Evolution*, 97: 1–16.

Lohman, D.J., Bruyn, M. de, Page, T., Rintelen, K. von, Hall, R., Ng, P.K.L., Shih, H.-T., Carvalho, G.R. and Rintelen, T. von (2011) Biogeography of the Indo-Australian Archipelago. *Annual Review of Ecology, Evolution, and Systematics*, 42: 205–226.

Louys, J., Price, G.J. and O'Connor, S. (2016) Direct dating of Pleistocene Stegodon from Timor Island, East Nusa Tenggara. *PeerJ*, 4: e1788, doi:10.7717/peerj.1788.

Mahirta, M. (2003) *Human Occupation on Roti and Sawu Islands, Nusa Tenggara Timur*. Ph.D. Thesis. Canberra: Department of Archaeology and Anthropology, Australian National University.

Marwick, B., Chris C., O'Connor, S. and Collins, S. (2016) Early modern human lithic technology from Jerimalai, East Timor. *Journal of Human Evolution*, 101: 45–61.

McBrearty, S. (1990) Consider the humble termite: termites as agents of post-depositional disturbance at African archaeological sites. *Journal of Archaeological Science*, 17: 111–143.

Meijer, H.J.M. and Due Awe, R. (2010) A new species of giant marabou stork (Aves: Ciconiiformes) from the Pleistocene of Liang Bua, Flores (Indonesia). *Zoological Journal of the Linnean Society*, 160: 707–724.

Mijares, A.S.B., Détroit, F., Piper, P., Grün, R., Bellwood, P. et al. (2010) New evidence for a 67,000-year-old human presence at Callao Cave, Luzon, Philippines. *Journal of Human Evolution*, 59: 123–132.

Morley, M.W., Goldberg, P., Sutikna, T., Tocheri, M.W., Prinsloo, L.C., Jatmiko, Wahyu Saptomo, E., Wasisto, S. and Roberts, R.G. (2017) Initial micromorphological results from

Liang Bua, Flores (Indonesia): site formation processes and hominin activities at the type locality of *Homo floresiensis*. *Journal of Archaeological Science*, 77: 125–142.
Morwood, M, and Oosterzee, P. van (2007) *A New Human: The Startling Discovery and Strange Story of the "Hobbits" of Flores, Indonesia*. New York: Smithsonian Books.
Morwood, M.J., O'Sullivan, P.B., Aziz, F. and Raza, A. (1998) Fission-track ages of stone tools and fossils on the east Indonesian island of Flores. *Nature*, 392: 173–176.
Morwood, M.J., Soejono, R.P., Roberts, R.G., Sutikna, T., Turney, C.S.M., Westaway, K.E., Rink, W.J., Zhao, J.-X., Bergh, G.D. van den, Due Awe, R., Hobbs, D.R., Moore, M. W., Bird, M.I. and Fifield, L.K. (2004) Archaeology and age of a new hominin from Flores in eastern Indonesia. *Nature*, 431: 1087–1091.
Norman, K., Inglis, J., Clarkson, Faith, J.T., Shulmeister, J. and Harris, D. (2018) An early colonisation pathway into northwest Australia 70–60,000 years ago. *Quaternary Science Reviews*, 180: 229–239.
O'Connell, J.F. and Allen, J. (2015) The process, biotic impact, and global implications of the human colonization of Sahul about 47,000 years ago. *Journal of Archaeological Science*, 56: 73–84.
O'Connell, J.F., Allen, J., Williams, M.A.J., Williams, A.N., Turney, C.S.M., Spooner, N., Kamminga, J., Brown, G. and Cooper, A. (2018) When did *Homo sapiens* first reach Southeast Asia and Sahul? *Proceedings of the National Academy of Sciences USA*, 115(34): 8482–8490, doi:10.1073/pnas.1808385115.
O'Connor, S. (2007) New evidence from East Timor contributes to our understanding of earliest modern human colonisation east of the Sunda Shelf. *Antiquity*, 81: 523–535.
O'Connor, S. and Aplin, K. (2007) A matter of balance: an overview of Pleistocene occupation history and the impact of the Last Glacial Phase in East Timor and the Aru Islands, eastern Indonesia. *Archaeology in Oceania*, 42: 82–90.
O'Connor, S., Barham, A., Aplin, K. and Maloney, T. (2017a) Cave stratigraphies and cave breccias: implications for sediment accumulation and removal models and interpreting the record of human occupation. *Journal of Archaeological Science*, 77: 143–159.
O'Connor, S., Barham, A., Spriggs, M., Veth, P., Aplin, K. and St Pierre, E. (2010) Cave archaeology and sampling issues in the tropics: a case study from Lene Hara, a 42,000 year old occupation site in East Timor, Island Southeast Asia. *Australian Archaeology*, 71: 29–40.
O'Connor, S., Louys, J., Kealy, S. and Carro, S.C.S. (2017b) Hominin dispersal and settlement east of Huxley's Line: the role of sea level changes, island size, and subsistence behavior. *Current Anthropology*, 58(Supplement 17): S567–S582.
O'Connor, S., Mahirta, S.C., Samper, S., Hawkins, S., Kealy, S., Louys, J. and Wood, R. (2017c) Fishing in life and death: Pleistocene fish-hooks from a burial context on Alor Island, Indonesia. *Antiquity*, 91: 1451–1468.
O'Connor, S. and Ono, R. (2013) The case for complex fishing technologies: a response to Anderson. *Antiquity*, 87: 885–888.
O'Connor, S., Ono, R. and Clarkson, C. (2011) Pelagic fishing at 42,000 years before the present and the maritime skills of modern humans. *Science*, 334: 1117–1121.
O'Connor, S., Robertson, G. and Aplin, K.P. (2014) Are osseous artefacts a window to perishable material culture? Implications of an unusually complex bone tool from the Late Pleistocene of East Timor. *Journal of Human Evolution*, 67: 108–119.
O'Connor, S., Spriggs, M. and Veth, P. (2002) Excavation at Lene Hara Cave establishes occupation in East Timor at least 30,000–35,000 years ago. *Antiquity*, 76: 45–50, doi:10.1017/S0003598X0008978X.

Ono, R., Soegondho, S. and Yoneda, M. (2009) Changing marine exploitation during Late Pleistocene in northern Wallacea: shell remains from Leang Sarru Rockshelter in Talaud Islands. *Asian Perspectives*, 48: 318–341.

Pawlik, A.F., Piper, P.J. and Mijares, A.S.B. (2014) Modern humans in the Philippines: colonization, subsistence and new insights into behavioural complexity. In R. Dennell and M. Porr (eds) *Southern Asia, Australia and the Search for Human Origins*, Cambridge: Cambridge University Press, pp. 135–147.

Perry, G.H. and Domini, N.J. (2009) Evolution of the pygmy human phenotype. *Trends in Ecology and Evolution*, 24: 218–225.

Robles, E.C. (2013) Estimates of Quaternary Philippine coastlines, land bridges, submerged river systems and migration routes: a GRASS GIS approach. *Hukay*, 18: 31–53.

Samper Carro, S.C., Louys, J. and O'Connor, S. (2017) Methodological considerations for icthyoarchaeology from the Tron Bon Lei sequence, Alor, Indonesia. *Archaeological Research in Asia*, 12: 11–22, doi:10.1016/j.ara.2017.09.006.

Samper Carro, S.C., O'Connor, S., Louys, J., Hawkins, S. and Mahirta, M. (2016) Human maritime subsistence strategies in the Lesser Sunda Islands during the Terminal Pleistocene–Early Holocene: new evidence from Alor, Indonesia. *Quaternary International*, 416: 64–79.

Schnetzler, C.C. and McHone, J.F. (1996) Source of Australasian tektites: investigating possible impact sites on Laos. *Meteoritics and Planetary Science*, 31: 73–76.

Selimiotis, H. (2006) *The Core of the Matter: Core Reduction in Prehistoric East Timor*. M.Phil. Thesis. Canberra: School of Archaeology and Anthropology, Australian National University.

Shea, J.J. (2011) *Homo sapiens* is as *Homo sapiens* was: behavioral variability versus "Behavioral Modernity" in Paleolithic Archaeology. *Current Anthropology*, 52(1): 1–35.

Siler, C.D., Oaks, J.R., Esselstyn, J.A., Diesmos, A.C. and Brown, R.M. (2010) Phylogeny and biogeography of Philippine bent-toed geckos (Gekkonidae: *Cyrtodactylus*) contradict a prevailing model of Pleistocene diversification. *Molecular Phylogenetics and Evolution*, 55: 699–710.

Simpson, G.G. (1977) Too many lines: the limits of the Oriental and Australian zoogeographic regions. *Proceedings of the American Philosophical Society*, 121(2): 107–120.

Smith, J.M.B. (2001) Did early hominids cross sea gaps on natural rafts? In I. Metcalf, J. Smith, I. Davidson and M.J. Morwood (eds) *Faunal and Floral Migrations and Evolution in SE Asia-Australasia*, The Netherlands: Swets and Zeitlinger, pp. 409–416.

Smith, M.A., Prescott, J.R. and Head, M.J. (1997) Comparison of ^{14}C and luminescence chronologies at Puritjarra rock shelter, central Australia. Quaternary Science Reviews, 16(3–5): 299–320.

Summerhayes, G.R. and Ford, A. (2014) Late Pleistocene colonisation and adaption in New Guinea: implications for modelling modern human behaviour. In R. Dennell and M. Porr (eds) *Southern Asia, Australia and the Search for Modern Human Origins*, Cambridge: Cambridge University Press, pp. 213–227.

Summerhayes, G.R., Leavesley, M., Fairbairn, A., Mandui, H., Field, J., Ford, A., Fullagar, R. (2010) Human adaptation and plant use in Highland New Guinea 49,000 to 44,000 years ago. *Science*, 330: 78–81.

Sutikna, T., Tocheri, M.W., Faith, J.T., Jatmiko, Due Awe, R., Meijer, H.J.M., Wahyu Saptomo, E. and Roberts, R.G. (2018) The spatio-temporal distribution of archaeological and faunal finds at Liang Bua (Flores, Indonesia) in light of the revised chronology for *Homo floresiensis*. *Journal of Human Evolution*, 124: 52–74.

Sutikna, T., Tochieri, M.W., Morwood, M.J., Wahyu Saptomo, E., Jatmiko et al. (2016) Revised stratigraphy and chronology for *Homo floresiensis* at Liang Bua in Indonesia. *Nature*, 532: 366–369.

Szabó, K., Brumm, A. and Bellwood, P. (2007) Shell artefact production at 32,000–28,000 BP in Island Southeast Asia: thinking across media? *Current Anthropology*, 48: 701–723.

Taçon, P.S.C., Tan, H.T., O'Connor, S., Xueping, J., Gang, L., Curnoe, D., Bulbeck, D., Hakim, B., Sumantris, I., Than, H., Sokrithy, I., Chia, S., Khun-Neay, K. and Kong, S. (2014) The global implications of the early surviving rock art of greater Southeast Asia. *Antiquity*, 88: 1050–1064.

Tocheri, M.W. (2019) Unknown human species found in Asia. *Nature*, 568: 176–178.

Veth, P., Ward, I., Manne, T., Ulm, S., Ditchfield, K. et al. (2016) Early human occupation of a maritime desert, Barrow Island, north-west Australia. *Quaternary Science Reviews*, 168, 19–29.

Westaway, K.E., Zhao, J.-X., Roberts, R.G., Chivas, A.R., Morwood, M.J. and Sutkina, T. (2007) Initial speleothem results from western Flores and eastern Java, Indonesia: were climate changes from 47 to 5 ka responsible for the extinction of *Homo floresiensis*? *Journal of Quaternary Science*, 22 (5): 429–438.

Williams, A.N. (2013) A new population curve for prehistoric Australia. *Proceedings of the Royal Society B*, 280: 20130486, doi:10.1098/rspb.2013.0486.

Xhauflair, H., Revel, N., Vitales, T.J., Callado, J.R., Tandang, D., Gaillard, C., Forestier, H., Dizon, E. and Pawlik, A. (2017) What plants might potentially have been used in the forests of prehistoric Southeast Asia? An insight from the resources used nowadays by local communities in the forested highlands of Palawan Island. *Quaternary International*, 448: 169–189.

PART 2

Prologue

The northern dispersal across Asia

The northern dispersal across Asia

Starting around 50,000 years ago, our species colonised the Palearctic Realm of continental Asia from the Levant to the Japanese islands. Between 40,000 and 30,000 years ago, they also colonised the Palearctic European peninsula at the western end of Eurasia. This expansion across Asia brought them into the most challenging environments that humans had yet encountered, particularly in the Arctic, Central Asia, Siberia, Mongolia and north China, with their brutally cold winters (see Figure 3.8), and on the Tibetan Plateau 4,500 m above sea level.

Our main evidence for this expansion is a small amount of skeletal data indicating the presence of *H. sapiens* in various parts of continental Asia after 50,000 years ago, and a large amount of artefactual evidence that is described by an umbrella term as Initial Upper Palaeolithic, or IUP (Kuhn and Zwyns 2014). This term was first used to describe the assemblage in level 4 at Boker Tachit, Jordan (Marks and Ferring 1988; see Chapter 8) but has since widened to include all assemblages between 35 ka BP and 50 ka BP that show Levallois features in blade production as far east as southern Siberia, Mongolia (Chapter 9) and northern China (Chapter 10). These assemblages occasionally include items such as beads and other personal ornaments (perhaps evidence of symbolism) and bone tools. Collectively, IUP assemblages outside the Levant appear to span several millennia between 32 ka BP and 47 ka BP, but their dating needs much improvement: many dates are at the effective limit of radiocarbon dating and were made before AMS and rigorous pre-treatment techniques. Across continental Asia, the appearance of the Initial Upper Palaeolithic and subsequent Early Upper Palaeolithic are often taken as a proxy indicator of our species, and the preceding Middle Palaeolithic assemblages are usually attributed to indigenous populations of Neandertals and, in parts of Siberia, to Denisovans. We need caution here because IUP assemblages are rarely unambiguously associated with the remains of *H. sapiens* (Kuhn and Zwyns 2014). In any case, given the genetic evidence for interbreeding between humans, Neandertals and Denisovans (Chapter 9), any of these, or hybrids of them, may have made IUP assemblages. Nevertheless, there is skeletal evidence of *H. sapiens* in Mongolia at ca. 35 ka BP (Chapter 9), North China ca. 40 ka BP and 36 ka BP (Chapter 10), and in the Arctic at ca. 29 ka BP (Chapter 11), so the expansion of the IUP across continental Asia clearly included the actual dispersal of our species as well as other processes such as technological convergence and cultural transmission across social networks. Here, I will take the risk of assuming that the IUP was the product of *H. sapiens* unless there is clear evidence to the contrary.

The foundation on which the expansion of our species across continental Asia was based would have been demographic. After 50,000 years ago, populations of our species increased to the point that some groups could colonise adjacent areas, and they continued to do so until virtually the whole of the Palaearctic Realm of Eurasia had been colonised. In this process of expansion, the indigenous populations of Neandertals and Denisovans became extinct, although a small amount of their DNA was taken up by our species through interbreeding. The two main ways in which a population can increase without immigration is through an increase in female fertility – the number of infants born per reproductive female, or more accurately, the number of infants per female that survive into adulthood – and through a decrease in mortality – either by people living longer, or because fewer died through accident, disease or some other misfortune.

So, what might have caused these demographic changes in fertility and mortality? Several explanations have been offered but none provides a "silver bullet" whereby one single change resulted in human supremacy. One is that genetic changes allowed language to develop, thereby greatly increasing human abilities to communicate complex ideas and to share these within and between groups. The gene FoxP2 has often been mentioned in this context (see e.g. Enard et al. 2002). Although the idea of a "magic switch", similar to turning on a light in a dark room, is attractive, it is currently impossible to demonstrate, and there is also no obvious reason why Neandertals and Denisovans lacked language. In any case this gene is also found in Neandertal DNA (Krause et al. 2007). Another explanation is technological, and in part rests on the assumed superiority of an Upper Palaeolithic tool-kit over a Middle Palaeolithic one. Yet it is not immediately obvious how blade tools might have allowed fertility to rise and mortality rates to fall. As pointed out many years ago, there is no "big deal" about blades (Bar-Yosef and Kuhn 1999); as an example, blades were used extensively at Qesem cave, Israel, 400,000 years ago without triggering a marked increase in population. This is not to say that technological changes were irrelevant: for example, John Shea and Matthew Sisk (2010) have argued that the appearance in the Levant ca. 45–35 ka BP of stone-tipped projectiles, with greater penetrating power, would have made hunting large animals more successful and probably safer as lethal wounds could have been inflicted at a safe distance. However, projectile points are fairly rare east of the Levant, and may have had only a regional significance. Bone points might have served the same purpose, as might have spear throwers, although these seem to appear after the Initial and Early Upper Palaeolithic. Arguably, developments in making cordage (string, twine, rope, thread, etc.) and knowing how to loop and knot effectively may have been more significant as a technological breakthrough that underpinned daily life (see Hardy 2008). In continental Asia, where winters were sub-freezing, warm, insulated clothing would have been essential (see Figure 3.10), so hide- and fur-working and evidence of tailoring (such as needles) would have been factors that aided dispersal. The deep antiquity of clothing is shown by the split between head lice and body lice (which infect clothing) between 80,000 and 170,000 years ago (Toups et al. 2011). Evidence from the use wear on scrapers shows that hide scraping was common after 400,000 years ago, and Neandertals certainly could not have survived in Siberia without warm clothing (although they may have used animal fats as a means of body insulation; d'Errico et al. 2018). As they used bone polishers (lissoirs) (see Chapter 3), our species may first have learned these skills by copying them. Although needles allow fine stitch-work, they are not essential for sewn clothing as this can be done using an awl to make perforations that are then threaded with, for example, sinew. Nevertheless, their widespread use after 45,000 years ago across continental Asia must have been a qualitative leap forward in technological know-how in keeping warm.

Symbolism is one innovation often associated with *Homo sapiens* that has been invoked as an important indication of "modern human behaviour" (see Chapter 1). The earliest examples are from southern Africa and comprise incised pieces of ochre, beads and perforated teeth that could be made into necklaces. Ornaments of this kind could have served as markers of the wearer's age, gender, status and/or group affinity. This may be so, but they could have been unconnected with increased fertility or declining mortality; alternatively, the need to symbolise one's identity could have been a consequence but not a cause of population growth. If, for example, group size increased, or groups were living at higher densities, there might have been a need to show one's identity in a more visual manner. (Equally, if population densities were very low [as in Mongolia, Chapter 9], effective social networks would have been essential, and shared types of ornaments may have been one way of demonstrating membership.) Unfortunately, the evidence for Palaeolithic symbolism is very patchy across Asia, and absence of its evidence need not indicate evidence of absence – people can use visual symbols that would not survive archaeologically – for example, by tattoos, scarification, body paint, hair style and much else besides.

Pat Shipman (2015) has advanced the interesting argument that the "silver bullet" that led to the extinction of Neandertals was the domestication of wolves. As we will see later (Chapter 11), the evidence is equivocal but the earliest domestic dogs may date to ca. 30,000 BP in Siberia. These could have been used when hunting, for defending groups against wild wolves or other human groups, and/or as pack animals. Dogs would have been useful in all these roles, although perhaps without being the game-changer as argued by Shipman.

One demographic explanation for the replacement of Neandertals by our species may simply be that we lived longer and needed less. Judging from the incidence of broken bones and other forms of trauma among Neandertals, Trinkaus and Zimmerman (1982) concluded that they sustained as many injuries as modern rodeo riders. This is not surprising if, as seems probable, they killed large herbivores such as bison at close range – and it also implies that adult male Neandertals were unlikely to die quietly at a ripe old age. The mortality rate of *Homo sapiens* hunters in that type of environment may have been lower if they had devised less dangerous ways of killing their prey. An additional potential benefit of living longer is that the chances of becoming a grandparent increases. Inactive grandparents are of course an additional burden on a group as an extra, unproductive mouth to feed. On the other hand, they do bring benefits. One is that they can help with bringing up grand-children (which is often delegated to them in many modern societies); another is that they have longer memories, which can be an asset in environments that experience frequent downturns or periodic runs of good and bad years.

An additional important factor may be that Neandertals were rugged, led energetic lifestyles and had higher food requirements than our species requires in similar environments (see Froehle and Churchill 2009): to quote Kuhn and Stiner (2006), "They were 'living fast', with very high caloric intake from high-yield but risky subsistence resources". Fred Smith (2015) has argued that Neandertals solved the problems of survival by using "brawn, not brain" – that is to say, brute strength was one of their main assets. He suggests that an adult Neandertal may have had a BMR (basal metabolic rate) that was 14% higher than that needed by a contemporary African. Stiner cites an estimate that adult Neandertal women needed 10% more food per day than adult *H. sapiens* women. These differences have two consequences. One is that when food ran short, Neandertals may have been more vulnerable than *H. sapiens*: a 20% reduction in food for someone needing 5,000 calories and now reduced to 4,000 calories a day may cause them to suffer more than one whose food is cut by the same proportion from 3,000 to 2,400 calories. The second is that if 11 *H. sapiens* women required the same amount of food as ten Neandertal women, they were already at a competitive advantage in terms of reproductive capacity.

One argument for human success in northern latitudes is offered by Steven Kuhn and Mary Stiner (2006). They argue that in the Mediterranean basin, the main change that occurred after ca. 50,000 years ago in hunting patterns involved small game. Prior to that, and best shown by Neandertals, was an almost exclusive focus on medium to large animals such as deer, horse, bison and *Bos primigenius* – consistent with Neandertals being a top predator. Smaller packages of protein, such as tortoises or shellfish, were often used but were probably little more than snacks. Tortoises and limpets are obviously easy to obtain and risk-free, but are also easily over-harvested and not therefore a viable long-term resource. After 50,000 years ago, there appears to have been a greater emphasis on capturing small, quick creatures such as hare, rabbits, fish and birds that could fly (and not flightless ones, as at some Neandertal sites). Animals such as hare or rabbit mature faster than tortoises, and have higher reproductive rates, and can therefore withstand a higher degree of predation (Stiner and Munro 2002). They also need to be caught with traps, snares or nets, and require delayed methods of hunting. Apart from nets, most forms of traps and snares are easy to make, but trapping is a skilled activity that requires a detailed understanding of the local terrain and the behaviour of the prey. Kuhn and Stiner argue that the shift towards trapping small, agile game implies a more flexible use of labour and a less exclusive reliance on hunting medium to large game. As they also point out, we should not ignore the role of plant foods, which are rarely preserved (see Hardy 2018 for an excellent review). We have already seen that these trends towards a greater use of small, agile game and plant foods were already in evidence in the Oriental Realm of Southeast Asia by 50,000 years ago – as examples, the hunting of monkeys in the rainforests of Sri Lanka after 36,000 years ago (Chapter 5), the hunting of monkeys in high-canopy rainforest and the detoxification of starch plants at Niah Cave, Borneo ca. 40,000 years ago (Chapter 6), and the evidence for offshore fishing on Timor (Chapter 7).

We need to be cautious here because Neandertals were also capable of trapping small game such as rabbit and eating plant foods. As examples, at least 225 rabbits are represented in the Mousterian site of Pié Lombard, France, dated to MIS 5 (Pelletier et al. 2019), and a variety of plant foods were cooked and eaten at Spy, Belgium and Shanidar, Iraq (Henry et al. 2011; see Chapter 8) and various sites in Iberia (Salazar-García et al. 2013). The difference between Neandertals and humans may be that Neandertals were largely opportunistic over dietary breadth but humans were more systematic over trapping small game and harvesting plant foods. Instead of being a top predator like Neandertals, humans became the top omnivore (Hockett and Haws 2005).

The main point here is that when our species began to enter the Palearctic regions of continental Asia after 50,000 years ago, they were probably more flexible and systematic in how they organised their activities than the indigenous Neandertals and Denisovans. Kuhn and Stiner (2006) posed the question, what did Neandertal women and children do apart from having babies and growing up? There is little evidence among Neandertals of any marked sexual division of labour on the lines seen in extant and historical hunter-gatherers: hunting appears to have been a group activity in which all adults took part, with the men perhaps engaged in the close-quarter killing and the women acting as scouts, drivers (to try to force a prey animal towards an ambush) or as blockers (in trying to restrict the routes open to a prey). In contrast, *Homo sapiens* women (and the older children) could have engaged in a wider set of activities than big-game hunting alone.

This is not to say that when our species colonised continental Asia, women were routinely and dutifully collecting nuts and berries while men focused on hunting large game. The main point is that human groups were flexible in how they organised their subsistence. One aspect that would have been crucial in

Palearctic Eurasia was in making warm, insulated clothing. As mentioned already (Chapter 3), one fundamental technological change that occurred in the colonisation of Asia was the shift from plant-based to hide- and fur-based clothing. As noted for the human colonisation of southern Asia, clothing might have been optional, or rudimentary unless it involves spun plant fibres. Hide-based clothing requires a considerable investment in time and labour: the hide or skin has to be cleaned of its grease and often hair, depending on its intended use; it has to be made supple, it may need tanning by vegetable dyes or smoke; and then has to be cut and tailored. This is not to imply that men cannot do this type of work, only that women and older children could have undertaken this vital, time-consuming task, thus allowing males to concentrate on obtaining meat and the hides required for clothing and other items. (In these environments with prolonged sub-freezing winters, fuel gathering [including kindling and tinder] would also have been an important activity that may have been organised by age and gender.)

One additional factor that affected both the indigenous Neandertals and Denisovans in continental Asia as well as the immigrant groups of our species was that the environment more or less closes down in winter, and plant foods are only available seasonally. The main responses by animals for over-wintering are either to hibernate (an option not available to humans), to store food for winter consumption (as with squirrels), or to move to areas where plant food for a prey species is still available in winter (as with many herbivores). The last two strategies are risky: stored food might rot or be stolen; in severe winters, animal populations that migrate are still vulnerable to death by starvation or exposure. The scale and extent of food storage by our species and the indigenous residents in continental Asia is unknown but was probably commonplace. The easiest ways of storing meat is by drying, smoking or freezing, none of which would leave visible archaeological traces. Herbivores such as bison, horse and gazelle need large annual territories to obtain grazing year-round, and often migrate long distances between winter and summer grazing areas. As consequences of these factors, populations of *Homo sapiens* as well as Neandertals and Denisovans would have been small; groups of these would usually have needed large annual territories and encountered problems of maintaining effective social and mating networks: Mongolia (Chapter 9) provides a good example.

The next five chapters examine the dispersal of our species across continental Asia. Chapter 8 examines the area between the Levant and Central Asia. Humans were in Greece by 210 ka BP and the Levant by perhaps 177,000–194,000 years ago and (as seen in Chapter 4) were likely in Arabia by this time. Whether these new finds denote a "failed dispersal" (Shea 2008) that never proceeded east of the Levant or was the beginning of their southern dispersal across southern Asia has yet to be determined but appears to stretch definitions of failure. However, it was not until after 45,000 years ago that humans began to disperse eastwards and northwards across continental Asia north of the Himalayas and the Tibetan Plateau. They probably began to supplant the Neandertal populations in the Zagros

Mountains of Iraq and western Iran by ca. 45,000 years ago. From there, humans would have entered the cold desert regions of northern Iran and Central Asia, which were probably the most forbidding environments that they had encountered up to that time. Evidence from here is patchy, but the trail resumes in the Altai Mountains near the borders of Russia, Kazakhstan, Mongolia and China (Chapter 9). Here, both Neandertals and Denisovans were present before 50 ka BP, but these were probably replaced by our species ca. 45–40 ka BP. It is, however, likely that our species interbred with both Neandertals and Denisovans across large parts of continental Asia, just as Neandertals and Denisovans interbred with each other.

Around the same time that our species appeared in the Altai Mountains, humans were also entering northern Mongolia by 42,000 years ago and perhaps even Siberia as far north as the Arctic. Chapter 10 examines the evidence from China and Tibet. China is unique in Asia because it extends from the Palaearctic Realm in the north to the Oriental Realm in the south and needs to be treated as a single unit. Humans were in north China by 42,000 years ago, and on the Tibetan Plateau at over 4,500 m above sea level by 36,000 years ago but probably in southern China before 80,000 years ago. The Arctic is briefly visited in Chapter 11, along with the colonisation of Korea and the Japanese islands, which were likely colonised after 38,000 years ago by sea crossings. By this stage, humans had now colonised every major region of Asia (and were in the process of doing so in the European peninsula at the western end of Eurasia).

References

Bar-Yosef, O. and Kuhn, S. (1999) The big deal about blades: laminar technologies and human evolution. *American Anthropologist*, 101(2): 322–338.

D'Errico, F., Doyon, L., Zhang Shuangquan, Baumann, M., Lázničková-Galetová, M. et al. (2018) The origin and evolution of sewing technologies in Eurasia and North America. *Journal of Human Evolution*, 125: 71–86.

Enard, W., Przeworski, M., Fisher, S.E., Lai, C.S.L., Wiebe, V. et al. (2002) Molecular evolution of FOXP2, a gene involved in speech and language. *Nature*, 418: 869–872.

Froehle, A.W. and Churchill, S.E. (2009) Energetic competition between Neandertals and Anatomically Modern Humans. *PaleoAnthropology*: 96−116.

Hardy, K. (2008) Prehistoric string theory. How twisted fibres helped to shape the world. *Antiquity*, 82: 271–280.

Hardy, K. (2018) Plant use in the Lower and Middle Palaeolithic: food, medicine and raw materials. *Quaternary Science Reviews*, 191: 393–405.

Henry, A.G., Brooks, A.S. and Piperno, D.R. (2011) Microfossils in calculus demonstrate consumption of plants and cooked foods in Neanderthal diets (Shanidar III, Iraq; Spy I and II, Belgium). *Proceedings of the National Academy of Sciences USA*, 108(2): 486–491.

Hockett, B. and Haws, J.A. (2005) Nutritional ecology and the human demography of Neandertal extinction. *Quaternary International*, 137: 21–34.

Krause, J., Lalueza-Fox, C., Orlando, L., Enard, W., Green, R., Burbano, H., Hublin, J.-J., Hänni, C., Fortea, J., Rasilla, M. de la, Bertranpetit, J., Rosas, A. and Pääbo, S. (2007) The derived FOXP2 variant of modern humans was shared with Neanderthals. *Current Biology*, 17, 1908–1912.

Kuhn, S.L. and Stiner, M.C. (2006) What's a mother to do? the division of labor among Neandertals and modern humans in Eurasia. *Current Anthropology*, 47(6): 953–980.

Kuhn, S.L. and Zwyns, N. (2014) Rethinking the Initial Upper Palaeolithic. *Quaternary International*, 34, 29–38.

Marks, A.E. and Ferring, C.R. (1988) The Early Upper Palaeolithic of the Levant. In J.E. Hoffecker and C.A. Wolf (eds) *The Early Upper Palaeolithic: Evidence from Europe and the Near East*, Oxford: British Archaeological Reports International Series 437, pp. 43–72.

Pelletier, M., Desclaux, E., Brugal, J.-P. and Texier, P.-J. (2019) The exploitation of rabbits for food and pelts by last interglacial Neandertals. *Quaternary Science Reviews*, 224: 105972.

Salazar-García, D.C., Power, R.C., Sanchis Serra, A., Villaverde, V., Walker, M.J. and Henry, A.G. (2013) Neanderthal diets in central and southeastern Mediterranean Iberia. *Quaternary International*, 318: 3–18.

Shea, J.J. (2008) Transitions or turnovers? Climatically-forced extinctions of *Homo sapiens* and Neanderthals in the East Mediterranean Levant. *Quaternary Science Reviews*, 27: 2253–2270.

Shea, J.J. and Sisk, M.L. (2010) Complex projectile technology and *Homo sapiens* dispersal into Western Eurasia. *PalaeoAnthropology*: 100–122.

Shipman, P. (2015) *The Invaders: How Humans and their Dogs Drove Neandertals to Extinction.* Cambridge, Mass.: Harvard University Press.

Smith, F.H. (2015) Neanderthal adaptation: the biological cost of brawn. In N. Sanz (ed.) *Human origin sites and the World Heritage Convention in Eurasia*, Paris: UNESCO, pp. 206–219.

Stiner, M.C. and Munro, N.D. (2002) Approaches to prehistoric diet breadth, demography, and prey ranking systems in time and space. *Journal of Archaeological Method and Theory*, 9(2): 181–214.

Toups, M.A., Kitchen, A., Light, J.E. and Reed, D.L. (2011) Origin of clothing lice indicates early clothing use by anatomically modern humans in Africa. *Molecular Biology and Evolution*, 28: 29–32.

Trinkaus, E. and Zimmerman, M.R. (1982) Trauma among the Shanidar Neandertals. *American Journal of Physical Anthropology*, 57: 61–76.

8

SOUTHWEST ASIA

From the Levant to Iran

The Levant

The Levant has the most detailed Palaeolithic and human skeletal records in Asia for the last 200,000 years, as well as the earliest evidence for *Homo sapiens* outside Africa. Those specimens dated to after 50,000 years ago are associated with Upper Palaeolithic, blade-dominated assemblages which serve as a proxy indicator of our species across much of continental Asia. For many researchers, the primary interest of the Levant is that it may contain the source population of our species that eventually colonised Europe. Here, my interest lies in the opposite direction, in those groups that ventured eastwards and colonised continental Asia.

The Levant lies at the southern edge of the modern Palearctic Realm and adjacent to the Afro-Arabian Realm; in hominin terms, there was a Neandertal world to the north, and a *sapiens* world to the south. Depending on the prevailing climate, the boundaries of these two realms see-sawed north or south during the Pleistocene. As with most biological boundaries, the frontier between these two worlds was unlikely to have been impermeable; instead there was probably a considerable amount of overlapping, co-existing, inter-mixing and interbreeding. The Levant is also one of the few parts of Asia with a Mediterranean climate, with winters generally above freezing, and warm but rarely excessively hot summers (compared with, for example, Arabia). Most rain falls between October and May and can be as high as 1,000 mm a year along the coast of Lebanon, but falls sharply eastwards and southwards to a yearly average of only 50 mm in the deserts of the Negev, Syria and Iraq. It was in other words a "goldilocks zone" that was for the most part not too hot or cold, or too dry or too wet for both Neandertals and *H. sapiens*.

We begin with the skeletal and archaeological record for hominins in the Levant before considering their climatic background and possible demographic history. Figure 8.1 shows the location of relevant sites.

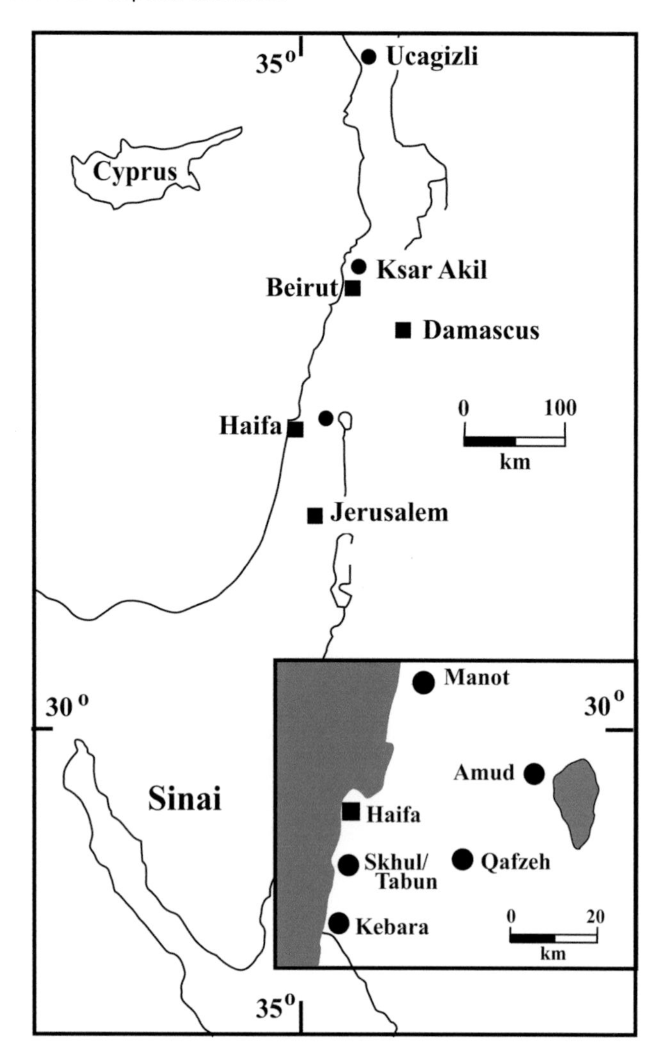

FIGURE 8.1 Middle and Upper Palaeolithic sites in the Levant
Source: The author.

The Levantine skeletal record

The Israeli skeletal record for the Upper Pleistocene is unique in Asia in two respects. First, many of the finds are from burials in caves, and thus there are several examples of partial and complete skeletons (see Table 8.1), in contrast with the much more fragmented evidence from south China and Southeast Asia. Second, the Israeli evidence has been exceptionally well dated; indeed (see Table 8.2), because there are so many dates and methods of dating, it is often difficult to assign precise ages to finds (see Millard 2008). Nevertheless,

TABLE 8.1 The fossil hominin evidence from the Levant, 250–30 ka.

Country	*Locality*	*Specimen*	*Context*	*Age (ka)*	*Identification*	*Comments*
Israel	Misliya	1 adult maxilla	Upper Terrace Unit 6	177–194 ka	*H. sapiens*	See also Table 8.2
	Tabun	Tabun I: almost complete skeleton Tabun II: mandible Molar and femur fragments	Layer C but probably intrusive from layer B Layer C Layer E	122 ± 16 ka if layer B (Grün and Stringer 2000) 165 ± 16 (Mercier and Valladas 2003) ca. 300 ka	Neandertal Neandertal, (*H. sapiens*: Howell 1999); neither (Harvati and Lopez 2017)	Indicates presence of Neandertal in MIS 6 in the Levant; see also Table 8.2 Indeterminate
	Skhul B	7 adults, 3 juveniles		119 ± 18 (TL)	*H. sapiens*	See also Table 8.2
	Qafzeh L	4 adults, 2 juveniles		92 ± 5 (TL)	*H. sapiens*	See also Table 8.2
	Qafzeh XV–XXII	2 adults, 5 juveniles, several isolated teeth		96 ± 13 (ESR EU) 115 ± 15 (ESR LU)	*H. sapiens*	
	Kebara F Kebara VII–XII	1 juvenile skeleton 1 partial adult skeleton, numerous isolated bones and teeth	Layer F Layers VII–XII	51.9 ± 3.5 to 59.9 ± 3.5 for layer XII by TL	All Neandertal	See also Table 8.2
	Amud B Amud B	2 adults, 2 juveniles 2 juveniles, numerous fragmentary remains	Layer B Layer B	(B1) 57.6 ± 3.7 (B2) 65.5 ± 3.5 (B4) 68.5 ± 3.4	Neandertal	See also Table 8.2
	'Ein Qashish[1]	Skull fragment, right M^3; leg bones and vertebra		60–70 ka	Neandertal	Been et al. (2017); Ekshtain et al. (2019)
	Manot	Partial cranium		54.7–65.5	*H. sapiens*	

(*Continued*)

TABLE 8.1 (Cont.)

Country	*Locality*	*Specimen*	*Context*	*Age (ka)*	*Identification*	*Comments*
Syria	Dederiyeh	3 infant burials various adult remains	Layers 3, 11 and 13 Layers 4 and 5		Neandertal	Akazawa et al. (1999)
Lebanon	Ksar Akil KS1	"Egbert", a skull and some post-cranial; now lost	Layer XXVII or XVIII	47 ± 9 43.7 ± 1.5 (^{14}C) 40.8–39.2 ka cal BP (Douka et al. 2013)	*H. sapiens*	See also Table 8.2
Lebanon	Ksar Akil KS2	"Ethelruda", a partial maxilla	Layer XXV	42.4–41.7 ka cal BP	*H. sapiens*	See also Table 8.2

1 'Ein Qashish is the first open-air site in the Levant where Neandertal remains have been recovered.
Source: Dennell 2014, except for 'Ein Qashish.

researchers disagree profoundly over the identification and explanation of much of the skeletal evidence. Before proceeding further, we need to bear in mind that specimens of early *H. sapiens* often retain "primitive" features (as example, the shape of the braincase at Jebel Irhoud; Chapter 1) that may also be found in Neandertals; and some Neandertals have "modern" features. We should also not expect Neandertals in the Levant to be identical to those from western Europe. In Europe, Neandertals were confined to a cul-de-sac but the Levant was, in Bar-Yosef's words, "the Central Bus Station" that was open to immigration from the south, east and north (see Tillier and Arensburg 2017). Additional considerations are that we know almost nothing about the Neandertal (and perhaps other) populations that inhabited Turkey and southeast Europe, and next to nothing about the populations in the Levant prior to *Homo sapiens*. (The recent report of a Neandertal cranium from the cave of Apidima in Greece, dated at 170 ka and a partial cranium identified as *H. sapiens* from the same cave but dated to 210 ka [Harvati et al. 2019] indicates that we might expect a complex demographic history for the region between Greece and the Levant – if the dating and identification of these specimens is confirmed.) Recent genetic evidence also indicates that non-African humans share a small amount of DNA with Neandertals, so interbreeding between the two is likely (Callaway 2015; Green et al. 2010), and we should therefore expect evidence of hybridisation.

TABLE 8.2 Dates (ka) for Levantine hominin skeletal remains

Context	TL	ESR EU	ESR LU	U-series	^{14}C	Comment	Source
H. sapiens							
Misliya	179 ± 48 ka; range 212 to 140 ka	174 ± 20		185 ± 8.0 on crust on maxilla			Hershkovitz et al. (2018)
Skhūl B	119 ± 18					Mean of 6	Mercier et al. (1993: 172)
Skhūl B		80.8 ± 12.6	101 ± 17.9			Mean of 7	Stringer et al. (1989: 757)
Skhūl B		59.7 ± 6.3	76.7 ± 8.2	49.0		Mean of 6	McDermott et al. (1993: 254)
Qafzeh XVII–XXIII	92 ± 5					Mean of 20	Valladas et al. (1988: 615)
Qafzeh XV–XXI		96 ± 13	115 ± 15			Mean of 16	Schwarcz et al. (1988: 735)
Qafzeh XIX		104 ± 10.5	120 ± 15	97.48		Mean of 2	McDermott et al. (1993: 254)
Neandertal							
Tabun B		102 ± 17	122 ± 16	104 + 33/ −18		Mean of 7	Tabun I skeleton, intrusive from layer C; Grün and Stringer (2000: 602)
Tabun B		76 ± 14	85 ± 18	50.69 ± 0.23			McDermott et al. (1993: 254)
Tabun C/Unit 1	165 ± 16					Mean of 7	Tabun II mandible; Mercier and Valladas (2003)
Tabun C		120 ± 16	140 ± 21	135 + 60/−30		Mean of 8	Grün and Stringer (2000: 602)

(*Continued*)

TABLE 8.2 (Cont.)

Context	*TL*	*ESR EU*	*ESR LU*	*U-series*	^{14}C	*Comment*	*Source*
Tabun C		117.6 ± 29.3	127 ± 34.3			Mean of 3	McDermott et al. (1993: 254)
Kebara VII	51.9 ± 3.5						Valladas et al. (1987: 159)
Kebara VII, Sq Q19					>44.8		Bar-Yosef et al. (1996: 301)
Kebara VIII	57.3 ± 4						Valladas et al. (1987: 159)
Kebara IX	58.4 ± 4						Valladas et al. (1987: 159)
Kebara X	61.6 ± 3.6						Valladas et al. (1987: 159)
Kebara X		60.4 ± 8.5	64.3 ± 9.2			Mean of 11	Schwarcz et al. (1988: 657)
Kebara XI	60 ± 3.5						Valladas et al. (1987: 159)
Kebara XII	59.9 ± 3.5						Valladas et al. (1987: 159)
Amud B1	57.6 ± 3.7					Mean of 6	Valladas et al. (1999: 265)
Amud B2	65.5 ± 3.5					Mean of 8	Valladas et al. (1999: 265)
Amud B4	68.5 ± 3.4					Mean of 5	Valladas et al. (1999: 265)
Amud B1/6–B1/7				53 ± 7			Rink et al. (2001: 713–714)
Amud B2				61 ± 9		Mean of 4	Rink et al. (2001: 713–714)
Amud B4				70 ± 11		Mean of 2	Rink et al. (2001: 713–714)

(*Continued*)

TABLE 8.2 (Cont.)

Context	*TL*	*ESR EU*	*ESR LU*	*U-series*	^{14}C	*Comment*	*Source*
Dederiyeh layers 2–4	50–70					Considered unsatisfactory	Akazawa and Nishiaki (2017)
Dederiyeh layers 8–9	60–90					Considered unsatisfactory	Akazawa and Nishiaki (2017)
Post-Neandertal *H. sapiens*							
Manot				54.7–65.5			Hershkovitz et al. (2015)
Ksar Akil, Egbert					40.8–39.2 ka cal BP; 43.2–42.9 ka cal BP		Douka et al. (2013) Bosch et al. (2015)
Ksar Akil, Ethelruda					42.4–41.7 ka cal BP; >45.9 ka cal BP		Douka et al. (2013) Bosch et al. (2015)
Ksar Akil XXVI				47 ± 9			Plicht et al. (1989)
Ksar Akil XXVI					43.7 ± 1.5		Mellars and Tixier (1989)
Ksar Akil XXXII				51 ± 4			Plicht et al. (1989)
Ksar Akil XXXII				49 ± 5			Plicht et al. (1989)

Levantine populations before the last interglacial

The only local hints of the indigenous population of the Levant that was absorbed, displaced or replaced by *H. sapiens* are eight teeth from Qesem cave (Hershkovitz et al. 2011), a few indeterminate ones (Harvati and Lopez 2017) from layer E at Tabun cave, and a partial cranium from Zuttiyeh cave. These are all dated to ca. 300 ka and are indeterminate. The Qesem teeth have Neandertal features, but cannot be assigned to any taxonomic group and might even be a new taxon (Fornai et al. 2016; Weber et al. 2016). Almost every opinion possible has been expressed about the Zuttiyeh cranium (Sohn and Wolpoff 1993).

Two important indications of when our species may have first appeared in the Levant come from Misliya Cave in Israel. Here, excavations have shown a sequence that spans the Late Lower Palaeolithic to the Early Middle Palaeolithic (Hershkovitz et al. 2018). The former comprises layers with late Lower Palaeolithic assemblages known as Acheulo-Yabrudian that include handaxes, thick side-scrapers and blades but lack Levallois technology, and the latter contain layers with assemblages known as early Levantine Mousterian, or Tabun-D type (named after the nearby Israeli cave site of Tabun) and characterised by the dominance of Levallois technology, laminar blade production and large elongate and retouched points (Meignen 2011). (A similar sequence was found at the cave of Dederiyeh in Syria; see Akazawa and Nishiaki 2017.) Thermoluminescense (TL) dating of burnt flints from Misliya place the boundary between these very different assemblages at ca. 250 ka, or at the end of Marine Isotope Stage (MIS) 8 or early MIS 7. This is broadly similar to the results from dating a similar sequence of deposits at Tabun. Because of the profound differences between Acheulo-Yabrudian and Early Middle Palaeolithic assemblages, Valladas and colleagues (2013) suggest that the change resulted from the influx of an immigrant population. As indicated below, this immigrant population is most likely *Homo sapiens*.

A recently discovered maxilla from Misliya that was dated to between 177,000 and 194,000 years old is identified as that of *Homo sapiens* and is associated with Tabun-D type Middle Palaeolithic assemblages (Hershkovitz et al. 2018). In climatic terms, the maxilla dates to the early part of MIS 6. As pointed out earlier (Chapter 1, Figure 1.10), the fossil record will always underestimate a taxon's first appearance, and so we will never pinpoint from skeletal evidence the date when a species first appeared in a region. Because the earliest records of our species are African and associated with Middle Stone Age assemblages (Chapter 2), the most likely source for the Misliya population is East or North Africa.

Analyses of the Middle Palaeolithic faunal remains at Misliya show that the commonest prey species were prime-aged fallow deer (*Dama mesopotamica*), followed by gazelle and aurochs (*Bos primigenius*). A few specimens of wild pig (*Sus scrofa*), red deer (*Cervus elaphus*), roe deer (*Capreolus capreolus*) and goat (*Capra* sp.) were also present. Small game animals were rare and limited to tortoises, ostrich eggs and partridge. After the evidence for cut-marked bone and burning were taken into account, the overall assessment was that the Middle Palaeolithic inhabitants at Misliya focused on prime-aged ungulates, transported whole or partial carcasses (depending on the prey size) back to camp, and were cracking bones for marrow (Yeshurun et al. 2007).

One important hominin find is an almost complete female skeleton (Tabun I) that was found in layer C at Tabun Cave, Mount Carmel, but was probably intrusive from the overlying deposit, layer B (Bar-Yosef and Callander 1999); this has been dated by ESR (late uptake) to 122 ± 16 ka (see Table 8.2). This age estimate and others from layer B would place the skeleton in MIS 5 or the end of MIS 6. Most researchers classify this skeleton as Neandertal. There is also a mandible (Tabun II) from layer C which has been dated to 165 ± 16 ka by TL, and slightly less by ESR (early and late uptake models) and U-series dating (see Table 8.2). Opinions differ widely over this specimen. Some researchers regard it as Neandertal (e.g. Schwartz and Tattersall 2003: 384), but Howell (1999: 217) placed it in the *H. sapiens* group of Skhūl and Qafzeh (see below). Others regard it as part of a late archaic lineage (Stefan and Trinkaus 1998), as neither Neandertal nor *H. sapiens* (Harvati et al. 2017), or as "modern-like" or a hybrid Neandertal–sapiens (Quam and Smith 1998). Seven upper teeth from layer B at Tabun were also identified as from a Neandertal individual (BC7) with an estimated age of 90 + 30/–16 ka (Coppa et al. 2007).

Levantine populations between ca. 125,000 and 60,000 years ago

Our main source of skeletal evidence in the Levant between 125,000 and 60,000 years ago comes from the caves of Amud, Kebara, Manot, Qafzeh and Skuhl in Israel, and Dederiyeh in Syria. At Skuhl, the remains of seven adults (Skuhl II-VI and IX) and three juveniles (Skuhl I, VIII and X) were found in layer B, and these have been dated to 119 ± 18 ka by TL. On stratigraphic grounds, Skuhl IX might be older than Skuhl II and V (Grün et al. 2005), and perhaps as much as 131 ka (Tillier and Arensburg 2017). Two perforated shells of *Nassarius gibbosulus* were also found that could have been used as beads (Vanhaeren et al. 2006). At Qafzeh, the remains of four adults and two juveniles were discovered in layer L in front of the cave in 1933–1935, and another two adults and five juveniles in units XV–XXII during the excavations of 1965–1977 (Shea 2003; Table 8.1). Depending on which suite of dates is preferred (Table 8.2), these remains may be the same age as, or slightly younger than, those from Skuhl (excluding Skuhl IX). At least 84 lumps of ochre (Hovers et al. 2003) and ten sea shells of the marine bivalve *Glycymeris insubrica* from the coast 35 km distant were also found; some were stained with ochre and may have been used as strung ornaments (Bar-Yosef Meyer et al. 2009). At Kebara, a fragmented infant skeleton (K1) was found in layer F in 1964. In later excavations (1984–1991), this layer was subdivided into Units VII–XII, and a partial skeleton of an adult was found in level XII, and numerous isolated teeth and bones in levels VII–XII. These have been dated to ca. 55–65 ka. At Amud, the remains of two adults and four juveniles were found in layer B, for which several dates are available, mostly in the range of 50–70 ka (Table 8.2). In northern Syria, three infant burials and numerous adult remains of Neandertals associated with Tabun B, C and D assemblages were excavated at the cave of Dederiyeh (Akazawa and Nishiaki 2017). At Manot, a partial cranium attributed to *H. sapiens* was recently found and dated to ca. 55–65 ka (Hershkovitz et al. 2015; see below).

Researchers disagree profoundly over this material. A widely held view is that there were two distinct and separate populations: *H. sapiens* at Skuhl and Qafzeh, and Neandertals at Amud and Kebarah (see e.g. Shea 2003, 2008; Shea and Bar-Yosef 2005). On this binary model, humans entered the Levant in MIS 5, and were then replaced by Neandertals when the climate deteriorated in MIS 4 until they in turn were replaced by a further immigrant population of humans after 50,000 years ago. Drawing on concepts outlined in Chapter 1, each of these dispersals would have been a simple case of range extension: *H. sapiens* moving north under warm conditions, and Neandertals moving south when it became colder. Much depends on dating, which still needs improvement to show their relative chronological ordering (Grün et al. 2005) and to prove or disprove links between these occupations and climatic changes (Millard 2008).

Three aspects of this model are problematic. First, our species and Neandertals are regarded as biologically separate populations, despite evidence that they interbred. Second, it is not obvious how these supposedly separate populations used the same type of Middle Palaeolithic stone tools, ate the same animals and lived in the same area yet remained distinct. Third, given the adaptability of *H. sapiens*, it is hard to understand how modest falls in temperature in MIS 4 between 80 ka and 60 ka would have made the Levant uninhabitable for them when no other local animal became extinct.

Other researchers maintain that the Israeli caves show a single, polymorphic population, as was argued by McCown and Keith (1939) when they examined the Neandertal-like skeletal material from Tabun and sapiens-like material from Skuhl (Kramer et al. 2001), and proposed a single, variable taxon called *Palaeoanthropus palestinensis*. Many researchers accept that both humans and Neandertals were present in the Levant between 120 ka and 50 ka but emphasise the variability of both populations. Arensburg and Belfer-Cohen (1998) claim Neandertal traits in the Skuhl and Qafzeh populations. Thackeray and colleagues (2005) also regard Skuhl and Qafzeh as variable populations, and Qafzeh 6 as intermediate between late Neandertals and late AMH (anatomically modern humans); this variability is attributed to interbreeding. Trinkaus (2005) also regards the early moderns as having features that are either Neandertal or retentions from their previous ancestry. Opinions vary over the likely incidence of interbreeding. Tillier and Arensburg (2017) maintain that Kebara, Tabun, Amud and Dederiyeh are not biologically homogeneous; instead, they argue, there is considerable intra-site variation in the material from Tabun, Skuhl and Qafzeh as well as inter-site variation between Tabun, Amud and Kebara.

My own view is that the Levant between 125 ka and 60 ka ago was probably a contact zone between a Neandertal world to the north and *H. sapiens* one to the south. Unlike western Europe, Neandertal and *H. sapiens* were not mutually exclusive categories, and the boundary between them was likely porous. Consequently, the inhabitants were largely or partly Neandertal or *H.* sapiens (in terms of their cranial morphology), depending upon the degree of immigration, interbreeding and the relative size of the Neandertal and *sapiens* gene pools at any one time.

Homo sapiens *after 60 ka*

Two sites are important in documenting our species after 60 ka: Manot and Ksar Akil.

Manot

Here, speleologists discovered a cave chamber that could only be accessed through a chimney as the entrance was probably sealed by a roof-fall between 15,000 and 30,000 years ago. The cave deposits show that the cave was occupied in the Late Middle Palaeolithic and more intensively during the Early Upper Palaeolithic (Ahmarian) between 38,000 and 34,000 cal BP (Alex et al. 2017; Marder et al. 2013, 2017, 2018; see Figures 8.2a and 8.2b). A human skull was also found on a limestone ledge in a side chamber – unfortunately not associated with any archaeological material. As a result of sampling 11 points of the calcite patina around the skull, an average uranium-thorium age of 54.7–65.5 ka was obtained. It is described as *H. sapiens* but with a mixture of modern and archaic features. Three explanations of these traits have been proposed. One is that it could represent a hybrid between Neandertals and our species. A second is that it is a direct descendant of the Skuhl-Qafzeh populations but the long time gap between them and Manot perhaps makes this unlikely. The preferred interpretation of the investigators is that it probably represents an immigrant population of *H. sapiens* that arrived in the Levant from Africa after ca. 60,000 years ago (Hershkovitz et al. 2015; Hershkovitz and Arensburg 2017). This interpretation remains to be tested because, unfortunately, the nearest relevant African example of *H. sapiens* is a fragile and poorly preserved child's skeleton from Taramsa 1, Upper Egypt that has a weighted average burial age of 55.3 ± 3.7 k (with a range from 49.8 to 80.4 ka) (Vermeersch et al. 1998) but does not allow detailed comparison with the Manot specimen. If the Manot individual did arrive from Africa, it did so before the IUP was developed – presumably by its descendants.

Ksar Akil

Prior to the discovery of Manot, the oldest examples of a post-Neandertal *H. sapiens* were a maxillary fragment (a.k.a. "Ethelruda") from layer XXV and a skull ("Egbert", now lost) from layer XVII or XVIII in the cave of Ksar Akil, Lebanon. This cave was excavated in the 1930s at a lower standard than today. At this cave, there is a long sequence of Levallois-Mousterian in layers XXXVI to XXVI, IUP assemblages in layers XXV–XXI and EUP (Ahmarian) assemblages in layers XX–XVI (Bergman et al. 2017). Both specimens are therefore post-Middle Palaeolithic, but there is uncertainty over their dating. According to Bosch and colleagues (2015), the EUP from layers XX to XVI dates from

FIGURE 8.2 Recent excavations at the Upper Palaeolithic site of Manot, Israel

Recent excavations here have made this site one of the most important Upper Palaeolithic sites in Southwest Asia.

Source: Reproduced with kind permission of Mae Goder-Goldberg.

44,000 to 37,200 cal BP. Because of the lack of suitable dating material, the age of Ethelruda (and the associated IUP) is estimated as starting before 45,900 cal BP. Slightly younger dates are proposed by Douka and colleagues (2013), who propose that Egbert dates from between 40.8 and 39.2 ka cal BP and Ethelruda between 42.4 and 41.7 ka cal BP, with a 68.2% probability in each case. As a note of caution, these estimates need constraining by more dates from above and below these specimens (Bergman et al. 2017). Several isolated teeth attibuted to *H. sapiens* were also recovered from the cave of Üçağızlı[1] in southwest Turkey and dated by ^{14}C as from 41.4 $\pm$ 1.1 to 29.1–0.4 uncalibrated ka BP (Güleç et al. 2007) but at least one is more Neandertal-like (Kuhn et al. 2009: 108).

The Levantine Initial Upper Palaeolithic (IUP) and Early Upper Palaeolithic (EUP)

The Levantine IUP comprises a mixture of Middle and Upper Palaeolithic attributes, with Levallois flakes, blades and points as well as retouched Upper Palaeolithic blades, end-scrapers and burins (see Figure 8.3). Bone tools are also present, although rare. Perforated shells are also common at Ksar Akil and Üçağızlı (see below). It remains unclear if the IUP developed from the local Middle Palaeolithic or was introduced from outside (Kuhn 2003) – or, perhaps more likely, was a combination of local developments and interactions with incoming populations (Meignen 2012). The EUP comprises the Ahmarian and Levantine Aurignacian. The Ahmarian is similar to the IUP and probably developed from it (Kuhn 2003) but is characterised by a bipolar prismatic blade technology, as well as bone tools and shells ornaments (Kuhn 2002). At Manot, these bone tools include retouchers (Yeshurun et al. 2017), awls and antler projectile points (Tejero et al. 2015). Levantine Aurignacian assemblages are characterised by end scrapers, carinated burins, bladelets, as well as bone and antler tools.

The dating of the IUP and EUP in the Levant is problematic, partly because its likely age is at or beyond the effective limit of radiocarbon carbon dating, partly because of the type of dating material (e.g. carbon or shell) and partly because of differences between various calibration methods. The IUP is known from Ksar Akil (Lebanon), Üçağızlı (Turkey), Umm el Tlel (Syria) and Boker Tachit (Israel). Rebollo and colleagues (2011) estimate that the IUP spans 49,000–46,000 cal BP. At Boker Tachit, four IUP levels were each separated by a sterile layer. Level 1 (the oldest) was dated at 50.5–48.3 ka cal BP and level 4, the youngest, at ca. 39 ka cal BP. These layers show a transition from using the Levallois technique to make Levallois points to using points to make prismatic blades. The tools in all levels were Upper Palaeolithic types such as end scrapers and burins (Olszewski 2017). At Üçağızlı, layers F-I contain IUP assemblages and layers B, C and probably D and E have EUP ones. Layers G, H, H1–3 and I formed between about 35,000 and 41,400 radiocarbon years BP, and layers B, B1–3 and C between roughly 29,000 and 34,000 radiocarbon years BP, but Kuhn et al. (2009) warn that these dates are minima and may under-shoot true age by at least 3,500–5,000 years. As noted above, the dating of the IUP and EUP at Ksar Akil is problematic because it was excavated such a long time ago, and radiocarbon dates were mainly produced from shells which are generally less reliable than charcoal for dating (Alex et al. 2017).

The main dating evidence of the EUP comes from the caves of Üçağızlı, Ksar Akil, Kebara, Manot and Mughr el-Hamamah in the Jordan Valley. Sites in the arid southern Levant have fewer dates, and those were mostly produced in the 1970s and early 1980s with less reliable methods than now. At Kebara the Ahmarian began by 47.5 to 46 ka cal BP and ca. 46,000 cal BP at Manot, where its replacement by Levantine Aurignacian assemblages occurred between 38 and 34 ka cal BP and probably more precisely between 37 and 35 ka cal BP. The Ahmarian

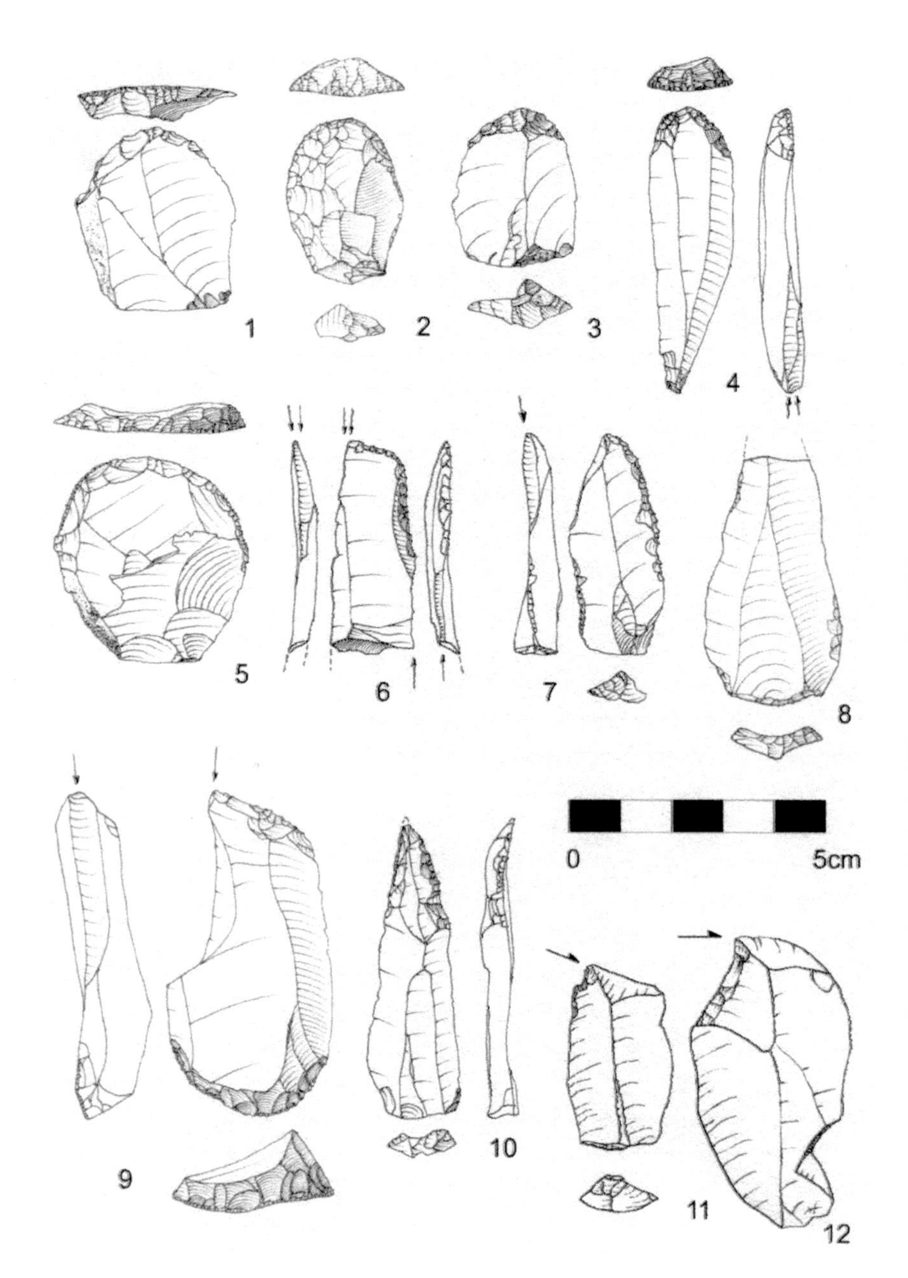

FIGURE 8.3 Initial Upper Palaeolithic (IUP) retouched tools from Üçağızlı, Turkey

Context: (1–2) layer F; (3, 8, 9) layer H; (4, 6) layer H2–3; (5) layer G; (7, 10) layer Fb–c; (11, 12) layer I.

Source: Kuhn et al. 2009, Fig. 10.

appears to be earlier at Kebara and Manot than at Ksar Akil and Üçağızlı where it dates from ca. 43 ka cal BP or 40 ka cal BP. This discrepancy is probably an artefact of the dating techniques used and the choice of dating materials. These concerns apply also to the dating of Ahmarian assemblages from the southern Levant, as the few dates available were made decades ago without modern treatment methods (Alex et al. 2017). At Mughr el-Hamamah, the EUP is dated to 45–39 ka cal BP (Stutz et al. 2015).

The features of the IUP and EUP are shown from the cave of Üçağızlı in southwest Turkey (Kuhn 2004; Kuhn et al. 2009). In both sets of stone tool assemblages, hide working is the most frequently identified activity from microwear. In a sample of 35 end-scrapers, almost all showed use wear consistent with working dry hide and the burins from layers G and H indicated use on fresh or wet hide. Bone and antler tools were present but never abundant. Ornaments were also very common, with over 2,000 perforated shells (mostly of the marine *Nassarius gibbosula*) that could be used as beads, pendants or other ornaments from the EUP layers (Stiner et al. 2013). A large number of perforated shells (perhaps a third of c. 3,600 specimens) were also recovered from the IUP layers at Ksar Alkil (Bosch et al. 2015, 2018).

The large animals that were most frequently hunted at Üçağızlı were *Dama mesopotamica* (fallow deer), goat (*Capra aegagrus*) and roe deer (*Capreolus capreolus*), as well as pig (*Sus scrofa*), red deer (*Cervus elaphus*) and aurochs (*Bos primigenius*). Some carcase parts may have been prepared for smoke-drying: at the junction of layers B and B1–3 and by the cave wall, there was a concentration of ash delineated by a row of limestone blocks, and this might have been part of a structure for smoking meat or hides. As noted earlier, small, agile prey were now part of IUP diet, which included hares (*Lepus capensis*), Persian squirrel (*Sciurus anomalous*) and birds such as partridge (*Alectoris*, also known as chukar). At Manot, the main animals in the EUP were gazelle followed by fallow deer (*Dama* sp.) and occasionally red deer, roe deer, tortoise, hare, fox and chukar partridge, and other birds in the archaeological layers of Area C; interestingly, fallow deer, aurochs, equid and caprids were the main animals in spotted hyena deposits in Area D. *Prunus amygdalus* (almond) was common in all levels (Marder et al. 2017). Yeshurun et al. (2019) suggest that human hunters might have preferentially targeted female gazelle herds because these tend ot be larger and more predictable than male ones; also, the use of spears with antler points would have made killing more effective.[2] These differences might indicate niche separation, with humans hunting gazelle in areas of open vegetation and spotted hyena hunting fallow deer in wooded areas (Orbach and Yeshurun 2019).

In summary, IUP and EUP inhabitants were like Neandertals in hunting big game but were also more flexible and included a wider range of small game such as birds and hare. They also used strung ornaments of perforated shells on a much larger scale than at Skuhl and Qafzeh as components of strung ornaments (Stiner et al. 2013).

The climatic background

The climate of the east Mediterranean and the Levant over the last 200,000 years is known from several sources, including studies of microfauna (e.g. Belmaker and Hovers 2011), large mammals (e.g. Tchernov 1992), offshore pollen (Langgut et al. 2011), marine sediments and sapropels (Cane et al. 2002; Emeis et al. 2000), cave speleothems (Vaks et al. 2006, 2007; Yasur et al. 2019), dental isotopes (Hallin et al. 2012) and lake levels (Torfstein 2019; Torfstein et al. 2013). For a small region such as the Levant, these studies are surprisingly difficult to synthesise. One reason is that each of these has its own methodological problems: for example, offshore pollen can be derived from several vegetation zones, or microfauna derived from owl predation probably came from a much smaller catchment area than the large mammals in the same cave fauna. Additionally, the Levant is complex physiographically, botanically and climatically. It comprises a Mediterranean coastal plain, the Galilee Mountains, the Dead Sea Valley, Golan Heights and Negev Desert. Botanically, these encompass a Mediterranean, Irano-Turanian and Saharan-Sindian[3] flora (see Figure 8.4) (Hallin et al. 2012). Grasses with a C3 signature[4] are the commonest plant group in the Mediterranean region, where most rain is regular and falls between November and April. In the Irano-Turanian and Saharan-Sindian regions, where annual temperatures are higher, and rainfall more intermittent and variable, grasses with a C4 signature are dominant. Climatically, the Levant in the Pleistocene was at the interface of two climate systems between a humid north and an arid south and southeast. One was driven by the extent of the European ice sheets; as these expanded, westerly storms were pushed southwards into the east Mediterranean, and the other was the African monsoon, which occasionally moved northwards into Arabia because of shifts in the Intertropical Convergence Zone (ITCZ) (Frumkin et al. 2011; Hallin et al. 2012; Vaks et al. 2006, 2007; see also Chapter 4). The transition between the two systems occurs over a distance of only 150 km from the Mediterranean climate of central Israel to the hyper-arid conditions of the southern Negev and Sinai Deserts (Vaks et al. 2007). (Even so, there are numerous perennial springs in the Negev that could have supported dispersing populations even if the climate was arid [Goder-Goldberg, pers. com].)

As shown in Chapter 4, northward shifts of the African monsoon system brought increased rainfall to the Arabian Peninsula, and also opened up a corridor between Northeast Africa and the Levant between 140 and 100 ka, thus allowing humans to disperse as far north as central Israel. At this time, the Levant became part of "Greater Africa", as proposed by Foley (2018; see also Chapter 4). This corridor between the Levant and Northeast Africa was closed when the summer monsoon moved offshore. This period was thought to have been drier than today (see Shea 2008) but now appears to have been wetter, judging from histories of lake levels (see Frumkin et al. 2011; Hallin et al. 2012). Torfstein (2019) and Torfstein et al. (2013) present a detailed account of changes in the level of the Dead Sea over the last 70,000 years and show that fluctuations in lake level closely match changes in the Greenland ice core and Chinese speleothem records

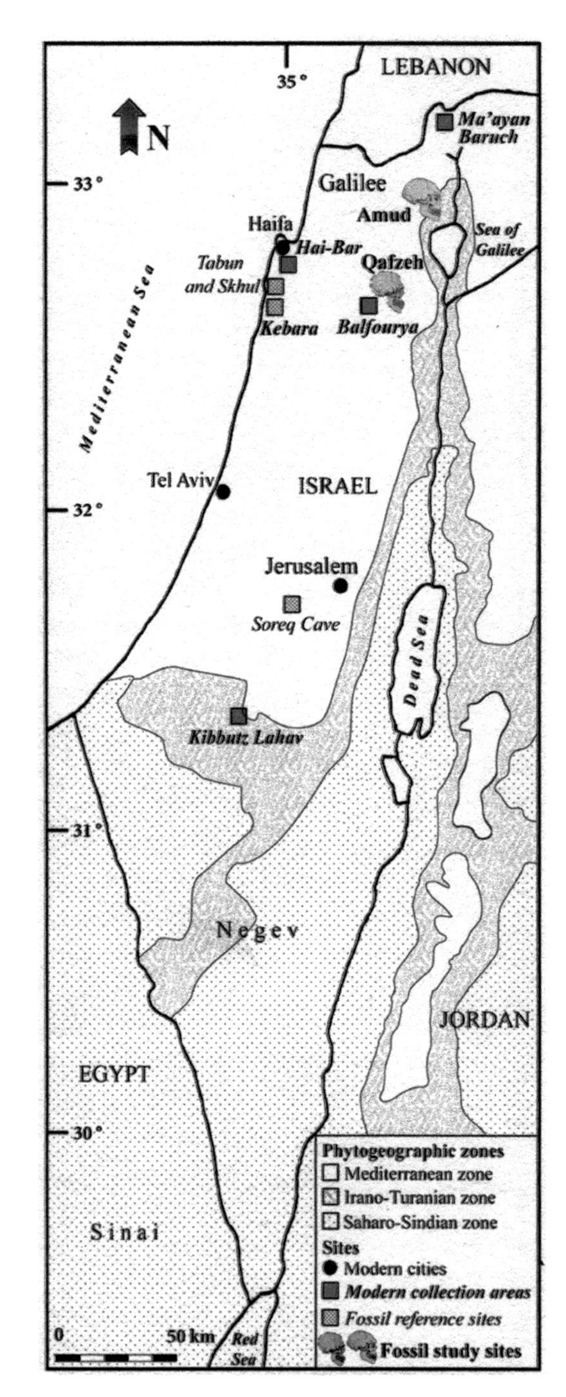

FIGURE 8.4 Modern vegetational zones of Israel

Source: Hallin et al. 2012, Fig. 1.

(Figure 8.5). They also show that there were rapid falls in lake level – along with temperatures and rainfall – during Heinrich Events, which were short but severe cold episodes in the North Atlantic. The overall regional picture is therefore one of a moist episode in the early part of MIS 5, and an increasingly arid Arabia in MIS 4, but a wet but variable central and northern Levant. Whether these changes were sufficient to cause Neandertal extinction in MIS 4 remains unclear. It is interesting, however, that Langgut and colleagues (2018) have identified a moist period in the east Mediterranean caused by an intensification of the African monsoon. This would have opened up a corridor through otherwise arid regions of North Africa, the Sinai and Negev Deserts, southern Jordan and the Arabian Peninsula that would have allowed the dispersal of humans into the Levant. This development may have been an important factor in the development of the Levantine IUP and EUP.

We are highly fortunate in having a detailed speleothem record from four stalagmites from Manot Cave that covers its occupation record to when it was abandoned after the cave entrance collapsed ca. 31,000 years ago (Yasur et al. 2019; see Figure 8.6). During the early Ahmarian (ca. 46–42 ka) there was a pronounced shift towards a drier, open grassy landscape (hence the preference for gazelle hunting), whereas wetter and more wooded conditions developed during the Levantine-

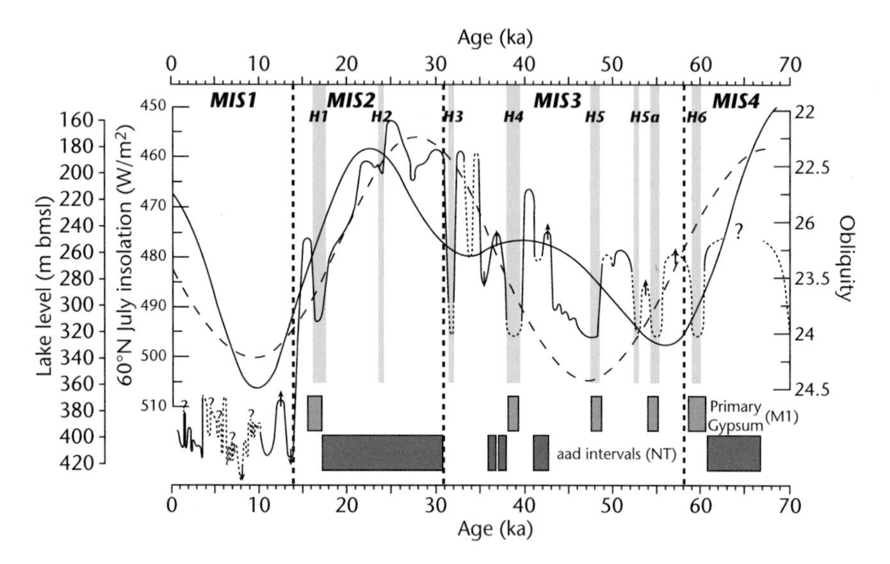

FIGURE 8.5 The climatic record from the Dead Sea

The figure shows the fluctuations in the depth of the Dead Sea over the last 70,000 years. These were highest in MIS 2, and considerably higher than in MIS 3. During MIS 3 between 30 ka and 60 ka, there were short but repeated falls in lake levels, with the largest associated with Heinrich (H) events and corresponding with the deposition of massive gypsum layers. Alternating aragonite-detritus (aad) sequences indicate a positive water balance in the lake system.

Source: Torfstein 2019, Fig. 2.

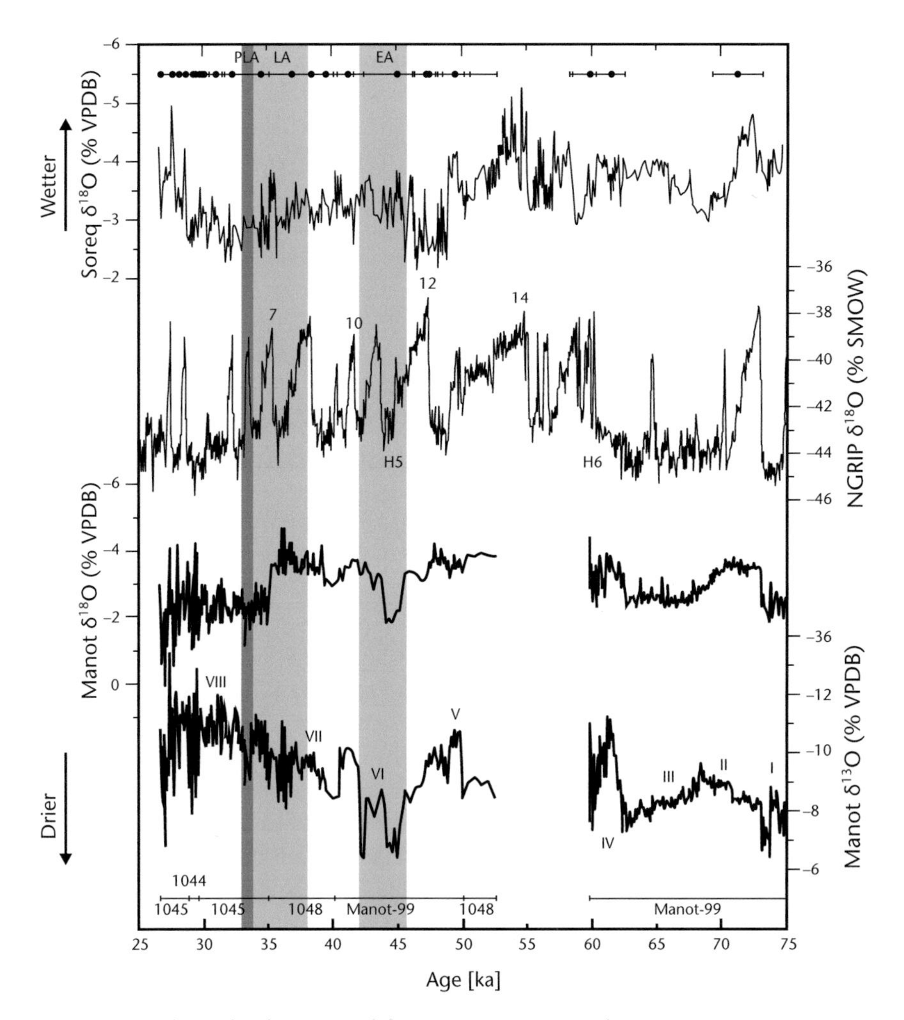

FIGURE 8.6 The speleothem record from Manot cave, Israel

Manot Cave speleothems $\delta^{13}C$ (bottom) and $\delta^{18}O$ (next to bottom) composite age records plotted with reference to the $\delta^{18}O$ records from Soreq Cave (top) and Greenland ice core NGRIP (second from top). Periods I–VIII are growth phases. The horizontal bars at the base of the diagram mark the portions of the speleothem records used to construct the composite profile. The duration of the Early Ahmarian (EA), Levantine Aurignacian (LA) and Post-Levantine Aurignacian (PLA) cultures are indicated by vertical bands. Cold Heinrich Events H5, H6 and warm-cold Dansgaard-Oeschger (D-O) cycles 7, 10, 12 and 14 are indicated adjacent to the NGRIP record. The circles with error bars at the top of the diagram indicate the measured U-Th ages of Manot Cave speleothems. Vertical arrows on the left side of the diagram show the wet and dry climatic trends for both isotope systems. (A colour version is shown in the original publication.)

Source: Yasur et al. 2019, Fig. 7.

Aurignacian (ca. 38–34 ka) and Post-Levantine Aurignacian layers (ca. 34–33 ka). The continuous growth of the speleothems indicates that water was always available throughout the cave's occupation, with average annual rainfall above 250–300 mm.

Towards a demographic history of the Levant

Virtually all aspects of the demographic history of the Levant before 40,000 years ago are fiercely contested. Researchers differ over whether there was one variable population, or two alternating populations of Neandertals and humans before 50,000 years ago; whether these interbred or even saw each other; whether climate played a role in the extinction of *H. sapiens* at the end of MIS 5 and Neandertals ca. 50,000 years ago; whether occupation was continuous or intermittent; and whether the eventual success of our species resulted from improved cognition, more flexible and diverse hunting practices, and/or climate change.

The Levant before 50,000 years ago

Three comments can be offered about the Levantine record before 50,000 years ago. First, it's a story of a small area with small populations. The perennial core area of settlement encompassing the Mediterranean woodland and its ecotone with the Irano-Turanian steppe was a narrow coastal strip bounded by the sea to the west, and deserts to the east and south. In this core area of settlement, the population was probably small. By reference to ethnographic examples, Shea (2008) suggests a maximum population of only 6,400. Size matters: small populations in a small region are vulnerable to local extinction through any of several factors, especially if these acted in combination, such as climatic downturns, random deaths through childbirth, fluctuations in infant mortality, deaths when hunting, inter- and intra-group violence and so on: there were numerous ways in which mortality could have outstripped fertility. An additional reason why populations were vulnerable was that they relied on a narrow resource base that focused on the hunting of medium and large animals, and the occasional intake of tortoises and other small and easily caught creatures. There is no indication of a fall-back strategy or any effective ecological buffering, unlike human populations after 50,000 years ago that did diversify to include small, agile prey such as hare or game birds.

Secondly, the Levant was the first area in Asia where the range of Neandertals and *H. sapiens* overlapped. I think it unlikely that they remained separate populations, and much more likely that they interacted and at times interbred. The Levantine skeletal record shows a series of highly variable populations, with some Neandertals showing traits of *H. sapiens*, and some *H. sapiens* individuals with Neandertal features. The relative strengths of the human and Neandertal gene pools varied over time, with Neandertals being commoner in some areas and periods, and vice versa with humans. For reasons still very poorly understood, Neandertals were no longer in the Levant after 50,000 years ago.

The Levant after 50,000 years ago

As noted above, some African populations may have been able to disperse into the Levant during the moist period between 56 ka and 44 ka. However, the origins of the IUP remain unclear but are unlikely to have been caused solely by the sudden influx of a new population with new ideas and a new technology. More likely perhaps is that it developed as a mosaic process driven over several generations by a series of small-scale developments among and between local populations, as suggested by Meignen (2012) and Kuhn and Zwyns (2014) as well as small-scale influxes of immigrants.

Two aspects of the IUP stand out as indicating major changes in the way people lived besides the shift to blade-based lithic assemblages. The first is the broadening of the subsistence base to include small agile prey and birds, and probably also plant foods, as seen at Üçağızlı. This denotes not just a more varied diet but a change in hunting technique to include traps, snares and other devices, and a change towards hunting being an individual activity as well as a group one. It could also denote the beginning of a gendered subsistence strategy, with more clearly defined roles for men, women as well as older children. Its origins need not have been tropical, let alone Africa: ecologically rich and diverse areas such as Galilee in northern Israel could easily have provided a suitable environmental context.

The second is the marked increase in personal ornaments, best illustrated by the hundreds of beads from Ksar Akil and Üçağızlı. It seems that personal identity began to matter in the way that people appeared to others. Although hard to demonstrate, social networks were becoming more formalised, with shared rules over how personal status and group identity were expressed. As seen in Chapters 9 and 10, the use of personal ornaments in the IUP forms a common strand that can be traced across continental Asia as far east as China: it obviously mattered to people how they looked and who they were.

On current estimates, the IUP lasted 3,000 years from 49,000 to 46,000 cal BP, or 1,200 generations if we assume a generational length of 25 years. What we see as a short period of rapid change (compared with the much slower rate of changes before 50,000 years ago) was probably one of profound stability to those who lived it. Cumulatively, however, over several centuries, populations began to grow (perhaps helped by immigration); in other words, fertility now began to outstrip mortality. The basis was forming for population expansion northwards into Europe, and eastwards into continental Asia. We can now begin to trace this eastward expansion.

Syria, Mesopotamia and Anatolia

Those hominins and later, humans, who dispersed eastwards from the Levant would have crossed what are now the deserts of modern-day Syria and Iraq, but were probably semiarid grasslands during MIS 3. The obvious water sources to follow would have been springs along the southern Taurus (Por 2004) and the Tigris-Euphrates river systems. Syria has a rich and largely untapped Palaeolithic record, particularly around the springs in the El Kowm region. In the long sequence of deposits that accumulated by an artesian spring at Hummal, there is an Ahmarian or Levantine Aurignacian occupation in Unit B Layer 4 that focused on blade and bladelet production of end scrapers and burins. At Umm el Tlel, there is another long sequence that includes no fewer than 29 layers with Ahmarian and Levantine Aurignacian assemblages (Le Tensorer 2015). In southwest Syria, the rock shelter of Baz contained a long sequence in which archaeological horizon VII was dated at 38,000–33,000 cal BP and contained an assemblage with blades and bladelets that was unlike the underlying Middle Palaeolithic horizons (Bretzke and Conard 2017; Deckers et al. 2009).

Little is known about the Palaeolithic of Anatolia in eastern Turkey. This is unsurprising because its winters are harsh, and the region was probably a barrier to dispersal in the Palaeolithic (Kuhn 2010). We know even less about the Palaeolithic of the Mesopotamian Plains, which is doubly unfortunate because these would have been used by groups that also settled in the adjacent Zagros Mountains.

The Zagros Mountains and Iran

Iran is defined geographically by the Iranian Plateau ca. 1,200 m a.s.l that is bounded by the Zagros Mountains in the west, the Alborz mountains in the north, and the Suleiman Range to the east. The Zagros Mountains begin in northern Iraq and extend as a series of parallel ranges 240 km (150 miles) wide for 1,800 km (1,000 miles) to the Strait of Hormuz at the southern end of the Arabian-Persian Gulf. After that, they become lower and sweep eastwards along the coast of the Makran, where they merge with the Suleiman Range near the border of Iran and Pakistan. Several mountains in the Zagros are over 3,000 m, and the highest point is Mount Dena, at 4,409 metres (14,921 ft), only a little lower than the Matterhorn (4,478 m). In the Alborz Mountains, Mount Damavand is Asia's highest volcano at 5,610 metres (18,403 ft). In the mountains, winters are severe and temperatures can fall to –25° C. (–13° F.). Snowfall can be heavy. Summers throughout Iran can be very hot and often exceed 40° C., especially in the desert regions of Central Iran.

Climatic and environmental background

Pleistocene climatic studies in Iran have progressed considerably in recent years. In northeast Iran, loess deposits are extensive east of the Caspian Sea, and sections at Agh Band and Toshan provide detailed histories of the last glacial-interglacial cycle (Lauer et al. 2017a, 2017b; Wang et al. 2017; Vlaminck et al. 2018). Sediment and pollen cores from lakes have also been studied. For many years, lake Zeribar in northwest Iran provided the best record of vegetational change in the region back to ca. 42 ka (Zeist and Bottema 1977; Wasylikowa 2005), but this has now been supplemented by a 100 m core from lake Urmia that covers the last 200,000 years (Djamali et al. 2008). Across the border in eastern Turkey, the PALAEOVAN project (Litt and Anselmetti 2014) established a vegetational and climate record for Lake Van for the last 600 ka (Litt et al. 2014) and a detailed record of the last 90 ka (Çatağay et al. 2014). Glaciers in Iran have also been studied (see Ferrigno 1993). These now cover only 20 sq km but were formerly more extensive in the Zagros and Alborz Mountains. Various studies show that summer snow lines that are currently at ca. 4,000 metres a.s.l. were depressed to 2,800–2,200 m in the northern Zagros (Wright, 1962), 2,400 m in the southeast Zagros (Kuhle 2007) and 3,000–2,500 m in the Taurus and eastern Turkey (Sarıkaya et al. 2011), with the consequence of some habitat loss and fragmentation (see Chapter 3), and greater restrictions on movement.

Some progress has been made in studying speleothems; as example, Rowe and colleagues (2012) present a speleothem record from Karaca Cave in northwest Turkey that extends from 5 ka to 80 ka and agrees closely with the record from Soreq Cave, Israel, Hulu and Sanbao Caves in China, and the east Mediterranean marine records (see Figure 8.7). In northwest Iran, Mehterian et al. (2017) document a speleothem record from Qal'e Kord that covers 73–125 ka and 6.5–7.5 ka; the intervening hiatus should be closed soon. The playas, or salt lakes, that cover much of central Iran have received some attention (see e.g. Abdi et al. 2018; Rahimpour-Bonab et al. 2012) but although cores from them indicate a change from a humid cold climate in the Late Pleistocene to an arid warm climate in the Holocene, they have so far proved difficult to date. Nevertheless, the presence of Middle Palaeolithic sites such as Mirak and Chah-i-Jam (Rezvani and Vahdati Nasab 2010; Vahdati Nasab and Hashemi 2016) on the edges of former lakes indicate that they were important foci for human/hominin groups. There is also an emerging field of faunal studies directed at environmental reconstruction by analyses of microfauna (e.g. Hashemi et al. 2006), macrofauna (Mashkour et al. 2009a) and pollen from hyena coprolites (Djamali et al. 2011).

Kehl (2009) provides a detailed overview of Quaternary climatic change in Iran. Its climate has been shaped by changes in the pressure systems of Siberia, the Mediterranean and the southwest monsoon. Today, ca. 75% of Iran has a semi-arid to arid climate, with ca. 350–50 mm of rainfall. The areas with the highest rainfall are the provinces of Mazandaran and Gilan along the south coast of the Caspian, and the high parts of the Alborz and the Zagros Mountains in northwest Iran: here, annual precipitation can exceed 1,000 mm. Most precipitation is delivered as rain and snow to the Zagros and Alborz Mountains by westerly winds from the Mediterranean and Black Sea, but little of this rainfall falls on the Plateau. Rainfall along the northern edge of the Alborz comes from the Caspian, and a small amount from the Indian monsoon reaches southern Iran. In cold periods, westerly airflows from the Mediterranean weakened and rainfall decreased. Average annual temperatures in the Zagros were probably depressed by 5° C.; by 5° to 8° C. in central Iran during the LGM; and by an average of 8° to 10° C. across Iran. In north and west Iran, the climate alternated from cold and dry conditions in glacial and stadial periods such as MIS 2 and MIS 4, and warmer and moister

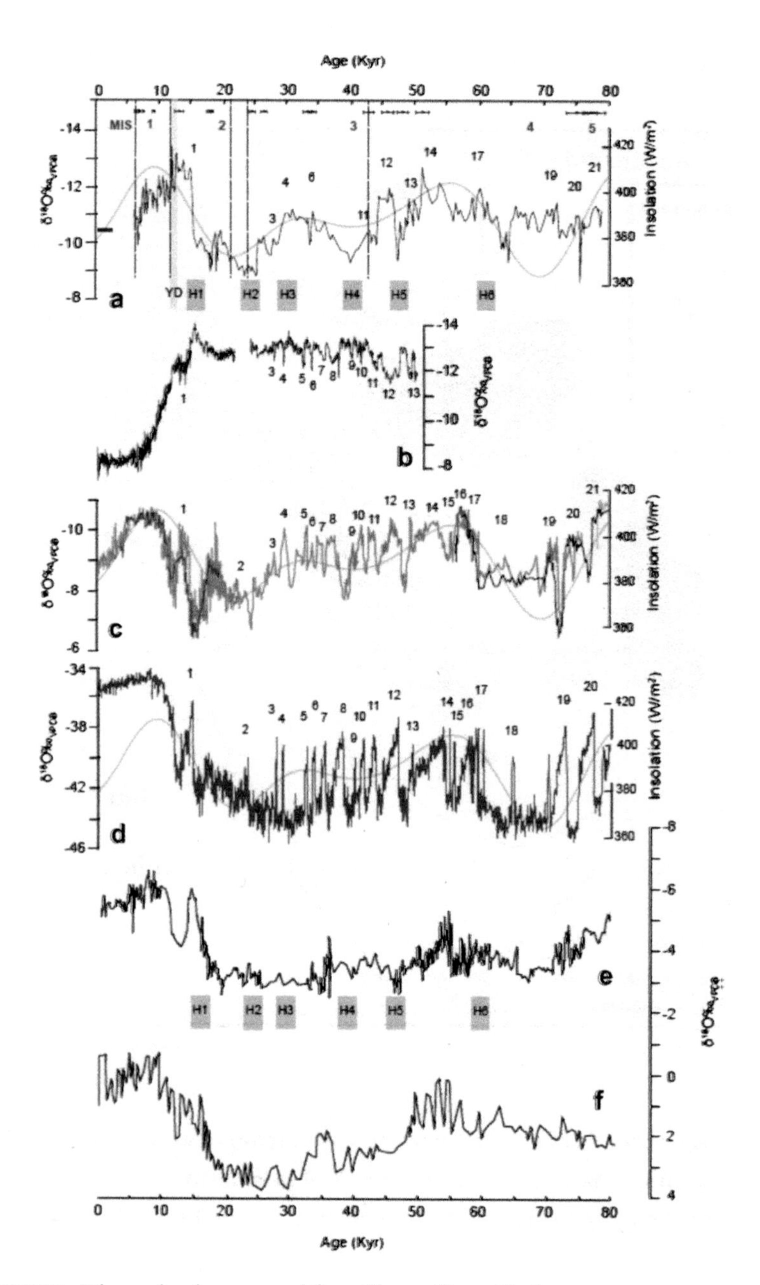

FIGURE 8.7 The speleothem record from Karaca Cave, Turkey

(a) The Karaca Cave K1 stalagmite $\delta^{18}O$ record. ^{230}Th dates and 2-sigma error bars are shown; YD denotes Younger Dryas; H1–H6 mark Heinrich Events; horizontal bar represents $\delta^{18}O$ value of calcite presently forming in nearby Akçakale Cave. Vertical dotted lines show positions of physical breaks or saw cuts in the stalagmite. (b) Sofular So-1 speleothem record, northwest Turkey. (c) Hulu/Sanbao (China) speleothem records; more negative values correlate with increased monsoon intensity. (d) Greenland NGRIP $\delta^{18}O$ record. (e) Soreq Cave speleothem record. (f) Sea surface $\delta^{18}O$ record of eastern Mediterranean surface waters from Core 9501. Numbers above (a) denote marine isotope stages; numbers 1–21 refer to Greenland Interstadials. Curved lines show July solar insolation at 65° N. (A colour version is shown in the original publication.)

Source: Rowe et al. 2012, Fig. 7.

conditions in MIS 3 and MIS 5. In cold periods, the dominant vegetation in the Zagros was steppe, dominated by *Artemisia*, Chenopods and Umbellifers,[5] with scattered stands of trees, most of which disappeared in the LGM (Zeist and Bottema 1977) because of aridity. The Iranian Plateau would have been dominated by northerly cold, dry winds from the Central Asian deserts, and increases in loess grain size is one indicator that wind strength may have increased (Wang et al. 2017). Maximum summer temperatures on the Plateau during the LGM may have been as low as 10° to 12° C. (Djamali et al. 2008).

The main challenges for humans in the Zagros and on the Plateau in cold periods would have been in over-wintering; lowered snow lines and winter snow would also have impeded movement in the Zagros and Elburz Mountains. On the Plateau, dust storms driven by strong winds would have been an additional hazard.

Palaeolithic research in Iran

Palaeolithic research in the Zagros Mountains of Iraq and Iran and in the rest of Iran has two distinct phases (see Conard et al. 2013). The first was led by European and American researchers and ended in 1979 with the Iranian Revolution. The earliest Palaeolithic research took place in Iraqi Kurdistan – Dorothy Garrod at Hazar Merd and Zarzi in 1928, followed by Henry Field's (1951) surveys and excavations in Iran (and Afghanistan), and Ralph Solecki's three seasons of fieldwork (1951, 1953, 1956/1957 [Ghasidian et al. 2017]) at the now-iconic Neandertal site of Shanidar (now under re-investigation by a Cambridge-led team). In 1949, Carlton Coon excavated Besitun in the Zagros, Tamtama on the northwest Iranian Plateau, and Kara Kamar in Afghanistan (Coon 1957). The main Palaeolithic research in Iran in the 1960s was in the Kermanshah and Khorramabad valleys of the central Zagros (Figure 8.8). These investigations were spin-offs from projects investigating the origins of agriculture. Robert Braidwood's team excavated the early Neolithic sites of Tepe Asiab and Tepe Sarab in the Kermanshah region but also the caves of Besitun and Kobeh and the rock shelter of Warwasi (Braidwood et al. 1961), and the team led by Frank Hole and Kent Flannery investigated the caves of Yafteh, Kunji, Pasangar and Gar Arjeneh in the Khorramabad Valley (Hole and Flannery 1967). Several other researchers also conducted Palaeolithic projects in various parts of Iran up to 1979. This phase of fieldwork was summarised by Smith (1986) and Olszewski and Dibble (1993).

After a long hiatus, Palaeolithic research resumed in Iran in the 2000s, but this time mostly led by a small but active group of Iranian archaeologists, sometimes working with European colleagues. These have continued to research in the Kermanshah Valley (e.g. Jaubert et al. 2009; Biglari and Shidrang 2016; Shidrang et al. 2016; Heydari-Guran and Ghasidian 2017) and the Khorrambad Valley (Mashkour et al. 2009b; Otte et al. 2009; Bazgir et al. 2014, 2017), but they have also opened up new areas, such as the southern Zagros (Ghasidian et al. 2017), the Caspian coastal plain (Biglari and Jahani 2011), the Alborz (Berillon et al. 2007); the Iranian Plateau (Biglari et al. 2009; Heydari-Guran et al. 2009, 2015; Rezvani and Vahdati

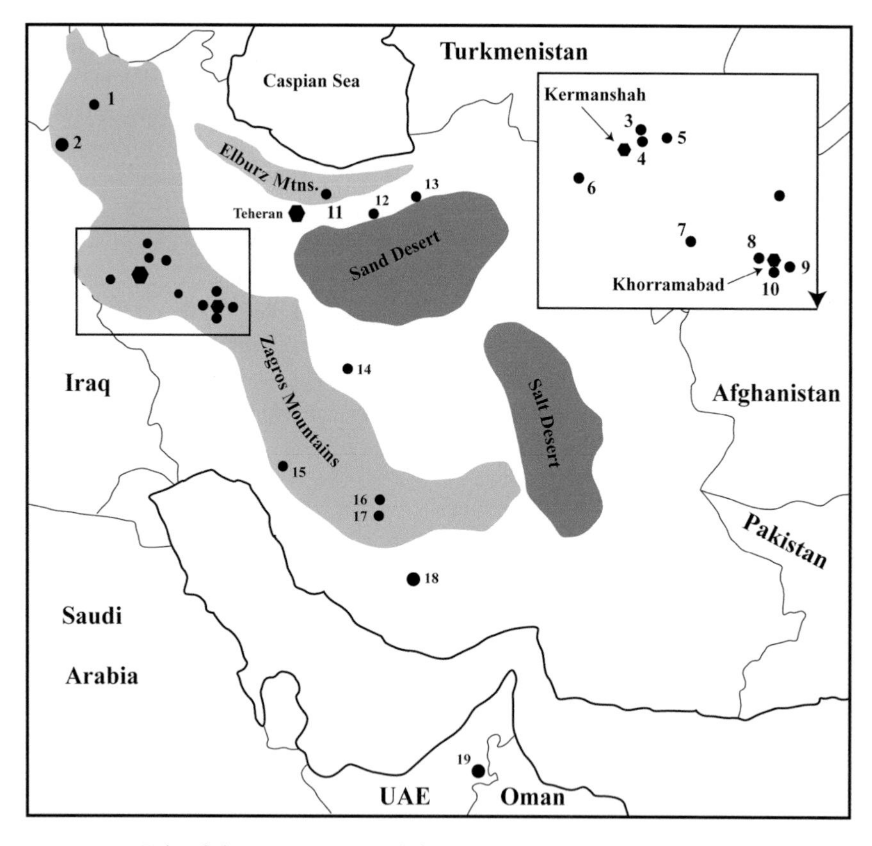

FIGURE 8.8 Palaeolithic sites in Iran and the Zagros Mountains

Iranian Palaeolithic sites: 1 Tamtama; 2 Shanidar; 3 Kobeh; 4Warwasi; 5 Besitun; 6 Wezmeh; 7 Houmian; 8 Yafteh; 9 Kunji; 10 Gar Arjeneh; 11 Garm Roud; 12 Mirak; 13 Chah-i-Jam; 14 Qaleh Bozi; 15 Ghar-i-Boof; 16 Eshkaft-i-Gavi; 17 Ghad-i-Barmeh; 18 Jahrom; 19 Jebel Faya.

Source: The author.

Nasab 2010; Vahdati Nasab and Hashemi 2016); southwest Iran (Bahramiyan and Shouhani 2016), and east Iran (Nikzad et al. 2015). Compared with 1979, there is now an impressive literature on the Palaeolithic of Iran. Nevertheless, what we don't know about the region still far outstrips what we do know.

When did Homo sapiens *appear in the Zagros region?*

Shanidar in Iraqi Kurdistan has a long Mousterian sequence that includes the remains of ten Neandertal individuals (Trinkaus 1983). In Iran, as in many other parts of continental Asia, the hominin skeletal evidence for either Neandertals or *Homo sapiens* is meagre: an indeterminate radius fragment from Besitun (Trinkaus and Biglari 2006), and Late Palaeolithic (and probably butchered) human remains

TABLE 8.3 Radiocarbon dates of Upper Palaeolithic sites in the Zagros

Site	*Period*	*C14 date*	*Cal BP*
Shanidar C	Upper Palaeolithic/Barodostian	33,300 ± 1,000	38,196–33,476
		33,900 ± 900	38,566–34,256
		34,000 ± 420	37,600–35,218
		35,440 ± 600	39,386–36,856
		34,540 ± 500	38,390–36,079
Yafteh	Upper Palaeolithic/Aurignacian	35,450 ± 600	40,510 ± 672
		32,770 ± 290	37,435 ± 491
		34,160 ± 360	39,220 ± 518
		33,430 ± 310	38,118 ± 471
		32,190 ± 290	36,755 ± 384
Ghar-I Boof	Upper Palaeolithic/Rostamian	35,950 ± 800	42,050–38,950
		36,030 ± 390/–370	41,355 ± 326
		33,850 ± 650	38,994 ± 1,419
		34,900 ± 600	39,949 ± 921
		31,620 ± 180	36,000–35,000
Kaldar	Upper Palaeolithic	33,480 ± 320	38,650–36,750
		39,300 ± 550	44,200–42,350
		49,200 ± 1,800	54,400–46,050

Source: Ghasidian et al. 2019.

from Eshkaft-e Gavi (Scott and Marean 2009). There is also a Late Pleistocene upper third premolar (P^3) from Wezmeh (see below and Table 8.4) that Trinkaus et al. (2008) decided was indeterminate but which is now identified as Neandertal (Zanolli et al. 2019) – the first positive indication of Neandertals in the Zagros. Unfortunately, its age is uncertain but probably lies somewhere in MIS 3 (Trinkaus et al., 2008: 376; and Zanolli et al. 2019: 4). The working assumption is necessarily that the Middle Palaeolithic of Iran was made by Neandertals (as at Shanidar) and the Upper Palaeolithic was made by our species. We need to be open-minded here: because *Homo sapiens* made Middle Palaeolithic assemblages in Arabia (Chapter 4) and the Levant, we cannot exclude the possibility that they also made them in the central and southern Zagros, southern Iran or on the Plateau. Nevertheless, the fact that there is skeletal evidence of *H. sapiens* in north China at 40 ka (Chapter 10) makes it likely that our species also made the Upper Palaeolithic in Iran after that date.

In the Zagros region, the big debates have been whether the Upper Palaeolithic developed from the local Mousterian; whether the Upper Palaeolithic can be regarded as Aurignacian; and whether the Aurignacian as known in Europe originated in the Zagros.

The cave of Shanidar in northeast Iraq has occupied a central role in these debates. There, Solecki identified an Upper Palaeolithic in layer C. This was flake-based, and contained burins, carinated pieces, side-scrapers and points, and was

TABLE 8.4 The fauna from Wezmeh Cave, western Iran

Taxon	*English name*	*NISP*	*MNI*
Carnivora			
Crocuta crocuta	Spotted hyaena	437	10
Ursus arctos	Brown bear	192	5
Panthera leo	Lion	16	2
Panthera pardus	Leopard	2	1
Caracal/Lynx/Felis chaus	Caracal/lynx/jungle cat	4	2
Felis sylvestris	Wildcat	8	2
Canis lupus	Wolf	176	6
Vulpes vulpes	Red fox	492	19
Meles meles	Badger	42	7
Mustela putorius	Polecat	7	2
Martes martes	Stone marten	2	1
Herpestes	Mongoose	5	3
Primate			
Homo sp.	Neandertal	1	1
Perissodactyla			
Dicerorhinus	Rhinoceros	2	1
Equus caballus	Horse	1	1
Hemionus/E. asinus	Hemione/onager/donkey	4	1
Artiodactyla			
Bos primigenius	Aurochs	13	2
Sus scrofa	Wild boar	97	4
Cervus elaphus	Red deer	59	3
Gazella sp.	Gazelle	17	1
Ovis orientalis	Mouflon, wild sheep	127	6
Capra aegagrus	Wild goat	31	2
Rodentia/Insectivora			
Erinacaeus sp.	Hedgehog	5	3
Hystrix indica	Indian porcupine	2	1
Lagomorpha			
Lepus capensis	Cape hare	145	17
Reptilia			
Testudo graeca	Spur-thighed tortoise	46	4
Laudakia	Agamid	1	1
Malpolon monspessulanus	Grass snake	14	1
Vipera lebetina	Viper	2	1

Comments: NISP = number of individual specimens; MNI = minimum number of individuals. Most taxa in this table are Palearctic and still extant in Iran, or were in historic times. We can add to this list the roe deer *Capreolus*, recorded at Kuldar and Gilvaran Caves, Khorramabad. *Malpolon* is found now only in Iberia and northwest Africa. Lions are now extinct in Asia except in Gir reserve, India, but were once widespread across southwest Asia, and are graphically shown in the Assyrian bas reliefs of the lion hunt of Ashurbanipal in the 7th century BCE. The Oriental leopard is now endangered and found only in northern Iran and the Caucasus. *Dicerorhinus* is now found only in Sumatra but was once widespread across southern Asia. Rhinos are now extinct in southwest Asia but were formerly distributed from Africa to southeast Asia. Bazgir et al. (2014) suggest that at Wezmeh, it may be the steppe rhino *Stephanorhinus*. *Lepus capensis* is mainly African but is found also in Arabia and Iran.

Source: Mashkour et al. 2009a; and Zanolli et al. 2019 for the identification of the Neandertal specimen.

dated by C14 as between 28,700 ± 700 and 35,440 ± 600 BP (uncalibrated). (These have since been re-modelled to ca. 40–45 ka cal BP by Becerra-Valdivia et al. [2017]; see below.) Although Solecki was struck by the general similarities of the Shanidar layer C assemblage with Early Upper Palaeolithic Aurignacian assemblages from Europe, he chose to name it Baradostian, after the local mountain, following advice from Dorothy Garrod (Ghasidian, pers. com.) Similar assemblages were reported but not fully published from other caves in the central Zagros such as Gar Arjeneh, Pasangar, Yafteh (Figure 8.9) and Warwasi, so it was clearly not

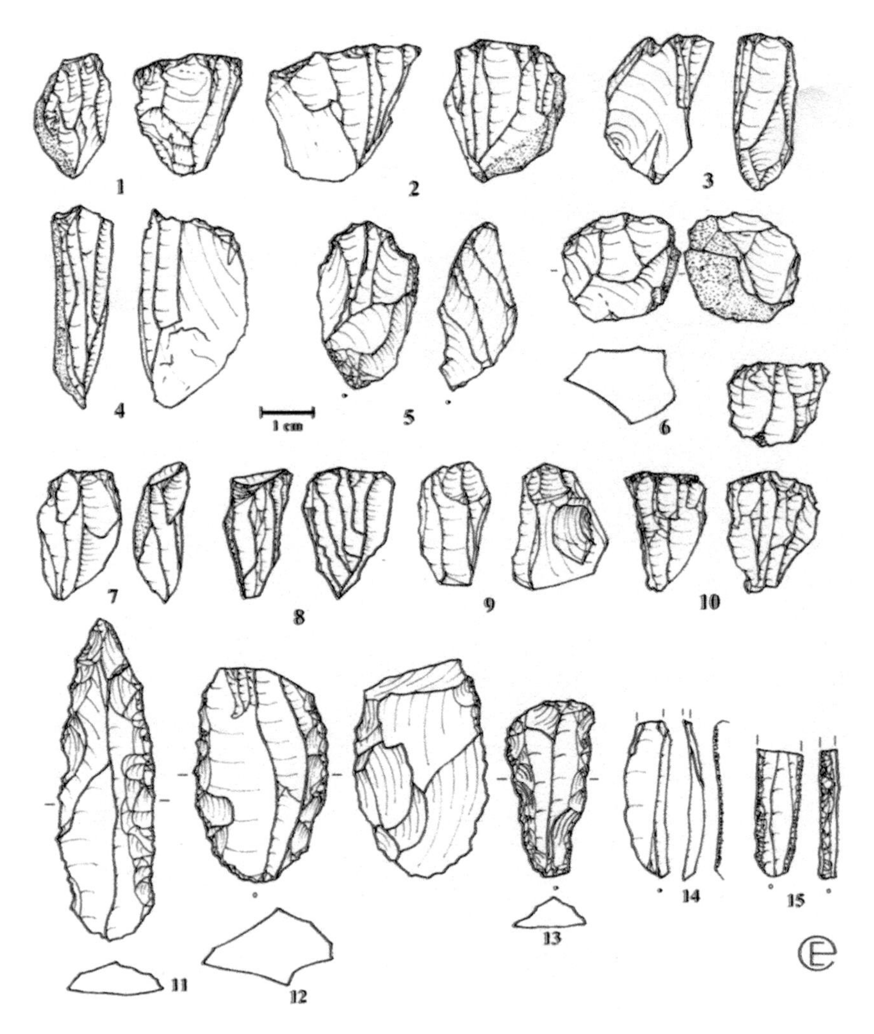

FIGURE 8.9 Early Upper Palaeolithic stone tools from Yafteh, western Iran

Key: 1–5, 7–10 bladelet cores; 6 discoidal flake core; 11 scraper; 12 retouched flake; 13 scraper on blade; 14 retouched bladelet; 15 backed bladelet.

Source: Ghasidian et al. 2019, Fig. 3.

unique to only one site. (The lithics from Pasangar and Yafteh have since been fully analysed and presented by Shidrang [2015] and Ghasidian [2019].) In the southern Zagros, Baradostian assemblages were also claimed from the caves of Ghad-i Barmeh Shour (Piperno 1974) and Eshkaft-e Gavi (Rosenberg 1985) in Fars Province, thus extending its distribution across the whole of the Zagros. Late Baradostian assemblages dated at (23,920 ± 160 ^{14}C BP and 28,486 ± 190 ^{14}C BP) were later recognised at the open-air site of Garm Roud 2 in the Alborz Mountains (Berillon et al. 2007).

Olszewski and Dibble (1993) analysed the Baradostian assemblages from Warwasi and proposed that these should be named Zagros Aurignacian, to emphasise their general similarities with the European Aurignacian but also their regional characteristics. These were emphasised even more strongly by Marcel Otte (Otte et al. 2009; Otte 2014), who argued that the Aurignacian actually originated in Iran and then spread into Europe. Olszewski and Dibble (1993) also argued that the presence of Mousterian type flakes in the early Baradostian layers suggested a local origin for the Baradostian. More recently, Conard and Ghasidian (2014) and Ghasidian and colleagues (2017) have argued that the Early Upper Palaeolithic assemblages from Ghār-e-Boof in the southern Zagros that date between 35 kyr cal BP and 42 kyr cal BP (Ghasidian et al. 2019) show substantial differences from the Baradostian of the north and central Zagros, and justify a separate naming as Roustamian. As shown below, this suggestion makes sense when considered against the topography of the Zagros. (However, see Shidrang [2018] for a contrary view.)

Resolution of these issues depends on secure chronologies and detailed assessments of cave stratigraphies that contain carefully excavated Late Middle and Early Upper Palaeolithic assemblages. Unfortunately, few caves contain both types. At Gar Arjeneh, the Middle and Upper Palaeolithic layers were disturbed (Shidrang 2018); Warwasi and Ghar-e-Khar were coarsely excavated in 20 cm and 10–30 cm spits respectively (Bazgir et al. 2017). Opinion is still divided over whether the Baradostian developed from the local Mousterian – as argued by Olszewski and Dibble (1993) and recently by Tsanova (2013). However, as Heydari-Guran (2015) points out, the lack of stratigraphic control makes this proposed local transition highly questionable; and we need more sites to consider than only Warwasi and Ghar-i-Khar (Shidrang 2018). My own view is that the Baradostian, or as some would prefer, the Zagros Aurignacian, is an intrusive development, as recently argued by Ghasidian and colleagues (2019). As the Zagros Upper Palaeolithic assemblages are now younger than the earliest Aurignacian in Europe (ca. 43 ka cal BP), it is also unlikely that they play any part in the origin of the European Aurignacian (see Ghasidian et al. 2019).

Current assessments

Excavations at the cave of Kaldar in the Khorramabad Valley showed a sequence of Middle to Upper Palaeolithic assemblages in layers 5 and 4 respectively (Bazgir et al. 2017). TL dates for layer 4 ranged from 23,100 ± 3300 to 29,400

± 2300 BP, and AMS dates from 38,650–36,750 cal BP, 44,200–42,350 cal BP, and 54,400–46,050 cal BP (all at the 95.4% confidence level). The excavators agreed that the lithics from this layer could be classed as Baradostian or Zagros Aurignacian. Recently, Becerra-Valdivia and colleagues (2017) have obtained new AMS dates for Kaldar Cave, Kobeh and Ghār-e Boof and modelled these (along with previously published dates from Yafteh and Shanidar) to indicate that the Upper Palaeolithic dates to 45,000–40,250 cal BP (with a 68.2% probability). On the assumption that *H. sapiens* made the Zagros Upper Palaeolithic, we can place our species in the Zagros ca. 45,000 calibrated years ago. This is slightly earlier than in the neighbouring Caucasus region where Upper Palaeolithic assemblages appear between 42 and 37 ka cal BP (Adler et al. 2008; Pinhasi et al. 2012).

Middle and Upper Palaeolithic subsistence in Iran

Wezmeh Cave is located in an inter-montane valley southwest of Kermanshah in the Central Zagros at 1,430 m and was primarily a carnivore den that was used from ca. 70 ka to sub-recent times (Mashkour et al. 2009a). As Table 8.4 shows, it contained 12 types of carnivore, including hyaena, bear, lion and wolf (one of the few recorded instances where both lion and wolf occupied the same area), nine types of herbivores (including a rare instance in Iran of rhinoceros) and various small mammals and reptiles. From this potential menu, Neandertals and humans mostly targeted wild goat, gazelle, horse/wild ass, aurochs and occasionally deer. Hesse (1989) suggests that there were two main hunting strategies in both the Middle and Upper Palaeolithic (see Table 8.4). At caves such as Shanidar, Ghar-e-Khar, Kobeh, Karim Shahir, Pasangar and Yafteh, wild sheep/goat[6] were the main prey, with goat more likely than sheep. In contrast, red deer, equids and aurochs were the main prey at Bisetun, Tamtama, Gar Arjeneh, Hazar Merd, Qaleh Bozi 2, Warwasi and the terminal Pleistocene site of Palegawra. Hunting these animals involves very different tactics. Wild goat inhabits rocky hilly terrain, and migrates vertically by going high in summer and descending to lower levels in winter. They are strongly territorial, live in small groups, have small annual territories and are "formidably elusive" (Marean and Kim 1998). Successful hunting would require either a skilled hunter with bow and arrow, or a combined effort to drive one or more off a cliff. However, despite their fecundity and ability to rebound after an episode of heavy mortality, over-hunting is a danger, and there will be a lag before migrants move into a new territory. Repeated, year-round hunting of wild goat is not therefore sustainable over long periods unless combined with other types of hunting. At Shanidar, for example, wild goat, boar and tortoise provided the dietary base, supplemented occasionally by wild sheep, red- and roe-deer (Evins 1982), and perhaps fish (Reynolds et al. 2018). The goats were mainly prime adults that may have been killed by, for example, being driven off a cliff. Analysis of phytoliths and starch grains in dental calculus shows that Neandertals at Shanidar also cooked grasses and legumes and ate dates (*Phoenix* sp.) (Henry et al. 2011). At Kobeh, wild goat was the main prey, followed by various types of horse and a small number of pig and gazelle (Marean and Kim 1998). At Yafteh Cave, sheep/goat were predominant but there were also numerous hare and fish bones[7] (Mashkour et al. 2009b) – hares are one of the few hints of hunting small, agile prey that are seen in both the Zagros and the Levant at Üçağızlı.

In contrast, horse, gazelle, deer and aurochs migrate between winter and summer pastures that can be far apart. These have much larger annual territories, and although deer, aurochs, onager and gazelle might be easier to hunt than wild goat, hunters need to be in the right place at the right time to intercept them. This might not be too difficult in the broad inter-montane valleys of the Zagros and Alborz Mountains, but would be harder on the wide expanses of the Iranian Plateau, where they can easily escape, and their movements may be harder to predict. Devès et al. (2014) point out ways in which hunters can lessen risk by their familiarity with the details of landscapes, such as soil quality, likely migration corridors, water sources and optimal places for ambushing (near water sources, for example, or in valleys in the foothills of mountains). Nevertheless, integrating the hunting of both goat in rocky terrain and onager, gazelle and deer in open landscapes would place considerable strains on hunters, whether Neandertal or human. It may be that humans replaced Neandertals because they were marginally more successful in hunting slightly more often than their Neandertal rivals. Yafteh provides a hint that humans may also have been more successful in widening their resource base by catching small elusive game such as hare and fish.

A biogeographic perspective

A number of points can be made about the human biogeography of Iran in MIS 3 between 50,000 and 30,000 years ago during a mild part of the last glacial cycle. The first is that Iran covers an enormous area of ca. 1.6 million sq km (or six times the size of Britain), and is also incredibly diverse, from the sand deserts in its centre to the glaciers on the highest peaks, from the dense forests along the Caspian coast to the salt deserts of southern Iran. Because of its size and diversity, it is most unlikely that the Middle and Upper Palaeolithic records from Iran will be uniform – we should expect to see regional and ecological diversity in its archaeological records. Secondly, its environments and climate were also dynamic: in the mountains snow- and tree-lines rose and fell, and in the deserts, lakes and streams expanded and contracted. As these changed, there would have been different opportunities and challenges for human and other animal populations. One key issue is the extent to which the Plateau was ever "green" like Arabia or the Sahara. Today, the deserts of central Iran are barriers to movement, especially across the sand desert of the Dasht-i-Kavir in the north, and the playas (salt lakes) of the Dasht-i-Lut in the south. The discovery of extensive tufa deposits associated with Palaeolithic artefacts (Heydari-Guran et al. 2009, Heydari-Guran et al. 2015) at several places in central Iran is a strong indication that parts of the desert plateau might have been green, and thus opened corridors for movement (and

TABLE 8.5 Prey taken at Zagros Caves

Site	*Goat/sheep*	*Bos*	*Cervus*	*Gazelle*	*Equid*	*NISP*
Goat-focused						
Shanidar Mousterian	98	<1	<1	<1	0	1,138
Shanidar Baradostian	98	0	2	0	0	???
Karim Shahir, Middle Palaeolithic	76	5	12	7	0	163
Houmian, Middle Pal.	Mainly goat/sheep with a pig, deer and onager					
Pa Sangar, Upper Pal.	Mostly goats					
Kobeh, Middle Pal.	213	2	8	1	72	302
Yafteh, Upper Pal.	614		14	23		1,181
Equid, deer focused						
Besitun, Middle Pal.	0	2	43	10	45	124
Tamtama Middle Pal.	3	3	78	<1	15	179
Warwasi, Middle Pal.	23	2	8	0	67	114
Gar Arjeneh Middle Palaeolithic	Mostly aurochs, onager and red deer					
Hazar Merd Middle Palaeolithic	Deer, gazelle, goat (in list order)					
Palegawra UP	27	1	20	4	48	163
Gar Arjeneh UP	Mostly aurochs, onager and red deer					
Qaleh Bozi 2, Middle Palaeolithic	Mostly horse; rhinoceros, aurochs, sheep/goat, gazelle also listed					

At Kobeh the other animals were pig (1), fox (3), hyaena (1) and wolf (1). At Yafteh, note the presence of 3 pig bones, 12 of carnivores; 86 bones of hare and 294 of fish. NISP = number of identified specimens.

Source: Hesse 1989 except for Kobeh (Marean and Kim 1998, Table 1), Yafteh (Mashkour et al. 2009b) and Qaleh Bozi 2 (Biglari et al. 2009).

colonisation). Likewise, sites such as Chah-i-Jam and Mirak on the edges of extinct lakes implies that the Plateau was considerably less arid than now.

An important aspect of Iran's diversity is that it encompasses two biological realms. Southern Iran along the coast of the Indian Ocean is the eastern part of the Afro-Arabian realm (Chapter 4), whereas the rest lies in the Palearctic Realm. In Central Iran, for example, the large mammals are Palearctic, but most of the reptiles are Saharo-Arabian, and the small mammals are a mixed group (Bazgir et al. 2017). These researchers raise an additional important point about the fauna of Iran in the last glacial cycle. As noted already, it was Palearctic – but, they note, it was an interglacial type of Palearctic fauna, with large mammals such as fallow deer, roe deer, pig and horse, and an absence of the cold-adapted Palearctic mammals such as reindeer, woolly mammoth and woolly rhinoceros. In other words, Iran was a fairly gentle introduction to the Palearctic world, especially regarding its winters, compared to what awaited them in Central Asia and Siberia.

Homo sapiens could have first entered Iran from several points and at several times along the 1,800 km length of the Zagros Mountains. Because of its topography, passage into, along and through the Zagros is restricted to a small number of routes, whether for early hominins, invading armies, nomads or modern buses. In "green" periods, access to the Plateau would also have been possible from the southern coastal regions (see Chapter 4). As Heydari-Guran (2015) has pointed out, the high mountains of the central Zagros, with peaks over 4,000 m, divide the northern from the southern Zagros. It would not therefore be surprising if the earliest Upper Palaeolithic assemblages in Iran showed regional variation, as implied by naming the Ghar-i-Boof assemblages from southwest Iran as Rostamian instead of Baradostian (Ghasidian 2014; Conard and Ghasidian 2011). Recently, Ghasidian and colleagues (2019) make an excellent case in arguing that the north, central and southern Zagros (i.e. Shanidar in the north, the Kermanshah and Khorrambad Valleys in the centre, and Ghār-e-Boof in the south) each show different cultural trajectories. Nor is it surprising that the Middle Palaeolithic on the Plateau is different from that in the northern Zagros (Bretzke and Conard 2017; Heydari-Guran et al. 2015). (As noted already, it could have been made by Neandertals, or *Homo sapiens*, or, indeed, both.) Given the size and diversity of Iran, it is most unlikely that we are witnessing a single major dispersal event of humans into and across Iran. Instead, it is far more likely that the record is a mosaic of small-scale movements into various areas (Heydari-Guran 2015). Humans would have probably overlapped with local Neandertals, whose own population was likely small as well. If, over a millennium or so, the demographic balance between fertility and mortality was only slightly more favourable for humans than for Neandertals, humans would have eventually outnumbered and replaced them. From this perspective, the "transition" from the Middle to the Upper Palaeolithic in Iran (and on a larger scale, Out of Africa 2) was not a single event but a protracted and complex process involving multiple points of entry and repeated migrations, intermingling and movements.

Notes

1 Pronounced *ewch-ah-zluh* (Kuhn et al. 2009).
2 Gazelle are not dangerous to kill as adults weigh only ca. 20 kg but they are very agile and alert. When I was doing fieldwork in central Iran in the 1970s, I tried stalking gazelle in sand desert and on open plain and never got within 50 metres of one. One or two hunters with spears, skill and much patience would probably have had the best chances of a successful kill.
3 The Irano-Turanian flora is centred on Iran but also found in Central Asia (Uzbekistan, Kazakhstan, Turkmenistan), Afghanistan, most of Pakistan, Anatolia and the inland part of the Levant. The Saharo-Sindian flora covers the Sahara, the Arabian Peninsula, southern Iran and southern Pakistan (Sind Province) and the Thar Desert. Both are primarily desert floras but the Irano-Turanian is better adapted to sub-freezing winters (see Manafzadeh et al. 2014).
4 Put simply, C3 and C4 refer to different ways that plants use to obtain carbon dioxide during photosynthesis.
5 *Artemisia* includes sagebrush; Chenopods, or species of *Chenopodium*, are cosmopolitan perennial shrubs and flowers; and Umbellifers, or Umbelliferae, include carrots and parsley and in the wild are mostly aromatic flowering plants. Together they form dominant parts of *Artemisia* steppe – dry, highly seasonal conditions.
6 Sheep and goat bones are difficult to distinguish from each other and are often grouped together. In the Zagros, most of the animals in the sheep/goat category that can be clearly identified were probably goat.

7 The recovery of these items shows the benefits of sieving excavated deposits through mesh sizes of 1 mm, 2.2 mm and 4.4 mm. It is likely that these were missed in excavations that did not screen deposits.

References

The Levant

Akazawa T., Muhesen, S., Ishida, H., Kondo, O. and Griggo C. (1999) New discovery of a Neanderthal child burial from the Dederiyeh Cave in Syria. *Paléorient*, 25(2): 129–142. doi:10.3406/paleo.1999.4691.

Akazawa, T. and Nishiaki, Y. (2017) The palaeolithic cultural sequence of Dederiyeh Cave, Syria. In Y. Enzel and O. Bar-Yosef (eds) *Quaternary of the Levant: Environments, Climate Change and Humans*, Cambridge: Cambridge University Press, pp. 307–314.

Alex, B., Barzilai, O., Hershkovitz, I., Marder, O., Berna, F., Caracuta, V., Abulafia, T. et al. (2017) Radiocarbon chronology of Manot Cave, Israel and Upper Paleolithic dispersals. *Science Advances*, 3(11): e1701450.

Arensburg, B. and Belfer-Cohen, A. (1998) Sapiens and Neandertals: rethinking the Levantine Middle Paleolithic hominids. In T. Akazawa, K. Aoki and O. Bar-Yosef (eds) *Neandertals and Modern Humans in Western Asia*, New York: Plenum Press, pp. 311–322.

Bar-Yosef Mayer, D.E., Vandermeersch, B. and Bar-Yosef, O. (2009) Shells and ochre in Middle Paleolithic Qafzeh Cave, Israel: indications for modern behaviour. *Journal of Human Evolution*, 56: 307–314.

Bar-Yosef, O., Arnold, M., Mercier, N., Belfer-Cohen, A., Goldberg, P. et al. (1996) The dating of the Upper Palaeolithic layers in Kebara Cave, Mount Carmel. *Journal of Archaeological Science*, 23: 297–306.

Bar-Yosef, O. and Callander, J. (1999) The woman from Tabun: Garrod's doubts in historical perspective. *Journal of Human Evolution*, 37: 879–885.

Been, E., Hovers, E., Ekshtain, R., Malinski-Buller, A., Agha, N. et al. (2017) The first Neanderthal remains from an open-air Middle Palaeolithic site in the Levant. *Scientific Reports*, 7(1): 2958.

Belmaker, M. and Hovers, E. (2011) Ecological change and the extinction of the Levantine Neanderthals: implications from a diachronic study of micromammals from Amud Cave, Israel. *Quaternary Science Reviews*, 30: 3196–3209.

Bergman, C., Williams, J., Douka, K. and Schyle, D. (2017) The palaeolithic sequence of Ksar 'Akil, Lebanon. In Y. Enzel and O. Bar-Yosef (eds) *Quaternary of the Levant: Environments, Climate Change, and Humans*, Cambridge: Cambridge University Press, pp. 267–276.

Bosch, M.D., Buck, L. and Strauss, A.S. (2018) Location, location, location: investigating perforation locations in *Tritia gibbosula* shells at Ksâr 'Akil (Lebanon) using micro-CT data. *PaleoAnthropology*, 2019: 52–63. doi:10.4207/PA.2019.ART123.

Bosch, M.D., Mannino, M.A., Prendergast, A.L., O'Connell, T.C., Demarchi, B., Taylor, S. M. et al. (2015) New chronology for Ksâr 'Akil (Lebanon) supports Levantine route of modern human dispersal into Europe. *Proceedings of the National Academy of Sciences USA*, 112: 7683–7688.

Callaway, E. (2015) Neanderthals gain human neighbour. Cranium discovery shows that *Homo sapiens* was living in Middle East 55,000 years ago. *Nature*, 517: 541.

Cane, T., Rohling, E.J., Kemp, A.E.S., Cooke, S. and Pearce, R.B. (2002) High-resolution stratigraphic framework for Mediterranean sapropel S5: defining temporal relationships between records of Eemian climate variability. *Palaeogeography, Palaeoclimatology. Palaeoecology*, 183: 87–101.

Coppa, A., Manni, F., Stringer, C., Vargiu, R. and Vecchi, F. (2007) Evidence for new Neanderthal teeth in Tabun Cave (Israel) by the application of self-organizing maps (SOMs). *Journal of Human Evolution*, 52: 601–613.

Dennell, R.W. (2014) Smoke and mirrors: the fossil record for Homo sapiens between Arabia and Australia. In R. W. Dennell and M. Porr (eds) *Southern Asia, Australia and the Search for Human Origins*, Cambridge: Cambridge University Press, pp. 33–50.

Douka, K, Bergman, C.A., Hedges, R.E.M., Wesselingh, F.P. and Higham, T.F.G. (2013) Chronology of Ksar Akil (Lebanon) and implications for the colonization of Europe by anatomically modern humans. *PLoS ONE*, 8(9): e72931, doi:10.1371/journal.pone.0072931.

Ekshtain, R., Malinsky-Buller, A., Greenbaum, N., Mitki, N., Stahlschmidt, M.C. et al. (2019) Persistent Neanderthal occupation of the open-air site of 'Ein Qashish, Israel. *PloS One*, 14(6): e0215668.

Emeis, K.C., Struck, U., Schulz, H.M., Rosenberg, R., Bernasconi, S. et al. (2000) Temperature and salinity variations of Mediterranean Sea surface waters over the last 16,000 years from records of planktonic stable oxygen isotopes and alkenone unsaturation ratios. *Palaeogeography, Palaeoclimatology, Palaeoecology*, 158: 259–280.

Foley, R. (2018) Evolutionary geography and the Afrotropical model of hominin evolution. *Bulletins et Mémoires de la Société d'Anthropologie de Paris*: 1–15, doi:10.3166/bmsap-2018-0001.

Fornai, C., Benazzi, S., Gopher, A., Barkai, R., Sarig, R. et al. (2016) The Qesem Cave hominin material (part 2): a morphometric analysis of dm2-QC2 deciduous lower second molar. *Quaternary International*, 398: 175–189.

Frumkin, A., Bar-Yosef, O. and Schwarcz, H.P. (2011) Possible paleohydrologic and paleoclimatic effects on hominin migration and occupation of the Levantine Middle Paleolithic. *Journal of Human Evolution*, 60: 437–451.

Green, R.E., Krause, J., Briggs, A.W., Maricic, T., Stenzel, U. et al. (2010) A draft sequence of the Neandertal genome. *Science*, 328: 710–722.

Grün, R. and Stringer, C. (2000) Tabun revisited: revised ESR chronology and new ESR and U-series analyses of dental material from Tabun C1. *Journal of Human Evolution*, 39(6): 601–612.

Grün, R., Stringer, C., McDermott, F., Nathan, R., Porat, N. et al. (2005) U-series and ESR analyses of bones and teeth relating to the human burials from Skhul. *Journal of Human Evolution*, 49: 316–334.

Güleç, E., Özer, I., Sağır, M. and Kuhn, S. (2007) Early Upper Paleolithic human dental remains from Ucagizli Cave (Hatay, Turkey). *American Journal of Physical Anthropology*, 132(S44): 122.

Hallin, K.A., Schoeninger, M.J. and Schwarcz, H.P. (2012) Paleoclimate during Neandertal and anatomically modern human occupation at Amud and Qafzeh, Israel: the stable isotope data. *Journal of Human Evolution*, 62: 59–73.

Harvati, K. and Lopez, E.N. (2017) A 3-D look at the Tabun C2 jaw. In A. Marom and E. Hovers (eds) *Human Paleontology and Prehistory: Contributions in Honor of Yoel Rak*, Springer, pp. 203–213.

Harvati, K., Röding, C., Bosman, A.M., Karakostis, F.A., Grün, R. et al. (2019) Apidima Cave fossils provide earliest evidence of *Homo sapiens* in Eurasia. *Nature*, doi:10.1038/s41586-019-1376-z.

Hershkovitz, I. and Arensburg, B. (2017) Human fossils from the Upper Palaeolithic through the Early Holocene. In Y. Enzel and O. Bar-Yosef (eds) *Quaternary of the*

Levant: Environments, Climate Change, and Humans, Cambridge: Cambridge University Press, pp. 611–619.

Hershkovitz, I., Marder, O., Ayalon, A., Bar-Matthews, M., Yasur, G., et al. (2015) Levantine cranium from Manot Cave (Israel) foreshadows the first European modern humans. *Nature*, 520: 216–219.

Hershkovitz, I., Smith, P., Sarig, R., Quam, R., Rodríguez, L. et al. (2011) Middle Pleistocene dental remains from Qesem Cave (Israel). *American Journal of Physical Anthropology*, 144: 575–592.

Hershkovitz, I., Weber, G.W., Quam, R., Duval, M. and Grün, R. (2018) The earliest modern humans outside Africa. *Science*, 359: 456–459.

Hovers, E., Ilani, S., Bar-Yosef, O. and Vandermeersch, B. (2003) An early case of color symbolism ochre use by modern humans in Qafzeh Cave. *Current Anthropology*, 44: 491–522.

Howell, F.C. (1999) Paleo-demes, species clades, and extinctions in the Pleistocene hominin record. *Journal of Anthropological Research*, 55: 191–243.

Kramer, A., Crummett, T.L. and Wolpoff, M.H. (2001) Out of Africa and into the Levant: replacement or admixture in Western Asia? *Quaternary International*, 75: 51–63.

Kuhn, S.L. (2002) Paleolithic archaeology in Turkey. *Evolutionary Anthropology*, 11: 198–210.

Kuhn, S.L. (2003) In what sense is the Levantine Initial Upper Paleolithic a "Transitional" Industry? In J. Zilhaõ and F. d'Errico (eds) *The Chronology of the Aurignacian and of the Transitional Technocomplexes. Dating, Stratigraphies, Cultural Implications*. Trabalhos de Arqueologia, 33: 61–70. Instituto Português de Arqueologia, Lisbon.

Kuhn, S.L. (2004) Upper Paleolithic raw material economies at Üçağızlı Cave, Turkey. *Journal of Anthropological Archaeology*, 23: 431–448.

Kuhn, S.L., Stiner, M.C., Güleç, E., Özer, I., Yılmaz, H., Baykara, I., Açıkkol, A., Goldberg, P., Molina, K.M., Ünay, E. and Suata-Alpaslan, F. (2009) The Early Upper Paleolithic occupations at Üçağızlı Cave (Hatay, Turkey). *Journal of Human Evolution*, 56: 87–113.

Kuhn, S. and Zwyns, N. (2014) Rethinking the Initial Upper Palaeolithic. *Quaternary International*, 347: 29–38.

Langgut, D., Almogi-Labin, A., Bar-Matthews, M., Pickarski, N. and Weinstein-Evron, M. (2018) Evidence for a humid interval at ~56–44 ka in the Levant and its potential link to modern humans dispersal out of Africa. *Journal of Human Evolution*, 124: 75–90.

Langgut, D., Almogi-Labin, A., Bar-Matthews, M. and Weinstein-Evron, M. (2011) Vegetation and climate changes in the south eastern Mediterranean during the Last Glacial-Interglacial cycle (86 ka): new marine pollen record. *Quaternary Science Reviews*, 30: 3960–3972.

Manafzadeh, S., Salvo, G. and Conti, E. (2014) A tale of migrations from east to west: the Irano-Turanian floristic region as a source of Mediterranean xerophytes. *Journal of Biogeography*, 41: 366–379.

Marder, O., Alex, B., Ayalon, A., Bar-Matthews, M. and Bar-Oz, G. (2013) The Upper Palaeolithic of Manot Cave, Western Galilee, Israel: the 2011–12 excavations. *Antiquity Project Gallery*, 087(337).

Marder, O., Barzilai, O., Abulafia, T., Hershkovitz, I. and Goder-Goldberger, M. (2018) Chrono-cultural considerations of Middle Palaeolithic occurrences at Manot Cave (western Galilee), Israel. In Y. Nishiaki and T. Akazawa (eds) *The Middle and Upper*

Paleolithic Archaeology of the Levant and Beyond, Springer, pp. 49–63, doi:10.1007/978-981-10-6826-3_4.

Marder, O., Hershkovitz, I. and Barzilai, O. (2017) The Early Upper Palaeolithic of Manot Cave, Western Galilee: chrono-cultural, subsistence, and palaeo-environmental reconstruction. In Y. Enzel and O. Bar-Yosef (eds) *Quaternary of the Levant: Environments, Climate Change, and Humans*, Cambridge: Cambridge University Press, pp. 277–284.

McCown, T. and Keith, A. (1939) *The Stone Age of Mount Carmel: The Fossil Human Remains from the Levalloiso-Mousterian*, Vol. II. Oxford: Clarendon Press.

McDermott, F., Grün, R., Stringer, C.B. and Hawkesworth C.J. (1993) Mass spectrometric dates for Israeli Neanderthal/early modern sites. *Nature*, 363: 252–255.

Meignen L. (2011) Contribution of Hayonim cave assemblages to the understanding of the so-called "Early Levantine Mousterian". In J.-M. Le Tensorer, R. Jagher and M. Otte (eds) *The Lower and Middle Palaeolithic in the Middle East and Neighbouring Regions. Proceedings of the Basel symposium (mai 8–10 2008)*, ERAUL 126: 85–101.

Meignen, L. (2012) Levantine perspectives on the middle to upper paleolithic "transition". *Archaeology Ethnology & Anthropology of Eurasia*, 40(3): 12–21.

Mellars, P. and Tixier, J. (1989) Radiocarbon accelerator dating of Ksar Aqil (Lebanon) and the chronology of the Upper Palaeolithic sequence in the Middle East. *Antiquity*, 63: 761–768.

Mercier, H. and Valladas, H. (2003) Reassessment of TL age estimates of burnt flints from the Paleolithic site of Tabun Cave, Israel. *Journal of Human Evolution*, 45: 401–409.

Mercier, N., Valladas, H., Bar-Yosef, O., Vandermeersch, B. and Stringer, C. (1993) Thermoluminescence data for the Mousterian burial site of Es-Skhul Mt. Carmel. *Journal of Archaeological Science*, 20(2): 169–174.

Millard, A.R. (2008) A critique of the chronometric evidence for hominid fossils: I. Africa and the Near East 500–50 ka. *Journal of Human Evolution*, 54: 848–874.

Olszewski, D. (2017) The Initial Upper Palaeolithic in the Levant. In Y. Enzel and O. Bar-Yosef (eds) *Quaternary of the Levant: Environments, Climate Change, and Humans*, Cambridge: Cambridge University Press, pp. 621–626.

Orbach, M. and Yeshurun, R. (2019) The hunters or the hunters: human and hyena prey choice divergence in the Late Pleistocene Levant. *Journal of Human Evolution*, doi:10.1016/j.jhevol.2019.01.005.

Plicht, J. van der, Wijk, A. van der and Bartstra, G.J. (1989) Uranium and thorium in fossil bones: activity ratios and dating. *Applied Geochemistry*, 4: 339–342.

Quam, R.M. and Smith, F.H. (1998) A reassessment of the Tabun C2 mandible. In T. Akazawa, K. Aoki and O. Bar-Yosef (eds), *Neandertals and Modern Humans in Western Asia*. New York: Plenum Press, pp. 405–421.

Rebollo, N.R., Weiner, S., Brock, F., Meignen, L., Goldberg, P. et al. (2011) New radiocarbon dating of the transition from the Middle to the Upper Paleolithic in Kebara Cave, Israel. *Journal of Archaeological Science*, 38: 2424–2433.

Rink, W.J., Schwarcz, H.P., Lee, H.K, Rees-Jones, J., Rabinovich, R. and Hovers, E. (2001) Electron spin resonance (ESR) and thermal ionization mass spectroscopic (TIMS) $^{23o}Th/^{234}U$ dating of teeth in Middle Paleolithic layers at Amud Cave, Israel. *Geochronology*, 16: 701–717.

Schwarcz, H.P., Buhay, W.M., Grün, R., Valladas, H., Tchernov, E., Bar-Yosef, O. and Vandermeersch, B. (1989) ESR dating of the Neanderthal site, Kebara Cave, Israel. *Journal of Archaeological Science*, 16: 653–659.

Schwarcz, H.P., Grün, R., Vandermeersch, B., Bar-Yosef, O., Valladas, H. and Tchernov, E. (1988) ESR dates for the hominid burial site of Qafzeh in Israel. *Journal of Human Evolution*, 17: 733–737.

Schwartz, J.H. and Tattersall, I. (2003) *The Human Fossil Record. Volume Two: Craniodental Morphology of Genus Homo (Africa and Asia)*. New York: Wiley-Liss.

Shea, J.J. (2003) The Middle Palaeolithic of the East Mediterranean Levant. *Journal of World Prehistory*, 17: 313–394.

Shea, J.J. (2008) Transitions or turnovers? Climatically-forced extinctions of *Homo sapiens* and Neanderthals in the East Mediterranean Levant. *Quaternary Science Reviews*, 27: 2253–2270.

Shea, J.J. and Bar-Yosef, O. (2005) Who were the Skhul/Qafzeh people? An archaeological perspective on Eurasia's oldest modern humans. *Journal of the Israel Prehistoric Society*, 35: 451–468.

Sohn, S. and Wolpoff, M.H. (1993) Zuttiyeh face: a view from the East. *American Journal of Physical Anthropology*, 91: 325–347.

Stefan V.H. and Trinkaus, E. (1998) Discrete trait and dental morphometric affinities of the Tabun C mandible. *Journal of Human Evolution*, 34: 443–468.

Stiner, M.C., Kuhn, S.L. and Güleç, E. (2013) Early Upper Paleolithic shell beads at Üçağızlı Cave I (Turkey): technology and the socioeconomic context of ornament life-histories. *Journal of Human Evolution*, 64: 380–398.

Stringer, C.B., Grün, R., Schwarcz, H. and Goldberg, P. (1989) ESR dates for the burial site of Es Skhul in Israel. *Nature*, 338: 756–758.

Stutz, A.J., Shea, J.J., Rech, J.A., Pigati, J.S., Wilson, J. et al. (2015) Early Upper Paleolithic chronology in the Levant: new ABOx-SC accelerator mass spectrometry results from the Mughr el-Hamamah Site, Jordan. *Journal of Human Evolution*, 85: 157–173.

Tchernov, E. (1992) Eurasian-African biotic exchanges through the Levantine corridor during the Neogene and Quaternary. In W. von Koenigswald and L. Werdelin (eds) Mammalian Migration and Dispersal Events in the European Quaternary. Courier Forschungsinstitut Senckenberg, 153: 103–123.

Tejero, J.-M., Yeshurun, R., Barzilai, O., Goder-Goldberger, M., Hershkovitz, I., Lavi, R., Schneller-Pels, N. and Marder, O. (2015) The osseous industry from Manot Cave (Western Galilee, Israel): technical and conceptual behaviours of bone and antler exploitation in the Levantine Aurignacian. *Quaternary International*, 403: 90–106.

Thackeray, J.F., Maureille, B., Vandermeersch, B., Braga, J. and Chaix, R. (2005) Morphometric comparisons between Neanderthals and "anatomically modern" *Homo sapiens* from Europe and the Near East. *Annals of the Transvaal Museum*, 42: 47–51.

Tillier, A.-M. and Arensburg, B. (2017) Neanderthals and modern humans in the Levant: an overview. In Y. Enzel and O. Bar-Yosef (eds) *Quaternary of the Levant: Environments, Climate Change, and Humans*, Cambridge: Cambridge University Press, pp. 607–610.

Torfstein, A. (2019) Climate cycles in the southern Levant and their global climatic connections. Quaternary Science Reviews, 221: 105881, 10.1016/j.quascirev.2019.105881.

Torfstein, A., Goldstein, S.L., Stein, M. and Enzel, Y. (2013) Impacts of abrupt climate changes in the Levant from Last Glacial Dead Sea levels. *Quaternary Science Reviews*, 69: 1–7.

Trinkaus, E. (2005) Early modern humans. *Annual Review of Anthropology*, 34: 207–230.

Vaks, A., Bar-Matthews, M., Ayalon, A., Halicz, L. and Frumkin, A. (2007) Desert speleothems reveal climatic window for African exodus of early modern humans. *Geology*, 35 (9): 831–834, doi:10.1130/G23794A.1.

Vaks, A., Bar-Matthews, M., Ayalon, A., Matthews, A., Frumkin, A., Dayan, U., Halicz, L., Almogi-Labin, A. and Schilman, B. (2006) Paleoclimate and location of the border between Mediterranean climate region and the Saharo-Arabian Desert as revealed by speleothems from the northern Negev Desert, Israel. *Earth and Planetary Science Letters*, 249: 384–399.

Valladas, H., Joron, J.-L., Valladas, G., Arensburg, B., Bar-Yosef, O. et al. (1987) Thermoluminescence dates for the Neanderthal burial site at Kebara in Israel. *Nature*, 330: 159–160.

Valladas, H., Mercier, N., Froget, L., Hovers, E. and Joron, J.-L. (1999) TL dates for the Neanderthal sites of the Amud cave, Israel. *Journal of Archaeological Science*, 26: 259–268.

Valladas, H., Mercier, N., Hershkovitz, I., Zaidner, Y., Tsatkin, A. et al. (2013) Dating the Lower to Middle Paleolithic transition in the Levant: a view from Misliya Cave, Mount Carmel, Israel. *Journal of Human Evolution*, 65: 585–593.

Valladas, H., Reyss, L. Joron, G., Valladas, H., Bar-Yosef, O. and Vandermeersch, B. (1988) Thermoluminescence dating of Mousterian Proto-Cro-Magnon remains from Israel and the origin of modern man. *Nature*, 331: 614–616.

Vanhaeren, M., d'Errico, F., Stringer, C., James, S.L., Todd, J.A. and Mienis, H.K. (2006) Middle Paleolithic shell beads in Israel and Algeria. *Science*, 312: 1785–1788.

Vermeersch, P.M., Paulissen, E., Stokes, S., Charlier, C., Peer, P. van, Stringer, C. and Lindsay, W. (1998) A Middle Palaeolithic burial of a modern human at Taramsa Hill, Egypt. *Antiquity*, 72: 475–484.

Weber, G.W., Fornai, C., Gopher, A., Barkai, R., Sarig, R. and Hershkovitz, I. (2016) The Qesem Cave hominin material (part 1): a morphometric analysis of the mandibular premolars and molar. *Quaternary International*, 398: 159–174.

Yasur, G., Ayalon, A., Matthews, A., Zilberman, T., Marder, O. et al. (2019) Climatic and environmental conditions in the western Galilee, during Late Middle and Upper Paleolithic periods, based on speleothems from Manot Cave, Israel, *Journal of Human Evolution*, doi: 10.1016/j.jhevol.2019.04.004.

Yeshurun, R., Bar-Oz, G. and Weinstein-Evron, M. (2007) Modern hunting behavior in the early Middle Paleolithic: faunal remains from Misliya Cave, Mount Carmel, Israel. *Journal of Human Evolution*, 53: 656–677.

Yeshurun, R., Schneller, N., Barzilai, O. and Marder, O. (2019) Early Upper Paleolithic subsistence in the Levant: zooarchaeology of the Ahmarian-Aurignacian sequence at Manot Cave, Israel. *Journal of Human Evolution*, doi:10.1016/j.jhevol.2019.05.007.

Yeshurun, R., Tejero, J.-M., Barzilai, O., Hershkovitz, I. and Marder, O. (2017) Upper Palaeolithic bone retouchers from Manot cave (Israel): a preliminary analysis of an (as yet) rare phenomenon in the Levant. In J. M. Hutson, A. García-Moreno, E.S. Noack, E. Turner, A. Villaluenga and S. Gaudzinski-Windheuser (eds) *The Origins of Bone Tool Technologies*, Mainz: Verlag des Römisch-Germanischen Zentralmuseums, pp. 287–295.

Yokoyama, Y., Falguères, C. and de Lumley, M.-A. (1997) Datation directe d'un crâne Proto-Cro-Magnon de Qafzeh par la spèctrometrie gamma non-destructive. *Comptes Rendues de l'Academie des Sciences, Paris*, 324 IIa: 773–779.

The Zagros and Iran

Abdi, L., Rahimpour-Bonab, H., Mirmohammad-Makki, M., Probst, J. and Rezaeian Langeroudi. S. (2018) Sedimentology, mineralogy, and geochemistry of the Late Quaternary Meyghan Playa sediments, NE Arak, Iran: palaeoclimate implications. *Arabian Journal of Geosciences*, 11: 589, doi:10.1007/s12517-018-3918-3.

Adler, D.S., Bar-Yosef, O., Belfer-Cohen, A., Tushabramishvili, N., Boaretto, E., Mercier, N., Valladas, H. and Rink, W.J. (2008) Dating the demise: Neanderthal extinction and the establishment of modern humans in the southern Caucasus. *Journal of Human Evolution*, 55: 817–833.

Bahramiyan, S. and Shouhani, L.A. (2016) Between mountain and plain: new evidence for the Middle Palaeolithic in the northern Susiana Plain, Khuzestan, Iran. *Antiquity*, 90: 354, doi:10.15184/aqy.2016.190.

Bazgir, B., Ollé, A., Tumung, L., Becerra-Valdivia, L. and Douka, K. (2017) Understanding the emergence of modern humans and the disappearance of Neanderthals: insights from Kaldar Cave (Khorramabad Valley, western Iran). *Scientific Reports*, 7: 43460, doi:10.1038/srep43460.

Bazgir, B., Otte, M., Tumung, L., Ollé, A., Deo, G. et al. (2014) Test excavations and initial results at the Middle and Upper Paleolithic sites of Gilvaran, Kaldar, Ghamari caves and Gar Arjene Rockshelter, Khorramabad Valley, western Iran. *Comptes Rendues Palévolution*, 13: 511–525.

Becerra-Valdivia, L., Douka, K., Comesky, D., Bazgir, B., Conard, N.J., Marean, C.W., Ollé, A., Otte, M., Tumung, L., Zeidi, M. and Higham, T.F.G. (2017) Chronometric investigations of the Middle to Upper Paleolithic transition in the Zagros Mountains using AMS radiocarbon dating and Bayesian age modelling. *Journal of Human Evolution*, 109: 57–69.

Berillon, G., Asghar Asgari Khaneghah, Antoine, P., Bahain, J.-J., Chevrier, B. et al. (2007) Discovery of new open-air Paleolithic localities in Central Alborz, northern Iran. *Journal of Human Evolution*, 52: 380–387.

Biglari, F. and Jahani, V. (2011) The Pleistocene human settlement in Gilan, south west Caspian Sea: recent research. *Eurasian Prehistory*, 8(1–2): 3–28.

Biglari, F., Javeri, M., Mashkour, M., Yazdi, M., Shidrang, S. et al. (2009) Test excavations at the middle paleolithic sites of Qaleh Bozi, southwest of central Iran, a preliminary report. In M. Otte, F. Biglari, F. and J. Jaubert (eds) *Iran Palaeolithic (Le Paléolithique d'Iran)*. British Archaeological Reports, 1968: 29–38.

Biglari, F. and Shidrang, S. (2016) New evidence of Paleolithic Occupation in the western Zagros foothills: preliminary report of cave and rockshelter survey in the Sar Qaleh Plain in the west of Kermanshah Province, Iran. In K. Kopanias and J. MacGinnis (eds) *The Archaeology of the Kurdistan Region of Iraq and Adjacent Regions*, Oxford: Archaeopress, pp. 29–47.

Braidwood, R.J., Howe, B.J. and Reed, C.A. (1961) The Iranian prehistoric project. *Science*, 133: 2008–2010.

Bretzke, K. and Conard, N.J. (2017) Not just a crossroad: population dynamics and changing material culture in Southwestern Asia during the Late Pleistocene. *Current Anthropology* 58 (Supplement 17): S449–S462.

Çatağay, M.N., Öğretman, N., Damcı, E., Stockhecke, M., Sancar, Ü., Eriş, K.K. and Özeren, S. (2014) Lake level and climate records of the last 90 ka from the Northern Basin of Lake Van, eastern Turkey. *Quaternary Science Reviews*, 104: 97–116.

Conard, N.J. and Ghasidian, E. (2011) Rostamian cultural group and the taxonomy of the Iranian Upper Paleolithic. In N. J. Conard, P. Dreschsler and A. Morales (eds) *Between Sand and Sea: The Archaeology and Human Ecology of Southwestern Asia*, Tubingen: Kerns Verlag, pp. 33–52.

Conard, N.J., Ghasidian, E. and Heydari-Guran, S. (2013) The Palaeolithic of Iran. In D. T. Potts (ed) *The Oxford Handbook of Ancient Iran*, Oxford: Oxford University Press, doi:10.1093/oxfordhb/9780199733309.013.0038.

Coon, C.S. (1957) *The Seven Caves: Archaeological Explorations in the Middle East*. New York: Alfred Knopf.

Deckers, K., Riehl, S., Jenkins, E., Rosen, A., Dodonov, A. et al. (2009) Vegetation development and human occupation in the Damascus region of southwestern Syria from the Late Pleistocene to Holocene. *Vegetation History and Archaeobotany*, 18: 329–340.

Devès, M., Sturdy, D., Godet, N., King, G.C.P. and Bailey, G.N. (2014) Hominin reactions to herbivore distribution in the Lower Palaeolithic of the southern Levant. *Quaternary Science Reviews*, 96: 140–160.

Djamali, M., Beaulieu, J.-L. de, Shah-Hosseini, M., Andrieu-Ponel, V., Ponel, P. et al. (2008) A Late Pleistocene long pollen record from Lake Urmia, NW Iran. *Quaternary Research*, 69: 413–420.

Djamali, M., Biglari, F., Abdi, K., Andrieu-Ponel, V., Beaulieu, J.-L. de, Mashkour, M. and Ponel, P. (2011) Pollen analysis of coprolites from a Late Pleistocene-Holocene cave deposit (Wezmeh Cave, west Iran): insights into the Late Pleistocene and late Holocene vegetation and flora of the central Zagros Mountains. *Journal of Archaeological Science*, 38: 3394–3401.

Evins, M.A. (1982) The fauna from Shanidar Cave: Mousterian wild goat exploitation in north-eastern Iraq. *Paléorient*, 8: 37–58.

Ferrigno, J.G. (1993) Glaciers of Iran. In R.S. Williams and J. Ferrigno (eds) *Satellite Image Atlas of Glaciers of the World*, US Geological Survey Professional Paper 1386-G-2: 31–47.

Field, H. (1951) Reconnaissance in Southwestern Asia. *Southwestern Journal of Anthropology*, 7(1): 86–102.

Ghasidian, E. (2014) *Early Upper Palaeolithic Occupation at the Ghār-e Boof Cave, a Reconstruction of Cultural Traditions in Southern Zagros Mountains of Iran*. Tübingen: Kerns Verlag.

Ghasidian, E. (2019) Rethinking the Upper Paleolithic of the Zagros Mountains. *PaleoAnthropology*, 2019: 240–310.

Ghasidian, E., Bretzke, K. and Conard, N. (2017) Excavations at Ghār-e Boof in the Fars Province of Iran and its bearing on models for the evolution of the Upper Palaeolithic in the Zagros Mountains. *Journal of Anthropological Archaeology*, 47: 33–49.

Ghasidian, E., Heydari-Guran, S. and Lahr, M.M. (2019) Upper Paleolithic cultural diversity in the Iranian Zagros Mountains and the expansion of modern humans into Eurasia. *Journal of Human Evolution*, 132: 101–118.

Hashemi, N., Darvish, J., Mashkour, M. and Biglari, F. (2006) Rodents and lagomorphs remains from Late Pleistocene and early Holocene caves and rockshelter sites in the Zagros region, Iran. *Iranian Journal of Animal Biosystematics*, 2(1): 25–33.

Henry, A.G., Brooks, A.S. and Piperno, D.R. (2011) Microfossils in calculus demonstrate consumption of plants and cooked foods in Neanderthal diets (Shanidar III, Iraq; Spy I and II, Belgium). *Proceedings of the National Academy of Sciences USA*, 108(2): 486–491.

Hesse, B. (1989) Paleolithic faunal remains from Ghar-i-Khar, western Iran. In P. J. Crabtree, D. Campana and K. Ryan (eds) *Early Animal Domestication and its Cultural Context*, MASCA Research Papers in Science and Archaeology, pp. 37–45.

Heydari-Guran, S. (2015) Tracking Upper Pleistocene human dispersals into the Iranian Plateau: a geoarchaeological model. In N. Sanz (ed) *HEADS 4: Human Origin Sites and the World Heritage Convention in Eurasia*, Paris: UNESCO, pp. 40–53.

Heydari-Guran, S. and Ghasidian, E. (2017) The MUP Zagros Project: tracking the Middle–Upper Palaeolithic transition in the Kermanshah region, west-central Zagros, Iran. *Antiquity*, 91: 355, e2: 1–7, doi:10.15184/aqy.2016.261.

Heydari-Guran, S., Ghasidian, E. and Conard, N.J. (2009) Iranian paleolithic sites on travertine and tufa formations. In M. Otte, F. Biglari and J. Jaubert (eds) *Iran Palaeolithic (Le Paléolithique d'Iran)*, British Archaeological Reports, 1968: 109–124.

Heydari-Guran, S., Ghasidian, E. and Conard, N.J. (2015) Middle Paleolithic settlement on the Iranian Central Plateau. In N.J. Conard and A. Delagnes (eds) *Settlement Dynamics of the Middle Paleolithic and Middle Stone Age*, Tübingen: Tübingen Publications in Archaeology, Kerns Verlag, pp. 171–204.

Hole, F. and Flannery, K.V. (1967) The prehistory of southwestern Iran: a preliminary report. *Proceedings of the Prehistoric Society*, 22: 147–206.

Jaubert, J., Biglari, F., Mourre, V., Bruxelles, L., Bordes, J.-G. et al. (2009) The Middle Palaeolithic occupation of Mar-Tarik, a new Zagros Mousterian site in Bisotun massif (Kermanshah, Iran). In M. Otte, F. Biglari and J. Jaubert (eds) *Iran Palaeolithic (Le Paléolithique d'Iran)*, British Archaeological Reports, 1968: 7–27.

Kehl, M. (2009) Quaternary climate change in Iran – the state of knowledge. *Erdkunde*, 63 (1): 1–17.

Kuhle, M. (2007) The Pleistocene glaciation (LGP and pre-LGP, pre-LGM) of SE Iranian mountains exemplified by the Kuh-i-Jupar, Kuh-i-Lalezar and Kuh-i-Hezar Massifs in the Zagros. *Polarforschung*, 77(2–3): 71–88.

Kuhn, S.L. (2010) Was Anatolia a bridge or a barrier to early hominin dispersals? *Quaternary International*, 223–224: 434–435.

Lauer, T., Frechen, M., Vlaminck, S., Kehl, M., Lehndorff, E., Shahriari, A. and Khormali, F. (2017a) Luminescence-chronology of the loess palaeosol sequence Toshan, northern Iran: a highly resolved climate archive for the last glacial-interglacial cycle. *Quaternary International*, 429: 3–12.

Lauer, T., Vlaminck, S., Frechen, M., Rolf, C., Kehl, M., Sharif, J., Lehndorff, E. and Khormali, F. (2017b) The Agh Band loess-palaeosol sequence – a terrestrial archive for climatic shifts during the last and penultimate glacial-interglacial cycles in a semiarid region in northern Iran. *Quaternary International*, 429: 13–30.

Le Tensorer, J.-M. (2015) Regional Perspective of early human populations in Syria: the case of El Kowm. In N. Sanz (ed.) *Human Origin Sites and the World Heritage Convention in Eurasia*, Paris: UNESCO, pp. 54–71.

Litt, T. and Anselmetti, F.S. (2014) Lake Van deep drilling project PALEOVAN. *Quaternary Science Reviews*, 104: 1–7.

Litt, T., Pickarski, N., Heumann, G., Stockhecke, M. and Tzedakis, P.C. (2014) A 600,000 year long continental pollen record from Lake Van, eastern Anatolia (Turkey). *Quaternary Science Reviews*, 104: 30–41.

Marean, C.W. and Kim, S.Y. (1998) Mousterian large-mammal remains from Kobeh cave. Behavioral implications for Neanderthals and early modern Humans. *Current Anthropology*, 39: 79–113.

Mashkour, M., Monchot, H., Trinkaus, E., Reyss, J.-L., Biglari, F., Bailon, S., Heydari, S. and Abdi, K. (2009a) Carnivores and their prey in the Wezmeh Cave (Kermanshah, Iran): a Late Pleistocene refuge in the Zagros. *International Journal of Osteoarchaeology*, 19: 678–694.

Mashkour, M., Radu, V., Mohaseb, A., Hashemi, N., Otte, M. and Shidrang, S. (2009b) The Upper Paleolithic faunal remains from Yafteh cave (Central Zagros), 2005 campaign. A preliminary study. In M. Otte, F. Biglari, and J. Jaubert (eds) *Iran Palaeolithic (Le Paléolithique d'Iran)*, British Archaeological Reports, 1968: 73–84.

Mehterian, S., Pourmand, A., Sharif, A., Lahijani, H.A.K., Naderi, M. and Swart, P.K. (2017) Speleothem records of glacial/interglacial climate from Iran forewarn of future water availability in the interior of the Middle East. *Quaternary Science Reviews*, 164: 187–198.

Nikzad, M., Sedighian, H. and Ghasemi, E. (2015) New evidence of Palaeolithic activity from South Khorasan, eastern Iran. *Antiquity*, http://antiquity.ac.uk/projgall/nikzad347.

Olszewski, D. and Dibble, H.L. (1993) *The Paleolithic Prehistory of the Zagros-Tauros*. University of Pennsylvania: The University Museum.

Otte, M. (2014) Central Asia as a Core Area: Iran as an origin for the European Aurignacian. *International Journal of the Society of Iranian Archaeologists*, 1: 27–32.

Otte, M., Biglari, F., Flas, D., Shidrang, S., Zwyns, N. et al. (2009) The Aurignacian in the Zagros region: new research at Yafteh Cave, Lorestan, Iran. *Antiquity*, 81: 82–96.

Pinhasi, R., Nioradze, M., Tushabramishvili, N., Lordkipanidze, D., Pleurdeau, D., Moncel, M.H., Adler, D.S., Stringer, C. and Higham, T.F.G. (2012) New chronology for the Middle Palaeolithic of the southern Caucasus suggests early demise of Neanderthals in this region. *Journal of Human Evolution*, 63: 770–780.

Piperno, M. (1974) Upper Palaeolithic Caves in southern Iran Preliminary Report. *East and West*, 24(1/2): 9–13.

Por, D. (2004) The Levantine waterway, riparian archaeology, paleolimnology, and conservation. In N. Goren-Inbar and J.D. Speth (eds) *Human Paleoecology in the Levantine Corridor*, Oxford: Oxbow Books, pp. 5–20.

Rahimpour-Bonab, H. and Abdi, L. (2012) Sedimentology and origin of Meyghan lake/playa deposits in Sanandaj–Sirjan zone, Iran. *Carbonates Evaporites*, 27: 375–393, doi:10.1007/s13146-012-0119-0.

Reynolds, T., Farr, L., Hill, E., Hunt, C., Jones, S. et al. (2018) Shanidar Cave and the Baradostian, a Zagros Aurignacian industry. *L'Anthropologie*, 122: 737–748.

Rezvani, H. and Vahdati Nasab, H. (2010) A major Middle Palaeolithic open-air site at Mirak, Semnan Province, Iran. *Antiquity*, 84: Issue 323.

Rosenberg, M.S. (1985) Report on the 1978 sondage at Eshkaft-e Gavi. Iran. *Journal of the British Institute of Persian Studies*, 23: 51–62.

Rowe, P.J., Mason, J.E., Andrews, J.E., Marca, A.D., Thomas, L. et al. (2012) Speleothem isotopic evidence of winter rainfall variability in northeast Turkey between 77 and 6 ka. *Quaternary Science Reviews*, 45: 60–72.

Sarıkaya, M.A., Çiner, A. and Zreda, M. (2011) Quaternary glaciations of Turkey. In J. Ehlers, P.L. Gibbard and P.D. Hughes (eds) *Developments in Quaternary Science*, 15: 393–403.

Scott, J.E. and Marean, C.W. (2009) Paleolithic hominin remains from Eshkaft-e Gavi (southern Zagros Mountains, Iran): description, affinities, and evidence for butchery. *Journal of Human Evolution*, 57: 248–259.

Shidrang, S. (2015) The Early Upper Paleolithic of Zagros: techno-typological assessment of three Baradostian lithic assemblages from Khar Cave (Ghar-e Khar), Yafteh Cave and Pa-Sangar rockshelter in the Central Zagros, Iran. Ph.D. dissertation, Bordeaux University.

Shidrang, S.Y. (2018) The Middle to Upper Paleolithic transition in the Zagros: the appearance and evolution of the Baradostian. In Y. Nishiaki and T. Akazawa (eds) *The Middle and Upper Paleolithic Archeology of the Levant and Beyond* (*Replacement of Neanderthals by Modern Humans Series*), Springer, pp. 133–156, doi:10.1007/978-981-10-6826-3_10.

Shidrang, S., Biglari, F., Bordes, J.-G. and Jaubert, J. (2016) Continuity and change in the Late Pleistocene lithic industries of the central Zagros: a typo-technological analysis of lithic assemblages from Ghar-e Khar cave, Bisotun, Iran. *Archaeology, Ethnology & Anthropology of Eurasia*, 44/1: 27–38.

Smith, P.E.L. (1986) *The Paleolithic of Iran*. Philadelphia: University Museum of Archaeology and Anthropology.

Trinkaus, E. (1983) *The Shanidar Neanderthals*. New York, London: Academic Press.

Trinkaus, E. and Biglari, F. (2006) Middle Palaeolithic human remains from Bisitun Cave, Iran. *Paléorient*, 32(2): 105–111.

Trinkaus, E., Biglari, F., Mashkour, M., Monchot, H., Reyss, J.-L., Rougier, H., Heydari, S. and Abdi, K. (2008) Late Pleistocene human remains from Wezmeh Cave, western Iran. *American Journal of Physical Anthropology*, 135: 371–380.

Tsanova, T. (2013) The beginning of the Upper Paleolithic in the Iranian Zagros. A taphonomic approach and techno-economic comparison of Early Baradostian assemblages from Warwasi and Yafteh (Iran). *Journal of Human Evolution*, 65: 39–64.

Vahdati Nasab, H. and Hashemi, M. (2016) Playas and Middle Paleolithic settlement of the Iranian Central Desert: the discovery of the Chah-e Jam Middle Paleolithic site. *Quaternary International*, 408: 140–152.

Vlaminck, S., Kehl, M., Rolf, C., Franz, S.O., Lauer, T., Lehndorff, S., Frechen, M. and Khormali, F. (2018) Late Pleistocene dust dynamics and pedogenesis in southern Eurasia. Detailed insights from the loess profile Toshan (NE Iran). *Quaternary Science Reviews*, 180: 75–95.

Wang, X., Wei, H., Khormali, F., Taheri, M., Kehl, M., Frechen, M., Lauer, T. and Chen, B. (2017) Grain-size distribution of Pleistocene loess deposits in northern Iran and its palaeoclimatic implications. *Quaternary International*, 429: 41–51.

Wasylikowa, K. (2005) Palaeoecology of Lake Zeribar, Iran, in the Pleniglacial, Lateglacial and Holocene, reconstructed from plant macrofossils. *The Holocene*, 15(5): 720–735.

Wright, H.E. (1962) Pleistocene glaciation in Kurdistan. *Eiszeitalter und Gegenwart*, 12: 131–164.

Zanolli, C., Biglari, F., Mashkour, M., Abdi, K., Monchot, H. et al. (2019) A Neanderthal from the central western Zagros, Iran. Structural reassessment of the Wezmeh 1 maxillary premolar. *Journal of Human Evolution*, 135: 102643.

Zeist, W. van and Bottema, S. (1977) Palynological investigations in western Iran. *Palaeohistoria*, 19: 19–85.

9

CENTRAL ASIA, SOUTHERN SIBERIA AND MONGOLIA

Introduction

Central Asia, southern Siberia and Mongolia cover a major part of continental Asia and were the harshest regions that our species had yet encountered. Central Asia is an enormous but poorly defined region: some define it as the "five stans" (Kazakhstan, Kyrgyzstan, Tajikistan, Turkmenistan and Uzbekistan) that cover an area almost the size of the EU[1], but Russian specialists include Mongolia, northwest China, Afghanistan and regions of Siberia south to the taiga belt.[2] Siberia – defined here as the landmass east of the Ural Mountains and north of Kazakhstan, Mongolia and China – covers ca. 12 million sq km, but here we are mainly concerned with only the southern part of the Altai Mountains and Transbaikalia east of Lake Baikal. Mongolia is vast[3] and provides a potential corridor from Siberia into north China. The whole of this region lies within the Palearctic Realm and is characterised by winter rain- and snowfall, prolonged sub-freezing winters and short, hot summers. Average winter temperatures drop steadily as one moves east from the Iranian Plateau: –5° C. in Uzbekistan, –10° C. in Kazakhstan, –20° C. in Irkutsk and –30° C. in Ulaan Baatar, Mongolia; minimum winter temperatures can fall below –45° C. Warm, insulated clothing, substantial winter settlements, food storage and controlled use of fire are among the obvious adaptations that were needed for survival.

Our species was not the first to enter this region, which was already inhabited at the time of contact by Neandertals and Denisovans. Because skeletal evidence for *H. sapiens* in these regions is so limited, the best indicator for the arrival of our species is the first appearance of blade-based lithic assemblages, known variously as Initial Upper Palaeolithic (IUP) or Early Upper Palaeolithic (EUP). Their timing is unclear in Central Asia, but likely to date from ca. 48 ka cal BP in the Altai Mountains, 45 ka cal BP in Transbaikalia and 42 ka cal BP in

Mongolia. These assemblages are sometimes found with personal ornaments such as perforated pendants and ostrich eggshell beads that provide some information about social networks. Herd animals such as bison, deer, horse and gazelle provided most of the subsistence: because some animals like horse and gazelle can have enormous annual territories, human groups must either have been extremely mobile or had developed effective food storage techniques to ensure successful over-wintering. Despite many likely failures, by 40,000 years ago our species had succeeded in adapting to the landscapes of Central Asia, southern Siberia and Mongolia despite the shortage of year-round plant foods and the brutally cold winters.

Central Asia

East of the Caspian Sea, the western part of Central Asia comprises arid and semi-arid plains and includes substantial deserts such as the Kara Kum and Kizyl Kum (see Figure 3.6). Undated surface assemblages of stone tools in the plains east of the Caspian show that the area was used in the Palaeolithic (Vishnyatsky 1999), perhaps when hunting horse and gazelle. Unpredictable rainfall, repeated periods of drought and dust-storms are major hazards in this region, and it is unsurprising that it contains little clear evidence of the Upper Palaeolithic. Most of the evidence for hominin and human settlement comes from the foothills and valleys of the great mountain ranges of the Hindu Kush, Pamirs and Tien Shan that rise up to 7,000 m. These are less arid, and streams are fed in spring and summer by melt-water. The Altai Mountains of southern Siberia have many deep valleys that were used in the Middle and Upper Palaeolithic and may have been a refugium. Transbaikalia at the head waters of the Angara and Selenga rivers has many EUP sites, and provides a corridor into northern Mongolia, where humans probably arrived ca, 42,000 cal BP years ago.

Archaeological evidence

Middle Palaeolithic assemblages from Central Asia can be described as Mousterian or Mousterian-like but have no close similarities to Mousterian assemblages further west. Although the Central Asian Middle Palaeolithic may have been derived from better-known areas such as the Zagros, Caucasus or eastern Europe to the west, prepared core technology in Central Asia could have originated to the east or south, or even have been a local development (see e.g. Glantz 2010). One inescapable consequence of naming Middle Palaeolithic assemblages in Central Asia as Mousterian is that any associated hominin remains have been automatically classed as Neandertal. As seen shortly, this may be an over-simplification. Few Upper Palaeolithic sites have been accurately dated or studied in detail; the main ones are shown in Figure 9.1. Sites such as Samarkandskaya and Kulbulak appear to retain Middle Palaeolithic

FIGURE 9.1 Location of Middle and Upper Palaeolithic sites in Central Asia

Key: 1 Anghilak cave; 2 Samarkandskaya; 3 Teshik Tash; 4 Kulbulak; 5 Shugnou; 6 Khudji; 7 Obi Rakhmat; 8: Ushbulak-1.

Source: The author.

features but the extent of stratigraphic mixing is unclear. The best-studied in this region is Shugnou, Tajikistan (Ranov et al. 2012). Here, numerous lithics (N=6,185) were recovered from four cultural layers. Ranov et al. (2012) interpreted these as part of a single Upper Palaeolithic tradition named the Kulbulak variant, within which the technology of producing bladelets with keeled cores was gradually evolving. In their view, this developed from the local Middle Palaeolithic assemblages such as those from Obi Rakhmat and Khudji (see below). They also claimed similarities in small tools with Baradostian assemblages in Iran (see Chapter 8). Other Russian specialists agree that the Upper Palaeolithic in Central Asia developed locally. As an example, Krivoshapkin et al. (2007) argue that Obi Rakhmat represents the local emergence of an Upper Palaeolithic by 44,000–42,000 years ago, similar to that seen at Kara Bom in the Altai (see below). On the other hand, the possibility of technological convergence by an intrusive tradition cannot be disregarded (Kuhn and Zwyns 2014). Ongoing research at the newly discovered site of Ushbulak-1 in Kazakhstan (Shunkov et al. 2017) may throw some light on these problems.

Central Asian skeletal evidence

The hominin skeletal record from this region is sparse. The best-known specimen is the skeleton of a young individual from the cave of Teshik Tash, (Uzbekistan) (Figure 9.1) that was associated with a Middle Palaeolithic assemblage but unfortunately cannot be securely dated. It was probably deliberately buried but later disturbed by carnivore activity (Gunz and Bulygina 2012). Until recently, it was regarded as the easternmost example of a Neandertal but remains classed as Neandertal are now recorded at Okladnikov Cave in the Russian Altai 2,000 km to the east (Krause et al. 2007; see Figure 1.1 and below). Although most researchers regard the Teshik Tash child as a Neandertal, some maintain that the mandible and cranium have modern rather than Neandertal features (Glantz et al. 2009). However, Gunz and Bylugina (2012) argue that the frontal bone is unequivocally Neandertal. Krause and colleagues (2007) also show by aDNA analysis that the Teshik Tash and Okladnikov individuals were genetically affiliated to European Neandertals; indeed, they are so similar that they might be recent immigrants into Central Asia from Europe. This suggestion is contradicted by recent evidence that indicates that Neandertals at Denisova Cave (see below) are probably earlier than Denisovans.

There are also six permanent maxillary teeth and 121 cranial fragments (all probably from a 9–12-year-old individual) from level 16 at Obi Rakhmat (Uzbekistan) that are associated with a Middle Palaeolithic assemblage; their age is ca. 60–90 ka by OSL (Glantz et al. 2008) and ca. 70 ka by ESR (Bailey et al. 2008). According to Glantz (2010), this individual has a mosaic of features: the teeth seem Neandertal but the left parietal is long and looks modern. At Anghilak Cave, also in Uzbekistan, a non-diagnostic fragment of an adult fifth metatarsal (AH-1) was found in stratum IV, for which there are radiocarbon dates of 43,900 ± 2,000 and 38,100 ± 2,100 uncalibrated years BP (Glantz et al. 2008: 225). There is also a second deciduous incisor from the open-air site of Khudji, Tajikistan. This was found in horizon 8 in a late Middle Paleolithic context, dated by radiocarbon to ca. 40 ka uncalibrated years BP (Trinkhaus et al. 2000) but is not particularly diagnostic. The only pertinent evidence from Afghanistan was a right temporal bone from Darra-i-Kur that was found in 1966, identified as Neandertal and dated by radiocarbon to 30,000 BP, but has been recently re-dated to only 4,500 BP (Douka et al. 2017) and thus now irrelevant to discussions of human evolution in Central Asia.

In summary, the quality of the Central Asian skeletal record is extremely poor; although the individuals from Teshik Tash and Obi Rahmat have Neandertal features, we should perhaps regard these as indicating a variable, Central Asian population that is genetically similar to European Neandertals but morphologically variable (Trinkhaus 2005), and thus showing a mosaic of Neandertal and "modern" features. Additionally, we have no indication of when *Homo sapiens* first appeared in the region (Dennell 2013).

Faunal data shows a pattern similar to that seen in Iran, with some sites showing an emphasis on hunting wild goat, presumably in rocky terrain, and others a focus on horse, deer, aurochs and other ungulates (see Table 9.1). Sites such as

TABLE 9.1 Fauna associated with Middle and Upper Palaeolithic sites in Central Asia

Site	*Elevation (m)*	*Dominant species*	
Ogzi Kichik Cave	1,200	Wild goat/sheep	*E. caballus, E. hydruntinus, Cervus, Coelodonta sp.* (woolly rhino), trace; tortoise abundant
Khudji, open-air site	1,200	Wild goat/sheep	
Obi Rahmat Cave	1,250	Wild goat (60%)	*Cervus elaphus* (30%), *Ovis sp., Sus scrofa*, marmot
Aman Kutan	1,300	Wild sheep	*C. elaphus, Capreolus, Capra sp., E. hemionus*, tortoise
Teshik Tash	1,800	Wild goat (83%)	*E. caballus, C. elaphus, Ursus, Felis-pardus, Hyena*
Amir Temir		Wild goat	
Kuturbulak, open air		*E. caballus* (49.3%)	*Bos/Bison, Equus, Cervus*
Aman Kutan		Wild sheep	*C. elaphus, Capreolus sp., Capra sp., E. hemionus*, tortoise; *Crocuta spelea*
Samarkandskaya		*E.* cf. *przewalskii*	*E. hemionus, E. hydruntinus, Bos primigenius, Camelus, C. elaphus, Ovis sp., Gazella sp., Sus scrofa*
Khudji	1,200	Wild goat/sheep	*Cervus sp., Equus, Alces* sp., *Ursus, Canis* sp., Tortoise
Selungir 1 and 2	2,000	Wild goat/sheep, *C. elaphus, C. elaphus, Ursus spelaeus*	
Selungir 3 and 4	2,000	*Bos primigenius, Dicerorhinus, Ovis, Capra*	
Kulbulak layers 45–12a	1,042	*Equus sp., Cervus sp., Bos primigenius, Capra sp., Sus scrofa, Lepus tolai*	
Karasu		*Equus sp.*	*Bison, Saiga, C. elaphus*

Notes: Aman-Kutan: at least 116 individual *Ovis orientalis* and 105 steppe tortoises (*Testudo horsfieldi*, 843 bones/carapace).Khudji: Trinkaus et al. (2000) give the altitude as 800 m.
Source: Vishnyatsky 1999.

Obi Rakhmat have been interpreted as a short-term hunting and butchery station of wild goat and to a lesser extent, red deer (Glantz et al. 2008).

The Altai and Transbaikalia

Two adjacent regions of southern Siberia contain the earliest evidence of the earliest Upper Palaeolithic. These are the Altai Mountains and Transbaikalia (see Figure 9.2). Because the Altai may have been a source region for the population that entered Transbaikalia, we can consider it first.

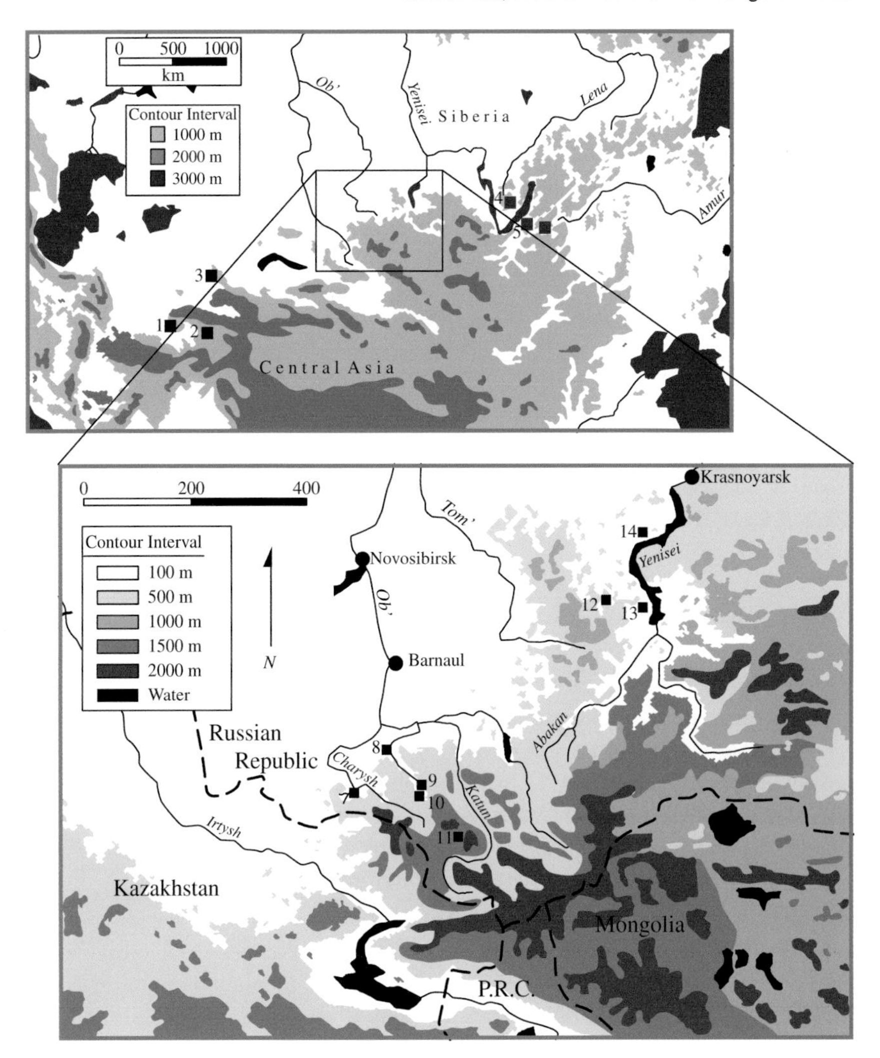

FIGURE 9.2 Location of sites in the Altai Mountains

Key: 1 Teshik Tash; 4 Makarovo; 5 Vavarina Gora; 6 Tolbaga; 7 Strashnaia cave; 8 Okladnikov cave; Denisova cave; 10 Ust'-Karakol; 11; Kara Bom; 12 Malaia Syia; 14 Kurtak

Source: Goebel et al. 1993, Fig. 1.

The Russian Altai Mountains

The Russian Altai Mountains lie near the borders of Kazakhstan, Mongolia and China, in the heart of continental Asia. They are a series of mountains up to 4,500 m high that extend ca. 2,000 km (1,200 miles) southeast to northwest from the Lake Baikal region and northern Mongolia to the West Siberian Plain, and are dissected by the valleys of several rivers that flow northwards into the rivers Irtysh and Ob. The climate is strongly continental, with average July temperatures of ca.

14–16° C., with highs of 30° C. and January temperatures averaging –15° to –20° C. (5° to –4° F.), excluding wind chill. As shown in Figure 9.2, the Russian Altai contains several cave and open-air sites that document hominin occupation of the regions over the past 200,000–300,000 years. Of these, the best-known is Denisova.

There were three types of resident hominins in the Altai before 40 ka – Neandertals, Denisovans and probably *H. sapiens*. Skeletal remains identified as Neandertal and dated to 37,750 ± 750 and 43,700 + 1,100/–1,300 years BP were found at Okladnikov Cave (Krause et al. 2007) (although these dates are probably too recent) and have also been found in recent excavations at Chagryskaya Cave (Buzhilova 2013; Mednikova, 2013). As noted above, aDNA (ancient DNA) analysis indicated that the mtDNA (mitochrondial DNA) of a subadult at Okladnikov was similar to that from Teshik Tash and European Neandertals. Without that analysis, the identity of the remains from Okladnikov would have been difficult to establish. Viola and colleagues (2011) observed that the teeth in general lacked the derived features seen in Neandertals, and the post-cranial specimens were hard to assign to any specific taxon.

The presence of Neandertals at Denisova Cave has recently been confirmed from aDNA analysis of a foot and hand phalange (Denisova specimens 5 and 9) and aDNA from an otherwise unidentifiable fragment (Denisova 15) (Viola et al. 2019) as well as from aDNA in the cave sediments (Slon et al. 2017). As at Okladnikov, the identification of Neandertals has been primarily on genetic, not morphological grounds. Denisova Cave became world-famous after the discovery of a previously unknown Denisovan population that was initially identified from the aDNA that was extracted from one tooth and part of a finger bone (Krause et al. 2010). These analyses showed that it was probably a sister clade, or near relation, of Neandertals.

Since then, Denisovan aDNA has been found in four other specimens (Denisova specimens 2–4, 8) (Viola et al. 2019). Denisovan DNA has a wide past and present distribution: it is present in the 430-ka-old population at Sima de los Huesos, Atapuerca, Spain (Meyer et al. 2014), and in modern populations in Melanesia (Reich et al. 2011). (We encounter a recent discovery of a claimed Denisovan mandible from Tibet in Chapter 10.) No specimens of *H. sapiens* have so far been recovered from an Early Upper Palaeolithic context in the Altai, but eight teeth of a 7–9-year-old are reported from Strashnaya Cave in a layer dated at 19,150 ± 80 cal BP (Krivoshapkin et al. 2018). There is also a *H. sapiens* cranium from Pokrovka in northeast Siberia that is dated to 27,740 ± 150 BP (Akimova et al. 2010). However, a modern male femur from Ust'-Ishim in western Siberia was dated to 46,880–43,210 cal BP, and its aDNA showed a small admixture of Neandertal (but no Denisovan) DNA (Fu et al. 2014). This helps us to place humans in the Altai region at ca. 47–43 ka cal BP.

Conditions in the Altai Mountains in the last glacial cycle appear to have been moderately temperate as there is little evidence of tundra in the faunal records, and much evidence for animals such as Siberian goat, red- and roe-deer that prefer woodlands. In the Late Pleistocene, the Altai had four faunal components (Agadjanian and Shunkov 2018): residents (those habitually living in the Altai), cosmopolitans (those with a wide Palearctic distribution), northern and southern migrants (species that appeared only incidentally and played a minor role in hominin subsistence) (see Table 9.2). The principal resident mammals were Siberian goat and red deer, and cosmopolitan species such as bison, woolly rhinoceros and mammoth. Most migrants came from the south, especially different types of horses, yak, gazelle and saiga. Northern migrants were rare but included occasional records of reindeer and arctic fox.

TABLE 9.2 Fauna of the Altai in the Late Pleistocene

Taxon	*English name*	*Abundance*
Resident		
Ovis ammon	Wild sheep, argali	√√
Capra sibirica	Siberian goat	√√√
Moschus moschiferus	Musk deer	√
Martes zibellina	Sable	√
Cervus elaphus	Red deer	√√√
Cosmopolitans		
Bison priscus	Bison	√√√
Capreolus pygargus	Roe deer	√√
Alces alces	Moose	√
Megaloceros giganteus	Giant elk	√
Equus ferus	Horse	√√
Coelodonta antiquitatis	Woolly rhinoceros	√√√
Mammuthus primigenius	Woolly mammoth	√√
Crocuta crocuta	Cave hyaena	√√√
Panthera leo spelaea	Cave lion	√√
Lynx lynx	Lynx	√
Canis lupus	Wolf	√√√
Southern migrants		
Equus hydruntinus	Pleistocene donkey	√√
Equus hemionus	Onager, kulan	√√
Equus przewalskii	Przewalski horse	√√
Poephagus sp.	Yak	√√
Procapra gutturosa	Dzereh, Mongolian gazelle	√√
Saiga tatarica	Saiga	√√
Cuon alpinus	Red dog	√√
Uncia uncia	Snow leopard	√√
Hystrix leucura	Porcupine	√
Northern migrants		
Rangifer tarandus	Reindeer	√√
Alopex lagopus	Arctic fox	√√

Notes: √ = present; √√ = rare; √√√ = common. Animals that avoid deep snow: *Cervus elaphus*, *Capreolus*, *Procapra*, *Saiga tatarica*; avoids mountains: mammoth; avoids low altitudes and warm conditions: yak. *Equus hemionus* can migrate 1,500 km.
Source: Agadjanian and Shunkov 2018.

The Initial (Early) Upper Palaeolithic

In the Altai, three sites are particularly important in showing the earliest Upper Palaeolithic.

Denisova

The cave of Denisova lies ca. 660 m above sea level in a narrow valley of the river Anui and is currently one of the most discussed Palaeolithic sites in Asia because of its long stratigraphy and records of Neandertals and Denisovans (Dennell 2019). The cave has three chambers – Main, East and South – that each has several metres of deposits which contain Middle Palaeolithic assemblages (layers 11–12), with an Upper Palaeolithic (layer 9) and later assemblages at the top of the sections. The Upper Palaeolithic assemblages include bone points and pendants made from perforated teeth of deer, none of which are found in Middle Palaeolithic assemblages in the region. Despite its long archaeological sequence, micro-stratigraphic analysis indicates that hominin use of the cave was intermittent and the main occupant of the cave was hyaena (Morley et al. 2019).

One of the biggest challenges in the investigations at Denisova has been its dating. This is difficult for several reasons: the strata in each chamber are difficult to correlate, and layers with the same number in different chambers are not necessarily of the same age; many layers have been disturbed by solifluxion, cryoturbation and bioturbation, particularly the burrowing actions of animals such as hyaena. Because of these factors, it is not always clear if small items – such as hominin teeth or beads – are in their original layer or have been incorporated into older or younger deposits. Fortunately, the chronology is much clearer now because of two recent dating programmes. The first, led by Katerina Douka from Oxford (Douka et al. 2019), took 50 new radiocarbon determinations using the most up-to-date techniques, and the other led by Zenobia Jacobs (Jacobs et al. 2019) from Wollongong, Australia, dated over 100 sediment samples by OSL. The parts of those programmes that are of most interest here concern the age of the most recent Neandertals and Denisovans at the cave, and the age of the artefacts in the IUP layers.

The most recent Neandertal specimens from Denisova are Denisova specimens 5 (a phalange fragment in layer 11.4) and 15 (a long bone fragment in layer 11.4[4]), with estimated ages of 90,000–130,000 years. Denisovans are more recent: the youngest are Denisova 3 (a phalange fragment) in layer 11.2 (51,600–76,200 BP) and Denisova 4 (an upper molar) in layer 11.1 (55,200–84,100 BP). A hybrid, Denisova 11 (a long bone fragment in layer 12.3, but possibly intrusive into that layer), was a young girl with a Neandertal mother and Denisovan father (Slon et al. 2018), and was dated at 79,300–118,000 BP. These all predate the Upper Palaeolithic material in layer 9. Four pendants made from red deer (*Cervus elaphus*) and elk (*Alces alces*) teeth were radiocarbon dated at ~32,000, ~40,000 and ~45,000 cal BP, and two bone points were dated to 42,660–48,100 and 41,590–45,700 cal BP.

These dates place the IUP at Denisova at ca. 43,000–48,000 cal BP. As seen in Figure 9.3, these dates match those for the *H. sapiens* femur from Ust'-Ishim, and strongly imply that our species was the manufacturer of the IUP. They also match the dates for the IUP at Kara Bom and Strashnaya Cave (see below). This is not to dismiss, of course, the possibility that it might have interbred with Denisovans or even Neandertals at the time of contact and we might expect evidence of that when eventually an IUP skeletal part is found in Denisova or elsewhere in the Altai.

Kara Bom

The open-air site of Kara Bom is the type site of the Kara Bom variant of the Upper Palaeolithic in the Altai (Derevianko et al. 2000). The site has 11 lithological layers divided into three depositional phases (Derevianko et al. 2000; Brantingham et al. 2001). Strata 4–6 contain Upper Palaeolithic assemblages:

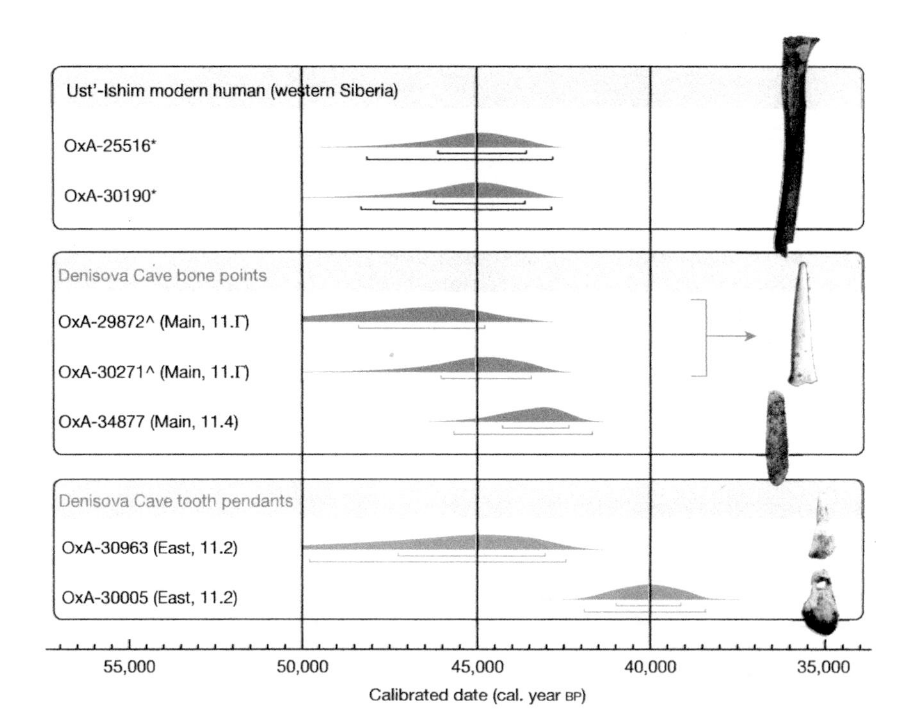

FIGURE 9.3 Dating the appearance of *H. sapiens* at Denisova in Siberia

This figure compares age determinations obtained for the oldest bone points and tooth pendants from Denisova Cave with the two direct age estimates for the Ust'-Ishim modern human femur. For each measurement, the lab code is shown and, for the Denisova artefacts, the chamber and stratigraphic context in brackets. Marked ages (⋆ †) were obtained on the same sample. Artefacts and human bone are not to scale.

Source: Douka et al. 2019, Fig. 2.

layer 4 has occupation horizon (OH) 1, layer 5a contains OH2, and layer 5b OH3, and layer 6 contains OH4. Below that, layers 7 and 8 are sterile, and layer 9a and 9b contain Middle Palaeolithic horizons MPH1 and MPH2 respectively. The earliest Upper Palaeolithic assemblage in layer 6 have radiocarbon ages of 43,300 ± 1,500 and 43,300 ± 1,600 BP; these are the earliest dates in Central Asia for the Upper Palaeolithic. Layers 5a and 5b are dated to 30,990 ± 460 BP and 34,180 ± 640 BP respectively, and layer 4 to 38,080 ± 910 BP. These dates place the beginning of the Upper Palaeolithic in layers 6 and 5 at around 45,000–42,000 BP. The underlying Middle Palaeolithic is poorly dated, with two infinite C14 dates of >42,000 for MPH1 and dubious EPR ages of 62–72 ka for MPH2.

Occupation levels 5 and 6 contained 1,472 lithic artefacts, of which 224 were tools. In level 6, 19.5% of tools were made on flakes and 70% on blades, some of which were very large. There were several new tool types, such as end-scrapers on blades, multifaceted burins and various points on blades. One interesting aspect of the initial, or Early Upper Palaeolithic at this site is that there was a greater emphasis on blade production, but these were produced on Middle Palaeolithic Levallois cores. Brantingham and colleagues (2001) suggest this may have been a way of maximising cutting edge length and the number of end products whilst minimising waste. Because there are sterile layers between the Middle and Upper Palaeolithic at this site, there is no possibility that material could have been mixed together during excavation.

Kara Bom is one of several Upper Palaeolithic sites in south Siberia with indications of symbolism and personal ornaments similar to those seen in the Levant (Chapter 8), Transbaikalia and Mongolia (see below). Occupation level 5 contained numerous lumps of ochre, a striped pebble with traces of ochre on one end, and three pendants (one of bone with traces of ochre, one pierced ungulate tooth and one drilled piece of bone).

Ust Karakol

This open-air site lies at confluence of Karakol and Anui rivers at an ideal location for a seasonal hunting site. The site comprises a 6.5-metre sequence of alluvial deposits at its base (strata 19–12) and loess-like sandy loams and palaeosols (strata 11–4). There is a long sequence of artefacts from Early Middle to Late Upper Palaeolithic: Early Middle Palaeolithic artefacts were found in stratum 19, Middle Palaeolithic ones in strata 18–13, and Early Upper Palaeolithic ones in strata 11–8. Stratum 5 contained a Middle Upper Palaeolithic, and a Late Upper Palaeolithic was present in strata 4–2. A radiocarbon date of 35,100 ± 2,850 BP was obtained for a hearth in the upper part of stratum 10, and ones of 33,400 ± 1,285 BP, 29,860 ± 355 BP and 29,720 ± 435 BP for hearths in stratum 9. There is also an RTL date of 50 ± 12 ka for a hearth lens in the base of stratum 9. Because older dates for the Early Upper Palaeolithic have been obtained from Denisova, Kara Bom and Strashnaya (see below), the Ust Karakol ^{14}C dates probably under-shoot true age and these layers should be re-dated. Derevianko and colleagues (2000: 369) suggest that the tentative age for stratum 9 is ca. 38 ka.

Regarding the Early Upper Palaeolithic assemblages, core reduction was aimed at detaching elongated blanks from single and double platform cores and prismatic cores. Microblade production from wedge- and cone-shaped cores was important. The tool-kit consisted mainly of longitudinal racloirs (side scrapers), carinated- and end-scrapers, backed knives, burins, awls and notches (Derevianko et al. 2003: 368–371).

Strashnaya Cave

Recent excavations at this cave show seven Middle Palaeolithic layers (10–4) and one Upper Palaeolithic layer (layer 3, with three sub-units). The earliest of these, horizon 3_3, is described as the Denisovan phase of the Early/Initial Upper Palaeolithic, with Levallois flaking techniques, blade cores, and bone tools and ornaments. Horizon 3_1b is described as an Early Upper Palaeolithic blade-based tradition with bone points, similar to that at Kara Bom, and is dated to 43,650 ± 650 cal BP and 44,050 ± 700 cal BP. Above that, horizon. 3_1a and dating to 20 kyr BC, contains a late Upper Palaeolithic with bladelet technology, personal ornaments and bone tools as well as eight teeth of a young *H. sapiens* (Krivoshapkin et al. 2018).

The Altai Middle and Upper Palaeolithic

The main type of Middle Palaeolithic assemblages in the Altai in the Upper Pleistocene are those with a Levallois technology. This group includes Kara Bom, Ust-Karakol-1, Anui-3, Ust-Kanskaya Cave and possibly Strashnaya Cave, and forms the "Kara Bom variant of the Altai Middle Palaeolithic". These show a dominance of Levallois reduction strategies, a developed method of blade detachment, large numbers of tools on blades, and few Mousterian forms or notched and denticulate tools. There is also the Denisova variant of the Altai Middle Palaeolithic, known only from Denisova, in which the primary reduction strategy is parallel and radial, with an insignificant use of the Levallois technique. This variant contains numerous Middle Palaeolithic types of scrapers, notched and denticulate tools made on medium-sized flakes. A third variant, known as the Sibiryachikha, derived from the caves of Okladnikov (formerly Sibiryachikha Cave) and Chagyrskaya. Most blanks are produced by radial flaking. A major feature of this variant are bifaces, and scrapers with different types of backing as well as déjeté, double and triple scrapers that are present at other Altai sites but not at such high frequencies (Shunkov 2017). It has been suggested that the inhabitants of these caves were Neandertal immigrants from Central Asia (Derevianko et al. 2013a, 2013b, 2018).

According to Russian specialists, the Early (or Initial) Upper Palaeolithic in the Altai developed between 50 ka and 40 ka from the local Middle Palaeolithic industries. Two variants are recognised, the Ust-Karakol and Kara Bom variants. The Ust-Karakol variant includes Ust-Karakol-1, Denisova, Anui-3 and possibly Strashnaya Cave. This variant is characterised by the serial production of elongated blanks from prismatic, cone-like and end-facetted cores, including wedge-shaped cores that are used alongside Levallois and simply prepared cores. Tools still include Middle Palaeolithic types of denticulates and racloirs, but few Levallois ones. Upper Palaeolithic tool types include bladelets, end- and carinated-scrapers, dihedral burins and large blades, and bone implements such as needles, piercers, cylindrical beads, rings of mammoth tusk, and pendants from ungulate teeth. The most characteristic tools are those of Aurignacian type (end-scrapers on blades, large blades with retouched sides, backed microblades) and bifacial foliate points, as well as bone and tooth ornaments. The Kara Bom variant includes Kara Bom, and possibly Maloyalomanskaya Cave: the emphasis here was on the repetitive bi-directional production of large blades and blade-based tools, leaf-shaped and oval bifaces, and also end-scrapers, dihedral burins, elongated points, and knives with retouched backs. There are also occasional Aurignacian forms, bifacially worked pieces and personal ornaments, but these are not common Derevianko et al. 2003; Shunkov 2017).

The Early Upper Palaeolithic in the Altai: indigenous or invasive?

There are two main ways of looking at the Middle and Upper Palaeolithic of the Altai. Russian researchers – particularly Professor Anatoly Derevianko from Novosibirsk – have argued that it developed indigenously from the local Middle Palaeolithic Kara Bom and Ust Karakol traditions. On this argument, the substrate of lithic reduction methods remained Middle Palaeolithic, but new types of artefacts appeared, as well as bone tools and personal ornaments. He further argues that *Homo sapiens* evolved locally; as one who believes in multi-regional evolution, he proposes that our species evolved from four sub-species, namely *Homo sapiens africanensis* in Africa, *Homo sapiens neanderthalensis* in Europe, *Homo sapiens orientalensis* in East and Southeast Asia, and *Homo sapiens altaiensis* in southern Siberia and Central Asia (see Derevianko 2011; Derevianko and Shunkov 2011). (We see in Chapter 10 that the East Asian fossil record has been similarly interpreted.) Most non-Russian researchers reject this notion of multi-regional evolution; instead, as seen in Chapter 1, most now agree that our species originated in Africa, and was an immigrant or invasive species in Eurasia. On the same argument, the Upper Palaeolithic is also seen as a largely, if not wholly, intrusive development that marks a break with the preceding Middle Palaeolithic – a view with which I concur.

We might bear in mind two points, however. The first is that we know now from genetic evidence that Neandertals interbred with Denisovans – as seen in the Denisova 11 hybrid – and also with our species; hence the presence of a small amount of Neandertal DNA in modern non-African populations. Second, we know that in Europe, Neandertals overlapped with humans for around 2,000–5,000 years, which provides ample time for some degree of interbreeding. We have no way of course of knowing whether sexual encounters between Neandertals, Denisovans and humans were long-term, caring and consensual, or short-term, violent and non-consensual (or all shades in between), and probably they were as varied as within *Homo sapiens* populations. It is, however, likely that at the time of contact, humans may have copied or learnt local methods of tool-use, and this is one way of explaining the persistence of Middle Palaeolithic core reduction techniques in Early Upper Palaeolithic assemblages. (We can also include the possibility of technological convergence, as argued in Chapter 5.) We have noted already (Chapter 1) the suggestion that humans may have learnt hide-working techniques from Neandertals, and the need for warm, insulated clothing was an obvious essential adaption that humans needed to survive prolonged sub-freezing winters after they had entered Central Asia and southern Siberia. My own view is that the Altai Upper Palaeolithic and *Homo sapiens* were intrusive but probably involved some degree of interbreeding with the local population and the incorporation of local methods of stone tool production. All this remains speculation until we have well-preserved, diagnostic skeletal specimens from the Altai in an Early Upper Palaeolithic context.

Transbaikalia

East of Denisova, the Sayan Mountains extend 1,500 km east of the Altai Mountains towards Lake Baikal, the world's deepest and largest by volume. This region contains the headwaters of three of Russia's mightiest rivers – the Yenesei, Angara and Lena – that drain into the Arctic Ocean 2,000 km to the north. Cis-Baikal is the area west of Lake Baikal, and Trans-Baikal lies to the east (see Figure 9.4); the Mongolian Plateau begins further south (see below). The regional climate is strongly continental: at Irkutsk, the average July temperature is 17.5° C., and in January –19.3° C., with average minimum temperatures of –27° C. Clearly, over-wintering would have challenged any Palaeolithic group in this region.

Boreal and mountain taiga, with *Pinus sylvestris* (Scots pine), *Larix sibirica* (Siberian larch) and *Betula* (birch) trees, with some *Populus tremula* (aspen) and *Alnus fruticosa* (shrubby alder) (Bezrukova et al. 2010) form the present-day vegetation around Lake Baikal. In the early part of MIS 3, open forests of *Picea*, *Pinus* and *Betula* covered the southern region but steppe increased in the latter part (Shichi et al. 2007). According to data from a sediment core from Lake Kotokel near Lake Baikal, the vegetation between 47 and 30 ka was largely tundra-steppe with *Artemisia*, Poaceae and Cyperaceae and some wood cover, indicating a harsh and unstable climate during the MIS 3 (Karginskiy) interstadial (see Figure 9.5).

Although little is known of the Palaeolithic in Cis-Baikal, the Transbaikal contains several open-air Early Upper Palaeolithic sites (see Figure 9.4). Their features are summarised in Table 9.3. The dating of these sites has been problematic, in part because they are in complicated depositional settings of alluvial,

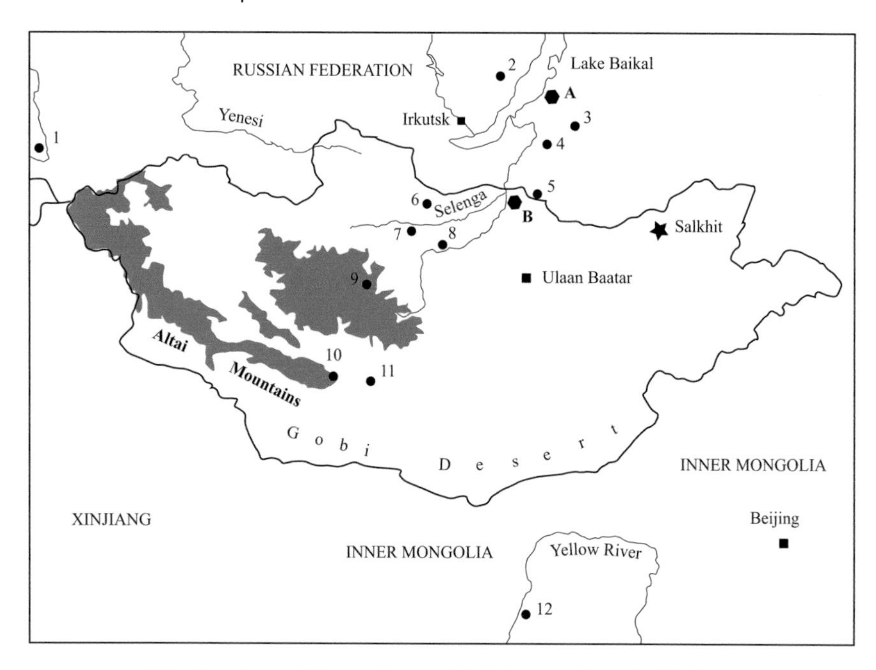

FIGURE 9.4 The location of the main stratified sites with early blade assemblages in Transbaikalia and Mongolia

Key: 1 Kara-Bom; 2 Makarovo-4; 3 Khotyk; 4 Kamenka A–C; 5 Podzvonkaya; 6 Dörölj-1; 7 Tolbor sites (T4, T15, T16, T21); 8 Orkhon sites (Orkhon-1, Orkhon-7); 9 Tsatsyn Ereg-2; 10 Tsagaan-Agui; 11 Chikhen-2; 12 Shuiddongou-1 and -2. Palaeo-environmental sites: A Lake Kotokel; B Shaamar.

Source: Redrawn and adapted from Zwyns et al. 2014a.

colluvial and/or aeolian sediments in which archaeological units have often been disturbed by cryoturbation, and also because of problems of excluding contamination from carbon-14 (Graf and Buvit 2017). Nevertheless, the available dates indicate that the Early Upper Palaeolithic in the Transbaikal is probably slightly younger than in the Altai (Rybin 2014). In the Cis-Baikal, there are three infinite C14 dates of >45,000 years BP for Makarovo 4 (in which the artefacts have been re-deposited [Rybin 2015]), and stratum 4 at Malaia Syia has two dates on bone of 40,800–38,600 cal BP (Graf 2014). In Transbaikalia, dates from Kamenka, Tolbaga, Varvarina Gora suggest that occupation began ca. 45,000–44,500 cal BP and continues to ca. 35,000 cal BP (see Table 9.4 and Buvit et al. 2015; Graf and Buvit 2017). Unlike the Altai, where the case can be made that the IUP develops from the local Middle Palaeolithic, the earliest Upper Palaeolithic in the Transbaikal appears to be an intrusive development that "indicates migrations within a limited territory and restricted home ranges" (Rybin 2015: 481). As seen in Figure 9.5, the environment between 46 ka and 30 ka was a cold tundra steppe with patches of tree cover, and less

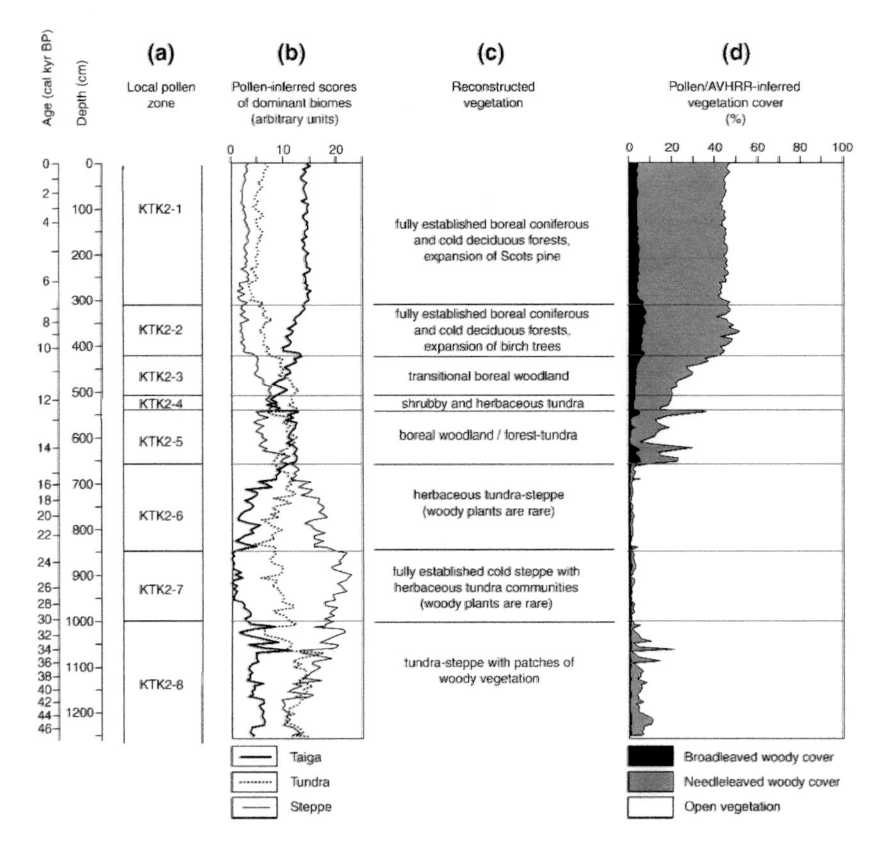

FIGURE 9.5 The environmental record of Lake Kotokel over the last 46,000 years
Source: Bezrukova et al. 2010, Fig. 5.

severe than the cold and largely treeless steppe that developed between 30 ka and 24 ka. Compared with the Altai, the conditions in the Transbaikal were less predictable and more severe; the steppe was also less diverse than the variety of ecozones in the Altai, and there were therefore fewer alternatives when conditions worsened (Rybin 2015).

Nevertheless, the impression is that human groups were learning to adapt to a new type of landscape (Graf 2014) and achieving more than mere survival (see Table 9.3). There are also several examples of personal ornaments (Derevianko and Rybin 2003; Lbova 2019) that are summarised in Table 9.4: these include drilled eggshell and bone beads and a drilled pendant at Podzvonkaya, two shell and bone beads from Khotyk layer 2, three talc beads with drilled holes and two stone rings from layer 3; bone pendants from Tolbaga and Varvarina Gora; and ochre at Tolbaga, Podzvonkaya and Khotyk layer 2, where traces were found on a battered grinding tool (Buvit et al. 2015). Other bone artifacts include a whistle from Kamenka A (Kuhn and Zwyns 2014) and a bird long-bone with

TABLE 9.3 Summary of the Early Upper Palaeolithic in the Trans-Baikal

Behaviour	*Evidence*	*Sites where present*	*Earliest evidence (cal BP)*
Technology	Blade-based lithic assemblages dominated by flat-based cores with diverse toolkits	Tolbaga, Chitkan, Varvarina Gora, Khotyk, Kamenka, Sapun, Sanmyi Mys	Kamenka A (47,60–41,200); Varvarina Gora (>40,030); Tolbaga (41,300–40,000)
	Bone and antler tools	Tolbaga layer 4, Varvarina Gora, Kamenka A and B	Kamenka A (47,760–41,200); Varvarina Gora (>40,030); Tolbaga (41,330–40,000)
	Stone-slab lined dwellings	Tolbaga layer 4, Varvarina Gora, Kamenka A and B, Khotyk layer 3	Kamenka A (47,760–41,200); Varvarina Gora (>40,030); Tolbaga (41,330–40,000)
	Stone-lined hearths	Tolbaga, Varvarina Gora, Sanmyi Mys, Podzvonkaya	Varvarina Gora (>40,030); Tolbaga (41,330–40,000)
	Storage pits	Varvarina Gora, Khotyk layer 2	Varvarina Gora (>40,030)
Ritual/ symbolic behaviour	Carved and incised eggshell, bone and stone ornaments	Tolbaga, Varvarina Gora, Khotyk layers 2 and 3, Podzvonkaya	Tolbaga (41,330–40,000); Varvarina Gora (>40,030)
	Carved bone mobiliary art	Tolbaga	Tolbaga (41,330–40,000)
	Ritual pits or hearths	Varvarina Gora, Podzvonkaya, Khotyk layer 3	Varvarina Gora (>40,030)
Subsistence	Resource-rich areas transitional between several environmental zones	All sites	Kamenka A (47,760–41,200)
	Repeatedly occupied and long-term dwellings	Tolbaga, Varvarina Gora, Kamenka A and B	Kamenka A (47,760–41,200)

Source: Buvit et al. 2015, Table 33.3: 502.

a drilled hole that might have been a flute/whistle from Khotyk, layer 3 (Buvit et al. 2015) which may be similar to the flutes found at Hohe Fels and Vogelherd, Germany (Conard et al. 2009). A selection of IUP examples is shown in Figure 9.6.

TABLE 9.4 Initial Upper Palaeolithic sites in Siberia and Mongolia with personal ornaments

Area	*Site, complex, stratum*	*Dating, years BP*[a]	*Artifacts*	*Material*
Mongolia	Tolbor – 4 stratum 5	>41,000	beads	ostrich egg-
	Tolbor – 15 stratum 7	29,000-	beads	shell
	Dorolg – 1 (Egiin-Gol)	34,000	beads	ostrich egg-
		29,000-		shell
		31,000		ostrich egg-
				shell
Trans	Podzvonkaya	37,000-	beads, rings	ostrich egg-
Baikal	Khotyk stratum 2	>44,000	beads, pendants, rings,	shell
	stratum 3	28,000-	bracelet	stone, bone,
	kamenka (A)	32,000	beads, fragments of	see-shell
	Varvarina Gora	>40,000	a diadem, other articles	stone, bone,
	stratum2	>40,000	beads, pendant	ostrich-
	Tolbaga stratum 3,4	28,000-	pendants, sculpture	shells, ivory
		35,000		stone, tooth
		25,000-		ostrich egg-
		34,000		shell(?),
				bone
Cis –	Pereselencheskyi punct	25,000-	pendant, unknown	stone
Baikal	(Gerasimovo-1)	34,000	articles	bone, ivory
	Military Hospital	20,000-	beads, other articles	stone
	Mamony	30,000	unknown articles	stone
	Schapovo	28,000-	unknown articles	ivory
	Ust-Kova low stratum	30,000	beads, sculpture	
		28,000-		
		30,000		
		22,000-		
		32,000		
Mountain	Denisova Cave, stratum	>37,000-	beads, pendants, brace-	stone, bone,
Altai	11(Central Hall, South	>50,000	let, other articles with	tooth, sea-
	and East Gallery)	27,000-	cuts	shells,
	Strashnaya Cave Stratum	30,000	beads, pendant	ivory
	2	30,000-	pendants	ivory
	Kara Bom stratum 5 UP	44,000	pendant	tooth
	1	29,000-	pendant	soft stone
	Ust-Karakol, stratum 9.2	34,000	pendant,	tooth
	Maloyaloman's Cave	30,000-	beads, pendant,	bone
	Ust-Kanskaya Cave	34,000	unknown articles	soft stone
	(entrance area)	MIS 3		
	Anuy-2 stratum13.2	MIS 3		
Kuznetskiy	Malaya Syya	29,000-	beads, pendants, other	soft stone,
Alatau		>36,000	articles	bone
Arctic	Yana site	28,000-	Beads, pendants, frag-	ivory, soft
Siberia		32,000	ments of a diadem,	stone, bone
			bracelets, other articles	

Source: Lbova 2019, Table 1.

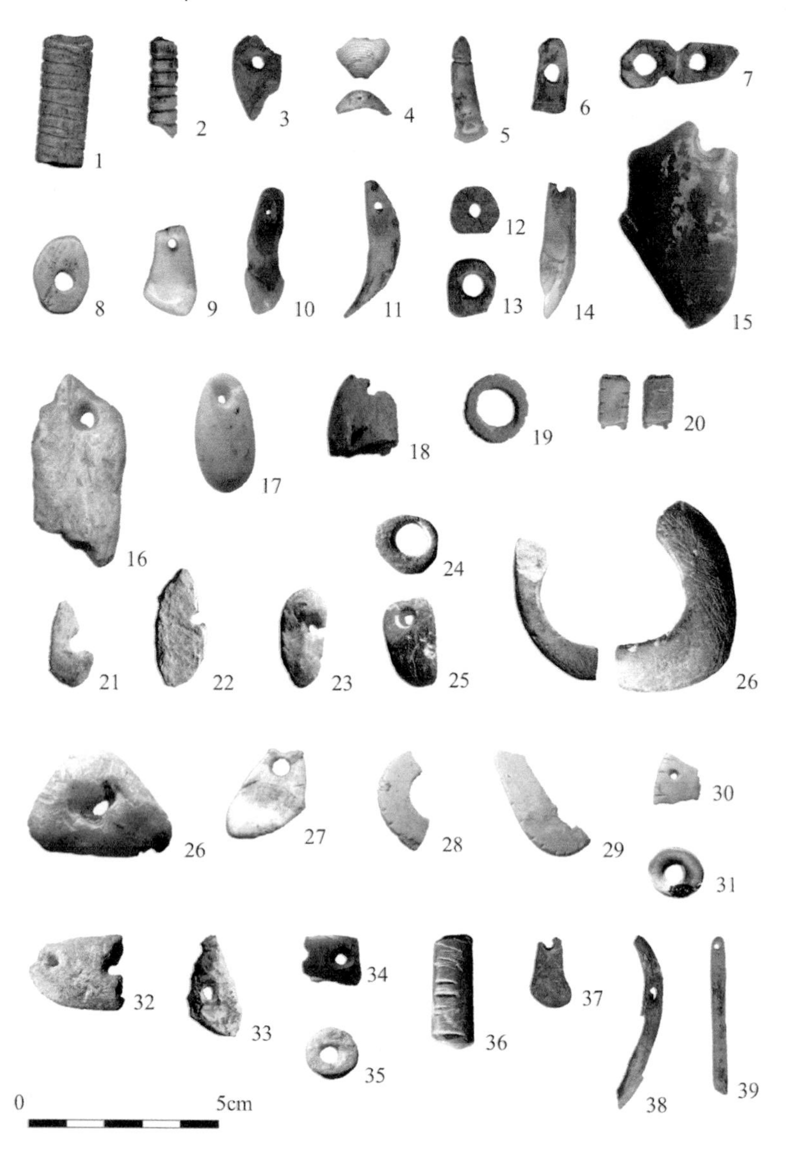

FIGURE 9.6 Personal ornaments in Initial Upper Palaeolithic (IUP) sites in Siberia

Key: Denisova Cave (1–21), Khotyk (22–31), Kara-Bom (32, 33), Podzvonkaya (34–36), Kamenka (38, 39), Maloyalomanskaya (37). Materials: bone and teeth (1, 2, 5, 10, 11, 14, 20, 27, 36–39); ivory (6, 7, 12, 13, 34); shell (4, 31); stone raw materials (3, 8, 9, 15–19, 21–26; 29–30, 32, 33, 35).

Source: Lbova 2019, Fig. 2.

There are a few hints that social networks in both the Altai and Transbaikalia were more extensive than during the Middle Palaeolithic. Whereas Middle Palaeolithic stone was overwhelmingly local, some raw materials at Kamenka came from 40–60 km distant, and jasper at Kara Bom stratum 6 came from

80 km away; the ostrich eggshell at Denisova (and Podzvonkaya) came from no less than 500 km away (Rybin 2015). Rybin (2014: 39) also argues that the presence in Transbaikalia of the same types of stone tools and reduction technologies and personal ornaments that are seen in the IUP of the Altai indicates "the transfer of a whole set of a unified cultural tradition" over a distance of 1,800 km (and much longer by less direct routes through the Altai and Sayan Mountains), and into a substantially different environment.

Faunal lists from the Transbaikal sites indicate woolly mammoth, woolly rhinoceros, horse, steppe bison, auroch, Irish elk, argali sheep, Siberian mountain goat, saiga antelope, red deer, roe deer, reindeer, arctic fox, red fox and hare (Graf and Buvit 2017). At Tolbaga, the three main species were argali sheep, horse and rhinoceros, and at Varvarina Gora, Mongolian gazelle, horse and argali sheep; the latter two were also present at Podzvonkaya. A detailed view of hunting preferences comes from the site of Kamenka (Germonpre and Lbova 1996). Here, horse (*E. caballus*) and Mongolian gazelle (*Procapra gutturosa*) were the main prey and formed 33.3% each (by minimum number of individuals) of the assemblage; the remainder were woolly rhinoceros, camel, Irish elk, bison, *Spirocerus kiakhtensis* (*spiral-horned antelope*), wild sheep and a trace of lion, *Panthera leo*. Horse prefer windswept landscapes where grass is exposed in winter, and Mongolian gazelle today live in dry grass steppe and semi-deserts, avoiding deep snow and steep slopes. The age at death of horse and gazelle indicates that they were killed in late summer/autumn or early winter. The absence of fur-bearing animals such as arctic fox and hare (whose pelts are best in winter) may indicate that Kamenka was a seasonal camp. The winter coat of Mongolian gazelle is good for fur coats (Germonpre and Lbova 1996) but those killed in summer and autumn would not have been suitable for this purpose.

Recent re-investigations of the site of Tolbaga (Izuho et al. 2019) conclude that it was occupied during three periods between 42,970 and 26,010 cal yr BP. The first, associated with an IUP assemblage, occurred between 42,970 and 40,425 cal yr BP, and the second, associated with EUP assemblages, took place between 37,785 and 33,290 cal yr BP. The final phase, with a MUP tool-kit, occurred between 29,320 and 26,010 cal yr BP. Each phase occurred during periods of warming: phase 1 between Heinrich Events H5 (46,000 cal yr BP) and H4 (39,000 cal yr BP); phase 2 within the interval between H4 and H3 (30,000 cal yr BP); and phase 3 within the interval between H3 and H2 (24,000 cal yr BP). Tolbaga is identified as a residential base, with stone-lined dwellings and hearths, storage pits, and diverse lithic and bone/antler tool types. The local environment is described as predominantly steppe with some open forest in warmer periods, and steppe-tundra in colder ones. Hunting preferences varied with each occupation phase. During the IUP phase of steppe and steppe-tundra, the main prey were woolly rhinoceros (N=3) and horse (N=1), but in the second phase, eight species are present: Asiatic wild ass, argali sheep, Mongolian gazelle (two of each), and one each of spiral-horned antelope, red deer, wolf, Baikal yak (*Poephagus baikalensis*) and reindeer. This variety indicates a mosaic landscape of taiga (boreal forest), steppe, and mountain-steppe and steppe-tundra. Horse, bison and reindeer (one of each) are the only species found in the final phase of steppe and steppe-tundra conditions.

Although there is no skeletal evidence for the Early Upper Palaeolithic of the Transbaikal, we do have evidence of *H. sapiens* at 45 ka BP from Ust'-Ishim, west Siberia (Fu et al. 2014), ca. 34 ka BP at Salkhit (see below) and at ca. 40 ka BP from Tianyuandong, north China (see Chapter 10), and it is not unreasonable to assume that it was responsible for the EUP. This seems additionally plausible because the main features of the Transbaikal EUP are similar to those found in the EUP of western Europe after 40 ka, which can safely be attributed to our species.

Mongolia

Mongolia is one of the largest and least populated countries in Asia. The beauty of its landscapes is offset by the harshness of its climate, in which summer temperatures can exceed 30° C. and winter night-time temperatures can fall below –40° C. Its average annual temperature is only 2.9° C. and rainfall averages 310 mm pa. Mongolia's climate is also extremely unpredictable and marked by wide variations in rainfall, dust storms, blizzards and cold snaps. Dzuds that result when snow melts and then re-freezes to form an ice crust over vegetation are especially lethal to livestock (see Chapter 3). Despite these challenges, hominins were in Mongolia before the Upper Palaeolithic, even if fleetingly. Middle and even Early Palaeolithic artefacts have been found in surface collections, all sadly undated, but a terminal Middle Palaeolithic may be present at the site of Kharganyn Gol 5. Here, Middle Palaeolithic tool assemblages from horizons 6 and 7 show a combination of Levallois reduction technology and blade production, and resemble Middle Palaeolithic assemblages from Kara-Bom and Layers 11.1 and 11.2 in the Eastern Gallery of Denisova Cave in the Russian Altai Mountains (Khatsenovich et al. 2017).

Several Upper Palaeolithic sites indicate more systematic occupation after 40,000 cal years ago. The earliest sites are recorded in the valley of the river Tolbor (Figure 9.7) which flows into the Selenga, which in turn flows into Lake Baikal and thus provides a corridor into northern Mongolia from Transbaikalia. These sites typically lie ca. 20–40 m above the alluvial plain and ca. 200–1,000 m from the present river, which implies considerable difficulties in obtaining water (Khatsenovich et al. 2017), unless water containers made from animal skins or water-proof baskets were used. Today, the average temperature is between –0.9° C. and –1.6° C. (with a minimum of –46.2° C.), and rainfall averages 250–300 mm, mostly between June and October. There are several local resource areas. Western slopes in the valley are covered by steppe grassland, with boreal trees such as aspen (*Populus tremula*), alder (*Alnus incana*) and elm (*Ulmus pumila*) in canyons adjacent to the valley; larch (*Larix sibirica*), Scottish and Siberian pine (*Pinus sylvestris* and *P. sibirica*) and white birch (*Betula*

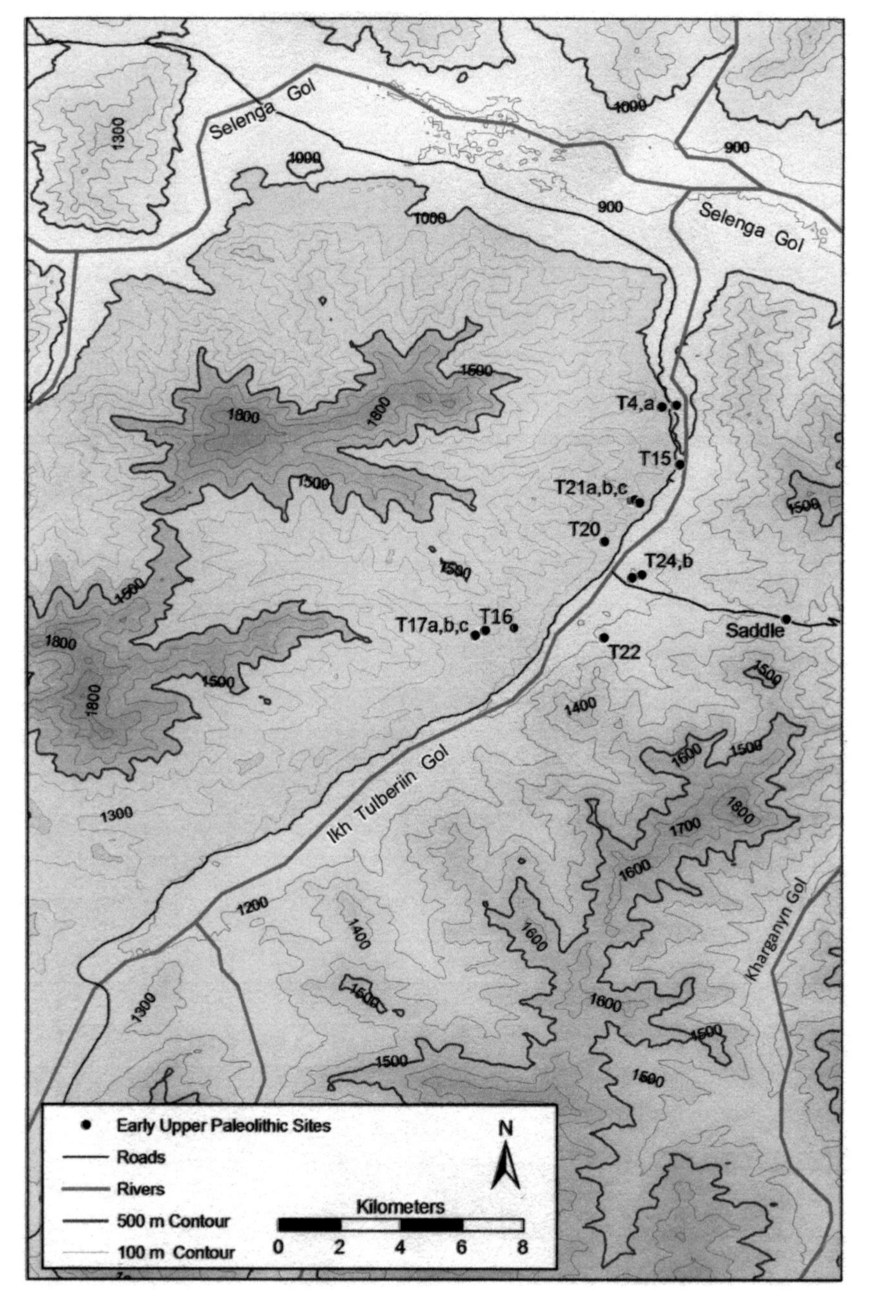

FIGURE 9.7 Location of Upper Palaeolithic sites in the Tolbor Valley, Mongolia
Source: Zwyns et al. 2014a, Fig. 2.

platyphylla) grow on northern slopes wherever the soils can provide sufficient moisture. Willow (*Salix mongolica*) grows along river banks (Gillam et al. 2012; Zwyns et al. 2014a).

Radiocarbon ages of Mongolian Upper Palaeolithic sites are shown in Table 9.5. As shown, the lowest archaeological horizons at Tolbor 4 (horizons 6 and 7), 15 (horizons 6 and 7) and Tolbor 16 have calibrated radiocarbon ages greater than 38,000 years. The earliest so far identified may be Tolbor 21, at which two radiocarbon dates on bone are 42,878–44,033 and 46,196–49,457 cal BP. These are classed by Zwyns et al. (2014a) as Initial Upper Palaeolithic (IUP). Some of the artefacts from Tolbor 16 are shown in Figure 9.8. The assemblages at Tolbor 4, levels 5 and 6, are dominated by blades and thick single or double platform cores. There is a low proportion of retouched tools; these are mostly carinated end-scrapers made on blades, side scrapers, notches and denticulates. (Zwyns et al. 2014b; Jaubert 2015). In the opinion of Zwyns and colleagues (2014a) as well as Derivianko and colleagues (2007), these are similar to those at Kara Bom horizons 5 and 6 as well as those in the Transbaikal such as Khotyk, Kamenka A–C and Podzvonkaya. Unsurprisingly, given the distances involved over this vast area from Mongolia to southern Siberia, there is considerable variation between assemblages but also important similarities. These include an emphasis on producing large, elongated blades by means of flat and subprismatic flaking techniques, including those aimed at making bladelets, bidirectional flaking resulting in pointed blades and flakes resembling Levallois points (Derevianko et al. 2013), as well as the occasional presence of ostrich eggshell beads (Rybin 2014).

Early Upper Palaeolithic (EUP) assemblages are known from several sites. The lithic tool-kits continued much as before, but blanks and tools became smaller, and a higher proportion of tools were made on flakes. Because of the absence of any major changes or innovations, it is assumed that there was continuity from the IUP to the EUP. The Middle Upper Palaeolithic is so far known only from after 25,000 radiocarbon years BP from the Orkhon Valley slightly east of the Tolbor Valley and is marked by the complete replacement of blade industries by flake industries, along with the parallel development of pressure-flaked microblades (Gladyshev et al. 2010, 2012). In this phase, Mongolia may have been largely depopulated as there is no record of occupation in the Tolbor or Kharganyn Gol Valleys between 30,000 and 19,000 years ago (Khatsenovich et al. 2017).

Useful background information on the environmental history of northern Mongolia of the past 38,000 years is provided by a study of a loess-palaeosol section at Shaamar (Ma et al. 2013; see Figure 9.9). Between 38,000 and 30,000 years ago, the local vegetation was a taiga forest dominated by *Picea*, *Abies*, and *Pinus* cf. *sibirica*, but from 30,000 to 23,000 BP was mainly steppe dominated by *Artemisia* and Poaceae denoting colder and drier conditions. This is in broad agreement with a study by Kolomiets and colleagues (2009) of the valley sediments at Tolbor 4 that indicate gradual but continuous

TABLE 9.5 Upper Palaeolithic dates from sites in Mongolia

Site	*Layer*	*C14 date*	*Date cal BP*	*Lab number*	*Industry*	*Dating sample*
Tolbor 4	Level 4	26,700 ± 300	30,984–31,764	AA-84135	EUP	*Struthio* eggshell
Tolbor 4	Level 5	31,210 ± 410	34,777–35,712	AA-93140	IUP	*Struthio* eggshell
Tolbor 4	Level 6	35,230 ± 680	39,137–41,120	AA-93141	IUP	*Struthio* eggshell
Tolbor 4	Level 6	37,400 ± 2,600	38,878–43,829	AA-79314	IUP	Bone
Tolbor 15	Level 5	28,460 ± 310	32,448–33,345	AA-84137	EUP	*Struthio* eggshell
Tolbor 15	Level 5	32,200 ± 1,400	35,182–38,664	AA-93136	EUP	bone
Tolbor 15	Level 7	29,150 ± 320	33,194–33,972	AA-84138	EUP	*Struthio* eggshell
Tolbor 15	Level 7	33,200 ± 1,500	36,047–40,129	AA-93137	EUP	Bone
Tolbor 15	Level 7	34,010 ± 200	38,656–40,537	MAMS-14934	EUP	Bone
Tolbor 15	Level 7	33,470 ± 190	37,426–40,318	MAMS-14935	EUP	Bone
Tolbor 15	Level 7	34,340 ± 210	38,903–40,532	MAMS-14937	EUP	Bone
Tolbor 16	T16 Pit 1 unit 7	33,320 ± 180	37,205–38,500	MAMS-14932	EUP?	Bone
Tolbor 16	Test pit (Pit 1) unit 7	AA-93134	>48,612	>45,400	IUP?	Bone
Tolbor 21	Test pit 1, level 4	44,640 ± 690	46,196–49,457	MAMS-14933	IUP?	Bone
Tolbor 21	Test pit 2, level 3	39,240 ± 360	42,878–44,033	MAMS-14936	IUP?	Bone
Dörölj-1		29,910 ± 310		GifA-99560; bone		Jaubert (2015)
Dörölj-1,		29,540 ± 390	38,220–34,456	GifA-99561; bone		Gladyshev et al. (2010, 2012)
Dörölj-1,		31,880 ± 800		GifA-11664; ostrich eggshell		Gladyshev et al. (2012)

(*Continued*)

TABLE 9.5 (Cont.)

Site	*Layer*	*C14 date*	*Date cal BP*	*Lab number*	*Industry*	*Dating sample*
Chikhen-2	Horizon 2.5:	30,550 ± 410		AA-31870		Gladyshev et al. (2012)
Chikhen Agui Rockshelter		27,432 ± 872		AA-26580		Charcoal in hearth; Gladyshev et al. (2010, 2012)
Orkhon-1	Trenches 1 and 2	29,465 ± 445				Jaubert (2015)
Orkhon-1	Trenches 1 and 2	34,400 ± 800				Jaubert (2015)
Orkhon-7	Trenches 1 and 3	33,295 ± 500				Jaubert (2015)
Orkhon-7	Trenches 1 and 3	31,490 ± 310				Jaubert (2015)
Kharganyn Gol 5	Layer 5, 1.4 m	38,716 ± 150	42,500–42,850	UGAMS-23064, NSKA-1503	Bone, IUP	Khatsenovich et al. (2017)
Kharganyn Gol 5	Layer 6, 1.71 m	46,180 ± 1,100	50,000–48,560	MAMS-21715	M. Pal?	Khatsenovich et al. (2017)
Kharganyn Gol 5	Layer 6, 1.76 m	43,340 ± 790	47,450–45,750	MAMS-21716	M. Pal?	Khatsenovich et al. (2017)

Dates for sites in the Tolbor Valley are from Zwyns et al. (2014a).

desiccation of the area. Although we need more detailed studies and better chronological resolution, human occupation of Mongolia in the Upper Palaeolithic may have been limited to short episodes of increased rainfall and higher temperatures. During these windows of opportunity, the main prey appears to have been Mongolian gazelle and horse, much as in the Transbaikal. These may have been hunted seasonally from locations in the Tolbor and neighbouring valleys by groups that relied on other resources when they were not in the vicinity. If they were hunted year-round, groups would have had to be extremely mobile and with large annual ranges. Mongolian gazelle can form enormous herds of several hundred animals (Ito et al. 2013) and have annual ranges that can cover 30,000 sq km, with distances of over 300 km between winter and summer ranges (Olson et al. 2010; Zwyns and Lbova 2019); their seasonal movements are also irregular, as calves rarely revisit their birth places. Onager (*Equus hemionus*) can migrate even further, up to 1,500 km (Agadjanian and Shunkov 2018). High mobility combined with high unpredictability of herd movements and climate would have

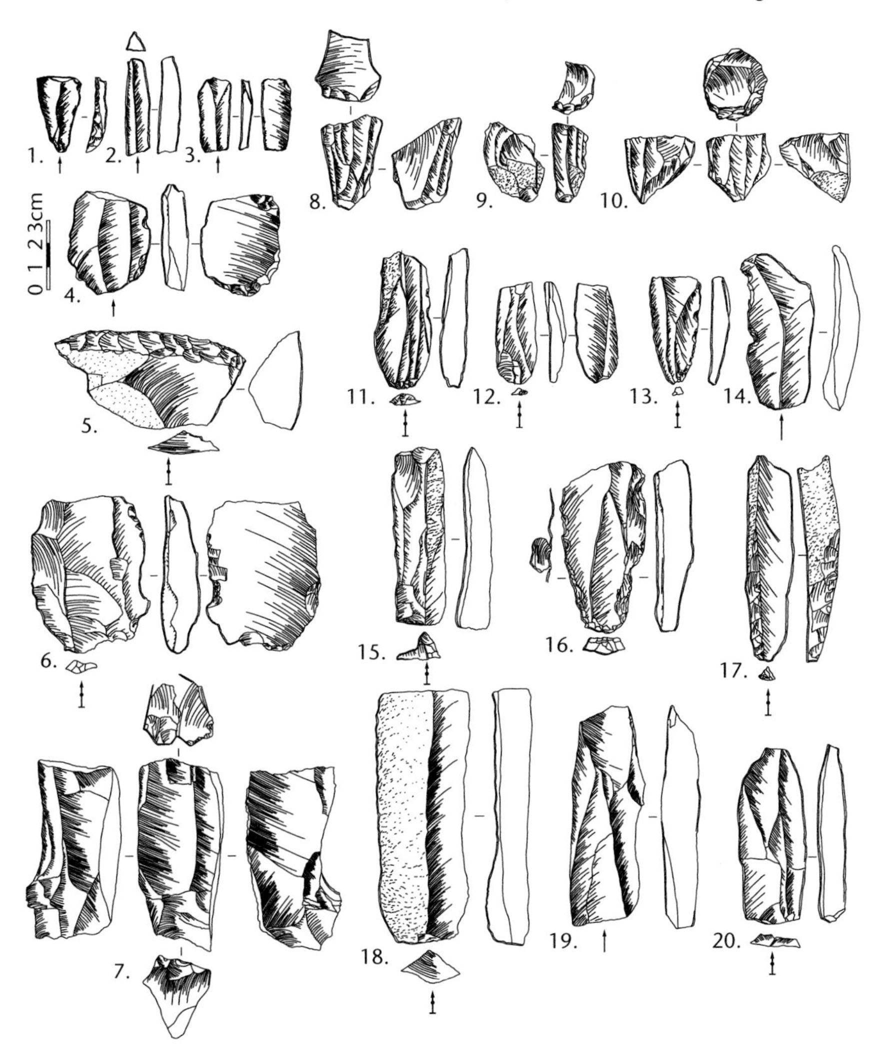

FIGURE 9.8 Initial Upper Palaeolithic (IUP) artefacts from Tolbor 16

Example of lithics from Unit 5/6 and Unit 7: 1. Blade with proximal retouch. 2–3 Small laminar blanks, 4 Blanks with inverse proximal retouch, 5 Transversal convex scraper, 6 Flake with orthogonal dorsal pattern, 7 Blade core, 8–10 Bladelet cores, 11–13 Blades with platform bludgering and trimming, 14–16, 19–20 Blade with bidirectional dorsal pattern, 17 Neo-crested blade, 18 Large cortical blade.

Source: Zwyns et al. 2014a, Fig. 7.

resulted in very low human population densities over an enormous area, but would help explain the similarities of the assemblages from northern Mongolia, and Shuidonggou in north China over 1,000 km to the south (see Chapter 10).

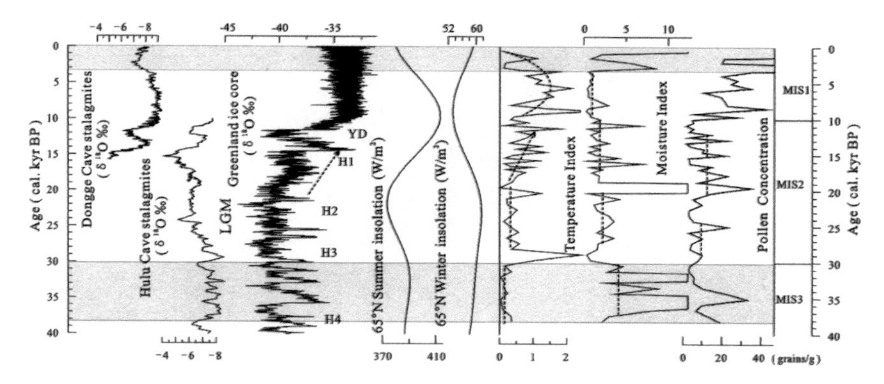

FIGURE 9.9 Climatic and environmental history of northern Mongolia since 40 ka
This shows the reconstructed temperature and moisture indices from the Shaamar eolian-paleosol section with monsoon records from East Asia, the Greenland ice-core record and changes in solar insolation.

Source: Ma et al. 2013, Fig. 6.

In a land of unpredictable resources and weather, maintaining viable social networks for obtaining mates, information and scarce materials, and for providing potential safety nets if and when resources failed would have been a major challenge. There are a few hints of their extent. One is the presence of ostrich eggshell beads on sites as far apart as Shuidonggou (see Chapter 10), Mongolia, the Transbaikal and Altai (Rybin 2014); Mongolian examples are beads at Tolbor 16 (Zwyns et al. 2014a), Dörölj-1 (Jaubert 2015) and Kharganyn Gol 5, layer 3 (Khatsenovich et al. 2017). Muscovite (a type of mica) was found at the latter site in layer 5 and was probably obtained from over 500 km away (Khatsenovich et al. 2017). There is also a remarkable cache of 57 mostly unused flakes that were probably deposited in a bag and were found near a path linking two side valleys of the Tolbor river valley. Although undated, the cache is most likely upper palaeolithic in age. Tabarev and colleagues (2013) interpret this find as a symbolic placement in a conspicuous part of the landscape. An alternative viewpoint is that the cache was placed (and perhaps with other perishable items) as an emergency resource and presumably signposted by a cairn or a similar landmark in the landscape.

Although human skeletal remains are absent at all the Upper Palaeolithic Mongolian sites that have so far been investigated, it has seemed reasonable to attribute them to *Homo sapiens*. This assumption is confirmed by the latest assessment of the cranial remains from Salkhit in northeast Mongolia. Here, a skull cap was found in 2006 in mining operations. Initially, it was ascribed an age of 800,000 years on the basis of a claimed association with woolly rhinoceros and was even assigned to a new palaeo-species *Mongolanthropus*. Subsequent study showed that its archaic features did not indicate that it was an archaic species; instead, these are regional features of an otherwise modern type of human (Lee

2015; Tseveendorj et al. 2016). In the latest assessment (Devièse et al. 2019), the skull cap has now been radiocarbon dated to 30,430 ± uncalibrated 300 BP, which when calibrated is 34,950–33,900 cal BP. Genetic analysis also shows that it has haplotype N, which is widespread across Asia. (Interestingly, there is no mention of any Denisovan content). On the basis of this study, the Upper Palaeolithic of Mongolia, and probably also north China and the Transbaikal, can be attributed to our species, *Homo sapiens*.

Overview

Immediately prior to human expansion, there were resident populations in Central Asia and the Altai, and probably also in Mongolia and north China. In Central Asia and the Altai, there were populations that were genetically similar to European Neandertals, but they may have been morphologically more variable. Denisova was also inhabited by a previously unknown population – the Denisovans – that was identified from its aDNA. Both are known to have interbred with each other, and both are likely to have interbred with our species when they expanded eastwards. Human expansion across Central Asia, the Altai Mountains, Transbaikalia and Mongolia and then into north China probably occurred between 48 ka and 42 ka cal BP. If we assume a generational length of 20 years, the Initial and Early Upper Palaeolithic spanned 300 generations over 6,000 years. During this time, because of the harshness of the landscape, and the severity and unpredictability of the climate, there were doubtless numerous setbacks and failures. It seems most improbable that the colonisation of this vast region was accomplished in a single wave of expansion. From the little that we know about subsistence, wild sheep/goat and horse were the main prey in Central Asia, and horse and gazelle in Transbaikalia and Mongolia. In the Altai Mountains, there was a much wider choice of prey that included wild sheep/goat, horse, bison, deer, woolly rhinoceros and many other species. In Central Asia and Mongolia especially, human groups that were dependent on horse and gazelle would have required large annual ranges and numerous short-term camps throughout the year, and probably effective food storage techniques for surviving winters. Social networks across this vast region were probably loose rather than tight (see Chapter 1) and are evidenced by similarities in stone artefacts and the use of perforated pendants and ostrich eggshell beads.

Notes

1 The "five stans" cover ca. 4 million sq km, and the EU (less the UK) cover 4.1 million sq km.
2 I thank Arina Khatsenovich for pointing this out.
3 Mongolia's area is ca. 1.6 million sq km – roughly the size of Britain, France, Germany, Spain and Italy combined.
4 Specimens Denisova 11 and 15 were identified by ZooMS (zoological mass spectrometry) and confirmed by its mitchrondrial DNA (mtDNA).

References

Central Asia

Bailey, S., Glantz, M., Weaver, T.D. and Viola, B. (2008) The affinity of the dental remains from Obi-Rakhmat Grotto, Uzbekistan. *Journal of Human Evolution*, 55: 238–248.

Dennell, R.W. (2013) Early Homo fossil records of west and Central Asia. In C. Smith (ed.) *Encyclopedia of Global Archaeology*, New York: Springer Science+Business Media, doi:10.1007/978-1-4419-0465-2.

Douka, K., Slon, V., Stringer, C., Potts, R., Hübner, A. et al. (2017) Direct radiocarbon dating and DNA analysis of the Darra-i-Kur (Afghanistan) human temporal bone. *Journal of Human Evolution*, 107: 86–93.

Glantz, M. (2010) Chapter 8: The history of hominin occupation of Central Asia in review. In C.J. Norton and D.R. Braun (eds) *Asian Paleoanthropology: From Africa to China and Beyond*, New York: Springer Science+Business Media B.V., pp. 101–112, doi:10.1007/978-90-481-9094-2_8.

Glantz, M., Athreya, S., Ritzman, T. (2009) Is Central Asia the eastern outpost of the Neandertal range? A reassessment of the Teshik-Tash child. *American Journal of Physical Anthropology*, 138: 45–61.

Glantz, M., Viola, B., Wrim, P., Chikisheva, T., Derevianko, A. et al. (2008) New hominin remains from Uzbekistan. *Journal of Human Evolution*, 55: 223–237.

Gunz, P. and Bulygina, E. (2012) The Mousterian child From Teshik-Tash is a Neanderthal: a geometric morphometric study of the frontal bone. *American Journal of Physical Anthropology*, 149: 365–379.

Krivoshapkin, A.I., Anoikin, A.A. and Brantingham, P.J. (2007) The lithic industry of Obi-Rakhmat grotto, Uzbekistan. *Bulletin of the Indo-Pacific Prehistory Association*, 26: 5–19, doi:10.7152/bippa.v26i0.11989.

Kuhn, S. and Zwyns, N. (2014) Rethinking the Initial Upper Palaeolithic. *Quaternary International*, 347: 29–38.

Ranov, V.A., Kolobova, K.A. and Krivoshapkin, A.I. (2012) The Upper Paleolithic assemblages of Shugnou, Tajikistan. *Archaeology Ethnology & Anthropology of Eurasia*, 40(2): 2–24.

Shunkov, M., Anoikin, A., Taimagambetov, Z., Pavlenok, K., Kharevich, V. et al. (2017) Ushbulak-1: new Initial Upper Palaeolithic evidence from Central Asia. *Antiquity*, 91: 1–7, doi:10.15184/aqy.2017.208.

Trinkaus, E. (2005) Early modern humans. *Annual Review of Anthropology*, 34: 207–230.

Trinkhaus, E., Ranov, V.A. and Lauklin, S. (2000) Middle Paleolithic human deciduous incisor from Khudji, Tajikstan. *Journal of Human Evolution*, 38: 75–583.

Vishnyatsky, L.B. (1999) The Paleolithic of Central Asia. *Journal of World Prehistory*, 13(1): 69–122.

The Altai

Agadjanian, K. and Shunkov, M. V. (2018) Paleolithic Man of Denisova Cave and Zoogeography of Pleistocene mammals of northwestern Altai. *Paleontological Journal*, 52(1): 66–89.

Akimova, E., Higham, T., Stasyuk, I., Buzhilova, A., Dobrovolskaya, M. and Mednikova, M. (2010) A new direct radiocarbon AMS date for an upper Palaeolithic human bone from Siberia. *Archaeometry*, 52: 1122–1130.

Brantingham, P.J., Krivoshapkin, A., Jinzeng, L. and Tserendagva, Y. (2001) The Initial Upper Paleolithic in northeast Asia. *Current Anthropology*, 42(5): 735–746.

Buzhilova, A.P. (2013) Dental Remains from the Middle Paleolithic Layers of Altai Cave Sites. *Archaeology, Ethnology and Anthropology of Eurasia*, 41(1): 55–65.

Callaway, E. (2016) Evidence mounts for interbreeding bonanza in ancient human species. *Nature*, doi:10.1038/nature.2016.19394.

Dennell, R.W. (2019) Dating of hominin discoveries at Denisova. *Nature*, 565: 571–572.

Derevianko, A.P. (2011) The origin of anatomically modern humans and their behavior in Africa and Eurasia. *Archaeology, Ethnology and Anthropology of Eurasia*, 39 (3): 2–31.

Derevianko, A.P., Markin, S.V., Kolobova, K.A., Chabai, V.P., Rudaya, N.A., Viola, B., Buzhilova, A.P., Mednikova, M.B., Vasilev, S.K., Zykin, V.S., Zykina, V.S., Zazhigin, V.S., Volvakh, A.O., Roberts, R.G., Jakobs, Z. and Li, B. (2018) *Multidisciplinary Studies of Chagyrskaya Cave – A Middle Paleolithic Site in Altai*. Novosibirsk: IAET SB RAS Press.

Derevianko, A.P., Markin, S.V. and Shunkov, M.V. (2013a) The Sibiryachikha Facies of the Middle Paleolithic of the Altai. Archaeology, Ethnology and Anthropology of Eurasia, 41(1): 89–103.

Derevianko, A.P., Markin, S.V., Zykin, V.S., Zykina, V.S., Zazhigin, V.S. et al. (2013b) *Chagyrskaya Cave: A Middle Paleolithic Site in the Altai. Archaeology, Ethnology and Anthropology of Eurasia*, 41(1): 12–27.

Derevianko, A.P., Petrin, V.T. and Rybin, E.P. (2000) The Kara-Bom site and the characteristics of the Middle-Upper Paleolithic transition in the Altai. *Archaeology, Ethnology and Anthropology of Eurasia*, 2(2): 33–52.

Derevianko, A.P. and Rybin E.P. (2003) The earliest representations of symbolic behaviour by paleolithic humans in the Altai Mountains. *Archaeology, Ethnology and Anthropology of Eurasia*, 3(15): 27–50.

Derevianko, A.P. and Shunkov, M.V. (2011) Anthropogenesis and colonization of Eurasia by archaic populations. Formation of anatomically modern human. In A.P. Derevianko and M.V. Shunkov (eds) *Characteristic Features of the Middle to Upper Paleolithic Transition in Eurasia: Development of Culture and Evolution of Homo Genus*, Novosibirsk: Institute of Archaeology and Ethnography, pp. 50–74.

Derevianko, A.P., Shunkov, M.V., Agadjanian, A.K., Baryshnikov, G.F. Maleva, E.M. et al. (2003) Paleoenvironment and paleolithic human occupation of Gorny Altai. In A. P. Derevianko and M.V. Shunkov (eds) *Paleoenvironment and Paleolithic Human Occupation of Gorny Altai*, Novosibirsk: Institute of Archaeology and Ethnography, pp. 361–393.

Douka, K., Slon, V., Jacobs, Z., Bronk Ramsey, C., Shunkov, M. et al. (2019) Age estimates for hominin fossils and the onset of the Upper Palaeolithic at Denisova Cave. *Nature*, 565: 640–644.

Fu, Q., Li, H., Moorjani, P., Jay, F., Slepchenko, S.M. et al. (2014) Genome sequence of a 45,000-year-old modern human from western Siberia. *Nature*, 514: 445–449.

Goebel, T., Derevianko, A.P. and Petrin, V.T. (1993) Dating the Middle-to-Upper-Paleolithic transition at Kara-Bom. *Current Anthropology*, 34(4): 452–458.

Jacobs, Z., Li, B., Shunkov, M., Kozlikin, M.B. and Bolikhovskaya, N.S. (2019) Timing of archaic hominin occupation 1 of Denisova Cave in southern Siberia. *Nature*, 565: 594–599.

Krause, J., Fu, Q., Good, J.M., Viola, B., Shunkov, M.V., Derevianko, A.P., Pääbo, S., (2010) The complete mitochondrial DNA genome of an unknown hominin from southern Siberia. *Nature*, 464: 894–897, doi:10.1038/nature08976.

Krause, J., Orlando, L., Serre, D., Viola, B., Prüfer, K., Richards, M.P., Hublin, J.-J., Hänni, C., Derevianko, A.P. and Pääbo, S. (2007) Neanderthals in central Asia and Siberia. *Nature*, 449: 902–904.

Krivoshapkin, A., Shalagina, A., Baumann, M., Shnaider, S. and Kolobova, K. (2018) Between Denisovans and Neanderthals: Strashnaya Cave in the Altai Mountains. *Antiquity*, 92: 365, e1: 1–7, doi:10.15184/aqy.2018.221.

Mednikova, M.B. (2013) An archaic human ulna from Chagyrskaya Cave Altai: morphology and taxonomy. *Archaeology, Ethnology and Anthropology of Eurasia*, 41(1): 66–77.

Meyer, M., Fu, Q., Aximu-Petri, A., Glocke, I., Arsuaga, J.-L., Gracia, A., Bermúdez de Castro, J.M., Carbonell, E. and Pääbo, S. (2014) A mitochondrial genome sequence of a hominin from Sima de los Huesos. *Nature*, 505: 403–406, doi:10.1038/nature12788.

Morley, M.W., Goldberg, P., Uliyanov, V.A., Kozlikin, M.B., Shunkov, M.V. et al. (2019). Hominin and animal activities in the microstratigraphic record from Denisova Cave (Altai Mountains, Russia). *Scientific Reports*, 9: 13785, doi:10.1038/s41598-019-49930-3.

Reich, D., Patterson, N., Kircher, M., Delfin, F., Nandineni, M.R. et al. (2011) Admixture and the first modern human dispersals into Southeast Asia and Oceania. *American Journal of Human Genetics*, 89: 516–528.

Shunkov, M.V. (2017) Denisova cave and the palaeolithic of the Altai. In *The Third Man: The Prehistory of the Altai*, Musée National de Préhistoire, Les Eyzies-de-Tayac, pp. 19–29.

Shichi, K., Kawamuro, K., Takahara, H., Hase, Y., Maki, T. and Miyoshi, N. (2007) Climate and vegetation changes around Lake Baikal during the last 350,000 years. *Palaeogeography, Palaeoclimatology, Palaeoecology*, 248(3–4): 357–75.

Slon, V., Hopfe, C.W., Weiß, C.L., Mafessono, F., Rasilla, M. de la, et al. (2017) Neandertal and Denisovan DNA from Pleistocene sediments. *Science*, 356: 605–608.

Slon, V., Mafessoni, F., Vernot, B., Filippo, C. de, Grote, S. et al. (2018) The genome of the offspring of a Neanderthal mother and a Denisovan father. *Nature*, 561: 113–116.

Viola, B., Markin, S.V., Zenin, A., Shunkov, M.V. and Derevianko, A.P. (2011) Late Pleistocene hominins from the Altai Mountains, Russia. In A. P. Derevianko and M. V. Shunkov (eds) *Characteristic Features of The Middle to Upper Paleolithic Transition in Eurasia*. Novosibirsk: Institute of Archaeology and Ethnography SB RAS, pp. 207–213.

Viola, B., Douka, K., Higham, T., Shunkov, M. and Kelso, J. (2019) Catalogue of hominin (or putative hominin) remains from Denisova cave. *Nature*, 565: 640–644, Supplementary Information, pp. 27–34.

Zwyns, N., Rybin, E.P., Hublin, J.-J. and Derevianko, A.P. (2012) Burin-core technology and laminar reduction sequences in the initial Upper Paleolithic from Kara-Bom (Gorny-Altai, Siberia). *Quaternary International*, 259: 33–47.

Transbaikalia

Bezrukova, E.V., Tarasov, P.E., Solovieva, N., Krivonogov, S.K. and Riedel, F. (2010) Last glacial–interglacial vegetation and environmental dynamics in southern Siberia: Chronology, forcing and feedbacks. *Palaeogeography, Palaeoclimatology, Palaeoecology*, 296: 185–198.

Buvit, I., Terry, K., Izuho, M. and Konstantinov, M.V. (2015) The emergence of modern behaviour in the Trans-Baikal, Russia. In Y. Kaifu, M. Izuho, H. Sato and A. Ono (eds) *Emergence and Diversity of Modern Human Behavior in Paleolithic Asia*. College Station: Texas A&M University Press, pp. 490–505.

Conard, N., Malina, M. and Münzel, S.C. (2009) New flutes document the earliest musical tradition in southwestern Germany. *Nature*, 460: 737–740.

Germonpre, M. and Lbova, L. (1996) Mammalian Remains from the Upper Palaeolithic site of Kamenka, Buryatia (Siberia). *Journal of Archaeological Science*, 23: 35–57.

Graf, K.E. (2014) Siberian odyssey. In K.E. Graf, C.V. Ketron and M.R. Waters (eds) *Paleoamerican Odyssey*. College Station: Texas A&M University Press, pp. 65–80.

Graf, K.E. and Buvit, I. (2017) Human dispersal from Siberia to Beringia: assessing a Beringian Standstill in light of the archaeological evidence. *Current Anthropology*, 58 (Supplement 17): S583–S603.

Izuho, M., Terry, K., Vasil'ev, S., Konstantinov, M. and Takahashi, K. (2019) Tolbaga revisited: scrutinizing occupation duration and its relationship with the faunal landscape during MIS 3 and MIS 2. *Archaeological Research in Asia*, 17: 9–23.

Kuhn, S. and Zwyns, N. (2014) Rethinking the Initial Upper Palaeolithic. *Quaternary International*, 347, 29–38.

Lbova, L.V. (2019) Personal ornaments as markers of social behavior, technological development and cultural phenomena in the Siberian early upper Paleolithic. *Quaternary International*, doi:10.1016/j.quaint.2019.07.037.

Rybin, E.P. (2014) Tools, beads, and migrations: specific cultural traits in the Initial Upper Paleolithic of southern Siberia and Central Asia. *Quaternary International*, 347: 39–52.

Rybin, E.P. (2015) Middle and Upper Palaeolithic interactions and the emergence of "modern behaviour" in southern Siberia and Mongolia. In Y. Kaifu, M. Izuho, T. Goebel, H. Sato and A. Ono (eds) *Emergence and Diversity of Modern Human Behavior in Paleolithic Asia*. College Station: Texas A&M University Press, pp. 470–489.

Mongolia

Derevianko, A.P. (2017) The early palaeolithic of Mongolia. In A.P. Derevianko (ed) *Three Global Migrations in Eurasia*, Novosibirsk: IAET SB RAS Publishing, pp. 619–688.

Derevianko, A.P., Rybin, E.P., Gladyshev, S.A., Gunchinsuren, B., Tsybankov, A.A. and Olsen, J.W. (2013) Early upper Paleolithic stone tool technologies of northern Mongolia: the case of Tolbor-4 and Tolbor-15. *Archaeology Ethnology & Anthropology of Eurasia*, 41(4): 21–37.

Derevianko, A.P., Zenin, A.N., Rybin, E.P., Gladyshev, S.A., Tsybankov, A.A., Olsen, J.W., Tseveendorj, D. and Gunchinsuren, B. (2007) The technology of Early Upper Paleolithic lithic reduction in northern Mongolia: the Tolbor-4 site. *Archaeology, Ethnology and Anthropology of Eurasia*, 1: 16–38.

Devièse, T., Massilani, D., Yi, S., Comeskey, D., Nagel, S. et al. (2019) Compound-specific radiocarbon dating and mitochondrial DNA analysis of the Pleistocene hominin from Salkhit Mongolia. *Nature Communications*, doi:10.1038/s41467-018-08018-8.

Gillam, J.C., Gladyshev, S.A., Tabarev, A.V., Gunchinsuren, B. and Olsen, J.W. (2012) Halfway to Mörön: shedding new light on paleolithic landscapes of northern Mongolia. *Legacy*, 16(2): 14–17.

Gladyshev, S.A., Olsen, J.W., Tabarev, A.V. and Jull, A.J.T. (2012) The Upper Paleolithic of Mongolia: recent finds and new perspectives. *Quaternary International*, 281: 36–46.

Gladyshev, S.A., Olsen, J.W., Tabarev, A.V. and Kuzmin, Y.V. (2010) Chronology and periodization of Upper Palaeolithic sites in Mongolia. *Archaeology Ethnology & Anthropology of Eurasia*, 38(3): 33–40.

Ito, T.Y., Tsuge, M., Lhagvasuren, B., Buuveibaatar, B., Chimeddorj, B., Takatsuki, A., Tsunekawa, A. and Shinoda, M. (2013) Effects of interannual variations in environmental conditions on seasonal range selection by Mongolian gazelles. *Journal of Arid Environments*, 91: 61–68.

Jaubert, J. (2015) The Palaeolithic peopling of Mongolia. In Y. Kaifu, M. Izuho, T. Goebel, H. Sato and A. Ono (eds) *Emergence and Diversity of modern Human Behavior in Paleolithic Asia*, College Station: Texas A&M University Press, pp. 453–469.

Khatsenovich, A.M., Rybin, E.P., Gunchinsuren, B., Olsen, J.W., Shelepaev, R.A. et al. (2017) New evidence for Paleolithic human behavior in Mongolia: the Kharganyn Gol 5 site. *Quaternary International*, 442B: 78–94, doi:10.1016/j.quaint.2016.10.013.

Kolomiets, V.L., Gladyshev, S.A., Bezrukova, E.V., Rybin, E.P., Letunova, P.P. and Abzaeva, A.A. (2009) Environment and human behaviour in northern Mongolia during the Upper Pleistocene. *Archaeology Ethnology & Anthropology of Eurasia*, 37(1): 2–14.

Lee, S.-H. (2015) *Homo erectus* in Salkhit, Mongolia? *HOMO – Journal of Comparative Human Biology*, 66: 287–298.

Ma, Y., Liu. K.-B., Feng, Z., Meng, H., Sang, Y., Wang, W. and Zhang, H. (2013) Vegetation changes and associated climate variations during the past ~38,000 years reconstructed from the Shaamar eolian-paleosol section, northern Mongolia. *Quaternary International*, 311: 25–35.

Olson, K.A., Fuller, T.K., Mueller, T., Murray, M.G., Nicolson, C., Ondokhuu, D., Bolortsetseg, S. and Schaller, G.B. (2010) Annual movements of Mongolian gazelles: nomads in the eastern steppe. *Journal of Arid Environments*, 74(11): 1435–1442.

Rybin, E.P. (2014) Tools, beads, and migrations: specific cultural traits in the Initial Upper Paleolithic of southern Siberia and Central Asia. *Quaternary International*, 347: 39–52.

Tabarev, A.V., Gillam, J.C., Kanomata, Y. and Gunchinsuren, B. (2013) A Paleolithic cache at Tolbor (northern Mongolia). *Archaeology Ethnology & Anthropology of Eurasia*, 41(3): 14–21.

Tseveendorj, D., Gunchinsuren, B., Gelegdorj, E., Yi, S. and Lee, S.-H. (2016) Patterns of human evolution in northeast Asia with a particular focus on Salkhit. *Quaternary International*, 400: 175–179.

Zwyns, N. and Lbova, L. V. (2019) The Initial Upper Paleolithic of Kamenka site, Zabaikal region (Siberia): a closer look at the blade technology. *Archaeological Research in Asia*, 17: 24–49.

Zwyns, N., Gladyshev, S.A., Gunchinsuren, B., Tsedendorj Bolorbat, Flas, D. et al. (2014a) The open-air site of Tolbor 16 (northern Mongolia): preliminary results and perspectives. *Quaternary International*, 347: 53–65.

Zwyns, N., Gladyshev, S., Tabarev, A. and Gunchinsuren, B. (2014b) Mongolia: Paleolithic. *Encyclopedia of Global Archaeology*, pp. 5025–5032. Springer.

10

CHINA

Introduction

The evidence for the earliest appearance of our species in China is arguably the most complex in the whole of Asia. This is partly because of China's size and environmental diversity, partly because it straddles both the Oriental and Palearctic Realms, and partly because of the complexity of its fossil and archaeological records. The demographic history of south China is clearly linked to neighbouring Southeast Asia (Chapter 6), but in north China, the main developments probably originated in Siberia and Mongolia (Chapter 9). For many years, discussions of the skeletal evidence for early *Homo sapiens* in China was polarised between two diametrically opposed models. Most Chinese and some western researchers argued for a multi-regional model of human evolution, whereby we evolved from an indigenous background in Africa, Europe and Asia (see e.g. Thorne and Wolpoff 1992). Opponents of this model such as Chris Stringer (2002, 2016) argue that our species originated in Africa (Chapter 2), and then dispersed into Asia and Europe and replaced all indigenous populations, without (in the extreme version of this model) any interbreeding: replacement was total.

The discovery that all modern non-Africans have a small amount of Neandertal DNA (Green et al. 2008, 2010) was a major stimulus in closing the gap between these two models. Subsequent analyses of Denisovan aDNA (Chapter 9) showed that it too had interbred with both Neandertals (Slon et al. 2018) and *H. sapiens* (Reich et al. 2011). Multi-regionalists now recognise that some immigration with interbreeding by *H. sapiens* in East Asia was likely, and those arguing for an African origin of our species also concede that there had been some interbreeding with indigenous populations. Most researchers now recognise that the appearance of our species in China was a more complex (and interesting) process than was envisaged under either a multi-regional view of human evolution, or one in which all indigenous non-Africans were replaced by immigrant populations of *H. sapiens* without any interbreeding or hybridisation. Instead, both local evolution and immigration were involved in the eventual emergence of our species in East Asia. The phrase "continuity with hybridisation" (Liu et al. 2010a: 19205) provides a convenient way of summarising the emerging picture of the Chinese skeletal record.

An additional factor in discussions of the Chinese evidence is the long shadow cast by Hallam Movius's (1948: 411) pronouncements that East China was a marginal and backward "region of cultural retardation" that was characterised by "monotonous and unimaginative assemblages of choppers, and chopping tools". This judgement was based on ignorance rather than any profound insight, as almost nothing was known of its Palaeolithic record in 1948. As seen below, it is in fact far more complex and interesting than it appeared when Movius was writing.

Before discussing the Chinese human skeletal and Palaeolithic records, the environmental and climatic background to later human evolution in China and neighbouring regions needs reviewing. Figure 10.1 shows the location of relevant sites in China.

Environmental and climatic background

China is an enormous country covering almost 10 million sq km and spanning 30° of latitude, from ca. 20° N. to 52° N., which is equivalent in latitude to Europe from Gibraltar at 35° N. to Iceland at 65° N., or North America from Miami at 24° N. to Vancouver at 54° N. Its basic division is between the north and south, with the Qinling Mountains forming the modern boundary (see Figure 10.1). China north of the Qinling Mountains lies in the Palearctic Realm and includes the Loess Plateau, Xinjiang, Inner Mongolia and the Northern Arid Area of China (NAAC), as well as Chinese Siberia and the Tibetan Plateau. Here, annual rainfall is low (usually <500 mm, and <50 mm in northwest China), and winters are usually sub-freezing, often for several months, with January temperatures averaging –20° C. near the Mongolian border and in –30° C. in Chinese Siberia. Today, wheat and millet are the main cereal crops, and the fauna is Palearctic. In the Pleistocene, this included woolly mammoth, woolly rhinoceros, horse and gazelle. South of the Qinling Mountains, China lies within the Oriental Realm.[1] The vegetation is largely sub-tropical (with rice as today's main crop), and the fauna includes panda (*Ailuripoda melaneuca*), tapir, and in the Pleistocene, the giant ape *Gigantopithecus*, *Megatapirus*, *Stegodon orientalis*, and oriental rhinoceros (*Rhinoceros sinensis*). Annual rainfall is high, and typically >1 m (and > 2m in the southeast), and winters are rarely sub-freezing. One major contrast between north and south China is that fresh plant foods are available year-round in the south, but are largely absent in winter in the north.

There are excellent palaeoclimatic records for China and its neighbouring regions that cover the last glacial cycle. For north China, the standard framework is the loess-palaeosol sequence of the Chinese Loess Plateau (Liu and Ding 1998), supplemented by studies of the desert regions of northern China (Fang et al. 2002a, 2002b; Guan et al. 2011). In southern China, there are excellent

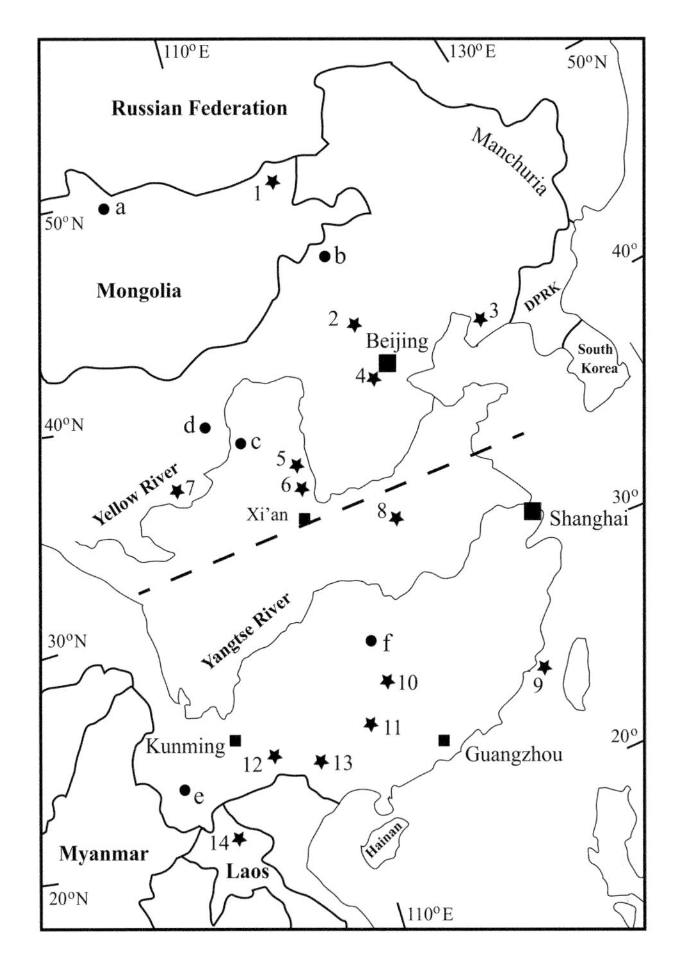

FIGURE 10.1 Fossil hominin and early Upper Palaeolithic sites in China and neighbouring countries

The dashed line indicates the present boundary between the Palearctic Realm of north China and the Oriental Realm of south China. This boundary shifted considerably during the Pleistocene, depending upon the relative strengths of the winter and summer monsoons. The Qinling Mountains that divide north from south China lie south of Xian and the Yellow River, and approximately along the modern boundary between the Palearctic and Oriental biogeographic realms.

Key: Sites with hominin remains: 1 Salkhit; 2 Xujiayao; 3 Jinnuishan; 4 Zhoukoudian Upper Cave and Tianyuandong; 5 Huanglong; 6 Dali; 7 Xiahe; 8 Xuchang (Linjing); 9 Penghu; 10 Fuyan; 11 Liujiang; 12 Longtanshan; 13 Luna; 14 Tam Pa Ling (Laos).

Archaeological sites: a Tolbor; b Jinsitai; c Shuidonggou; d South Temple Canyon; e Xiaodong; f Ma'anshan

Source: The author.

speleothem records from caves such as Fengyu (Li et al. 2017), Hulu (Wang et al. 2001, 2008), Wulu and Donge (Yuan et al. 2004; Liu et al. 2016) that cover the last glacial cycle in great detail (see Figure 10.2). These are supplemented by marine cores from the South China Sea (Wei et al. 2003), and Lakes Baikal and Kotokel (Chapter 9) in central Siberia provide other high-quality climate archives that are relevant to when hominins might have entered China from the north and west (Shichi et al. 2007; Bezrukova et al. 2010).

Rainfall, temperatures and vegetation

In cold, dry periods that are the equivalent of glaciations in Europe, the boundary of the summer monsoon in China was pushed southwards by a strengthened winter monsoon, and both temperatures and rainfall decreased. Consequently, the deserts of northern China expanded and blocked entry from Central Asia, where loess deposition rates increased and deserts such as the Kara Kum and Kyzyl Kum also expanded (Yang and Ding 2006; Dennell 2013). Maher and Thompson (1995) and Maher et al. (1994) estimated rainfall variations over the past 1.1 million years in north China at the Xifeng section in the central Loess Plateau from fluctuations in magnetic susceptibility, which they argue was more strongly controlled by precipitation than by temperature. Their estimated variations in precipitation over the last 125,000 years are shown in Figure 10.3. A similar study, based on four sections in different parts of the Loess Plateau, concluded that rainfall was slightly higher than today during the last interglacial, and ca. 30% lower in the last glaciation in the northern Loess Plateau (Florindo et al. 1999). These estimates come with a health warning, however, as Bloemendal and Liu (2005) argue against reliance on any single parameter in attempting to quantify past rainfall patterns.

A more detailed study of rainfall over the last 140,000 years in north China was also based on the Xifeng section and used magnetic susceptibility as a proxy indicator of rainfall (Liu et al. 1995). They suggested that annual precipitation varied from 150 to 250 mm during MIS 2, MIS 4, and MIS 6; during interstadials, from 300 to 500 mm; and in the last interglacial, 550–700 mm, that is, ca. 100 mm higher than today (see Figure 10.4). Their estimates imply a much greater reduction of 40–75% in precipitation than Maher and Thompson's, with levels as low as 100–200 mm over the central part during the Last Glacial Maximum and previous stadial. Such reductions in rainfall during cold arid periods across north China would clearly have had major consequences for hominin populations.

Similar trends are implied by analyses of spores, pollen, molluscs and magnetic susceptibility at Milanggouwan in the Mu Us Desert north of the Loess Plateau (Li et al. 2000). Here, temperatures might have been 2–6° C. higher in the last interglacial than before or after, and rainfall between 40 and 120% higher. In warm, moist periods such as MIS 5e, MIS 5a and MIS 3, the boundary of the summer monsoon moved northwards, rainfall increased by up to 20% (Liu et al. 1995) or even 80% (Maher and Thompson 1995) in the last interglacial, and the desert margin would

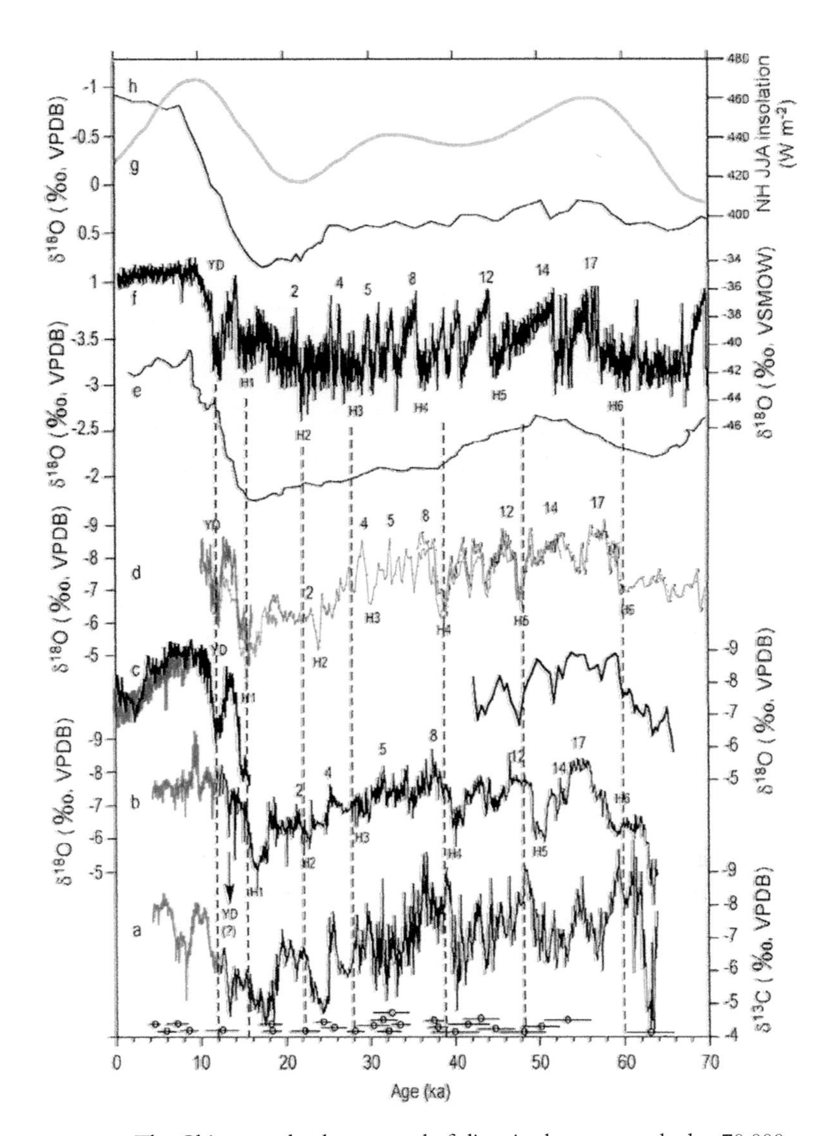

FIGURE 10.2 The Chinese speleothem record of climatic change over the last 70,000 years

The figure shows the climatic records of Chinese speleothems and other global climate records of the last 70,000 years. a) the δ^{13}C and b) the δ^{18}O records of Fengyou-1 speleothem; c) the Dongge Cave stalagmites; d) the Hulu Cave δ^{18}O stalagmite record; e) the δ^{18}O record of marine sediment core MD972142 from the South China Sea; f) the δ^{18}O record of the Greenland GRIP ice core; g) northern hemisphere insolation at 65° N. and h) the SPECMAP (SPECtral MApping) marine record of climate change. YD denotes the Younger Dryas; H1–H6 are the short, cold episodes known as Heinrich Events, and the numerical numbers are D–O (Dansgaard–Oeschger) Events; these are rapid climatic fluctuations that occurred at least 25 times in the last glacial cycle. The symbols of a circle with a horizontal bar on the bottom of the figure represent ^{230}Th/U dates with uncertainty.

Source: Li et al. 2017, Fig. 9.

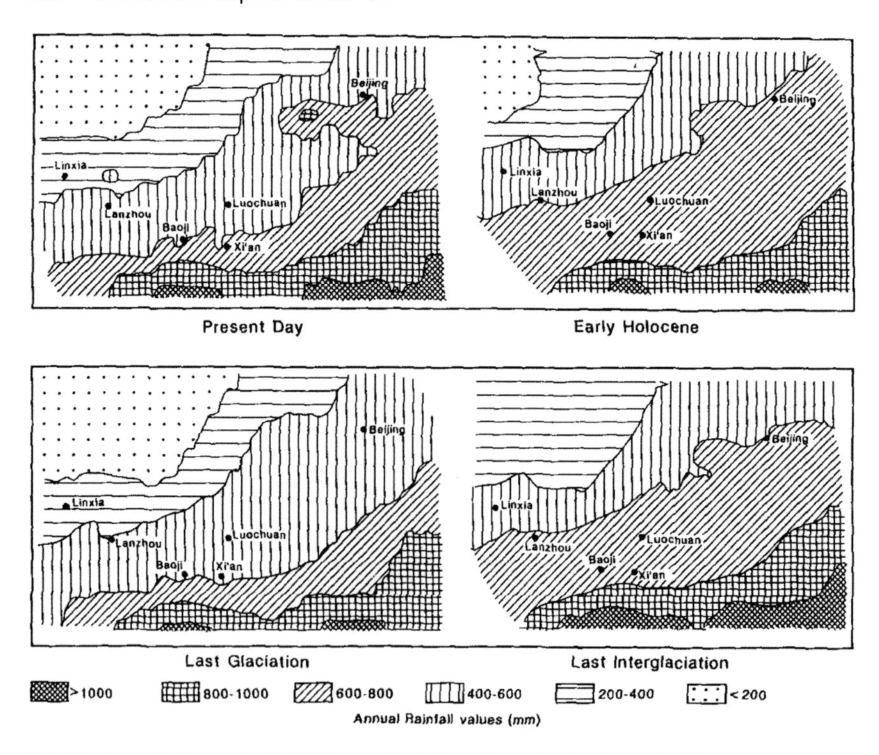

FIGURE 10.3 Estimated rainfall in north China since the last interglacial

The key feature of these figures in the extent of desert in north China. In the coldest and driest parts of the last glacial cycle, the desert boundary (defined by <200 mm of precipitation per year) shifted south in northwest China, but retreated northwards in the last interglacial. This boundary was critical in determining the northern limits of human occupation in north China.

Source: Maher and Thompson 1995, Fig. 4.

have contracted. As in Arabia, there were numerous palaeolakes in areas of north China (see Yu et al. 2001); those in the Tengger Desert were largest in MIS 5a (80–90 ka) (Long et al. 2012), and not between 42–40 ka during MIS 3, as previously thought (Zhang et al. 2004).[2]

Sun et al. (1997) studied pollen from a profile at Weinan, which lies on the southern edge of the Loess Plateau and covers the last glacial-interglacial cycle. Their assessment was that the dominant vegetation was *Artemisia* and chenopods, alternating with shrubs such as Cyperaceae, Ranunculaceae and Liliaceae; in other words, an alternation of steppe and meadow-steppe environments. Forest vegetation existed for only short intervals: *Ulmus* forest at 90.7–95.1 ka; *Corylus* forest at 25.0–21.1 ka; and *Tsuga* at 13.7–11.8 ka. Otherwise, the Loess Plateau has been largely treeless in the last 100,000 years. Further north, Cai and colleagues (2019) show that in the Hetao Basin of Inner Mongolia, conditions deteriorated from forest steppe and open forest steppe in MIS 5, to desert steppe in MIS 4, steppe in MIS 3, and desert steppe in MIS 2 (see Figure 10.5).

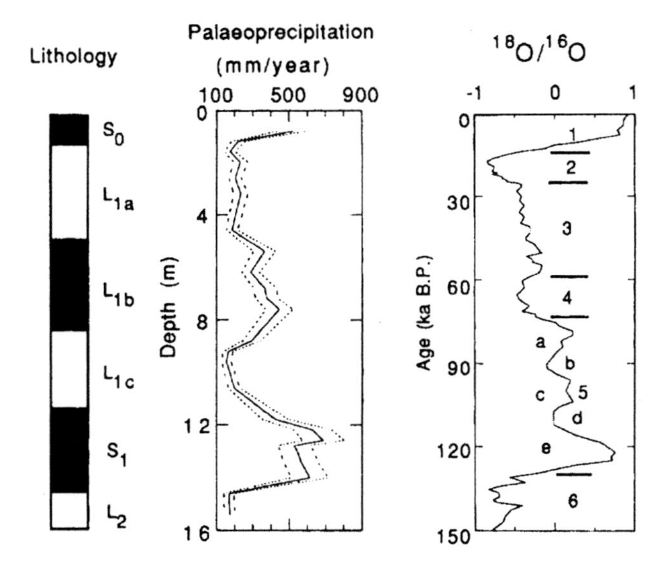

FIGURE 10.4 Estimated rainfall over the last 140,000 years in the Loess Plateau

L_{1a}–L_{1c} = the subdivisions of Loess 1 (L1) in the Chinese loess-palaeosol record; S_o is the Holocene soil and S_1 is the palaesol from the last interglacial. A composite $\delta^{18}O$ deep-sea record of climate change is shown on the right.

Source: Liu et al. 1995, Fig. 4.

Fauna

During cold, dry periods of the Late Pleistocene, the fauna of north China was characterised by horse (*Equus przewalskyi* and *E. hemionus*), gazelle (*Gazella przewalskyi* (=*Procapra przewalskyi*), woolly rhinoceros (*Coelodonta antiquitatis*), *Bison* and *Bos primigenius*. During cold and dry episodes in MIS 3, woolly mammoth grazed in Inner Mongolia and Shandong Province south of Beijing (see Larramendi 2014; Takahashi et al. 2007), and even as far south as the Qinling Mountains in the LGM. In warmer, moister periods, these mega-herbivores would have been replaced by herbivores such as *Bubalus* (buffalo), *Cervus nippon* (sika) and pig (Li et al. 2012).

The boundary between the Oriental and Palearctic faunas was not static but fluctuated frequently and often considerably during the Pleistocene, particularly north of the Qinling Mountains (Norton et al. 2010). However, cold-adapted animals rarely moved south of the Qinling Mountains, so there was greater faunal stability in south China. Several animal taxa did move north of the Qinling Mountains in warmer and moister periods. Tong (2007) notes that *Hystrix* (porcupine), *Macaca* (macaque), *Palaeoloxodon* (straight-tusked elephant), *Dicerorhinus* (Oriental rhinoceros) and *Bubalus* (water buffalo) are the most frequently recorded warm-adapted elements that appeared in north China. Because *Hystrix* has been found at 22 localities in north China, it is probably one of the best faunal indicators of a warm climate[3] in this region (Tong 2008).

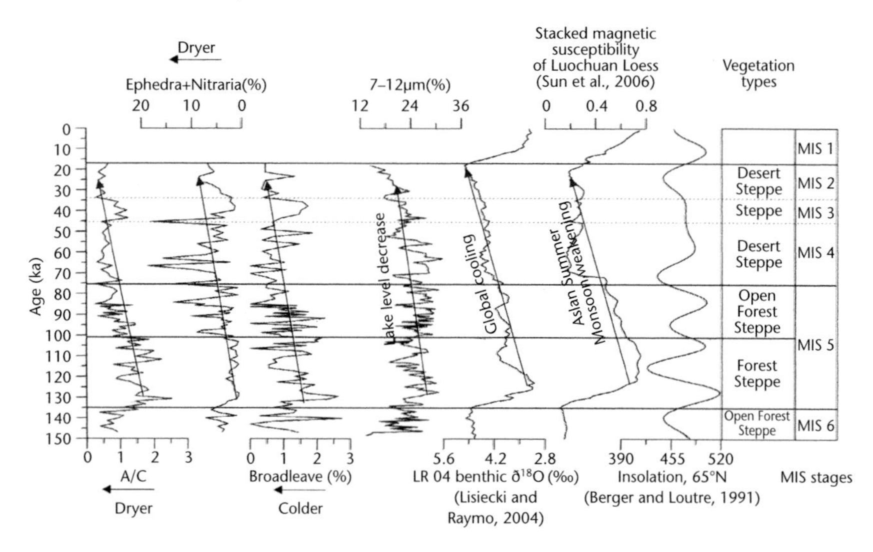

FIGURE 10.5 Changes in vegetation in Inner Mongolia over the last 140,000 years

The A/C ratio of *Artemisia* to Chenopodiaceae is used as an indicator of humidity in arid and semi-arid regions. The LR 04 benthic $\delta^{18}O$ marine record shows the long-term cooling between the last interglacial (MIS 5) and the Last Glacial Maximum (MIS 2). The main trend in Inner Mongolia are the changes from forest steppe to open steppe and desert steppe between MIS 5 and MIS 2

Source: Cai et al. 2019, Fig. 4.

The last glacial cycle (MIS 6–MIS 3): the Chinese coastal shelf

The penultimate ice age, MIS 6, appears to have been very severe in East Asia (Dennell 2009: Table 7.3). In north China, rainfall decreased by 25% to >50%; in south China, the montane tree line was lowered by >600 m and temperatures were 3–6° C. cooler (Zheng and Lei 1999). An additional development during MIS 6 was the formation of the northern part of the Chinese coastal shelf (see Figure 10.6) as a result of tectonic changes that affected offshore bathymetry (Sun et al. 2003) and increased the area of the Chinese coastal shelf to ca. 2 million sq km (Sun et al. 2000). As a result, when sea levels dropped in MIS 6, MIS 4 and MIS 2, the coastline of northern China moved up to 1,000 km eastwards (Liu and Ding 1998: 140), thereby joining the Korean peninsula to the Chinese mainland when sea levels were at their lowest. Pollen data indicates that the northern part of this coastal shelf was covered in *Artemisia*-dominated grassland (Sun et al. 2000). Because rainfall decreased from the southeast coast to the northwest, the eastward shift of coastline would also have made the interior of north China even more arid (Liu and Ding 1998: 140).

Population dynamics of China

Two factors need considering when thinking about the population dynamics of China during the Pleistocene. The first is how humans (and other animals) responded

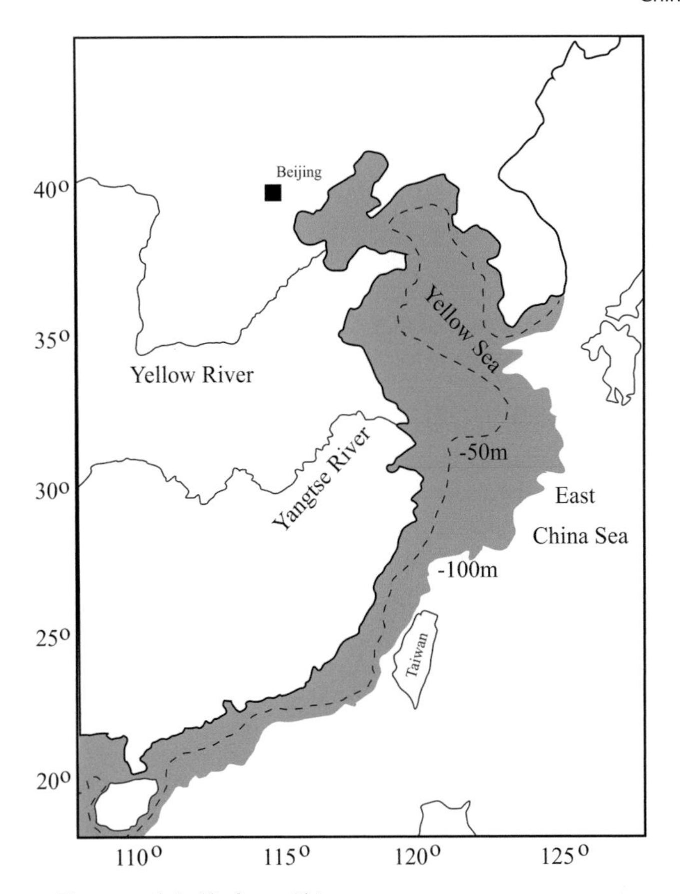

FIGURE 10.6 The coastal shelf of east China

The northern part did not emerge until MIS 6, ca. 150–130 ka ago, and would have reduced the amount of precipitation in inland areas.

Source: Redrawn from Li et al. 2018c, Fig. 1a.

to climatic change, particularly the repeated shifts in the respective strengths of the winter and summer monsoon that determined the severity or mildness of winter temperatures (particularly in the north of China) and the degree of aridity. The second factor is immigration into China – specifically, when, from where and under what climatic conditions.

Population dynamics in response to climate change in China

The response of warm-adapted faunas to the repeated climatic shifts in China between cold, dry climates and ones that were warmer and moister can be envisaged in a simple source-sink model (see Chapter 3, Figure 3.12). Maria Martinón--Torres and colleagues (2016) applied this type of model to China and showed how the Chinese hominin record could be modelled in terms of population contraction, expansion, coalescence and fragmentation.

The basins along the Luohe and Hanjiang Rivers in the Qinling Mountains are potential areas of glacial refugia and source populations for those that expanded northwards in warm, moist periods (Sun et al. 2018). The Qinling Mountains provided some shelter from the winter monsoon, and although some loess was deposited in these basins, it was much less than on the Loess Plateau, and winters appear to have been milder. Basins such as the Luonan and Lushi have been systematically surveyed for several years by Chinese teams and contain ca. 270 Palaeolithic sites that date from 1.2 Ma to the last glacial cycle (see e.g. Lu et al. 2011a, 2011b; Sun et al. 2012, 2017a, 2018). What makes this area especially interesting is that these basins were occupied in cold periods of loess deposition (equivalent to glaciations in Europe) as well as in periods of palaeosol development (equivalent to European interglacials) – as one would expect for glacial refugia. The Yangtse Valley was also another likely glacial refugia for southern populations no longer able to inhabit northern China.

These models consider hominins as part of a warm-adapted, Oriental fauna that would have expanded northwards when conditions became warmer, and retreated southwards towards glacial refugia when conditions became colder. Cold-adapted animals such as lemmings or mammoth can also be modelled in terms of source-sink populations, but these would have expanded southward when conditions were cold, and retreated northwards into interglacial refugia when conditions were at their warmest. This point is especially relevant to Pleistocene China in that there were repeated opportunities in cold periods for immigration by cold-adapted faunas into north China from Siberia, Mongolia and Manchuria (Li et al. 2012). Elephants provide a clear example, with the expansion southwards of the cold-adapted woolly mammoth when the climate of north China was extremely cold, and its replacement by the warm-adapted *Palaeoloxodon* when the climate became milder.

The fascinating aspect of China's population dynamics in the Pleistocene is that humans belonged to both the cold-adapted Palaearctic and warm-adapted Oriental Realms. A key factor here is the extent to which a hominin or human population was cold-adapted in terms of its ability to survive extremely cold conditions. As discussed in Chapter 9, the most important adaptation would have been warm, insulated clothing and footwear as well as effective control over fire. Recent evidence from north China indicates that these adaptations were likely in place by at least 45 ka.

Immigration into China

Figures 10.7 and 10.8 present two simple models of the climatic and environmental conditions that would facilitate or impede immigration. The first (Figure 10.7) assumes conditions were similar to those in the equivalent of an interglacial, when it was warm and the summer monsoon could penetrate much of northern China and stronger westerly winds could bring some rain into Xinjiang in northwest China.

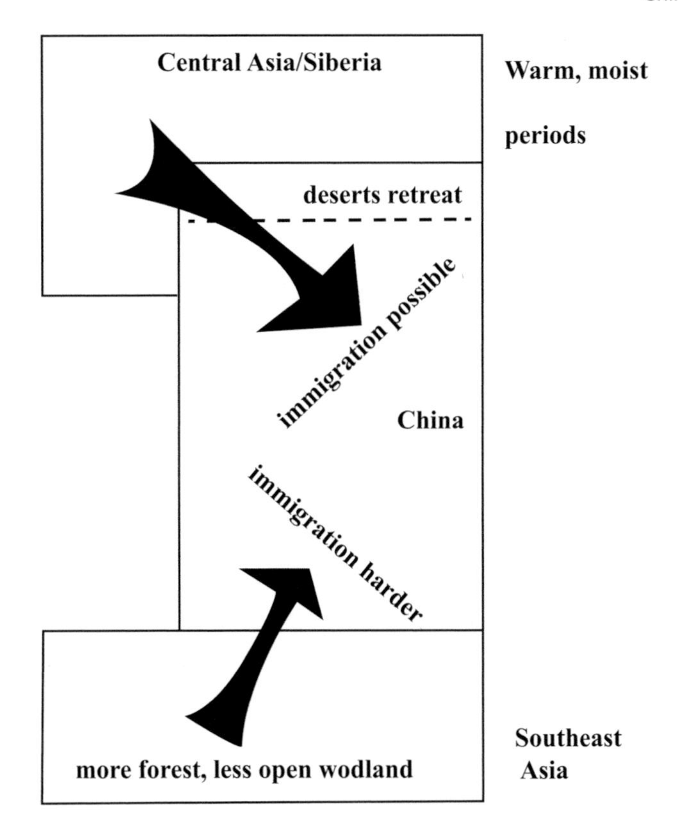

FIGURE 10.7 Model of immigration into China in warm periods

In warm, moist periods, the desert boundary retreated and immigration into north China was possible from regions to the north and west. In south China, immigration from Southeast Asia might have been more difficult because of the expansion of rainforest.

Source: The author.

These conditions would be the most optimal for immigration by a warm-adapted mammal into north China from the west and north from Siberia and Mongolia: the desert and steppe margins would have retreated northwards; there would have been numerous lakes in the Gobi and neighbouring deserts; and winters would have been less severe. Li Feng and others (2019b) have recently suggested that there may have been a desert corridor through a "green Gobi" at times of higher rainfall.[4] However, interglacial conditions were not necessarily optimal for migrations into southern China because warmer temperatures and higher rainfall would have led to an expansion in tropical and sub-tropical forest at the expense of open woodland and savannah grassland that would provide corridors through which incoming populations could disperse. Although immigration might not have been impossible, it would have been more difficult. (We need to bear in mind that increases in the extent of tropical forest might not have been a problem for populations that could colonise rainforest, as seen in Chapters 5 and 6 for South and Southeast Asia.)

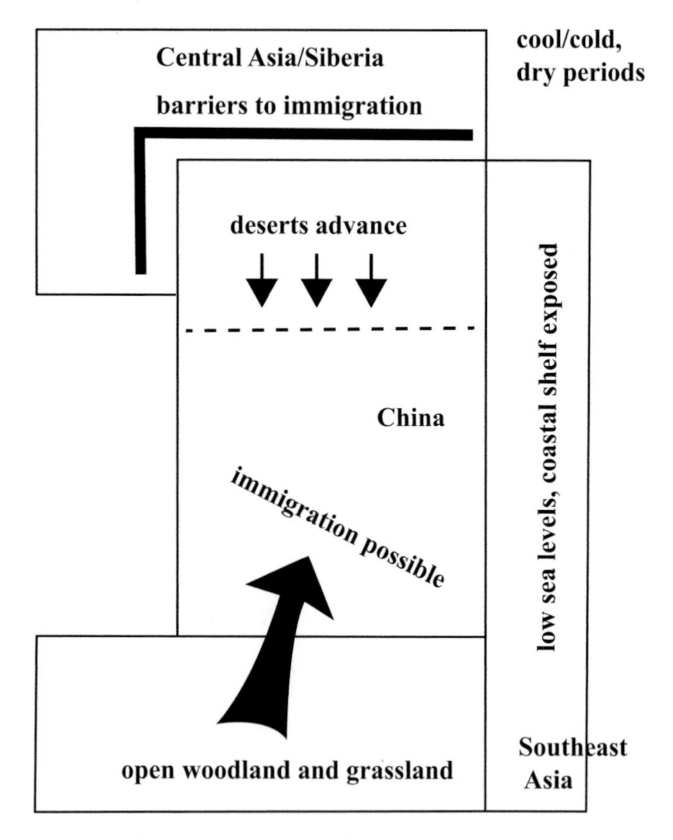

FIGURE 10.8 Model of immigration into China in cold periods

In cold, dry periods, the desert boundary shifted southwards in north China, and immigration from regions to the north and west would have become more difficult. On the other hand, the contraction of rain forest and expansion of open woodland and grassland in south China would have facilitated immigration from Southwest Asia.

Source: The author.

Figure 10.8 summarises conditions at the height of a glacial maximum. In north China, desert and steppe would have been at their greatest extent. With the fall in sea levels, the emergence of the east China coastal shelf would have moved the coastline several hundred kilometres eastwards, and this would further have reduced the amount of precipitation inland. Winters would also have been much more severe. As an extreme example, the average annual temperature on the Loess Plateau during the deposition of loess (L)15 1.2 Ma ago may have been as low as 1.3–3.0° C., similar to that of a polar desert (Guo et al. 1998). Under these conditions, immigration into China by a warm-adapted mammal from the north or west would have been difficult, if not impossible. In contrast, conditions that were cooler and drier in southern China would have facilitated immigration because corridors of savannah grassland and open woodland would have provided corridors for immigration, as argued previously (Chapter 6) for mainland Southeast Asia.

At this point, we can consider the skeletal and archaeological evidence for immigration into China by humans and their predecessors in the Middle and Upper Pleistocene.

Evidence for immigration into China

The Chinese skeletal record is consistent with evidence that animals (including hominins) were able to disperse into China during the Middle and Late Pleistocene. A few specimens suggest that immigration occurred from Southeast Asia but most indicate immigration from the north and west.

Immigration from the Oriental Realm of Southeast Asia

Two cave sites in south China have featured prominently in recent discussions of when our species first appeared in southern China.

Fuyan/Daoxian

This cave has a flowstone covering the entire floor that is dated by ^{230}Th analyses and faunal evidence to ca. 80–100 ka. Underneath were 47 teeth attributed to *H. sapiens*. The associated fauna includes five extinct taxa, so a minimum age of ca. 80 ka seems reasonable (Liu et al. 2015; Martinón-Torres et al. 2017). One interpretation is that these teeth indicate an immigration event of *H. sapiens* from Southeast Asia during late MIS 5/early MIS 4 (Dennell 2015), i.e. at a time when sea levels were lower than today (i.e. MIS 5d, MIS 5b or MIS 4), thus providing a wide corridor for entry when the vegetation was more open, as outlined above. Although some have reasonably asked for more data on the dating and taphonomic history of the cave deposits (Michel et al. 2016), nothing at present invalidates the dating of the evidence from Fuyan.

Zhirendong

At this cave, a toothless mandible and two molars were recovered from the upper part of layer 2 of Unit B and associated with a Late Middle or Early Upper Pleistocene fauna (Jin et al. 2009; Liu et al. 2010a). These were dated by reference to the lowest of a series of flowstones in the cave profile that was dated to 106.2 ± 6.7 ka. Cai and colleagues (2017) considered the faunal, palaeomagnetic and U-series dates and suggest that the human fossils are between 116 and 106 ka in age. These flowstones, however, were tightly packed within a thin stratigraphic zone with a vertical thickness of only 10 cm between one sample at ca. 51 ka and another at ca. 106 ka (see Figure S6 in Liu et al. 2010a), which suggests a "number of small-scale sedimentation or repeated erosional and depositional events at this portion of the cave sediments" (Kaifu and Fujita 2012: 3). Without knowing how the flowstones extended across the cave floor from the section, it is not possible to identify the one that relates to the human mandible, and it could therefore be younger than 110 ka. Dating remains a problem, and further research is needed into the formation processes of the cave sediments and their age.

The mandible shows signs of gracilisation, which was interpreted as implying "early modern human dispersal or gene flow across at least southern Asia sometime before the age of the Zhiren Cave human remains" (i.e., before 100 ka, the age of the oldest dating sample, and also implying that "any 'dispersal' involved substantial admixture between dispersing early modern human populations or gene flow into regional populations" (Liu et al. 2010a: 19205). As such, this evidence is consistent with a model of "continuity with hybridisation". An alternative explanation is that it indicates a *H. erectus* population that had become more gracile over time (Dennell 2010).

Other caves in south China have produced teeth attributed to *H. sapiens*, but the dental samples are small, the stratigraphic context is often unclear and the dating is often problematic. Luna Cave produced two teeth with a likely minimum age of 70 ka (Bae et al. 2014). Seven teeth of *H. sapiens* were recovered from layer 3 at Huanglong Cave (Liu et al., 2010b), but their age is unclear. U-series dating of two rhinoceros teeth gave an age range of 94.7 ± 12.5 ka and 79.4 ± 6.3 ka. Speleothem dating by U-series indicated ages of 103.7 ± 1.616 ka and 103.1 ± 1.348 ka, but ESR dating of a rhinoceros tooth produced an age of 44.18 ± 4.54 ka on a late uptake model and 34.78 ± 3.28 ka on an early uptake model. It is also unclear whether the speleothem sample from the section came from above or below the hominin teeth. This point highlights the difficulty of correlating speleothems exposed in a cave section with material found in the cave floor. Two teeth from the cave of Longtanshan, Yunnan Province, are identified as *H. sapiens* and dated to 60–83 ka, although further dating work may be required (Curnoe et al. 2016). There is also a partial skeleton from the cave at Liujiang that was found when local famers were digging for fertiliser. This find has caused much controversy. The deposits in which it is thought to have been found are dated to between 68 ka and >153 ka, with a probable age of 111–139 ka (Shen et al. 2002), but there is no way of knowing how these dates relate to the skeleton. It could be much younger – for example, it might represent an individual who was buried in a grave cut into those deposits. Until the skeleton is dated directly, it cannot be used as evidence of an early presence of *H. sapiens* in south China.

To summarise: until the dating of the teeth from Fuyan is disproved, this site shows the presence of our species in south China by at least 80 ka (early MIS 4), and perhaps earlier. Because the teeth are associated with a fauna that includes five extinct taxa, a recent age is improbable. Zhirendong might represent either a late *Homo erectus* (if older than e.g. 100 ka), or a hybrid resulting from some interbreeding with an early incoming population of *H. sapiens*. The evidence from other caves suggests that *H. sapiens* may have been present in south China before 80 ka, but not conclusively. The most likely source of these populations was likely to have been in Southeast Asia, where we know that

H. sapiens is present between 63–73 ka at Lida Ajer, Sumatra and 46–63 ka at Tam Pa Ling, Laos (Chapter 6).

The archaeological evidence from south China is currently insufficient to show how and when immigrant populations might have dispersed into south and central China, or how they adapted to local conditions. The little we do know shows that southern China was most unlikely to have been a "region of cultural retardation", as suggested by Movius (1948); see Qu Tongli et al. (2013) and Wang Youping (2017) for a summary of recent investigations in the Yangtse Valley.

Immigration from the Palearctic Realm of western Eurasia: skeletal evidence

Some Middle Pleistocene Chinese fossils indicate that immigration from western Eurasia may have occurred during warm periods. Rightmire (2001), for example, classified as *H. heidelbergensis* the specimens from Dali (ca. 240–280 ka old, Sun et al. 2017b) and Jinnuishan (ca. 280 ka old, Lu et al. 2003), which implies that "[the] spread of some populations of *Homo heidelbergensis* into the Far East cannot be ruled out" (Rightmire, 1998: 225). Others have commented on the overlap between *H. erectus* and non-erectus Middle Pleistocene *Homo*. Under the "continuity with hybridisation model" of hominin evolution in East Asia (see e.g. Liu et al. 2010a), "the suite of traits exhibited by Dali could be indicative of a local transition between *H. erectus* and *H. sapiens* that included some influence from western Eurasian populations during the Middle Pleistocene" (Wu and Athreya 2013: 154). Bae (2010: 90) further comments "even small amounts of gene flow from dispersing *H. heidelbergensis* groups into eastern Asia during the Middle Pleistocene is probably the most parsimonious explanation as to why similar morphological features occasionally appear among penecontemporaneous western and eastern Old World hominins".

One of the most intriguing sets of specimens are from Xujiayao (=Hujiayao), in the western part of the Nihewan basin in north China. This site was excavated in the 1970s. Stratigraphic level II contained a large lithic and faunal assemblage as well as several hominin remains that include several skull fragments, a mandible, a maxilla with six teeth in situ, and four isolated teeth (Wu and Poirier 1995; Wu and Trinkaus 2014; Xing et al. 2015). The lithic assemblage contained over 13,000 core and flake tools that included polyhedral cores, scrapers, points, and burins, most of which were made from local quartz. The faunal remains indicate a cool, temperate open grassland or steppe environment that was dominated by Przewalksi's horse (*Equus przewalski*), *E. heminonus* (onager) and gazelle (*Procapra przewalskyi*), which are mainly distributed in cold dry steppe, desert and semi-desert areas of northwest China (Norton and Gao 2008a). The presence of forest- or woodland-dwelling animals such as *Cervus* and *Sus* indicates either that the site lay on the ecotone of two environments or that the fauna is mixed. The presence of woolly rhino (*Coelodonta*) does not necessarily imply extremely cold conditions as it was distributed further south than mammoth and appears to have been tolerant of a wider range of temperatures (Li et al. 2012). The age of this material is uncertain. One study indicates an age older than 140 ka and likely between 160–220 ka (Tu et al. 2015), but another study implies an age range of ca. 260–370 ka (Ao et al. 2017). Here, I agree with Norton and Gao (2008a) that the fossil and archaeological material from stratigraphic level II is probably Late Middle or Early Late Pleistocene in age.

The hominin specimens are particularly interesting for their mosaic of primitive and derived dental features because these suggest a population of unclear taxonomic status with regard to other groups such as *H. sapiens* and *H. neanderthalensis*. The scarcity of fossil information available from Denisova (Chapter 9) prevents a detailed comparison with the Xujiayao specimens. However, Maria Martinón-Torres and colleagues (2017) point out that there are intriguing parallels between the morphological and genetic mosaic of the Xujiayao and the Denisova samples, respectively, which point in both cases to a hominin lineage that is different from *H. sapiens*; shares features with *H. neanderthalensis*; and preserves the heritage of a "mysterious" and primitive hominin, which in the Xujiayao fossils could be reflected in the preservation of some typical Asian Early and Middle Pleistocene features.

Overall, the morphology of the Xujiayao hominin specimens implies that their main similarities lie with hominin populations at Denisova in the Altai Mountains of Siberia and not with populations further south.

This assessment is consistent with their faunal context in a steppe or semi-arid environment. It is also consistent with an intriguing recent report of a Denisovan mandible from Xiahe on the edge of the Tibetan Plateau, ca. 3,000 m above sea level (Chen et al. 2019). The mandible is claimed to share features with a mandible recovered in fishing nets in the Penghu Channel between Taiwan and the mainland, and dated by the type of fauna occasionally found by fishermen to ca. 10–190 ka (Chang et al. 2015). Protein molecules extracted from the Xiahe mandible are claimed to be Denisovan. This might hint at an emerging Denisovan population extending from Denisova to the edge of the Tibetan Plateau and across northern and central China. If confirmed, this emerging population would provide a source population for the presence of Denisovan DNA in Southeast Asia and Melanesia (Reich et al. 2011).

Lingjing/Xuchang

Another major recent discovery are two hominin crania from layer 11 of the site of Lingjing in Xuchang County in Central China that are dated by OSL to ca. 105–125 ka and are thus dated to MIS 5e or MIS 5d. The fauna is diverse and includes grassland or steppe animals such as *Equus caballus and E. hemionus* (onager) and *Procapra* (gazelle), and open woodland or grassland such as *Cervus*, *Bos* and *Coelodonta*. The high proportion of limb bones and cutmarks indicates that the site was probably a kill and butchery site. Interestingly, two bones had been engraved with several parallel lines; one of the bones also had traces of ochre (Li et al. 2019).

The overall shape of the crania indicates continuity with earlier Middle Pleistocene hominins from East Asia, but they also show some Neandertal features. Together, these features "argue both for substantial regional continuity in eastern Eurasia into the early Late Pleistocene and for some level of east-west population interaction across Eurasia" (Li et al. 2017: 971). An intriguing possibility is that these crania have belonged to Denisovans, as suggested by Chris Stringer and Maria Martinon-Torres: "something with an Asian flavor but closely related to Neandertals" (see Gibbons 2017: 899). The Xiahe mandible strengthens this proposal.

Tianyuandong

The earliest unambiguous indication of *H. sapiens* in north China is from the cave of Tianyuandong near Beijing. Excavations in 2003 produced 34 remains of *H. sapiens*: most of the front and right side of a mandible, and parts of upper and lower limbs, all most likely from one individual (Shang et al. 2007), but unfortunately no archaeological evidence. The human fossils were found clustered an unconsolidated breccia in layer III. The associated faunal assemblage comprises 39 taxa (mostly rodents) but the commonest large mammals are cervids, including *Cervus elaphus*, sika (*Cervus nippon*) and *Siberian musk deer (Moschus moschiferus)*. Of these, the original range of sika extended from east Siberia to Vietnam, and musk deer is now distributed in Inner Mongolia, Mongolia, Manchuria, Siberia and the Korean Peninsula, so is further south at Tianyuandong than now. As it prefers mountainous terrain and forested slopes, it could have been hunted locally at Tianyuandong. Six bones from layer III were dated by AMS ^{14}C to between 30,500 ± 370 and 39,430 ± 680 BP (uncalibrated). A sample from the human femur was dated to 34,430 ± 510 ka BP uncalibrated, or 40,328 ± 816 ka BP calibrated. The lack of wear on the toe bones indicates that Tianyuan individual may also provide indirect – and so far the earliest – evidence for the habitual use of footwear (Trinkaus and Shang 2008).

Dentally, the Tianyuan individual shared similarities with teeth from European Upper Palaeolithic humans from Arene Candide (Italy), Dolní Vestonice and Mladeč (Czech Republic) (Shang et al. 2007). Analysis of the nuclear DNA of this individual showed that it was derived from a population that was ancestral to many present-day Asians and Native Americans but had already diverged from the ancestors of present-day Europeans. The proportion of Neandertal or Denisovan DNA sequences in its genome was also no larger than in present-day people in the region (Fu et al. 2013). Another study showed that the Tianyuan individual was more related to Asians than to past and present Europeans, but the 35,000-year-old individual from Goyet, Belgium, shared more alleles with Tianyuan than any other west European. The similarity of the Tianyuan individual to this specimen and some South American populations implies the persistence of a structured populations that lasted until the colonisation of the Americas (Yang et al. 2017). Interestingly, isotopic analysis of the human humerus indicated the consumption of freshwater fish year-round and on a scale sufficient to leave an isotopic signature (Hu et al. 2009).

Upper Cave, Zhoukoudian

Upper Cave was excavated by Pei Wen Zhong in 1933 and 1934, and produced faunal material, a small artefact assemblage (N=25), seven stone beads and 125 perforated teeth (Li et al. 2018a), and the remains of eight individuals of *H. sapiens*, including three crania. Weidenreich (1946) incorporated the Upper Cave remains into his multi-regional model of human evolution, and drew a direct link between *Sinanthropus* (now known as *H. erectus*) and modern Mongoloids, a term which he never rigorously defined. Sadly, most of this material – including the human crania – was lost in World War II, but thankfully casts of the crania and some other specimens that were made in the 1930s did survive. The faunal assemblage was re-examined by Norton and Gao (2008b) who found that only 20 of the 49 taxa in the original collection could be relocated. Of these, cervids and a wide range of carnivores were the main components, and they concluded that the lower recess in this cave was a natural bear trap, and humans hunted the local deer in the upper part. Kaminga and Wright (1988) studied one of the first-generation casts of Skull 101 (known as the "Old Man") and concluded that it was not Mongoloid. Later, Harvati (2009) showed that the Upper Cave crania 101 and 103 resembled Upper Palaeolithic Europeans and "may represent members of an as yet undifferentiated early modern human population that expanded across Eurasia in the Late Pleistocene" (Harvati, 2009: 761).

The dating of the Upper Cave material has been contentious, with age estimates ranging from 10 to 34 ka. Recently, a new set of accelerator mass spectrometry radiocarbon dating results from Upper Cave combined with a re-examination of previous research indicates that the archaeological layers in Upper Cave – including the crania – are dated to at least 35.1–33.5 cal ka and may be as old as 33,551–38,376 yr cal BP (Li et al. 2018a).

The most significant aspect of the Upper Cave material is that it shares more features with similar evidence from Siberia, Mongolia and western Eurasia than with south China. We have already noted Harvati's conclusion about the affinities to crania 101 and 103 to Upper Palaeolithic Europeans. The ornaments are most like those from Denisova, and both assemblages include perforated canine teeth of various types of deer and small carnivores, round beads, bone pendants, and perforated shells, none of which are found in south China but are widespread across Mongolia and southern Siberia (see Chapter 9). Li and colleagues (2018a: 170) conclude entirely reasonably that the inhabitants of Upper Cave "were part of dispersal events across northern Eurasia toward Siberia and eventually reaching into northern China".

Immigration from the Palearctic Realm of western Eurasia: archaeological evidence

There are archaeological indications of at least three immigration events into north China from Siberia and Mongolia in MIS 3 (60–25 ka), and a fourth immigration event into the Tibetan Plateau.

Jinsitai: a Neandertal intrusion?

There is a recent indication that Neandertals may have expanded eastwards into Inner Mongolia, ca. 2,000 km east of the previously recorded easternmost location in the Altai region of Siberia. At the cave of Jinsitai in Inner Mongolia, levels 7 and 8 contained assemblages classified as Mousterian that date from at least 47–42 ka and persisted until around 40–37 ka. It was thus a substantial occupation episode. Although the correspondence of Mousterian with Neandertals is not automatic – as in the Levant – most Mousterian sites are associated with them, and in Central Asia, Neandertals are the only hominin associated with Mousterian assemblages, so the same is likely at Jinsitai (Li et al. 2018b). However, we cannot discount the possibility that the occupants of Jinsitai were partly or even wholly Denisovan, given the evidence for interbreeding between Denisovans, Neandertals and our own species (see Chapter 9).

Shuidonggou and the Upper Palaeolithic of north China

Shuidonggou (SDG) is the most important Late Palaeolithic site cluster in north China. SDG 1 was discovered and first excavated in 1923 by the French missionary Emile Licent and the Jesuit Teilhard de Chardin (1925), and has since been re-excavated several times. SDG 2 lies opposite SDG 1 (Figure 10.9); in the same valley there is also SDG 7 (Niu et al. 2016) and SDG 9, and SDG 12, a late Pleistocene microblade site. Several methods have been used to date these sites, including U-series, AMS and OSL. Table 9.1 shows a selection of the dates thought most reliable by Li and colleagues (2019a). There are several excellent syntheses (e.g. Li et al. 2013a, 2013b, 2019a; Peng et al. 2018).

SDG 1 comprises a lower and an upper cultural layer. The lower cultural layer is 5 m thick and was excavated as a single assemblage; current re-excavation is likely to show multiple archaeological levels. It contains a blade-based technology produced by Levallois and prismatic methods of core reduction (Peng et al. 2014; see Figure 10.10). Similar assemblages are also known from South Temple Canyon (Madsen et al. 2014), SDG 2 cultural layers 5a and 7 (Li et al. 2014), SDG 7 and SDG 9 (Niu et al. 2016). Ostrich eggshell beads are also present, similar to those seen in Siberia and Mongolia (Peng et al. 2018), and an engraved limestone core was also found (Peng et al. 2012). OSL dating by Nian and colleagues (2014) indicates that the SDG 1 assemblage dates from ca. 43 ka (see Table 10.1), Overall, the SDG 1 assemblage is similar in age and composition to the IUP that is seen in Mongolia and the Altai (Peng et al. 2018; Li et al. 2019a; see Chapter 9).

FIGURE 10.9 Shuidonggou 1 and 2

Shuidonggou (SDG) 1 is in the section facing the camera. The low ridge in the background is a remnant of the Great Wall. The excavation of SDG2 is in the foreground.

Source: The author.

SDG 2 has seven archaeological layers (AL) (Li et al. 2013b, 2019a). The topmost AL1 is dated to 28.2–27.3 ka cal BP, and AL7 at the base is dated at 41.4–40.4 ka cal BP at the base of section (see Table 10.1). AL7 and AL5a contained the type of blade-based assemblage seen at SDG 1. Those from AL6, AL5b, and AL4–AL1 are flake-based with numerous side scrapers. The presence of a blade-based assemblage in layer 5a over the flake-based assemblages in AL5b and AL 6 indicates that the two traditions overlapped.

The flake-based assemblages at SDG 2 are similar to those from other sites in north China and indicate a local tradition that continued until the Last Glacial Maximum. These assemblages probably denote short-term camps. Several shallow hearths are found throughout the sequence. Studies of starch grains, phytoliths and plant tissue fragments on the artefacts from AL1–AL3 provide direct evidence of

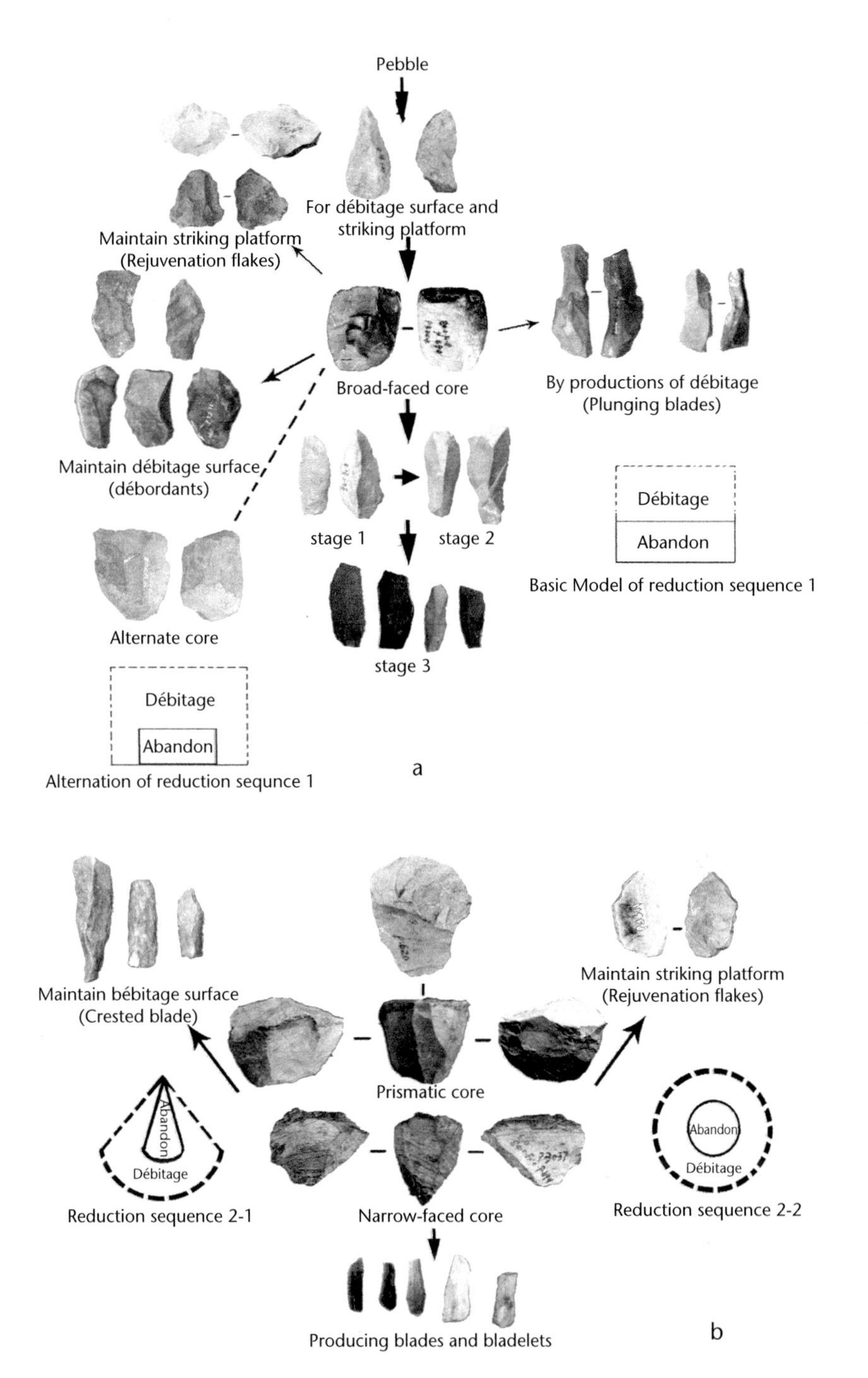

FIGURE 10.10 Blade production techniques at Shuidonggou (SDG) 1, lower cultural layer

Source: Peng et al. 2014, Fig. 3.

TABLE 10.1 Dates for Shuidonggou (SDG) 1 and SDG 2

Site	*Depth (m)*	*Dating material*	*Dating method*	*Age estimate BP*	*Age range cal yr BP (2-σ for 14C dates)*	*Source*
SDG 1 north section	6.1	Sediment	OSL	43,000 ± 3,000	46,000–40,000	Nian et al. (2014)
	6.5	Sediment	OSL	43,000 ± 3,000	46,000–40,000	Nian et al. (2014)
	6.5	Charcoal	AMS	36,200 ± 140	41,279–40,420	Morgan et al. (2014)
	7.0	Sediment	OSL	42,000 ± 3,000	45,000–39,000	Nian et al. (2014)
	7.6	Sediment	OSL	46,000 ± 3,000	49,000–43,000	Nian et al. (2014)
	7.6	Sediment	OSL	35,000 ± 3,000	38,000–32,000	Nian et al. (2014)
	8.1	Sediment	OSL	35,000 ± 3,000	38,000–32,000	Nian et al. (2014)
	8.6	Sediment	OSL	33,000 ± 3,000	35,000–30,000	Nian et al. (2014)
	9.1	Sediment	OSL	33,000 ± 2,000	35,000–30,000	Nian et al. (2014)
	9.6	Sediment	OSL	39,000 ± 3,000	42,000–36,000	Nian et al. (2014)
SDG 2: AL1a	3.5–3.64	Bone	AMS	23,450 ± 80	27,794–27,453	Li et al. (2019b)
	3.5–3.64	Bone	AMS	23,320 ± 70	27,720–27,383	Li et al. (2019b)
	3.5–3.64	Bone	AMS	23,270 ± 70	27,693–27,351	Li et al. (2019b)
	3.5–3.64	Bone	AMS	23,690 ± 70	27,928–27,618	Li et al. (2019b)
SDG 2: AL2	n/a	Charcoal	AMS	26,350 ± 190	31,012–30,203	Madsen et al. (2001)
	n/a	Charcoal	AMS	25,670 ± 140	30,339–29,414	Madsen et al. (2001)
	n/a	Ostrich eggshell	AMS	26,930 ± 120	31,207–30,818	Madsen et al. (2001)
	n/a	Charcoal	AMS	26,830 ± 200	31,220–30,699	Madsen et al. (2001)
SDG 2: AL3	5.76–5.86	Bone	AMS	28,290 ± 110	32,665–31,655	Li et al. (2013a)
SDG 2: AL6	10.4	Sediment	OSL	38,300 ± 3,500	41,800–34,800	Liu et al. (2009)
SDG 2: AL7	10.5–11.4	Wood	AMS	36,329 ± 215	41,475–40,441	Liu et al. (2009)

Notes: AL = archaeological layer. The Madsen et al. (2001) dating samples from SDG 2 were taken from the natural profile.

Sources: Li et al. 2019b; Li et al. 2013a; Liu et al. 2009, Madsen et al. 2001; Morgan et al. 2014; Nian et al. (2014

plant use activities during late MIS 3 (Guan et al. 2014). A grinding tool fragment in layer AL1 (Li et al. 2019a) may also have been used for plant processing. Ornaments and bone tools are present in AL3 and AL2: one freshwater shell bead fragment was found in AL3 (Wei et al., 2016), and more than 70 ostrich eggshell beads (including some ochre-coloured ones; see Martí et al. 2017) and a bone needle fragment were found in AL2, dating to ca. 30 ka cal BP (Li et al. 2014; Martí et al. 2017; Wei et al. 2017).

Who made the SDG 1 assemblage?

SDG 1 has a blade-based assemblage dated to ca. 43–46 ka cal BP but no hominin remains; Tianyuandong has evidence of *Homo sapiens* at ca. 40 ka but no archaeology: so who made the SDG 1 assemblage? By far the most likely candidate is *Homo sapiens* but until the crucial skeletal evidence is found, we should keep an open mind because we cannot exclude the possibility that the occupants of SDG 1 were Denisovan, or a sapiens-Denisovan hybrid (see Li et al. 2014), or even Neandertal. Because of the similarities of the lithic assemblages from Shuidonggou, Mongolia and southern Siberia, and the common element of ostrich eggshell beads, the SDG occupants were presumably part of an extensive network of groups that operated over an enormous territory in pursuit of highly mobile migratory prey such as horse and gazelle. A high degree of mobility would also involve a high degree of risk and uncertainty, thereby increasing the need to sustain long-distance social networks. As a long-term strategy, the risk may have been too great, and this may be why the initial blade-based tradition was short in north China compared with the succeeding flake-based occupations. It may be worth noting that there is no middle Upper Palaeolithic in Mongolia. On current evidence it is possible that the IUP groups in Mongolia went locally extinct, and those in north China were absorbed into groups using a flake-based lithic tool-kit.

Shuidonggou 12 and the microblade tradition

Although beyond the scope of this book, it is worth noting the late glacial microblade industry which probably originated in Siberia (Yi et al. 2013, 2016) and appeared in north China ca. 23 ka cal BP. This industry produced a superb lightweight toolkit by pressure-flaking minute blades that could be slotted into arrow shafts or handles, and is widely distributed across Siberia (see Chapter 9), Japan, the Korean Peninsula (Chapter 11), and in China as far south as Linxian in Central China (Li and Ma 2016). Shuidonggou 12 is a superb end-Pleistocene example of a microblade hunting camp focussed on trapping fur animals (Yi et al. 2014). As Yi and colleagues (2016: 138) point out, "microblade technology was a perfect solution to problems of provisioning through long, harsh winters when resources were less abundant and more difficult to access, and when failure to procure sufficient resources had fatal consequences". The diffusion of a microblade technology across Central Asia, Mongolia, and southern Siberia provides a clear demonstration of how successful humans had become at surviving some of the most hostile environments of the Pleistocene.

Tibet

One of the last places in Asia that was colonised was the Tibetan Plateau, which presented one of the most challenging and hostile environments of the northern hemisphere. With an average elevation of ca. 5,000 m and an average annual temperature close to freezing point, and with only half the amount of air as at sea level, it is not hard to see why it was such a challenge. Some researchers have suggested that it was only permanently colonised in the Holocene with the adoption of agriculture, and the domestication of barley and yak (Chen et al. 2015). Others suggest that it might have been initially colonised at the end of the last ice age when hunter-foragers began to hunt at higher altitudes during the summer months, and gradually learnt how to acclimatise to such harsh surroundings (Brantingham et al. 2013). Some genetic evidence from modern Tibetans suggests that the Plateau might have been colonised 30,000 years ago before the Last Glacial Maximum (Qin et al. 2010; Lu et al. 2016), but until now archaeological evidence for this has been lacking. For Palaeolithic hunters, the attractions of the Tibetan Plateau must have been the herds of gazelle, horse and yak (and perhaps other herbivores such as woolly rhinoceros) – the Plateau was undoubtedly harsh but was not barren.

For these reasons, recent investigations of a stratified Palaeolithic site on the Tibetan Plateau are of great interest in showing not only the earliest occupation of the "roof of the world" but the earliest record world-wide for humans living at high attitude. Zhang and colleagues (2018) report the excavation of a site called Nwya Devu (ND), which lies at 4,600 m in eastern Tibet. The site lies on the edge of a former lake and near a low ridge of black slate that provided a high-quality source of flakable stone for making stone tools. Unlike almost all other sites on the Tibetan Plateau that may have Palaeolithic-looking stone tools, this one is stratified and has almost two metres of deposits. The site has three strata, of which the lowest is a layer of fine-grained sands denoting gentle deposition. Over 3,600 stone artefacts were found, of which 300 are in the lowest layer, and 200 in a middle layer of sands and gravels and the rest were found in the top most layer of sand and silt. There are no obvious typological, technological or morphological differences in the assemblages from each layer, and the excavators argue that all the artefacts can be regarded as part of the same assemblage. They also argue that the artefacts were primarily associated with the lowest layer but through geological processes such as freeze-thawing and gelifluction, most were worked upwards.

Dating here is clearly critical. Because there was not enough material for radiocarbon dating, the layers were dated primarily by OSL. The 24 dates are largely internally consistent. The topmost layer is dated to ca. 4–13 ka and is largely Holocene in age, and the middle layer dates from ca. 18–25 ka, indicating deposition during the Last Glacial Maximum (LGM) when conditions were at their most severe. The oldest dates of ca. 30–40 ka are from the bottom layer in the deepest part of the stratigraphy and indicate the likely age of the stone tool assemblage. This estimate is consistent with a radiocarbon date from that layer of ca. 43–48 ka on a shell. Palaeo-environmental evidence from the Tibetan Plateau indicates that during this period, the climate was less severe than today. According to Shi and colleagues (2001), between 30 ka and 40 ka, annual average temperatures were 2–4° C. warmer than today, rainfall 40–100% higher, and there were numerous lakes resulting from a strengthened Indian summer monsoon.

A primary aim of those who stayed at Nwya Devu was to produce long, parallel-sided blades that could be used as knives or scrapers that were most likely hafted onto a bone or wooden handle. Some of these blades were over 20 cm long which testifies both to the quality of the raw material and the skill of the person flaking it. Because of the proximity of the site to a large source of flakable stone, the site is likely a workshop where tools were made that were then used at other locations. The assemblage is more or less unique in East Asia but the palaeolithic record of the vast area of northern China, Tibet, and Mongolia is very poorly known, so it is not surprising that there are no obvious parallels for the Nwya Devu assemblage.

We know from the cave site of Tianyuandong that our species was already in northern China by 40 ka, so the first Tibetans were also likely *Homo sapiens*. An interesting perspective on the colonisation of the Tibetan Plateau is that the adaptation that may have enabled humans to live at such a high altitude may have come from the Denisovans (see Chapter 9). Their relevance here is that they may have contributed our adaptation to high-altitude hypoxia through a gene region known as EPAS1. (If the Xiahe mandible is indeed Denisovan and the dating is correct, hominins were already living at 3,000 m above sea level 160,000 years ago.) The unusual haplotype structure of this gene is found only in Tibetans and Denisovans (Huerta-Sánchez et al. 2014). We know that Denisovans interbred with Neandertals, and Neandertals interbred with humans, and therefore it is likely that the Tibetan Plateau was colonised by humans that had interbred with Denisovans. This scenario is consistent with the suggestion that the Tibetan Plateau was colonised in a two-wave model (Lu et al. 2016). The first wave began ca. 40 ka by populations named as "SUNDers", or Siberian-Unknown-Neandertal-Denisovan", and the second after 9 ka of neolithic colonists. It was presumably this first wave of "SUNDers" who brought with them the adaptation to hypoxia.

Unfortunately, no bone was preserved at Nwya Devu, so we have no direct skeletal evidence about the first Tibetans, nor about the animals that they hunted. Nevertheless, the evidence from Nwya Devu that humans were living at 4,600 m above sea level 30–40,000 years provides a graphic example of how successful our species has been as a colonising animal (Zhang and Dennell 2018).

Relict populations

The demographic model proposed here envisages population fragmentation, isolation and recombination brought about by climate change. One example of an isolated population is from Yunnan Province, southwest China. Cranial and dental material from the caves of Longlin and Maludong (both dated to the interval 14.3–11.5 ka) have features that look archaic. These "may represent a late-surviving archaic population … or alternatively, East Asia may have been colonised during multiple waves during the Pleistocene, with the Longlin-Maludong morphology possibly reflecting deep population substructure in Africa prior to modern humans dispersing into Eurasia" (Curnoe et al. 2012: 1). Isolation in the mountain basins of Yunnan is entirely plausible as a mechanism for the retention of primitive features in these populations. Another example may be 15,000-year-old skeletal remains from the cave of Dushan in southern China, which may represent an isolated population that retained archaic dental features (Liao et al. 2019).

Discussion

The palaeodemography of Pleistocene China after 50 ka was based on two systems. The first was in response to climate change of humans (and their predecessors) as part of a warm-adapted Oriental fauna in south and central China. When conditions were favourable, they and their prey were able to move northwards into north China but then retreated southwards when necessary because of climatic deterioration. Their glacial refugia were probably in the basins in the Qinling Mountains and the Yangtse Valley. The second system involved humans who belonged to a cold adapted Palearctic fauna that moved into the steppe grasslands of north China from Mongolia and Siberia. After 50 ka, there would doubtless have been interactions between northern and southern groups along their zones of contact.

Immigration into China would have been from both the north and the south. *Homo heidelbergensis* probably entered from western Eurasia during warm periods. The fossil material from Xujiayao and Xuchang probably represent other northern immigrants prior to MIS 4. Neandertals might have been another northern immigrant if they made the Mousterian assemblages at Jinsitai. Our own species is first indicated in north China by the skeletal material from Tianyuandong, ca. 40 ka, and later by the inhabitants of Upper Cave, Zhoukoudian, ca. 33–35 ka. In both cases, their similarities to other populations lie more with regions to the west than to south China. The colonisation of the Tibetan Plateau between 30 and 40 ka was most likely another northern immigration, this time facilitated by the infusion from Denisovans of the EPAS1 gene that alleviates high-altitude apoxia. Modern humans may have moved into southern China before 80 ka, but considerably more research needs to be done on the dating and identification of fossil material before we can be certain.

The Palaeolithic record of China reflects this division between a southern and a northern demographic system. Mousterian, and the later blade-based assemblages are found only in north China. Microblade assemblages probably originated in Siberia and in China, are found primarily in north China but their users occasionally hunted as far south as central China. The archaeological record of southern China shows a much greater degree of stability. Most of these assemblages are based on unstandardised flakes and cores, but it would be uncritical to dismiss them as simple and monotonous (Qu et al. 2013; Wang 2017). Ma'anshan, for example, in Guizhou Province, has the earliest formal bone tools in China that date back to ca. 35 ka, and these include barbed bone points, dated at 23–28 ka (Zhang et al. 2016). Xiaodong in Yunnan Province in southwest China has the earliest Hoabinhian assemblage – another link to south east Asia – that is dated to 45 ka (Ji et al. 2015). Because this site lies in rainforest, this site may indicate the earliest colonisation of rain forest in mainland southeast Asia. There is probably far greater complexity and variety in the Palaeolithic of southern China than we currently recognise.

We can assume that the Neandertals at Jinsitai, and the *Homo sapiens* populations at Tianyuandong, Shuidonggou and Zhoukoudian had warm, wind-proof clothing, as did their counterparts in Mongolia and Siberia. This adaptation played a vital part in the late glacial colonisation of the North China Plain that is comparable to the colonisation of the North European Plain on the other side of Eurasia, and provides yet another example of our superb abilities as a colonising animal. In common with late Palaeolithic groups across continental Eurasia, status and social context within a network of groups was symbolised in ornaments made from beads and perforated teeth that were worn. In south China, where winters were never as severe as in north China, plant-based clothing would have sufficed, as suggested in Chapter 3, and visual statements about social status could have been made by using paint, scarification or tattoo to modify the appearance the body.

As an additional point, there is no need for conflict between a Recent Out of Africa (ROA) model that allows some hybridisation, and a model of continuous local evolution that allows some immigration. Both processes were involved, even if their relative contribution varied according to local climatic and environmental circumstance: most immigrant hominin and human populations into China interacted and hybridised with indigenous populations, and most indigenous populations are likely to have assimilated some immigrants. Both processes are consistent with a restated version of the Assimilation Model under which "The strong majority of modern human biology is clearly of African origin, but genomic and morphological data both demonstrate relatively small, but significant, archaic contributions from Neandertals, Denisovans and possibly other archaic human groups" (Smith et al. 2017: 127).

A second point that arises is that species' boundaries are inevitably blurred in models that allow a significant degree of hybridisation. Instead of "cherry picking" those traits that are deemed representative of *H. sapiens* or Neandertals, it is more productive to consider instead the overall morphology. A consequence of this shift of emphasis is that the term "archaic" has probably outlived its usefulness in discussions of human evolution in East Asia. It is not always clear if and how a specimen attributed to "archaic *H. sapiens*" was directly ancestral to "modern" *H. sapiens*. Additionally, the term "archaic *H. sapiens*" is a blanket term that conceals more than it reveals: it masks the complexity and variety of palaeodemes, morphs or metapopulations that constitute the demography of Pleistocene China. If we are to escape "the hybrid mess" in both Asia and Africa (Martinón-Torres et al. 2017), we need to develop methodologies that do not require us to assign specimens arbitrarily to one of a limited range of species.

Summary

China lies within both the Palearctic and Oriental biographic realms, and experienced frequent and often considerable environmental and climatic changes in the last glacial cycle.

The main points about the colonisation of China by *H. sapiens* are that i) it had experienced several immigration events before *H. sapiens*; ii) immigration from southeast China into the Oriental Realm of south China probably began by 80,000 years ago; iii) immigration into the Palearctic Realm of north China came from Mongolia and southern Siberia ca. 45,000 years ago by people who were already adapted to the cold with warm, wind-proof clothing; and iv) the subsequent history of China is one in which populations from both north and south developed locally but also mixed. Immigration of new groups and their assimilation by indigenous populations are the main features of its demographic history.

Notes

1 In a recent updating of Wallace's (1876) biogeographic regions, Holt and colleagues (2013) recognise a Sino-Japanese Realm in north China between the Qinling Mountains and Mongolia, based on the distribution and phylogenetic history of birds, amphibians and non-marine mammals. As we are concerned mainly with large mammals (including *Homo*), I retain here Wallace's divisions.
2 The discrepancy is probably the result of different dating techniques: OSL versus 14C.
3 *Hystrix* would also make a good convenience food. Once found, it is easy to kill and I understand that its flesh can be eaten raw, although I have not tried that.
4 Interestingly, Li, K. et al. (2018) report a grinding stone and evidence of seed processing dated to 13 ka in the Lop Nor region of the Tarim Basin, in an area now unrelenting desert with only ca. 20 mm rain – there may indeed have been a desert corridor.

References

Ao, H., Liu, C.-R., Roberts, A.P., Zhang, P. and Xu, X. (2017) An updated age for the Xujiayao hominin from the Nihewan Basin, north China: implications for Middle Pleistocene human evolution in East Asia. *Journal of Human Evolution*, 106: 54–65.

Bae, C.J. (2010) The Late Middle Pleistocene hominin fossil record of Eastern Asia: synthesis and review. *Yearbook of Physical Anthropology*, 53: 75–93.

Bae, C.J., Wang, W., Zhao, J., Huang, S., Tian, F. and Shen, G. (2014) Modern human teeth from Late Pleistocene Luna Cave (Guangxi, China). *Quaternary International*, 354: 169–183.

Bezrukova, E.V., Tarasov, P.E., Solovieva, N., Krivonogov, S.K. and Riedel, F. (2010) Last glacial–interglacial vegetation and environmental dynamics in southern Siberia: chronology, forcing and feedbacks. *Palaeogeography, Palaeoclimatology, Palaeoecology*, 296: 185–198.

Bloemendal, J. and Liu, X. (2005) Rock magnetism and geochemistry of two plio-pleistocene Chinese loess-palaeosol sequences – implications for palaeoclimatic reconstruction. *Palaeogeography, Palaeoclimatology, Palaeoecology*, 226: 149–166.

Brantingham, P.J., Xing, G., Madsen, D.B., Rhode, D., Perrault, C., Woerd, J. van der and Olsen, J.W. (2013) Late occupation of the high-elevation northern Tibetan Plateau based on cosmogenic, luminescence, and radiocarbon ages. *Geoarchaeology*, 28: 413–431.

Cai, Y., Qiang, X., Wang, X., Jin, C., Wang, Y. et al. (2017) The age of human remains and associated fauna from Zhiren Cave in Guangxi, southern China. *Quaternary International*, 434: 84–91.

Cai, M., Ye, P., Yang, X. and Li, C. (2019) Vegetation and climate change in the Hetao Basin (northern China) during the last interglacial-glacial cycle. *Journal of Asian Earth Sciences*, 171: 1–8.

Chang, C.-H., Kaifu, Y., Takai, M., Kono, R.T., Grün, R., Matsu'ura, S., Kinsley, L. and Lin, L.-K. (2015) The first archaic *Homo* from Taiwan. *Nature Communications*, 6: 6037. doi:10.1038/ncomms7037.

Chen, F.H., Dong., G.H., Zhang, D.J., Liu, X.Y., Jia, X. et al. (2015) Agriculture facilitated permanent human occupation of the Tibetan Plateau after 3600 B.P. *Science*, 347: 248–250.

Chen, F., Welker, F., Shen, C.-C., Bailey, S.E., Bergmann, I. et al. (2019) A late Middle Pleistocene Denisovan mandible from the Tibetan Plateau. *Nature*, 569: 409–412, doi:10.1038/s41586-019-1139-x.

Curnoe, D., Ji, X., Herries, A.I.R., Kanning, B., Taçon, P.S.C., Zhende, B., Fink, D., Yunsheng, Z., Hellstrom, J., Yun, L., Cassis, G., Bing, S., Wroe, S., Shi, H., Parr, W.C. H., Shengmin, H. and Rogers, N. (2012) Human remains from the Pleistocene-Holocene transition of southwest China suggest a complex evolutionary history for East Asians. *PLoS One*, 7: e31918.

Curnoe, D., Ji, X., Shaojin, H., Taçon, P.S.C. and Li, Y. (2016) Dental remains from Longtanshan cave 1 (Yunnan, China), and the initial presence of anatomically modern humans in East Asia. *Quaternary International*, 400: 180–186.

Dennell, R.W. (2009) *The Palaeolithic Settlement of Asia*. Cambridge: Cambridge University Press.

Dennell, R.W. (2010) Early *Homo sapiens* in China. *Nature*, 468: 512–513.

Dennell, R.W. (2013) Hominins, deserts, and the colonisation and settlement of continental Asia. *Quaternary International*, 300: 13–21.

Dennell, R.W. (2015) *Homo sapiens* in China by 80,000 years ago. *Nature*, 526: 647–648.

Fang, X.M., Lü, L. Q., Yang, S.L., Li, J.J., An, Z.S., Jiang, P.A. and Chen, X.L. (2002b) Loess in Kunlun Mountain and its implications on desert development and Tibetan Plateau uplift in west China. *Science in China (Series D)*, 45(4): 289–299.

Fang, X., Shi, Z., Yang, S., Yan, M., Li, J. and Jiang, P. (2002a) Loess in the Tian Shan and its implications for the development of the Gurbantunggut Desert and drying of northern Xinjiang. *Chinese Science Bulletin*, 47: 1381–1387.

Florindo, F., Zhu, R. and Guo, B., 1999. Low-field susceptibility and paleorainfall estimates. New data along a N-S transect of the Chinese Loess Plateau. *Physics and Chemistry of the Earth* (A), 24(9): 817–821.

Fu, Q., Meyer, M., Gao, X., Stenzel, U., Burbano, H.A., Kelso, J. and Pääbo, S. (2013) DNA analysis of an early modern human from Tianyuan Cave, China. *Proceedings of the National Academy of Sciences USA*, 110: 2223–2227.

Gibbons, A. (2017) Close relative of Neandertals unearthed in China: partial skulls may belong to elusive Denisovans, who are known almost exclusively by their DNA. *Science*, 355: 899.

Green, R.E., Krause, J., Briggs, A.W., Maricic, T., Stenzel, U. et al. (2010) A draft sequence of Neandertal Genome. *Science*, 328: 710–722.

Green, R.E., Malspinas, A.-S., Krause, J., Briggs, A.W., Johnson, P.L.F. et al. (2008) A complete Neandertal mitochondrial genome sequence determined by high-throughput sequencing. *Cell*, 134: 416–426.

Guan, Q., Pan, B., Li, N., Zhang, J. and Xue, L. (2011) Timing and significance of the initiation of present day deserts in the northeastern Hexi Corridor, China. *Palaeogeography, Palaeoclimatology, Palaeoecology*, 306: 70–74.

Guan, Y., Pearsall, D.M., Gao, X., Chen, F., Pei, S. and Zhou, Z. (2014) Plant use activities during the Upper Paleolithic in east Eurasia: evidence from the Shuidonggou Site, northwest China. *Quaternary International*, 347: 74–83.

Guo, Z., Liu, T., Fedoroff, N., Wei, L., Ding, Z., Wu, N., Lu, H., Jiang, W. and An, Z. (1998) Climate extremes in loess in China coupled with the strength of deep-water formation in the North Atlantic. *Global and Planetary Change*, 18: 113–128.

Harvati, K. (2009) Into Eurasia: a geometric morphometric re-assessment of the Upper Cave (Zhoukoudian) specimens. *Journal of Human Evolution*, 57: 751–762.

Holt, B.G. et al. (2013) An update of Wallace's zoogeographic regions of the world. *Science*, 339: 74–78.

Hu, Y. Shang, H., Tong, H., Nehlich, O., Liu, W., Zhao, C., Yu, J., Wang, C., Trinkaus, E. and Richards, M.P. (2009) Stable isotope dietary analysis of the Tianyuan 1 early modern human. *Proceedings of the National Academy of Sciences USA*, 106(27): 10971–10974, doi:10.1073/pnas.0904826106.

Huerta-Sánchez, E., Jin, X., Asan, E., Zhuoma, B., Peter, B.M. et al. (2014) Altitude adaptation in Tibetans caused by introgression of Denisovan-like DNA. *Nature*, 512: 194–197.

Ji, W., Kuman, K., Clarke, R.J., Forestier, H., Li, Y., Ma, J., Qiu, K., Li, H. and Wu, Y. (2015) The oldest Hoabinhian technocomplex in Asia (43.5 ka) at Xiaodong rockshelter, Yunnan Province, southwest China. *Quaternary International*, 400: 166–174.

Jin, C., Pan, W., Zhang, Y., Cai, Y., Xu, Q., Tang, Z., Wang, W., Wang, Y., Liu, J., Qin, D., Edwards, L. and Hai, C. (2009) The *Homo sapiens* cave hominin site of Mulan Mountain, Jiangzhou District, Chongzuo, Guangxi with emphasis on its age. *Chinese Science Bulletin*, 54: 3848–3856.

Kaifu, Y. and Fujita, M. (2012) Fossil record of early modern humans in East Asia. *Quaternary International*, 248: 2–11.

Kamminga, J. and Wright, R.V.S. (1988) The Upper Cave at Zhoukoudian and the origins of the Mongoloids. *Journal of Human Evolution*, 17: 739–767.

Larramendi, A. (2014) Skeleton of a Late Pleistocene steppe mammoth (*Mammuthus trogontherii*) from Zhalainuoer, Inner Mongolian Autonomous Region, China. *Paläontologische Zeitschrift*, doi:10.1007/s12542-014-0222-8.

Licent, E. and Teilhard de Chardin, P. (1925) Le Paléolithique de la Chine. *L'Anthropologie*, 25: 201–234.

Li, B., Zhang, D.D., Jin, H., Zheng, W., Yan, M., Wu, S., Zhu Y. and Sun, D. (2000) Paleo-monsoon activities of Mu Us Desert, China since 150 ka B.P. – a study of the stratigraphic sequences of the Milanggouwan Section, Salawusu River area *Palaeogeography, Palaeoclimatology, Palaeoecology*, 162: 1–16.

Li, F., Bae, C.J., Ramsey, C.B., Chen, F. and Gao, X. (2018a) Re-dating Zhoukoudian Upper Cave, northern China and its regional significance. *Journal of Human Evolution*, 121: 170–177.

Li, F., Chen, F.-Y. and Gao, X. (2014) "Modern behaviors" of ancient populations at Shuidonggou Locality 2 and their implications. *Quaternary International*, 347: 66–73.

Li, F., Gao, X., Chen, F., Pei, S., Zhang, Y., Zhang, X., Liu, D., Zhang, S., Guan, Y., Wang, H. and Kuhn, S.L. (2013a) The development of Upper Palaeolithic China: new results from the Shuidonggou site. *Antiquity*, 87: 368–383.

Li, F., Kuhn, S.L., Bar-Yosef, O., Chen, F.-Y., Peng, F. and Gao, X. (2019a) History, chronology and techno-typology of the Upper Paleolithic sequence in the Shuidonggou

Area, northern China. *Journal of World Prehistory*, 32: 111–141, doi:10.1007/s10963-019-09129-w.

Li, F., Kuhn, S.L., Chen, F., Wang, Y., Southon, J., Peng, F. et al. (2018b) The easternmost Middle Paleolithic (Mousterian) from Jinsitai Cave, north China. *Journal of Human Evolution*, 114: 76–84.

Li, F., Kuhn, S.L., Gao, X. and Chen, F. (2013b) Re-examination of the dates of large blade technology in China: a comparison of Shuidonggou Locality 1 and Locality 2. *Journal of Human Evolution*, 64: 161–168.

Li, F., Wang, J., Zhou, X., Wang, X., Long, H., Chen, Y., Olsen, J.W. and Chen, F. (2018c) Early Marine Isotope Stage 3 human occupation of the Shandong Peninsula, coastal north China. *Journal of Quaternary Science*, 33(8): 934–944.

Li, F., Vanwezer, N., Boivin, N., Gao, X., Ott, F., Petraglia, M. and Roberts, P. (2019b) Heading north: Late Pleistocene environments and human dispersals in central and eastern Asia. *PLoS ONE*, 14(5): e0216433.

Li, H.-C., Bar-Matthews, M., Chang, Y.-P., Ayalon, A., Yuan, D.-X., Zhang, M.-L. and Lone, M.A. (2017) High-resolution $\delta^{18}O$ and $\delta^{13}C$ records during the past 65 ka from Fengyu Cave in Guilin: variation of monsoonal climates in south China. *Quaternary International*, 441: 117–128.

Li, K., Qin, X., Yang, X., Xu, B., Zhang, L. et al. (2018) Human activity during the late Pleistocene in the Lop Nur region, northwest China: Evidence from a buried stone artifact. *Science China Earth Sciences*, 61: 1659–1668, doi:10.1007/s11430-017-9257-3.

Li, Y.-X., Zhang, Y.-X. and Xu, X.-X. (2012) The composition of three mammal faunas and environmental evolution in the Last Glacial Maximum, Guanzhong area, Shaanxi Province, China. *Quaternary International*, 248: 86–91.

Li, Z., Doyon, L., Li, H., Wang, Q., Zhang, Z., Zhao, Q. and d'Errico, F. (2019) Engraved bones from the archaic hominin site of Lingjing, Henan Province. *Antiquity*, 93: 886–900.

Li, Z. and Ma, H.H. (2016) Techno-typological analysis of the microlithic assemblage at the Xuchang Man site, Lingjing, central China. *Quaternary International*, 400: 120–129.

Li, Z., Wu, X., Zhou, L., Liu, W., Gao, X., Nian, X. and Trinkaus, E. (2017) Late Pleistocene archaic human crania from Xuchang, China. *Science*, 355: 969–972.

Liao, W., Xing, S., Li, D., Martinón-Torres, M., Wu, X. et al. (2019) Mosaic dental morphology in a terminal Pleistocene hominin from Dushan Cave in southern China. *Scientific Reports*, 9: 2347, doi:10.1038/s41598-019-38818-x.

Liu, D., Wang, Y., Cheng, H., Edwards, R.L., Kong, X., Li, T.-Y. (2016) Strong coupling of centennial-scale changes of Asian monsoon and soil processes derived from stalagmite $\delta^{18}O$ and $\delta^{13}C$ records, southern China. *Quaternary Research*, 85: 333–346.

Liu, D.-C., Wang, X.-L., Gao, X., Xia, Z.-K., Pei, S.-W., Chen, F.-Y. et al. (2009) Progress in the stratigraphy and geochronology of the Shuidonggou site, Ningxia, north China. *Chinese Science Bulletin*, 54(21): 3880–3886.

Liu, T. and Ding, Z. (1998) Chinese loess and the paleomonsoon. *Annual Review of Earth and Planetary Sciences*, 26: 111–145.

Liu, W., Jin, C., Zhang, Y., Cai, Y., Xing, S., Wu, X., Cheng, H., Edwards, R.L., Pan, W., Qin, D., An, Z., Trinkaus, E. and Wu, X. (2010a) Human remains from Zhirendong, south China, and modern human emergence in East Asia. *Proceedings of the National Academy of Sciences USA*, 107: 19201–19206.

Liu, W., Martinón-Torres, M., Cai, Y., Xing, S., Tong, H. et al. (2015) The earliest unequivocally modern humans in southern China. *Nature*, 526: 690–700.

Liu, W., Wu, X., Pei, S., Wu, X. and Norton, C.J. (2010b) Huanglong Cave: a Late Pleistocene human fossil site in Hubei Province, China. *Quaternary International*, 211 (1–2): 29–41.

Liu, X., Rolph, T., Bloemendal, J., Shaw, J. and Liu, T. (1995) Quantitative estimates of palaeoprecipitation at Xifeng, in the Loess Plateau of China. *Palaeogeography, Palaeoclimatology, Palaeoecology*, 113: 243–248.

Long, H., Lai, Z.-P., Fuchs, M., Zhang, J.-R. and Li, Y. (2012) Timing of Late Quaternary palaeolake evolution in Tengger Desert of northern China and its possible forcing mechanisms. *Global and Planetary Change*, 92–93: 119–129.

Lu, D., Lou, H., Yuan, K., Wang, X., Wang, Y. et al. (2016) Ancestral origins and genetic history of Tibetan highlanders. *American Journal of Human Genetics*, 99: 580–594.

Lu, H., Sun, X., Wang, S., Cosgrove, R., Zhang, H., Yi, S., Ma, X., Wei, M. and Yang, Z. (2011a) Ages for hominin occupation in Lushi Basin, middle of South Luo River, central China. *Journal of Human Evolution*, 60: 612–617.

Lu, H., Zhang, H., Wang, S., Cosgrove, R., Sun, X., Zhao, J., Sun, D., Zhao, C., Shen, C. and Wei, M. (2011b) Multiphase timing of hominin occupations and the paleoenvironment in Luonan Basin, Central China. *Quaternary Research*, 76: 1142–147.

Lu, Z. (2003) The Jinniushan hominind in anatomical, chronological and cultural context. In C. Shen and S.G. Keates (eds) *Current Research in Chinese Pleistocene Archaeology*, British Archaeological Reports (International Series), 1179: 127–36.

Madsen, D.B., Jingzen, L., Brantingham, P.J., Xing, G., Elston, R.G. and Bettinger, R.L. (2001) Dating Shuidonggou and the Upper Palaeolithic blade industry in north China. *Antiquity*, 75: 706–716.

Madsen, D.B., Oviatt, C.G., Zhu, Y., Brantingham, P.J., Elston, R.G., Chen, F., Bettinger, R.L. and Rhode, D. (2014) The early appearance of Shuidonggou core-and-blade technology in north China: implications for the spread of Anatomically Modern Humans in northeast Asia? *Quaternary International*, 347: 21–28.

Maher, B. and Thompson, R. (1995) Paleorainfall reconstructions from pedogenic magnetic susceptibility variations in the Chinese loess and paleosols. *Quaternary Research*, 44: 383–391.

Maher, B.A., Thompson, R. and Zhou, L.P. (1994) Spatial and temporal reconstructions of changes in the Asian palaeomonsoon: a new mineral magnetic approach. *Earth and Planetary Science Letters*, 125: 461–471.

Martí, A.P., Wei, Y., Gao, X., Chen, F. and d'Errico, F. (2017) The earliest evidence of coloured ornaments in China: the ochred ostrich eggshell beads from Shuidonggou Locality 2. *Journal of Anthropological Archaeology*, 48: 102–113.

Martinón-Torres, M., Wu, X., Bermúdez de Castro, J.M., Xing, S. and Liu, W. (2017) *Homo sapiens* in the Eastern Asian Late Pleistocene *Current Anthropology*, 58(Supplement 17): S434–S448.

Martinón-Torres, M., Xing, S. Liu, W. Bermúdez de Castro, J.M. (2016) A "source and sink" model for East Asia? Preliminary approach through the dental evidence. *Comptes Rendues Palévolution*, 17(1–2): 33–43, doi:10.1016/j.crpv.2015.09.011.

Michel, V., Valladas. H., Shen, G., Wang, W., Zhao, J.-X., Shen, C.-C., Valensi, P. and Bae, C.J. (2016) The earliest modern *Homo sapiens* in China? *Journal of Human Evolution*, 101: 101–104.

Morgan, C., Barton, L., Yi, M., Bettinger, R.L., Gao, X. and Peng, R. (2014) Redating Shuidonggou Locality 1 and implications for the initial Upper Paleolithic of East Asia. *Radiocarbon*, 56: 1–15.

Movius, H.L. (1948) The Lower Palaeolithic cultures of southern and eastern Asia. *Transactions of the American Philosophical Society*, 38(4): 329–420.

Nian, X., Gao, X. and Zhou, L. (2014) Chronological studies of Shuidonggou (SDG) Locality 1 and their significance for archaeology. *Quaternary International*, 347: 5–11.
Niu, D., Pei, S., Zhang, S., Zhou, Z., Wang, H. and Gao, X. (2016) The Initial Upper Palaeolithic in northwest China: new evidence of cultural variability and change from Shuidonggou locality 7. *Quaternary International*, 400: 111–119.
Norton, C.J. and Gao, X. (2008a) Hominin-carnivore interactions during the Chinese Early Paleolithic: Taphonomic perspectives from Xujiayao. *Journal of Human Evolution*, 55: 164–178.
Norton, C.J. and Gao, X. (2008b) Upper Cave revisited. *Current Anthropology*, 49: 732–745.
Norton, C.J., Jin, C., Wang, Y. and Zhang, Y. (2010) Rethinking the Palearctic-Oriental biogeographic boundary. In C.J. Norton and D.R. Braun (eds) *Asian Paleoanthropology from Africa to China and Beyond*, New York: Springer Business+Media, pp. 81–100.
Peng, F., Gao, X., Wang, H., Chen, F., Liu, D. and Pei, S.W. (2012) An engraved artifact from Shuidonggou, an Early Late Paleolithic site in northwest China. *Chinese Science Bulletin*, 57: 4594–4599.
Peng, F., Guo, J., Lin, S., Wang, H. and Gao, X. (2018) The onset of Late Paleolithic in north China: an integrative review of the Shuidonggou site complex, China. *L'Anthropologie*, 122(1): 74–86, doi:10.1016/j.anthro.2018.01.006.
Peng, F., Wang, H. and Gao, X. (2014) Blade production of Shuidonggou Locality 1 (northwest China): a technological perspective. *Quaternary International*, 347: 12–20.
Qin, Z., Yang, Y., Kang, L., Yan, S., Cho, K. et al. (2010) A mitochondrial revelation of early human migrations to the Tibetan Plateau before and after the Last Glacial Maximum. *American Journal of Physical Anthropology*, 143: 555–569.
Qu, T., Bar-Yosef, O., Wang, Y. and Wu, X. (2013) The Chinese Upper Paleolithic: geography, chronology, and techno-typology. *Journal of Archaeological Research*, 21: 1–73, doi:10.1007/s10814-012-9059-4.
Reich, D., Patterson, N., Kircher, M., Delfin, F., Nandineni, M.R. et al. (2011) Admixture and the first modern human dispersals into Southeast Asia and Oceania. *American Journal of Human Genetics*, 89: 516–528.
Rightmire, G.P. (1998) Human evolution in the Middle Pleistocene: the role of *Homo heidelbergensis*. *Evolutionary Anthropology*, 6: 218–227.
Rightmire, G.P. (2001) Comparison of Middle Pleistocene hominids from Africa and Asia. In L. Barham and K. Robson-Brown (eds) *Human Roots: Africa and Asia in the Middle Pleistocene*. Bristol: Western Academic and Specialist Press Ltd, pp. 123–133.
Shang, H., Tong, H., Zhang, S., Chen, F. and Trinkaus, E. (2007) An early modern human from Tianyuan Cave, Zhoukoudian, China. *Proceedings of the National Academy of Sciences USA*, 104: 6573–6578.
Shen, W., Wang, Q., Zhao, J., Collerson, K., Zhou, C. and Tobias, P.V. (2002) U-series dating of Liujiang hominid site in Guangxi, southern China. *Journal of Human Evolution*, 43: 817–29.
Shi, Y., Yu, G. Liu, S., Li, B. and Yao, T. (2001) Reconstruction of the 30–40 ka BP enhanced Indian monsoon climate based on geological records from the Tibetan Plateau. *Palaeogeography, Palaeoclimatology, Palaeoecology*, 169: 69–83.
Shichi, K., Kawamuro, K., Takahara, H., Hase, Y., Maki, T. and Miyoshi, N. (2007) Climate and vegetation changes around Lake Baikal during the last 350,000 years. *Palaeogeography, Palaeoclimatology, Palaeoecology*, 248(3–4): 357–375.
Slon, V., Mafessoni, F., Vernot, B., Filippo, C. de, Grote, S. et al. (2018) The genome of the offspring of a Neanderthal mother and a Denisovan father. *Nature*, 561: 113–116.

Smith, F.H., Ahern, J.C., Janković, I. and Karavanić, I. (2017) The Assimilation Model of modern human origins in light of current genetic and genomic knowledge. *Quaternary International*, 450: 126–136.

Stringer, C. (2002) Modern human origins: progress and prospects. *Philosophical Transactions of the Royal Society B*, 357: 563–579.

Stringer, C.B. (2016) The origin and evolution of *Homo sapiens*. *Philosophical Transactions of the Royal Society B*, 371: 20150237.

Sun, X., Li, Y.H., Feng, X.B., Lu, C.Q., Lu, H.Y., Yi, S.W., Wang, S.J. and Wu, S.Y. (2017a) Early human settlements in the southern Qinling Mountains, central China. *Quaternary Science Reviews*, 164: 168–186.

Sun, X., Li, X., Luo, Y. and Chen, X. (2000) The vegetation and climate at the last glaciation on the emerged continental shelf of the South China Sea. *Palaeogeography, Palaeoclimatology, Palaeoecology*, 160(3–4): 301–316.

Sun, X., Lu, H.Y., Wang, S.J. and Yi, S.W. (2012) Ages of Liangshan Paleolithic sites in Hanzhong Basin, central China. *Quaternary Geochronology*, 10: 380–386.

Sun, X., Lu, H., Wang, S., Xu, X., Zeng, Q., Lu, X., Lu, C., Zhang, W., Zhang, X. and Dennell, R. (2018) Hominin distribution in glacial-interglacial environmental changes in the Qinling Mountains range, central China. *Quaternary Science Reviews*, 198: 37–55.

Sun, X., Luo, Y., Huang, F., Tian, J. and Wang, P. (2003) Deep-sea pollen from the South China Sea: Pleistocene indicators of East Asian monsoon. *Marine Geology*, 201: 97–118.

Sun, X., Song, C., Wang, F. and Sun, M. (1997) Vegetation history of the Loess Plateau of China during the last 100,000 years based on pollen data. *Quaternary International*, 37: 25–36.

Sun, X.F., Yi, S.W., Lu, H.Y. and Zhang, W.C. (2017b) TT-OSL and post-IR IRSL dating of the Dali Man site in central China. *Quaternary International*, 434: 99–106.

Takahashi, K., Wei, G., Uno, H., Yoneda, M., Jin, C., Sun, C., Zhang, S. and Zhong, B. (2007) AMS ^{14}C chronology of the world's southernmost woolly mammoth (*Mammuthus primigenius* Blum.). *Quaternary Science Reviews*, 26: 954–957.

Thorne, A. and Wolpoff, M. (1992) The multiregional evolution of humans. *Scientific American*, 266(4): 76–83.

Tong, H. (2007) Occurrences of warm-adapted mammals in north China over the Quaternary period and their paleo-environmental significance. *Science in China D*, 50: 1327–1340.

Tong, H. (2008) Quaternary *Hystrix* (Rodentia, Mammalia) from north China: taxonomy, stratigraphy and zoogeography, with discussions on the distribution of *Hystrix* in Palearctic Eurasia. *Quaternary International*, 179: 126–134.

Trinkaus, E. and Shang, H. (2008) Anatomical evidence for the antiquity of human footwear: Tianyuan and Sunghir. *Journal of Archaeological Science*, 35: 1928–1933.

Tu, H., Shen, G., Li, H., Xie, F. and Granger, D.E. (2015) ^{26}Al/^{10}Be burial dating of Xujiayao-Houjiayao site in Nihewan Basin, northern China. *PLoS ONE*, 10(2): e0118315, doi:10.1371/journal.pone.0118315.

Wallace, A.R. (1876) *The Geographical Distribution of Animals, with a Study of the Relations of Living And Extinct Faunas as Elucidating The Past Changes of the Earth's Surface*. London: Macmillan.

Wang, Y. (2017) Late Pleistocene human migrations in China. *Current Anthropology*, 58 (Supplement 17): S504–S513.

Wang, Y.J., Cheng, H., Edwards, R.L., An, Z.S., Wu, J.Y., Shen, C.-C. and Dorale, J.A. (2001) A high-resolution absolute-dated late Pleistocene monsoon record from Hulu Cave, China. *Science*, 294: 2345–2348.

Wang, Y.J., Cheng, H., Edwards, R.L., Kong, X., Shao, X., Chen, S., Wu, J., Jiang, X., Wang, X. and An, Z. (2008) Millennial- and orbital-scale changes in the east Asian monsoon over the past 224,000 years. *Nature*, 451: 1090–1093.

Wei, K.-Y., Chiu, T.-C. and Chen, Y.-G. (2003) Toward establishing a maritime proxy record of the East Asian summer monsoons for the late Quaternary. *Marine Geology*, 201: 67–79.

Wei, Y., d'Errico, F., Vanhaeren, M., Li, F. and Gao, X. (2016) An early instance of upper Palaeolithic personal ornamentation from China: the freshwater shell bead from Shuidonggou 2. *PLoS ONE*, 11(5): e0155847, doi:10.1371/journal.pone.0155847.

Wei, Y., d'Errico, F., Vanhaeren, M., Peng, F., Chen, F. and Gao, X. (2017) A technological and morphological study of Late Paleolithic ostrich eggshell beads from Shuidonggou, north China. *Journal of Archaeological Science*, 85: 83–104.

Weidenreich, F. (1946) *Apes, Giants and Men*. Chicago: University of Chicago Press.

Wu, X. and Athreya, S. (2013) A description of the geological context, discrete traits, and linear morphometrics of the Middle Pleistocene hominin from Dali, Shaanxi Province, China. *American Journal of Physical Anthropology*, 150: 141–157.

Wu, X. and Poirier, F.E. (1995) *Human Evolution in China: A Metric Description of the Fossils and a Review of the Sites*. New York/Oxford: Oxford University Press.

Wu, X.J. and Trinkaus, E. (2014) The Xujiayao 14 mandibular ramus and Pleistocene *Homo* mandibular variation. *Comptes Rendues Palévolution*, 13: 333–341.

Xing, S., Martinón-Torres, M., Bermúdez de Castro, J.M., Wu, X. and Liu, W. (2015) Hominin teeth from the early Late Pleistocene site of Xujiayao, northern China. *American Journal of Physical Anthropology*, 156(2): 224–240.

Yang, S. and Ding, Z. (2006) Winter–spring precipitation as the principal control on predominance of C_3 plants in Central Asia over the last 1.77 Myr: evidence from $\delta^{13}C$ of loess organic matter in Tajikistan. *Palaeogeography, Palaeoclimatology, Palaeoecology*, 235: 330–339.

Yang, M.A., Gao, X., Theunert, C., Tong, H.-W., Aximu-Petri, A., Nickel, B., Slatkin, M., Meyer, M., Pääbo, S., Kelso, J. and Fu, Q.-M. (2017) 40,000-year-old individual from Asia provides insight into early population structure in Eurasia. *Current Biology*, 27: 3202–3208.

Yi, M., Barton, L., Morgan, C., Liu, D., Chen, F., Zhang, Y., Pei, S., Guan, Y., Wang, H., Gao, X. and Bettinger, R. (2013) Microblade technology and the rise of serial specialists in north-central China. *Journal of Anthropological Archaeology*, 32: 212–223.

Yi, M., Bettinger, R.L., Chen, F., Pei, S. and Gao, X. (2014) The significance of Shuidonggou Locality 12 to studies of hunter-gatherer adaptive strategies in north China during the Late Pleistocene. *Quaternary International*, 347: 97–104.

Yi, M., Gao, X., Li, F. and Chen, F. (2016) Rethinking the origin of microblade technology: a chronological and ecological perspective. *Quaternary International*, 400: 130–139.

Yu, G., Harrison, S.P. and Xue, B. (2001) Lake status records from China: Data Base Documentation. *Max Planck Institute for Biogeochemistry Technical Report*, 4: 1–247.

Yuan, D.X., Cheng, H., Edwards, R.L., Dykoski, C.A., Kelly, M.J., Zhang, M., Qing, J., Lin, Y., Wang, Y., Wu, J., Dorale, J.A., An, Z. and Cai, Y. (2004) Timing, duration, and transition of the last interglacial Asian monsoon. *Science*, 304: 575–578.

Zhang, J.-F. and Dennell, R.W. (2018) The last of Asia conquered by *Homo sapiens*: excavation reveals the earliest human colonization of the Tibetan Plateau. *Science*, 362: 992–993.

Zhang, S., d'Errico, F., Backwell, L.R., Zhang, Y., Chen, F. and Gao, X. (2016) Ma'anshan Cave and the origin of bone tool technology in China. *Journal of Archaeological Science*, 65: 57–69.

Zhang, H.C., Peng, J.L., Ma, Y.Z., Chen, G.J., Feng, Z.-D., Li, B., Fan, H.F., Chang, F. Q., Lei, G.L. and Wünnemann, B. (2004) Late Quaternary palaeolake levels in Tengger Desert, NW China. *Palaeogeography, Palaeoclimatology, Palaeoecology*, 211: 45–58.

Zhang, X.L., Ha, B.B., Wang, S.J., Chen, Z.J., Ge, J.Y. et al. (2018) The earliest human occupation of the high-altitude Tibetan Plateau 40 thousand to 30 thousand years ago. *Science*, 362: 1049–1051.

Zheng, Z. and Lei, Z.-Q. (1999) A 400,000 year record of vegetational and climatic changes from a volcanic basin, Leizhou Peninsula, southern China. *Palaeogeography, Palaeoclimatology, Palaeoecology*, 145(4): 339–362.

11

HUMANS ON THE EDGE OF ASIA

The Arctic, Korean Peninsula and the Japanese islands

Introduction

The colonisation of the Arctic and the Japanese islands effectively completed the colonisation of Asia by our species. Both are unmistakable examples of our prowess as a colonising species that proved itself adaptable to every climate or environment that it encountered. The colonisation of the Arctic and Japanese islands also provided the stepping stones by which the Americas were colonised at the end of the last ice.

Inside the Arctic Circle

The gateways to the Arctic would have been the great river systems of the Ob, Lena, Yenesi, Angara and others that flow northwards from southern Siberia and could be used by boat, or as an ice highway when frozen. Even now, in the present interglacial and with modern technology, the Far North and the Arctic (the region inside the Arctic Circle at 66° N.) are among the least hospitable regions of the planet. For Palaeolithic populations, the Far North provided a brutal test of their survival strategies. In addition to the extreme cold of an Arctic winter, humans also had to adapt to a world in which there was almost no daylight in mid-winter. Yet the Mammoth Steppe of the Far North was teeming with animals (Guthrie 1990), and probably comparable in biomass to the African savannah (see Zimov et al. 2012). It also offered an irresistible resource that would greatly increase the prospects for survival. This was the woolly mammoth, the most iconic ice-age mammal. Mammoths were the ultimate Pleistocene provider; they were a mobile super-store that could provide literally tons of meat, fat and marrow (Boschian et al. 2019); bones for making houses, handles or other tools, including ones that were normally made from stone (see Boschian and Saccà 2015); they provided dung and bone for fuel; hair for making felt, rope or snares; skin for tents; gut as a multi-purpose material; tallow for lamps; and they even provided a network of paths by their repeated use of regular routes. As with other herbivores, their stomach contents might have been a vitamin-rich delicacy (especially useful in winter when plant foods are otherwise unavailable), and the stomach itself could be used as a tough, water-proof bag (Buck and Stringer 2013). Above all, they provided ivory. In a treeless landscape, ivory became the ideal substitute for wood that could be used for making spears as well as the material of choice for making beads, sculptures and other non-utilitarian items. Although other animals such as woolly rhinoceros, bison and reindeer were hunted, none came near mammoth as an all-round provider: "in the elephant, everything is good" (Boschian and Saccà 2015: 288). Excuse the pun: Palaeolithic hunters in the Arctic ran a mammoth economy.

In European Russia, Neandertals probably ventured up to 60° N. (roughly the latitude of Perm), where modern winters average –19° C., with 170–185 days of snow per year. Those conditions probably marked the limits of their ecological tolerance (Pavlov et al. 2004). Sites in European Russia such as Mamontovaya Kurya (66° N., 36 ka) (Pavlov et al. 2001), Byzovaia[1] (65° N., ca. 28 ka) and Zaozer'e (65° N., 30–33 ka) were almost certainly created by our species in their pursuit of mammoths (Pavlov et al. 2004), which constitute most of the faunal remains at those sites.

In Siberia, there are at least three indications that humans were hunting mammoths inside the Arctic Circle before 40,000 years ago. At this time, when sea levels were 50 m to 80 m below present levels, the Arctic coastline extended beyond the New Siberian Islands. Summer temperatures may have been 4–5° C. warmer than now, and precipitation perhaps 50–100 m higher (see Figure 11.1), but winters would of course have been extremely severe. At this latitude, there was also no escaping the mid-winter months of near total darkness. The first site is on the Tamyr Peninsula and is known as the SK site (Sopochnaya Karga) (see Figure 11.2). Here, at 71° N., the skeleton of a mammoth was found that showed injuries before death and cut-marked bone (Pitulko et al. 2016, 2017). Two ribs had a notch-like lesion consistent with ones caused by spears, and post-mortem damage was also evident on the tip of one tusk and the mandible. The skeleton was dated at ca. 40–45 ka BP. In the light of what we know from European evidence (see Pavlov et al. 2004), it is most unlikely that Neandertals or Denisovans were hunting this far north, and it is therefore more probable that it was our species that killed the SK mammoth. There is also an assemblage from a site known as Bunge-Toll 1885, where remains of woolly mammoth, woolly rhinoceros and bison were found, as well as a wolf humerus that had a puncture-cut wound that was likely caused by a human projectile. The humerus was dated to 44,650 + 950/–700 BP, which is close in age to a date of 47,600 + 2,600/–2,000 BP for the mammoth remains (Pitulko et al. 2017). What we don't know

Stratigraphic units of Western Beringia				Palaeoenveronments	
Ages [kyr]	Nomenclature		Marine Isotope Stages (MIS)	Vegetation	
	Siberian	North American		West part Laptev Sea region, Yana-Idighirka lowland	East part Chukotka
11.0 - 0 ka BP	Holocene	Holocene	MIS 1	After ca. 10.3 kyr BP- shrub tundra and wetland vegetation; after 5 ^{14}C kyr BP -grass tundra, dwarf birches in protected places	After 10 ^{14}C kyr BP- mosaic of subshrubs-graminoid tundra and Betula forest-tundra; after ca 4.5 ^{14}C kyr BP- herb tundra
24.0 - 11.0 ka BP	Sartan	Late Wisconsinan	MIS 2	Open arctic pioneer and steppe herb-dominated tundra- and steppe-like environments, and after ca. 13 kyr BP- grass-sedge dominated habitats, shrubs in protected places	Herb-dominated tundra, and herbshrub tundra; abundant steppe communities
57.0 - 24.0 ka BP	Karginsk	Middle Wisconsinan	MIS 3	Open hemicryophyte tundra-steppe herb dominated vegetation, shrub willow stands in more protected places and wetland vegetation in riparian sites with a stable high water level	Mix of shrub and herb-shrub tundra that occasionally included tree Betula and Pinus pumila. River valleys supported forest-tundra with Betula, chosenia, and Larix

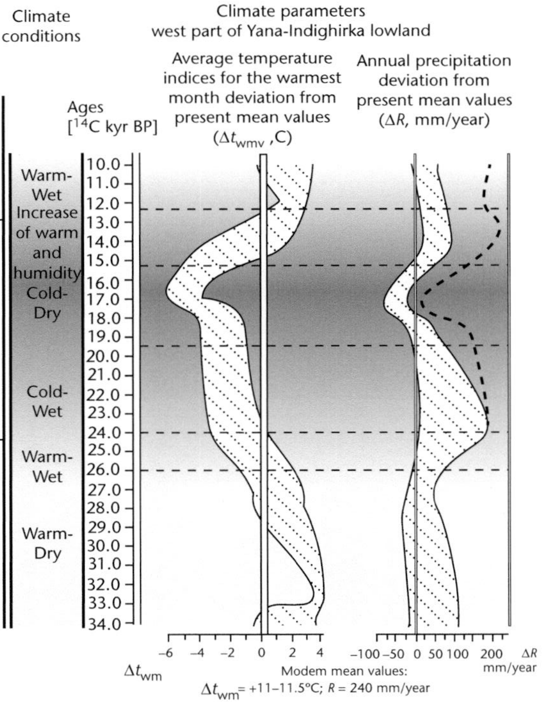

Climate conditions
Warm-Wet
Increase of warm and humidity
Cold-Dry
Cold-Wet
Warm-Wet
Warm-Dry

Ages [^{14}C kyr BP]
10.0 11.0 12.0 13.0 14.0 15.0 16.0 17.0 18.0 19.0 20.0 21.0 22.0 23.0 24.0 25.0 26.0 27.0 28.0 29.0 30.0 31.0 32.0 33.0 34.0

Climate parameters west part of Yana-Indighirka lowland
Average temperature indices for the warmest month deviation from present mean values (Δt_{wmv} ,C)
Annual precipitation deviation from present mean values (ΔR, mm/year)
−6 −4 −2 0 2 4
Δt_{wm}
−100 −50 0 50 100 200 ΔR mm/year
Modem mean values:
Δt_{wm}= +11–11.5°C; R = 240 mm/year

FIGURE 11.1 Stratigraphic units and palaeo-environments of Arctic Siberia and Western Beringia
Source: Pitulko et al. 2017, Fig. 2.

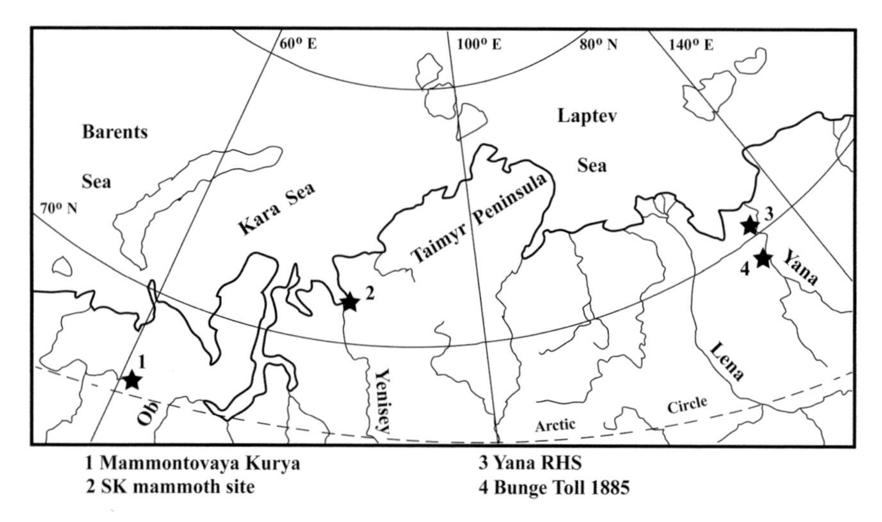

FIGURE 11.2 Palaeolithic sites in the Arctic
Source: The author.

is whether these two sites denote rare events or were already part of a regular pattern of mammoth hunting by humans in the Far North. What it probably does imply is that humans 40,000 years ago were regularly hunting north of latitude 60° N. – in other words, much further north than Neandertals appear to have managed.

Conclusive evidence of mammoth hunting comes from the site known as Yana RHS (rhino horn site). This also lies at 71° N. This site contains an enormous amount of faunal remains, stone tools, and ones made of bone and ivory (see Figure 11.3). Although there is no indication of any structures or hearths, a truly impressive number of ornaments were found. These include over 1,500 beads of ivory or hare bones, pendants made from perforated teeth, and numerous incised items. Also noteworthy are a series of eyed needles (among the oldest yet found) and some exquisite ivory vessels. There were also two ivory spear fore-shafts (onto which a stone or ivory tip could be attached) and one of rhinoceros horn, which is strong but also flexible and less likely to break. The presence of amber and anthraxolite (a soft jet-black mineral) that may have come from the New Siberian Islands 600 km north of Yana shows how extensive social networks and territories must have been (Pitulko et al. 2012). The stone tools at Yana were made from locally obtained pebbles that were mainly used for making chopping tools and scrapers. The site was used on at least three occasions between 27,000 and 29,000 years ago. Mammoth, horse, reindeer, bison, woolly rhinoceros, hare and birds (unspecified) were identified in the faunal assemblage from the cultural layer, and musk-ox, wolf, arctic fox, lion and a wolverine were present in adjacent assemblages that included bones that had been burnt or scraped. (These animals all provide good skins or fur but the

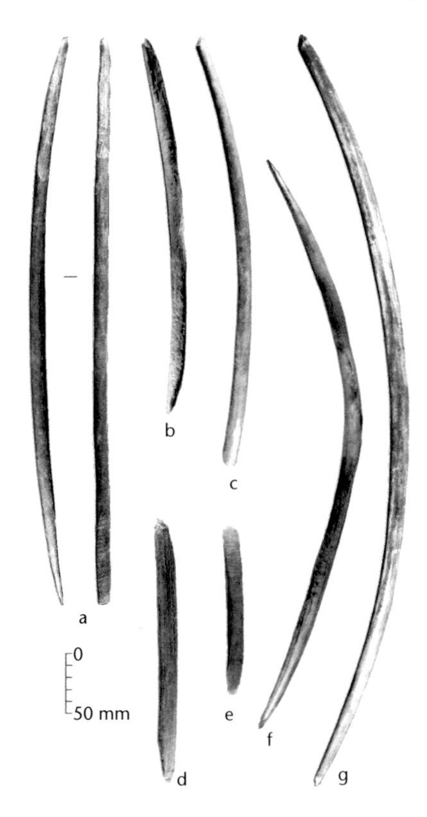

FIGURE 11.3 Ivory and bone tools from the Yana RHS site

Key: a, d and e – foreshafts; b, f, g – spear points; c – part of a long spear point. (a, d – woolly rhinoceros horn; b, c, e–g – mammoth ivory).

Source: Pitulko et al. 2017, Fig. 7.

best is the under-wool of musk-ox [or qiviut] which is the finest in the world). Of these animals, only reindeer and wolf still inhabit the area (Pitulko et al. 2004). Because mammoth bones formed only 3% of the faunal assemblage, the initial conclusion was that mammoth hunting was a negligible part of the group's subsistence. This view was abandoned after the discovery nearby of what is called the Yana Mass Accumulation of Mammoth (YMAM) site that was found by local miners searching for ivory (Basilyan et al. 2011). Only a small part could be salvaged, but the assemblage of ca. 1,000 bones that was recovered included the remains of 26 mammoths[2] and a few bones of woolly rhinoceros, bison, horse, reindeer and bear as well as side-scrapers, discoidal cores and flakes. Radiocarbon dating showed that this assemblage was the same age as the Yana RHS site. When the RHS and YMAM assemblages are combined, they show that mammoth was a prime target. Mammoth tongue bones (or hyoids) in the cultural layer show that mammoth tongue was eaten, presumably as a delicacy. Holes caused by spears in mammoth shoulder blades and pelves (and spear fragments embedded in some) leave no doubt that these animals were hunted. (And, presumably, large amounts of meat were preserved for winter consumption by smoking or freezing.) However, because mammoth survived into the Holocene in Siberia, with the last recorded on Wrangel Island dated to only 3,600 BP (uncalibrated) (see Vartanyan et al. 1993), its hunting was sustainable as a long-term strategy (Nikolskiy et al. 2011; Nikolskiy and Pitulko 2013). Large numbers of Arctic hare were also procured (presumably by nets, traps or snares) for their fur for items such as socks, gloves and undershirts. In the ethnographic present, the Thule Greenlanders would trap ca. 1,000–1,500 hares each year for their fur but rarely ate them (see Pitulko et al. 2004, 2017). Two human milk teeth confirm that our species was present; DNA extracted from these also provides clues about the first humans to reach North America (see Sikora et al. 2019 and below).

Siberian Palaeolithic dogs?

Before leaving Siberia and Mongolia, we should note an ongoing debate over whether dogs were already domesticated before 30,000 years ago in Siberia and eastern Europe. The claims rest on skeletal specimens from the cave of Razboinichya in the Rusian Altai Mountains at 33.5 ka (Ovodov et al. 2011) and others from Goyet, Belgium, at 36.5 ka BP; Predmosti, Czech Republic, at 31 ka; and Kostienki 8, Ukraine, at ca. 33.5–26.5 ka BP (Germonpré et al. 2009, 2012). The Razboinichya specimens are described as "incipient dogs", but the others are described as dogs, i.e. morphologically distinguishable from the wolves at those sites. The criteria used for these claims are primarily metrical (dogs tend to be smaller than wolves) and morphological (snouts are shorter, tooth rows are more crowded, for example). Ancient DNA analysis is also claimed to show that these dogs were closer to modern dogs and prehistoric New World canids than to modern wolves (Druzkhova et al. 2013). Shipman (2015a, 2015b) has developed these claims to argue that the combination of stone projectile technology and domestic dogs gave humans a major competitive edge over Neandertals in the hunting of large game (such as mammoth), and this was a key factor in Neandertal extinction.

These claims have attracted widespread criticism. The primary one is methodological: that the criteria used to distinguish dog from wolf are inadequate and fail to take into account the widespread variability in wolf populations (Perri 2016; Janssens et al. 2019). Much also depends upon the choice of comparatives of both dogs and modern wolves. The claimed domestic dogs might simply be, for example, short-faced wolves and fall within modern wolf variation (Crockford and Kuzmin 2012; Boudadi-Maligne and Escarguel 2014; Morey 2014). (In fairness, Germonpré et al. [2013, 2015] contest much of this criticism, so the debate is clearly not over.) Karen Lupo (2017) also questions the claimed advantages of hunting with dogs: citing ethnographic literature, she shows that much depends on the type of animal being hunted, the number of dogs involved, type of vegetation (whether dense or open for example), and whether prey is killed quickly or dies from its wounds some distance from the hunter.

More research is needed before we should accept claims that dogs were domesticated before 30,000 years ago. We might also consider the domestic context of these alleged domestic dogs. Razboinichya Cave, for example, has a rich faunal assemblage and appears to have been used primarily as a hyaena den, as well as by bear, fox and wolf. No artefacts are present, but traces of fire supposedly indicate that humans used the cave occasionally. It is nevertheless an odd context for an "incipient dog", especially when dogs/wolves are either absent or very rare at the Siberian sites that have been considered in Chapter 9. Too much attention has perhaps been paid to the role of dogs for hunting. In Mongolia, for example, it is hard to see how dogs would have made the hunting of gazelle any easier, and they might even have been a hindrance in encouraging them to stampede. Dogs have many other uses: in many regions, dogs are primarily sentries that deter unwelcome intruders, whether human or another predator; they can also be used as pack animals, or for pulling sledges. If they were used in that way, it might be useful to consider leg and shoulder strength instead of jaw shape as evidence of domestication. Depending upon preference, they can also, of course, be pets, or part of the menu. Until stronger evidence is forthcoming that the skull and mandibles from Razboinichya represent an "incipient dog", the possibility that human populations in Siberia and Mongolia were using domestic dogs before 30,000 years ago can probably be disregarded.

Korea and the Japanese islands

We can turn now to the Korean Peninsula and Japanese islands. The Korean Peninsula is interesting in several ways: first, its geographical position meant that it was open to immigration from modern-day China to the west and Russia to the north; second, the first colonists in PalaeoHonshu (the conjoined islands of Kyushu, Honshu and Shikoku) probably originated in its nearest neighbour across the Tsushima Strait; and third, Korea played a major role in the exchange networks of obsidian that developed in northeast Asia (Kuzmin 2017). Finally, Korea is interesting in showing that there is no "package" of new features that is equivalent to the Upper Palaeolithic in western Europe. Recent overviews of East Asia that include the Korean Peninsula are by Bae (2017), Lee (2015), Norton (2000) and Norton and Jin (2009).

The Korean Peninsula is cool and dry in the north, and warmer and wetter in the south. Modern vegetation reflects this gradation, with spruce (*Picea jezoensis*), pine (*Pinus* sp.) and birch (*Betula platyphlla* and *B. chinensis*) the main forest components in the north, temperate broad-leafed forest in the centre and subtropical evergreen broad-leaved forest in the south. In the last glaciation, the northern forests and steppe expanded southwards when the climate deteriorated. During the glacial maximum, annual average temperatures in central Korea were 5–6° C. cooler than now, and 40% drier. Conditions are thought to have been comparable to northern Mongolia today, with *Artemisia* steppe and steppe-forest of *Picea* (spruce), *Abies* (fir) and *Betula* (birch) as the main vegetation (see Chapter 9) (Chung et al. 2006; Yi and Kim 2010; Kim et al. 2015). The exposed floor of the Yellow Sea would also have been primarily a land of *Artemisia* steppe (see Chapter 10).

At present, there are ca. 100 Palaeolithic sites in Korea, most of which are in the southern part (Bae 2010). There is a faint but definite hominin presence in Korea that precedes human occupation, and is best represented by bifaces at sites such as Jeongok-ri (Chongnokni) that may date to the late Middle Pleistocene (Norton et al. 2006; Bae 2010). The Late (or Upper) Palaeolithic in Korea (see Box 11.1) developed after 40 ka. Some of the best-known sites are summarised in Table 11.1. The main features in Korea are that blades appear ca. 40,000 years ago, and microblades after 24,000 BP. Core and flake tools that were used before 40 ka continued in use (Bae 2017; Lee 2015). As in Japan, ground edge axes and tanged points were used in the Late Palaeolithic; the earliest ground axe is from Galsanri, dated to 30–40 ka BP, and the earliest tanged point is from Yongho-Dong, dated at 38.5 ka BP (Bae 2017).

TABLE 11.1 Radiocarbon-dated Late Palaeolithic sites from the Korean Peninsula

Site	*N. lithics*	*Assemblage*						*Age BP*
		Ct	Fl	Bt	Mbt	Tp	Ot	
Yongho-dong 2	662	x	x	x		x		38,500 ± 1,000
Yongho-dong 3	975	x	x	x		x		31,200 ± 900
								30,000 ± 1,400
Hwadae-ri 2	3,709	x	x	x		x		31,200 ± 900
Hopyeong-dong 1	3,023	x	x	x		x		30,000–27,000
Hopyeong-dong 2	4,761	x	x	x			x	24,000–16,000
Sinbuk	ca. 31,000	x	x	x	x	x	x	25,500–18,500
Jangheung-ri	664	x	x	x	x		x	24,400–24,200
Yongsan-dong	2,228	x	x	x		x		24,430–19,310
Jingeunul	ca. 12,000	x	x	x	x	x	x	22,850–17,310
Seokjang-ri	??	x	x	x	x	x	x	20,830 ± 1,880
Haga	ca. 27,000	x	x	x	x	x		19,700–19,500
Suyanggae	ca. 27,000	x	x	x	x	x	x	18,600–15,350
Hahwagye	2,267	x	x		x		x	13,390 ± 60

Ct = core tool; Fl = flake; Bt = blade tool; Mbt = micro-blade tool; Tp = tanged point; Ot = Obsidian tool.

Source: Lee 2015, Table 20.1.

Box 11.1 The problem with names

One unfortunate legacy of the French dominance of Palaeolithic archaeology in the early 20th century was the way its framework was copied in regions such as East Asia (Bae 2017). Palaeolithic sites in western Europe are still classed as Lower, Middle or Upper Palaeolithic but these periods are inappropriate for China, Korea or Japan. Gao and Norton (2002) argued that there is no equivalence in China of the Middle Palaeolithic and Seong and Bae (2016) have suggested that there is no convincing Middle Palaeolithic in Korea. The situation in Japan is somewhat confusing as some researchers regard assemblages that are more than 30 ka as Middle Palaeolithic but most regard those younger than 40,000 BP as Upper Palaeolithic (Bae 2017). The debate is not wholly concluded as Jinsitai in Inner Mongolia (Li et al. 2018) contains assemblages identified as Mousterian, and therefore as Middle Palaeolithic; and Kei (2012) maintains that the notion of a Chinese Middle Palaeolithic is still valid.

The concept of an Upper Palaeolithic is problematic in East Asia if defined by the appearance of prismatic blades and shaped bone tools. In north China, there are grounds for identifying an Initial Upper Palaeolithic (as at Shuidongdou 1), but in south China, lithic traditions continue as before. In Japan, the earliest assemblages are better defined by trapeze flakes, ground edge axes and pit traps, and in Korea, there is no sudden break with preceding traditions. My own view is it makes sense to think about an Early and Late Palaeolithic in East Asia, with a boundary at ca. 40 ka for south and central China, Korea and Japan that probably coincides with the appearance of our species. It is somewhat unfortunate that the Late Palaeolithic is sometimes divided into an Early Late Palaeolithic and a Late Late Palaeolithic, divided by the appearance of microblades. These are tolerable providing that no further subdivisions arise (such as Early Middle Late Palaeolithic).

Source: Bae 2017; Gao and Norton 2002; Kei 2012; Li et al. 2018.

There have been two contrasting interpretations of the Korean Late Palaeolithic. Some researchers have argued that there were four separate lithic traditions in Korea based on flake tools with heavyweight tools, flake tools, blades and microblades whereas others such as Lee (2015) prefer a single unilinear tradition that involved a variety of tool types of techniques. This interpretation seems reasonable given what we know about how hunter-foragers use different tool-kits according to season, location, local conditions, raw material availability, type of site and so on. Various views have emerged on how and from where the Late Palaeolithic in Korea developed. Some have suggested that it was an in-situ development, but this suggestion is at odds with the evidence from north

China, Mongolia and Transbaikalia that it involved immigration (Chapter 9). Immigration from the north is likely, especially when the climate deteriorated as that would have forced people to move into more favourable areas. Immigration from the south across the exposed floor of the Yellow Sea seems improbable as people are unlikely to move into a harsher environment.

Bae and Bae (2012: 33) make an important point about the Korean Late Palaeolithic: "if modern humans swept into eastern Asia carrying with them a superior stone tool industry and replacing the indigenous peoples, they do not appear to have done much sweeping in the Korean peninsula". As they point out, the Early Late Palaeolithic sites contain only a few blades and tanged points, and lithic assemblages continued to contain large numbers of the type of core and flake tools of the Early Palaeolithic. It was not until after 30,000 BP that blades, tanged points and microblades become important. This raises three issues. First, there is no equivalent here of an "Upper Palaeolithic revolution" of the kind claimed for western Europe around this time but instead, only a few minor changes to an existing technology (Norton and Jin 2009). Second, we do not know if any change in population occurred. The skeletal record of the Korean Peninsula is poor – of the nine sites in the Korean Peninsula with hominin fossils (Norton 2000), the best are probably the cranial and post-cranial remains of *Homo sapiens* from layers 9–12 of Yonggok (Ryonggok) Cave near Pyongyang, North Korea, which was dated to 44,300 ± 2000 and 49,900 ± 2000 cal BP (Bae and Guyomarc'h 2015; Park et al. 2018), and fragments of hominins from the Rangjeonggol cave that were found associated with some crude stone artefacts and dated to between 52,000 BP and 43,000 BP by several different methods (Bae 2010). There is also a large first metatarsal (foot bone) from Gunang Cave, South Korea, dated to 40,900 and 44,900 BP (Park et al. 2018). Although it is probable that *Homo sapiens* appeared in Korea ca. 40,000 BP, definite evidence is so far lacking. The third problem is that Korea is the most likely source population for the colonisation of PalaeoHonshu, but it scarcely gives the impression that it had a population sufficiently large enough for some to undertake this. One answer may simply be that the bulk of the population lived on the now-submerged floor of the Yellow Sea – or the impression that population levels were very low might simply result from a lack of survey and excavation.

Japan

Japan has been overlooked in most western narratives about the expansion of our species across Asia. This is partly because of an obvious language barrier, but there is now a large literature in English on the Japanese Palaeolithic, including overviews (e.g. Bae 2017). There are several reasons why it deserves far more attention than it has so far received outside Japan. First, the Japanese islands are second only to Australia as a Pleistocene example of maritime colonisation. Secondly, there are over 10,000 Palaeolithic sites in Japan, of which at least 500 are Early Upper Palaeolithic and relate to its initial colonisation (Izuho and

Kaifu 2015). Thirdly, Japan has some of the best examples of long-distance Palaeolithic exchange networks in which obsidian was a prime commodity. Finally, Okinawa in the Ryuku Islands may provide one of the earliest examples of humans deliberately trans-locating an animal to a new environment.

The Japanese Islands

Japan is roughly twice the size of the United Kingdom, and comprises the four main islands (Kyushu, Honshu, Shikoku and Hokkaido) as well as the Ryuku Islands, which are a 1,200 km-long chain that extends southwards from Kyushu to Taiwan. During the Early and Middle Pleistocene, Kyushu, Honshu and Shikoku were occasionally joined to the mainland (see Norton and Jin 2009) but were isolated for the last 130,000 years because of the 130–200 m-deep Tsushima Strait between Kyushu and Korea, and the 130 m-deep Tsugaru Strait between Hokkaido and Honshu (Chen et al. 2017) which narrowed to less than 12 km during the LGM (Iwase et al. 2015). In the Late Pleistocene, Shikoku, Honshu and Kyushu formed one large island known as PalaeoHonshu, and Hokkaido was joined to Sakhalin Island (PalaeoSakhalin–Hokkaido, or PSHK) which in turn was joined to the Siberian mainland (see Figure 11.4; Morisaki 2012; Nakazawa and Bae 2018). In the last glacial cycle, montane and tundra vegetation expanded across Sakhalin and northern Hokkaido as well as mountainous parts of northern Honshu, and boreal coniferous and temperate deciduous forest spread over the southern part of PalaeoHonshu at the expense of evergreen broadleaved forest (Morisaki 2012; see Figure 11.5). Fine-tuned vegetational records of the last 40,000 years are provided by Hayashi et al. (2010) and Chen et al. (2017). On PalaeoHonshu, the fauna was temperate Palearctic and included Naumann's elephant (*Palaeoloxodon naumanni*), *Sinomegaceros* (a giant deer), *Cervus nippon*, *Ursus arctos*, *Macaca fuscata*, wolf and fox, whereas on Hokkiado and Sakhalin, the Palearctic fauna included *Mammuthus primigenius Bison priscus*, *E. hydruntinus*, *C. elephus*, *Megaceros giganteus*, musk ox, woolly rhino, arctic fox, reindeer and saiga antelope (*Saiga tatarica*) (Iwase et al. 2015), as in Siberia at that time. Unfortunately, faunal remains are rarely preserved on Japanese Palaeolithic sites because the soils are acidic. As now, it is likely that fish played a major role in the diet of the first inhabitants. The Ryuku Islands (see below) have a generally depauperate fauna that includes some island endemics, none of which is larger than a small deer. The northern Ryukus have a Palaearctic fauna like Kyushu, but the fauna in the central and southern Ryukus is Oriental (Kawamura et al. 2016). All these islands have been isolated since the Middle Pleistocene.

The Japanese Palaeolithic

Japan was most likely colonised around 38,000 years ago. There have been occasional claims of earlier settlement but these have not been accepted by most researchers (see Norton and Jin 2010). PalaeoHonshu could have been colonised

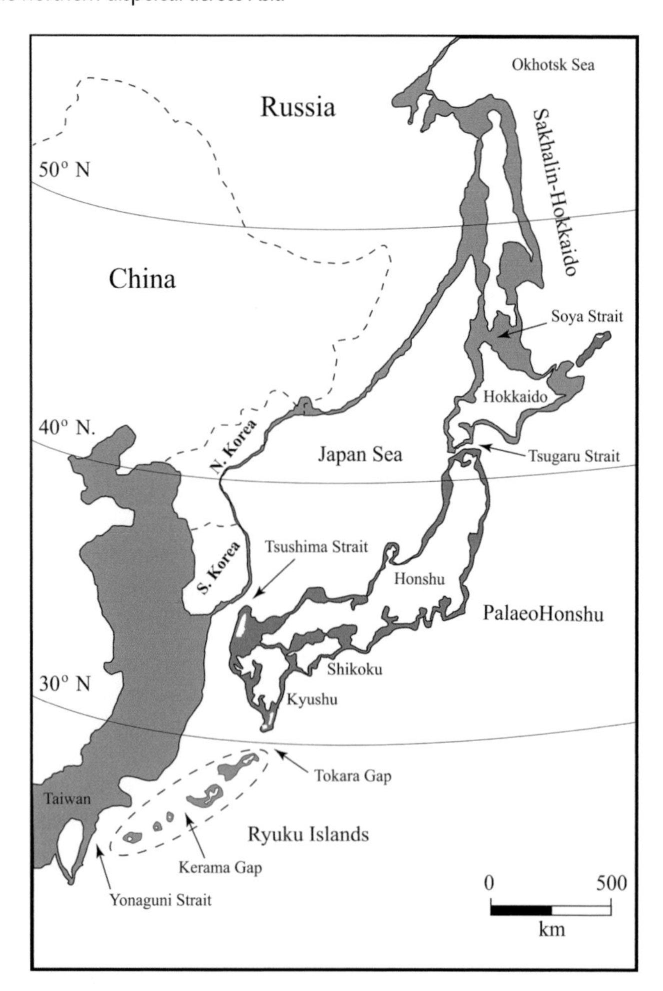

FIGURE 11. 4 Coastlines of Japanese islands during the Last Glacial Maximum (LGM)
Source: Redrawn by the author from Nakazawa and Bae 2018, Fig. 1.

from several directions (see Figure 11.6), all of which involve a marine crossing (Nakazawa 2017). The shortest and probably most likely was across the Tsushima Strait which was reduced from ca. 100 km in width to a narrow channel only 20 km wide in the Last Glacial Maximum (LGM) (Park et al. 2000).

Many Japanese archaeologists divide the Palaeolithic of Japan into an early and late phase. The Early Upper Palaeolithic (EUP) begins ca. 38 cal ka BP with the initial colonisation and ends at ca. 30 ka cal BP, which is the age of a volcanic ash known as the A-T (Aira-Tanzawa) tuff that forms a widespread marker horizon across southern Japan (see Machida 2002). The earliest assemblages are flake-based, ca. 38,000–37,000 cal BP and the first blade technology dates from ca. 36,000 cal BP. The nearest parallels on the Asian mainland are

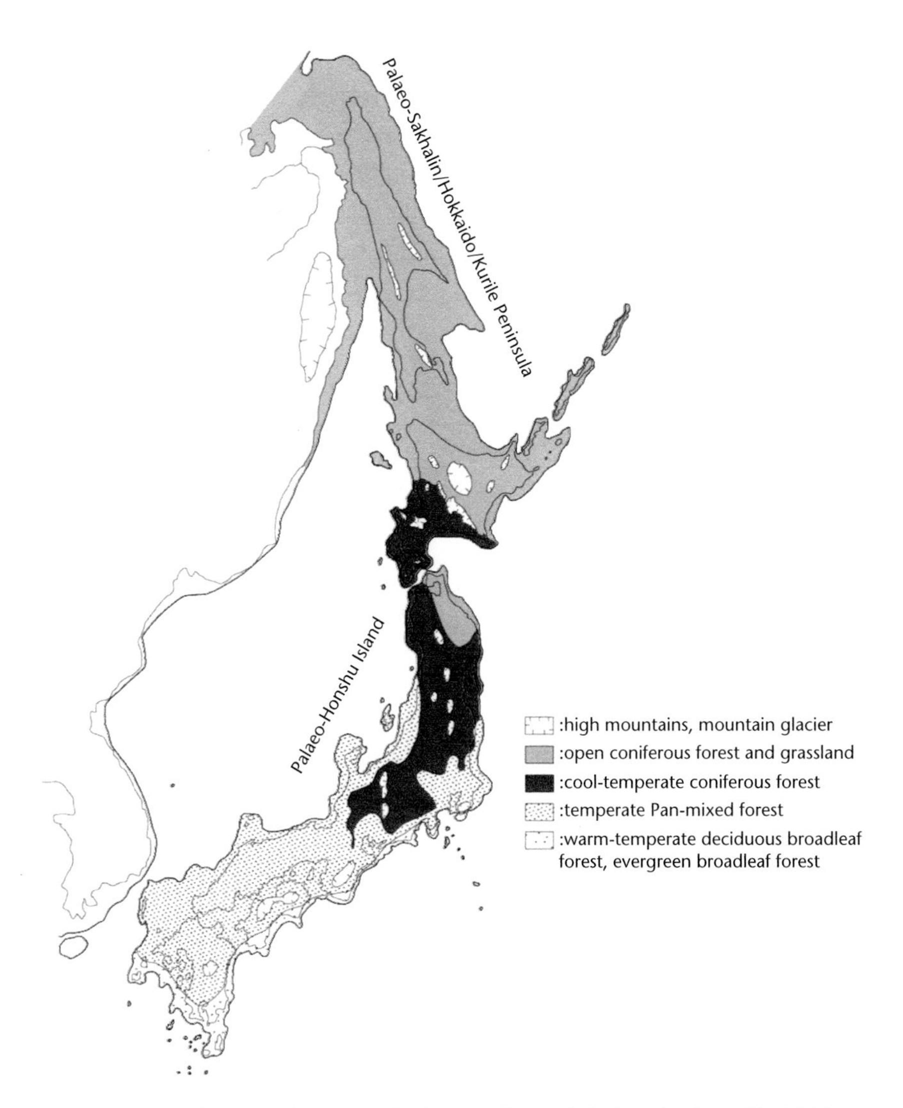

FIGURE 11.5 Main vegetation zones in Japan today and during the Last Glacial Maximum (LGM)

Source: Morisaki 2012, Fig. 4.

with Korea – as might be expected given the distance between them – and not with China (Morisaki et al. 2019). The Late Upper Palaeolithic (LUP) includes flake and blade assemblages from 29 ka cal BP onwards, pointed-tool assemblages from 22 to 19 ka cal BP and microblade assemblages from 20 to 16 ka cal

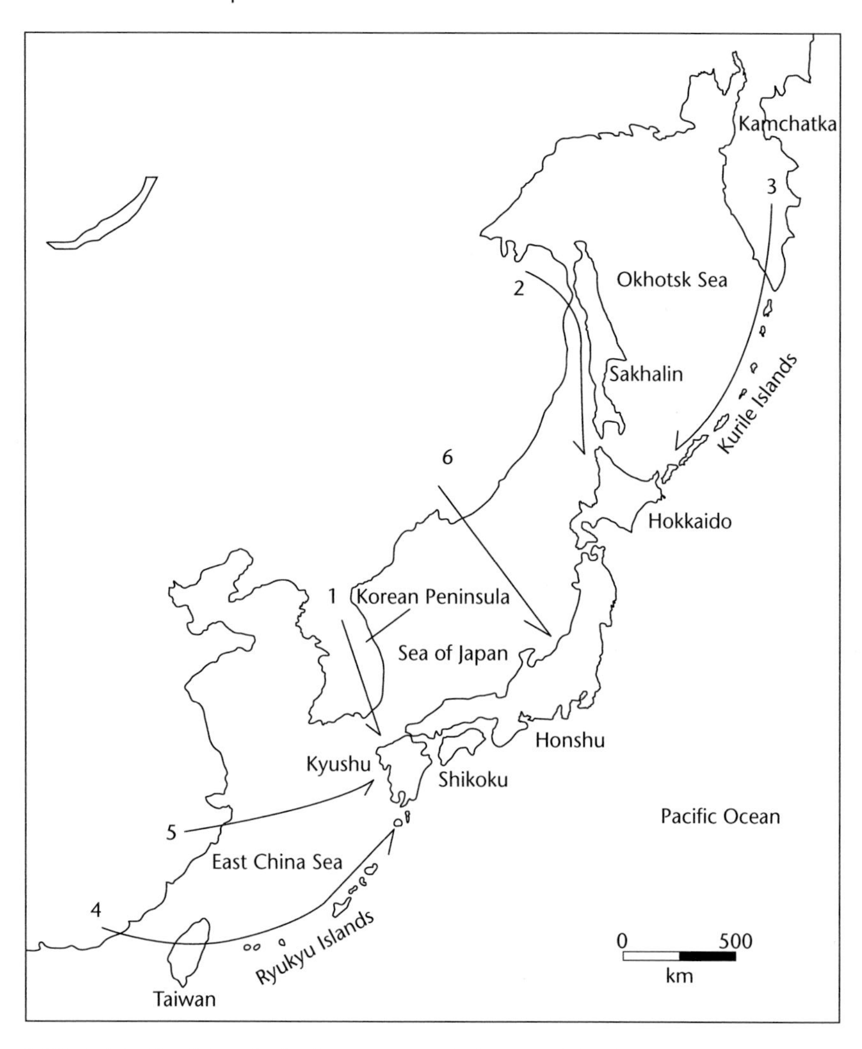

FIGURE 11.6 Possible colonisation routes to the Japanese islands
Source: Nakazawa 2017, Fig. 1.

BP (Kudo and Kumon 2012) that are similar to those found in Korea, North China, Mongolia and Siberia (see Chapter 10).

At present the earliest site in Japan is Idemaruyama on Honshu, dated to ca. 38,000 cal BP. EUP sites are characterised by trapezoids, ground edge axes (see Figures 11.7 and 11.8) and pit traps which are three unique features not seen on the mainland, as well as pointed-shaped backed blades, burins, end- and side-scrapers, and circular concentrations of charcoal and cobbles that may denote structures (Kudo and Kumon 2012; Nakazawa 2017). Watercraft can also be included. Trapezoids were designed for hafting and were probably used as

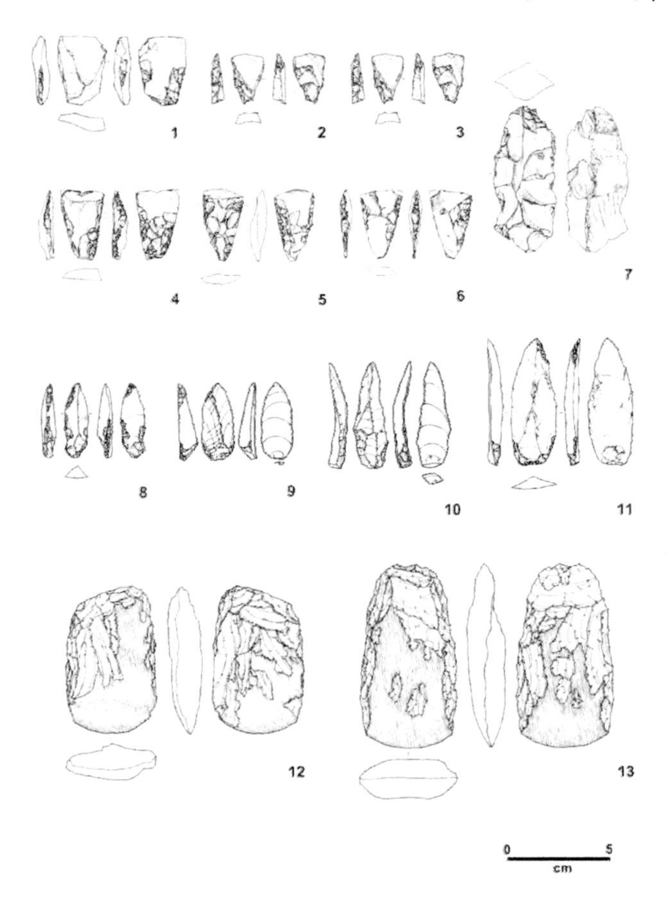

FIGURE 11.7 Examples of the major stone tool types from early Upper Palaeolithic assemblagesin Japan

1–6 trapezoids; 7 a flake core with small flake scars served for blanks of trapezoids; 8–11 knife-shaped tools (backed blades); 12–13 edge-ground axes.

Source: Nakazawa 2017, Fig. 3.

projectile points, quite possibly in bows and arrows (Yamaoka 2012). Because bow and arrow technology is complicated[3] and difficult to make, it was probably introduced by the first settlers as a proven technology (Sano 2016).

Hokkaido was probably colonised ca. 30 ka, several millennia after PalaeoHonshu. The earliest assemblages are based on small flakes. These were followed by blade and flake-based assemblages dating to between 27,000 and 25,000 cal BP, and then various microblade assemblages from 26,000 to 15,000 cal BP (Izuho et al. 2018). No comparable information is available on Sakhalin, which may reflect the lack of research.

Pit traps are peculiarly Japanese. Typically, a pit trap is ca. 1 m deep, 2 m wide and barrel-shaped in profile in order to make it harder for an animal

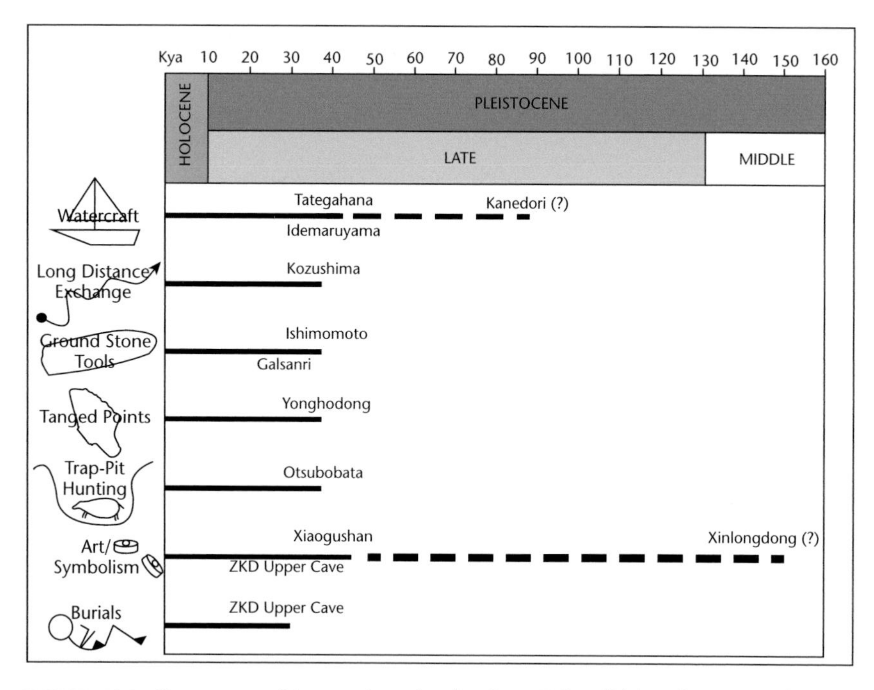

FIGURE 11.8 Summary of innovations in the Late Palaeolithic of Japan and north China

Source: Bae 2017, Fig. 2.

to climb out of. They were often dug in a row along a hillside, and the gaps between each trap were probably filled with a fence. They presumably were used in communal drives which would urge game to run downhill, and are commonest in southern Kyushu, which is an area that would have been rich in plant foods. Sato (2015) suggests that pit traps were used by sedentary hunter-foragers who had small territories, and a local fauna to hunt. The most likely prey species were medium-sized animals such as deer and pig. The oldest pit traps were excavated at the Otsu-bobata site on Tageshima Island in the northern Ryukus, and are dated to ca. 35–38,000 cal BP (Sato 2015).

The extinction of the megafauna

As in continental Asia, large herbivores such as Naumann's elephant, mammoth and *Sinomegaceros* are now extinct, and the timing and cause of their extinction are worth investigating. At the time of colonisation, the fauna on PalaeoHonshu would have been naïve (Chapter 1) in that it had never encountered human predators. One might therefore expect that it rapidly became extinct, as happened in North America within a few centuries of human arrival. Interestingly, there was a lengthy period between human arrival and megafaunal extinction of ca. 12,000 years for PalaeoHonshu and ca. 7,000 years for Sakhalin and Hokkaido. The main factor was probably climatic, and caused by vegetational changes (see Norton et al. 2010; Iwase et al. 2012, 2015).

Obsidian

One of the most fascinating aspects of the Japanese Palaeolithic is the way in which obsidian was quarried and distributed over considerable distances and often across the sea. Over 70 obsidian sources are known in Kyushu, Honshu and Hokkaido (Izuho and Kaifu 2015), and the resulting distribution has been mapped by using NAA (neutron activation analysis) and XRF (X-ray fluorescence); as of 2011, over 80,000 items had been analysed (Ikeya 2015). The earliest sources were used almost immediately after colonisation. Kozushima is a small island currently 50 km south of Kyushu, and probably 45 km distant when humans first arrived. Sailing there would have taken at least 6–7 hours and involved great skill, but people kept returning because of its excellent obsidian. Japanese researchers have shown that this was being used at Idemaruyama at 38 ca BP (Ikeya 2015). Other obsidian sources were used at an early date. As example, obsidian from Koshidake on Kyushu was brought across the Tsushima Strait to southern Korea, 350 km away (Kuzmin 2012).

The Ryuku Islands

The colonisation of the Ryuku Islands was a major challenge for the first settlers. This is because the Kurishio Current, which flows northwards along the coast of mainland China and southeast Japan (see Figure 11.9), is one of the strongest ocean currents in the world as it flows at 2–3 knots. Additionally, voyaging necessitated long passages across open sea and out of sight of land (Kaifu et al. 2015; Nakazawa and Bae 2018). The most likely starting points would have been Kyushu in the north and Taiwan in the south, which was joined to the mainland in MIS 3 when sea levels fell by 70 m (Kawamura et al. 2016). Neither route would have been easy. Sailing north from Taiwan necessitated long voyages out of sight of land. In the southern Ryukus, the nearest island to Taiwan is Yonaguni, 105 km distant.[4] A 60-km voyage was then needed to reach the conjoined islands of Iriomote and Ishigaki, and then two more shorter 30-km journeys to reach Miyako at the northern end of the southern Ryukus. If colonists then set sail to the Central Ryukus, they would have had to cross 220 km of open water to reach the Okinawa Islands. Alternatively, Okinawa could have been reached from the north via a series of 10–50 km crossings along the row of small active volcanoes in the Tokara Group, but this meant sailing against the Kurishio Current and using stopping points that were difficult to land at because of their steep cliffs. Another possibility may have been to use small southward currents, but without the benefit of a visible land target. The

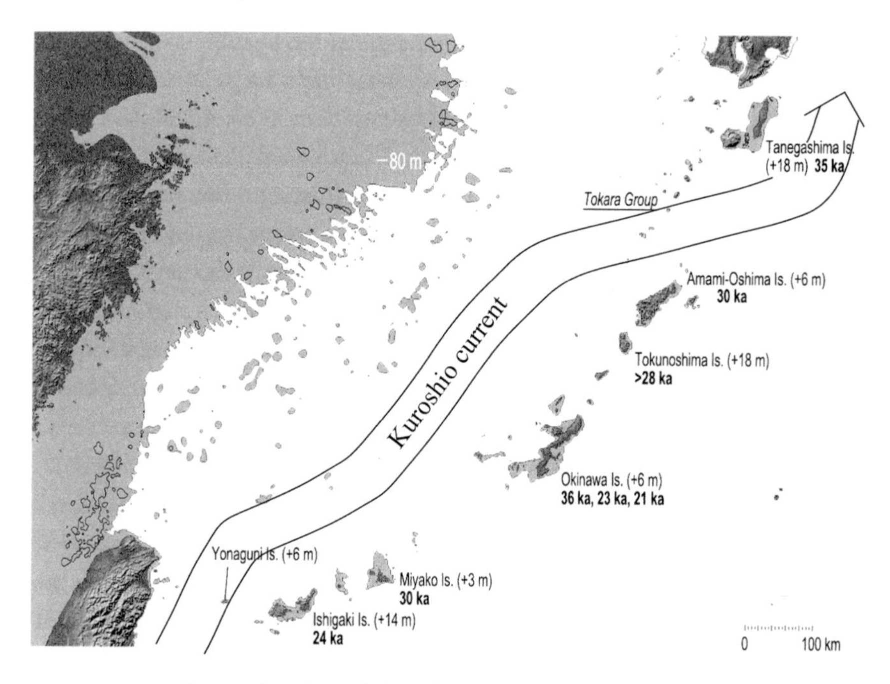

FIGURE 11.9 Earliest archaeological dates from the Ryuku Islands
Kuroshio current added by the author from Kaifu 2015, Fig. 1.
Source: Kaifu 2015, Fig. 24.3.

Osumi Group in the northern Ruyukus were the easiest to reach, as Tanegashima would have been only 22 km from Kyushu. It is thus not surprising that it was colonised in the Early Upper Palaeolithic ca. 35 ka cal BP by people with the same type of edge-ground axes seen on Kyushu (Kaifu et al. 2015).

It is therefore impressive that Okinawa, in the centre of the Ryukus, was colonised as early as 36 ka: charcoal from immediately above human remains at the cave fissure of Yamachita-cho are dated to 32,100 ± 1,000 (=34,210–38,500 cal BP). At that time, Okinawa was three times larger than its present area of ca. 1,200 sq km, but was poor in flakable stone and animal resources as the largest was a dwarf deer (*C. astylodon*). Nevertheless, people have lived there for 35,000 years. The Palaeolithic occupation of Okinawa has recently been illuminated by the excavation of Sakitari Cave (Fujita et al. 2016). Here, a long sequence of near-continuous deposition from 35,000 to 13,000 cal BP shows the consumption of fresh-water crabs and snails, and occasional hunting of an extinct dwarf deer (*Cervus astylodon*) and, in and after layer II (dated at 19,635 to 23,425 cal BP), pig (*Sus scrofa*). According to Kawamura and colleagues (2016), pig was probably a deliberate introduction to Okinawa. If so, this may be the earliest example of the deliberate relocation of an animal species by humans. (We need to bear in mind that only four pig bones were found in layer II [see Fujita et al. 2016, Table S4] so it is too early to say that pigs were absent on the island before 23 ka cal BP.) Also, interestingly, shell was used for making beads, scrapers and other tools; these include fish-hooks dated to 23 ka cal BP (see Figure 11.10), comparable in age or older than those from Timor (Chapter 7). Other islands such as Osumi, Amami and Miyako were reached by at least 30 ka, and probably earlier (see Figure 11.9). As way of comparison, the sea crossing to Buka from New Ireland at 32.5 ka involved a direct voyage of 170 km, comparable to the distances involved in reaching the Central Ryukus (Kaifu et al. 2015).

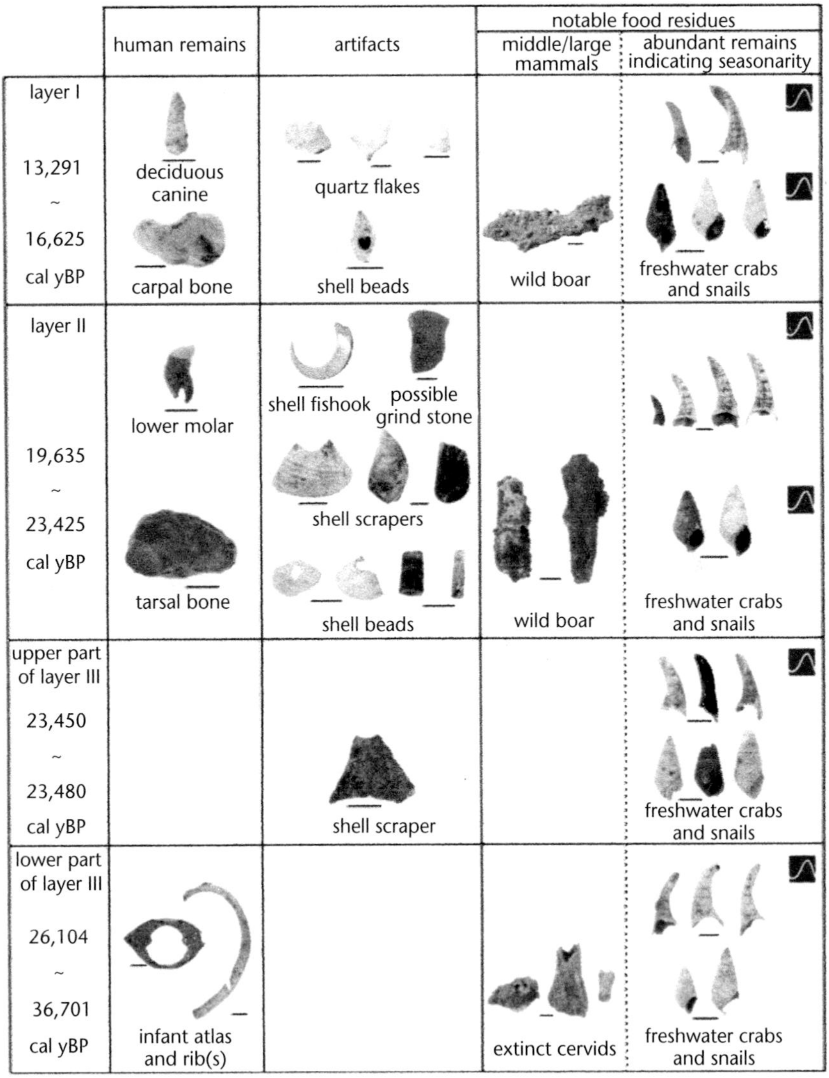

	human remains	artifacts	notable food residues: middle/large mammals	notable food residues: abundant remains indicating seasonarity
layer I 13,291 ~ 16,625 cal yBP	deciduous canine; carpal bone	quartz flakes; shell beads	wild boar	freshwater crabs and snails
layer II 19,635 ~ 23,425 cal yBP	lower molar; tarsal bone	shell fishook; possible grind stone; shell scrapers; shell beads	wild boar	freshwater crabs and snails
upper part of layer III 23,450 ~ 23,480 cal yBP		shell scraper		freshwater crabs and snails
lower part of layer III 26,104 ~ 36,701 cal yBP	infant atlas and rib(s)		extinct cervids	freshwater crabs and snails

FIGURE 11.10 The sequence of Sakitari Cave, Okinawa

Source: Fujita et al. 2016, Fig. 4.

Getting to America by land or sea

The colonisation of the Americas probably occurred around or shortly after 16,000 years ago, and thus lies outside the scope of this book. Nevertheless, a few comments are appropriate on the background to this event. There have been numerous suggestions over how the Americas were reached. Stanford and Bradley (2012) proposed that Upper Palaeolithic (Solutrean) groups from Spain and southwest France could have crossed the Atlantic by following the edge of the Atlantic pack-ice, using driftwood as fuel and seals and fish as the main food. Faught (2017) suggested that in the absence of any hard evidence from northeast Asia that a Pacific route from southeast Asia was more likely. Montenegro and others (2006) simulated ocean crossings by drifting and paddling, and showed that North America could have been reached from Europe or Northeast Asia, and South America from Africa – but they also noted that a successful sea crossing would not necessarily result in colonisation.

In most accounts, people reached North America from northeast Siberia by crossing the huge land shelf of Beringia that was exposed when sea levels fell. If so, the key question is when people first moved into Beringia. Some researchers have suggested that the first Beringians were there before the Last Glacial Maximum, and perhaps as early as 30,000 years ago (e.g. Bonatto and Solzano 1997). Archaeological evidence does not support this view. As seen, the Yana site complex allows us to place humans inside the Arctic Circle 29,000 cal years ago. There is then a gap in occupation records in northeast Siberia until ca.14,500 cal BP, when the site of Duiktai in northeast Siberia is recorded (Graf 2013). The glacial maximum in northern Siberia was almost certainly exceptionally harsh, with winter temperatures perhaps 20° C. colder than now. Although researchers are still debating whether Siberia was occupied throughout the glacial maximum (see e.g. Kuzmin and Keates 2005 versus Graf 2009, 2015; Graf and Buvit 2017), it is likely that much of it was depopulated, and most unlikely that people headed into Beringia as a pleasanter alternative. We should also bear in mind the sheer scale of the Siberian and Beringian landscapes. Yana to Fairbanks, Alaska, is ca. 3,000 km (1,800 miles), which is about the same as London to Istanbul. At present, there are no good archaeological reasons for assuming that people were already on Beringia before 20,000 years ago.

In the absence of archaeological data, most of the scenarios for the colonisation of North America via Beringia have been by geneticists. Several now suggest that the colonisation of the Americas was complex and involved more than one population (see Graf and Buvit 2017). Raghavan and colleagues (2014) analysed the ancient DNA from a 24,000-year-old skeleton from Mal'ta and another, 17,000 years old, from Afontova-Gora, both in Siberia. They concluded that both were very similar to each other and to a 36,000-year-old skeleton from Kostienki in the Ukraine, and that between 14% and 38% of Native American genome is derived from this population. (This homogeneity across Eurasia implies very extensive and robust mating networks.) The remaining 86%

to 62% is attributed to an incoming group of East Asians. In another study, Skoglund and colleagues (2015) also concluded that Amazonian Native Americans descended from a Native American population that was more closely related to indigenous Australians, New Guineans and Andaman Islanders than to any present-day Eurasians or Native Americans. Recently, high quality genomes have been recovered from two milk teeth from the Yana site complex (Sikora et al. 2019). These were similar to that seen at the 34,000-year-old burial site of Sunghir near Moscow, 4,500 km to the west. (This is yet another powerful demonstration of the robustness of social and mating networks across the Eurasian landmass at this time.) The descendants of this population called Ancient North Siberians later mixed (perhaps between 25 and 10 ka) with an incoming population with East Asian ancestry, resulting in a new population of Palaeo-Siberians and Ancient Beringians who have affinities with present-day groups on both sides of the Bering Strait. Later still, Alaska was reached by Neo-Siberians after 10 ka. This study is important in implying that the Yana-Sunghir population formed the reservoir which later formed part of a founding population that crossed Beringia into Alaska at the end of the last ice age.

A Kelp Highway?

Asian people may also have reached North America by sea. We've seen already (Chapter 7) that humans were sailing around Wallacea and navigating their way to Australia by 55,000 years ago, and had also colonised Palaeo-Honshu 38,000 years ago and the Ryuku Islands shortly after. Jon Erlandson and colleagues (2007, 2015) have argued that people could have sailed northwards from Japan along the Kurile and Aleutian Islands, and then down the North American coast (see Figure 11.11) and indeed all the way down to Chile. The incentive would have been the "Kelp Highway", as these coasts are among the richest marine environments in the world, with an abundance or even super-abundance of marine mammals, birds, fish, shellfish, crustaceans and much more besides. On environmental grounds, this model makes a great deal of sense, and it is also plausible because of the deep antiquity of seafaring in East Asia – humans in East Asia certainly had the experience and capability by 20,000 years ago to undertake this kind of voyaging. Actual archaeological evidence is weak on the Asian side, however, but this is hardly surprising for such a vast and remote region. Additionally, as Erlandson et al. (2015) point out, there would be numerous gaps in the Kelp Highway – some coasts are difficult to access and relatively unproductive, for example, or have few inland resources. For "bold" colonists, these can be bypassed in a series of "jump dispersals", as mentioned earlier (Chapter 1). In this light it is not surprising that the earliest evidence from the Kurile Islands is from the Holocene (at the Yankito cluster on Iturup Island, ca. 6000–7500 cal BC [Yanshina and Kuzmin 2010]) as these are a string of small volcanic islands

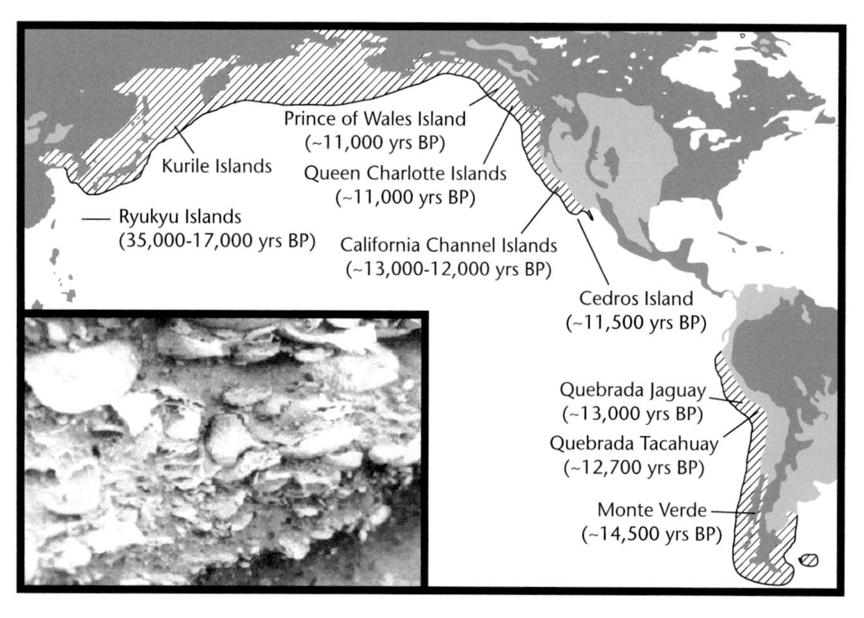

FIGURE 11.11 The Kelp Highway and relevant archaeological sites along the Pacific Rim.
Source: Erlandson 2007, Fig. 2.

with little to offer. The best supporting evidence is from the Kamchatka Peninsula, which had numerous obsidian sources on Kamchataka that were exploited from the Late Palaeolithic onwards and widely distributed over hundreds of kilometres (Grebennikov et al. 2010). The microblade site of Ushki is dated to ca. 13 ka cal BP (Goebel et al. 2010) and according to Coutouly and Ponkratova (2016), this site shares features with Swan Point in Alaska. There are also archaeological sites on the Russian side of the Bering Strait but their dating is uncertain because of cryoturbation, solifluction and erosion (Slobodin 1999). The recent discovery of the Cooper's Ferry site in Idaho, dated to between 16,560 and 15,280 years ago, indicates that humans were present in the United States before the development of an ice-free corridor through Alaska. This supports the idea that the first immigrants arrived via a Pacific route. Because the projectile points resemble those found at the Japanese Late Palaeolithic site of Kamishirataki 2 on Hokkaido (Davis et al. 2019), the "Kelp Highway" emerges as the most likely route.

Summary

Humans were the first hominin to inhabit the Arctic. The earliest evidence is from the RHS site and is ca. 45,000 years old, and the richest evidence is the Yana site, dated to ca. 29,000 years ago. Our species probably entered the

Korean Peninsula ca. 40,000 years ago, and main Japanese islands of Kyushu and Honshu (PalaeoHonshu) were colonised by sea ca. 38,000 years ago. Okinawa in the Ryuku Islands were reached ca. 36,000 years ago and involved several voyages between islands. People may also have trans-located pigs to that island ca. 19–23,000 years ago. Hokkaido was probably colonised from the Sakhalin Peninsula in eastern Siberia ca. 30,000 years ago. North America could have been reached via Beringia and an ice-free corridor through Alaska, or by a coastal route. At present, the latter seems more likely because the earliest evidence from the northern United States predates the formation of an ice-free corridor.

Notes

1 Contrary to Slimak et al. (2011), Byzovaia can be regarded as Upper Palaeolithic, not late Mousterian (Zwyns et al. 2012).
2 The total is now 31 mammoths – see Nikolskiy and Pitulko (2013).
3 The bow has to be strong but supple; the string needs to be strong but thin and inelastic; arrows need to be straight and provided with flights and in this case, with arrow heads.
4 In July 2019, a reed boat was successfully paddled in 45 hours by a crew of five from Taiwan to Yonaguni in a crowd-funded project directed by Professor Yosuke Kaifu at the National Museum of Nature and Science in Tokyo (Normile 2019).

References

The Arctic

Basilyan, A.E., Anisimov, M.A., Nikolskiy, P.A. and Pitulko, V.V. (2011) Woolly mammoth mass accumulation next to the Paleolithic Yana RHS site, Arctic Siberia: its geology, age, and relation to past human activity. *Journal of Archaeological Science*, 38: 2461–2474.

Boschian, G., Caramella, D., Saccà, D. and Barkai, R. (2019) Are there marrow cavities in Pleistocene elephant limb bones, and was marrow available to early humans? New CT scan results from the site of Castel di Guido (Italy). *Quaternary Science Reviews*, 215: 86–97.

Boschian, G. and Saccà, D. (2015) In the elephant, everything is good: carcass use and reuse at Castel di Guido (Italy). *Quaternary International*, 361: 288–296, doi:10.1016/j.quaint.2014.04.030.

Buck, L.T. and Stringer, C.B. (2013) Having the stomach for it: a contribution to Neanderthal diets? *Quaternary Science Reviews*, 96: 161–167.

Guthrie, R.D. (1990) *Frozen Fauna of the Mammoth Steppe*. Chicago: University of Chicago Press.

Nikolskiy, P.A. and Pitulko, V.V. (2013) Evidence from the Yana Palaeolithic site, Arctic Siberia, yields clues to the riddle of mammoth hunting. *Journal of Archaeological Science*, 40: 4189–4197.

Nikolskiy, P.A., Sulerzhitsky, L.D. and Pitulko, V.V. (2011) Last straw versus Blitzkrieg overkill: climate-driven changes in the Arctic Siberia mammoth population and the Late Pleistocene extinction problem. *Quaternary Science Reviews*, 30: 2309–2328.

Pavlov, P., Roebroeks, W. and Svendsen, J.I. (2004) The Pleistocene colonization of northeastern Europe: a report on recent research. *Journal of Human Evolution*, 47: 3–17.

Pavlov, P., Svendsen, J.I. and Indrelid, S. (2001) Human presence in the European Arctic nearly 40,000 years ago. *Nature*, 413: 64–67.

Pitulko, V.V., Nikolskiy, P.A., Girya, E.Y., Basilyan, A.E., Tumskoy, V.E., Koulakov, S. A., Astakhov, S.N., Pavlova, E.Y. and Anisimov, M.A. (2004) The Yana RHS site: humans in the Arctic before the last glaciation. *Science*, 303: 52–56.

Pitulko, V., Pavlova, E. and Nikolskiy, P. (2017) Revising the archaeological record of the Upper Pleistocene Arctic Siberia: human dispersal and adaptations in MIS 3 and 2. *Quaternary Science Reviews*, 165: 127–148.

Pitulko, V.V., Pavlova, E.Y., Nikolskiy, P.A. and Ivanova, V.V. (2012) The oldest art of Eurasian Arctic. *Antiquity*, 86: 642–659.

Pitulko, V.V., Tikhonov, A.N., Pavlova, E.Y., Nikolskiy, P.A., Kuper, K.E. and Polozov, R.N. (2016) Early human presence in the Arctic: evidence from 45,000-year-old mammoth remains. *Science*, 351: 260–263.

Sikora, M., Pitulko, V.V., Sousa, V.C., Allentoft, M.E., Vinner, L. et al. (2019) The population history of northeastern Siberia since the Pleistocene. *Nature*, 570: 182–188.

Slimak, L., Svendsen, J.I., Mangerud, J., Plisson, H., Presthus Heggen, H., Brugère, A. and Pavlov, P.Y. (2011) Late Mousterian Persistence near the Arctic Circle. *Science*, 332: 841–845, doi:10.1126/science.1203866.

Vartanyan, S., Garrut, V. and Sher, A. (1993) Holocene dwarf mammoths from Wrangel Island in the Siberian Arctic. *Nature*, 382: 337–340.

Zimov, S.A., Zimov, N.S., Tikhonov, A.N. and Chapin, F.S. III. (2012) Mammoth steppe: a high-productivity phenomenon. *Quaternary Science Reviews*, 57: 26–45.

Zwyns, N., Roebroeks, W., McPherron, S.P., Jagich, A. and Hublin, J.-J. (2012) Comment on "Late Mousterian Persistence near the Arctic Circle". *Science*, 335: 167–b, doi:10.1126/science.1209908.

Siberian dogs and wolves

Boudadi-Maligne, M. and Escarguel, G. (2014) A biometric re-evaluation of recent claims for Early Upper Paleolithic wolf domestication in Eurasia. *Journal of Archaeological Science*, 45: 80–89.

Crockford, S.J. and Kuzmin, Y.V. (2012) Comments on Germonpré et al. *Journal of Archaeological Science*, 36, 2009 "Fossil dogs and wolves from Palaeolithic sites in Belgium, the Ukraine and Russia: osteometry, ancient DNA and stable isotopes", and Germonpré, Lázkicková-Galetová, and Sablin, *Journal of Archaeological Science*, 39, 2012 "Palaeolithic dog skulls at the Gravettian Predmostí site, the Czech Republic". *Journal of Archaeological Science*, 39: 2797–2801.

Druzkhova, A.S., Thalmann, O., Trifonov, V.A., Leonard, J.A., Vorobieva, N.V. et al. (2013) Ancient DNA analysis affirms the canid from Altai as a primitive dog. *PLoS ONE*, 8(3): e57754, doi:10.1371/journal.pone.0057754.

Germonpré, M., Lázničková-Galetova, M. and Sablin, M.V. (2012) Palaeolithic dog skulls at the Gravettian Predmostí site, the Czech Republic. *Journal of Archaeological Science*, 39: 184–202.

Germonpré, M., Sablin, M.V., Després, V., Hofreiter, M., Lázkičková-Galetová, M. et al. (2013) Palaeolithic dogs and the early domestication of the wolf: a reply to the comments of Crockford and Kuzmin (2012). *Journal of Archaeological Science*, 40: 786–792.

Germonpré, M., Sablin, M.V., Lázničková-Galetova, M., Despres, V., Stevens, R.E. et al. (2015) Palaeolithic dogs and Pleistocene wolves revisited: a reply to Morey (2014). *Journal of Archaeological Science*, 54: 210–216.

Germonpré, M., Sablin, M.V., Stevens, R.E., Hedges, R.E.M., Hofreiter, M. et al. (2009) Fossil dogs and wolves from Palaeolithic sites in Belgium, the Ukraine and Russia: osteometry, ancient DNA and stable isotopes. *Journal of Archaeological Science*, 36: 473–490.

Janssens, L., Perri, A., Crombe, P., Van Dongen, S. and Lawler, D. (2019) An evaluation of classical morphologic and morphometric parameters reported to distinguish wolves and dogs. *Journal of Archaeological Science Reports*, 23: 501–533.

Lupo, K. (2017) When and where do dogs improve hunting productivity? The empirical record and some implications for early Upper Paleolithic prey acquisition. *Journal of Anthropological Archaeology*, 47: 139–151.

Morey, D.F. (2014) In search of Paleolithic dogs: a quest with mixed results. *Journal of Archaeological Science*, 52: 300–307.

Ovodov, N.D., Crockford, S.J., Kuzmin, Y.V., Higham, T.F.G., Hodgins, G.W.L. et al. (2011) A 33,000-year-old incipient dog from the Altai Mountains of Siberia: evidence of the earliest domestication disrupted by the Last Glacial Maximum. *PLoS ONE*, 6(7): e22821, doi:10.1371/journal.pone.0022821.

Perri, A. (2016) A wolf in dog's clothing: initial dog domestication and Pleistocene wolf variation. *Journal of Archaeological Science*, 68: 1–4.

Shipman, P. (2015a) How do you kill 86 mammoths? Taphonomic investigations of mammoth megasites. *Quaternary International*, 359–360: 38–46.

Shipman, P. (2015b) *The Invaders: How Humans and Their Dogs Drove Neandertals to Extinction*. Cambridge, Mass.: Harvard University Press.

Korea

Bae, C.J. (2017) Late Pleistocene human evolution in Eastern Asia: behavioral perspectives. *Current Anthropology*, 58(Supplement 17): S514–S526.

Bae, C.J. and Bae, K. (2012) The nature of the Early to Late Paleolithic transition in Korea: current perspectives. *Quaternary International*, 281: 26–35.

Bae, C.J. and Guyomarc'h, P. (2015) Potential contributions of Korean Pleistocene hominin fossils to palaeoanthropology: a view from Ryonggok Cave. *Asian Perspectives*, 54: 31–57.

Bae, K. (2010) Origin and patterns of the Upper Paleolithic industries in the Korean Peninsula and movement of modern humans in East Asia. *Quaternary International*, 211: 103–112.

Chung, C.-H., Lim, H.S. and Yoon, H.I. (2006) Vegetation and climate changes during the Late Pleistocene to Holocene inferred from pollen record in Jinju area, South Korea. *Geosciences Journal*, 10: 423–431.

Gao, X. and Norton, C.J. (2002) A critique of the Chinese "Middle Paleolithic". *Antiquity*, 76: 397–412.

Kei, Y.M. (2012) The Middle Palaeolithic in China: a review of current interpretations. *Antiquity*, 86: 619–626.

Kim, S.-J., Kim, J.-W. and Kim, B.-M. (2015) Last Glacial Maximum climate over Korean Peninsula in PMIP3 simulations. *Quaternary International*, 384: 52–81.

Kuzmin, Y.V. (2017) Obsidian as a commodity to investigate human migrations in the Upper Paleolithic, Neolithic, and Paleometal of Northeast Asia. *Quaternary International*, 442: 5–11.

Lee, G. (2015) The characteristics of Upper Palaeolithic industries in Korea. In Y. Kaifu, M. Izuho, T. Goebel, H. Sato and A. Ono (eds) *Emergence and Diversity of Modern Human Behavior in Paleolithic Asia*, College Station: Texas A&M University Press, pp. 270–286.

Lee, H.W., Bae, C.J. and Lee, C. (2017) The Korean early Late Paleolithic revisited: a view from Galsanri. *Archaeological and Anthropological Sciences*, 9: 843–863, doi:10.1007/s12520-015-0301-0.

Li, F., Kuhn, S.L., Chen, F., Wang, Y., Southon, J., Peng, F. et al. (2018) The easternmost Middle Paleolithic (Mousterian) from Jinsitai Cave, North China. *Journal of Human Evolution*, 114: 76–84.

Norton, C.J. (2000) The current state of Korean paleoanthropology. *Journal of Human Evolution*, 38: 803–825.

Norton, C.J., Bae, K., Harris, J.W.K. and Lee, H. (2006) Middle Pleistocene handaxes from the Korean Peninsula. *Journal of Human Evolution*, 51: 527–536.

Norton, C.J. and Jin, J.J.H. (2009) The evolution of modern human behaviour in East Asia. *Evolutionary Anthropology*, 18: 247–260.

Park, S.-J., Kim, J.-Y., Lee, Y.-J. and Woo, J.-Y. (2018) A Late Pleistocene modern human fossil from the Gunang Cave, Danyang county in Korea. *Quaternary International*, 519: 82–91.

Seong, C. and Bae, C.J. (2016) The eastern Asian 'Middle Palaeolithic' revisited: a view from Korea. *Antiquity*, 90: 1151–1165, doi:10.15184/aqy.2016.141.

Yi, S. and Kim, S.-J. (2010) Vegetation changes in western central region of Korean Peninsula during the last glacial (ca. 21.1−26.1 cal kyr BP). *Geosciences Journal*, 14(1): 1–10, doi:10.1007/s12303-010-0001-9.

Japan

Bae, C. (2017) Late Pleistocene human evolution in eastern Asia. *Current Anthropology*, 58 (Supplement 17): S514–S526.

Chen, J., Liu, Y., Shi, X., Suk, B.-C., Zou, J. and Yao, Z. (2017) Climate and environmental changes for the past 44 ka clarified by pollen and algae composition in the Ulleung Basin, East Sea (Japan Sea). *Quaternary International*, 441: 162–173.

Fujita, M., Yamasakia, S., Katagiria, C., Oshiro, I., Sano, K. et al. (2016) Advanced maritime adaptation in the western Pacific coastal region extends back to 35,000–30,000 years before present. *Proceedings of the National Academy of Sciences USA*, 113: 11184–11189.

Hayashi, R., Takahara, H., Hayashida, A. and Takemura, K. (2010) Millennial-scale vegetation changes during the last 40,000 yr based on a pollen record from Lake Biwa, Japan. *Quaternary Research*, 74: 91–99.

Ikeya, N. (2015) Maritime transport of obsidian in Japan during the Upper Palaeolithic. In Y. Kaifu, M. Izuho, T. Goebel, H. Sato and A. Ono (eds) *Emergence and Diversity of Modern Human Behavior in Paleolithic Asia*, College Station: Texas A&M University, pp. 362–375.

Iwase, A., Hashizume, J., Izuho, M., Takahashi, K. and Sato, H. (2012) Timing of megafaunal extinction in the late Late Pleistocene on the Japanese Archipelago. *Quaternary International*, 255: 114–124.

Iwase, A., Takahashi, K. and Izuho, M. (2015) Further study on the Late Pleistocene megafaunal extinction in the Japanese archipelago. In Y. Kaifu, M. Izuho, T. Goebel, H. Sato and A. Ono (eds) *Emergence and Diversity of Modern Human Behavior in Paleolithic Asia*, College Station: Texas A&M University Press, pp. 325–344.

Izuho, M. and Kaifu, Y. (2015) The appearance and characteristics of the Early Upper Palaeolithic in the Japanese archipelago. In Y. Kaifu, M. Izuho, T. Goebel, H. Sato and A. Ono (eds) *Emergence and Diversity of Modern Human Behavior in Paleolithic Asia*, College Station: Texas A&M University Press, pp. 289–313.

Izuho, M., Kunikita, D., Nakazawa, Y., Oda, N., Hiromatsu, K. and Takahashi, O. (2018) New AMS Dates from the Shukubai-Kaso Site (Loc. Sankakuyama), Hokkaido (Japan): refining the chronology of small flake-based assemblages during the Early Upper Paleolithic in the Paleo-Sakhalin-Hokkaido-Kurile Peninsula. *Paleoamerica*, 2018, doi:10.1080/20555563.2018.1457392.

Kaifu, Y., Fujita, M., Yoneda, M. and Yamasaki, S. (2015) Pleistocene seafaring and colonization of the Ryuku Islands, southwestern Japan. In Y. Kaifu, M. Izuho, T. Goebel, H. Sato and A. Ono (eds) *Emergence and Diversity of Modern Human Behavior in Paleolithic Asia*, College Station: Texas A&M University Press, pp. 345–361.

Kawamura, A., Chang, C.-H. and Kawamura, Y. (2016) Middle Pleistocene to Holocene mammal faunas of the Ryukyu Islands and Taiwan: an updated review incorporating results of recent research. *Quaternary International*, 397: 117–135.

Kudo, Y. and Kumon, F. (2012) Paleolithic cultures of MIS 3 to MIS 1 in relation to climate changes in the central Japanese islands. *Quaternary International*, 248: 22–31.

Kuzmin, Y.V. (2012) Long-distance obsidian transport in prehistoric northeast Asia.*Bulletin of the Indo-Pacific Prehistory Association*, 32: 1–5.

Machida, H. (2002) Volcanoes and tephras in the Japan area. *Global Environmental Research*, 6: 19–28.

Morisaki, K. (2012) The evolution of lithic technology and human behavior from MIS 3 to MIS 2 in the Japanese Upper Paleolithic. *Quaternary International*, 248: 56–69.

Morisaki, K., Sano, K. and Izuho, M. (2019) Early Upper Paleolithic blade technology in the Japanese Archipelago. *Archaeological Research in Asia*, 17: 79–97.

Nakazawa, Y. (2017) On the Pleistocene population history in the Japanese Archipelago. *Current Anthropology*, 58(Supplement 17): S539–S552.

Nakazawa, Y. and Bae, C.J. (2018) Quaternary paleoenvironmental variation and its impact on initial human dispersals into the Japanese Archipelago. *Palaeogeography, Palaeoclimatology, Palaeoecology*, 512: 145–155.

Normile, D. (2019) Update: explorers successfully voyage to Japan in primitive boat in bid to unlock an ancient mystery. *Science*, doi:10.1126/science.aay6005.

Norton, C.J. and Jin, J.J.H. (2009) The evolution of modern human behaviour in East Asia: current perspectives. *Evolutionary Anthropology*, 18: 247–260.

Norton, C.J., Kondo, Y., Ono, A., Zhang, Y. and Diab, M.C. (2010) The nature of megafaunal extinctions during the MIS 3–2 transition in Japan. *Quaternary International*, 211: 113–122.

Park, S.-C., Yoo, D.-G., Lee, C.-W. and Lee, E.-I. (2000) Last glacial sea-level changes and palaeogeography of the Korea (Tsushima) Strait. *Geo-Marine Letters*, 20: 64–71.

Sano, K. (2016) Evidence for the use of the bow-and-arrow technology by the first modern humans in the Japanese islands. *Journal of Archaeological Science Reports*, 10: 130–141.

Sato, H. (2015) Trap-pit hunting in Japan. In Y. Kaifu, M. Izuho, T. Goebel, H. Sato and A. Ono (eds) *Emergence and Diversity of Modern Human Behavior in Paleolithic Asia*, College Station: Texas A&M University Press, pp. 389–405.

Yamaoka, T. (2012) Use and maintenance of trapezoids in the initial Early Upper Paleolithic of the Japanese Islands. *Quaternary International*, 248: 32–42.

Getting to America

Bonatto, S.L. and Salzano, F.M. (1997) A single and early migration for the peopling of the Americas supported by mitochondrial DNA sequence data. *Proceedings of the National Academy of Sciences USA*, 94: 1866–1871.

Coutouly, Y.A.G. and Ponkratova, I.Y. (2016) The Late Pleistocene microblade component of Ushki Lake (Kamchatka, Russian Far East). *PaleoAmerica*, 1–29, doi:10.1080/20555563.2016.1202722 2016.

Davis, L.G., Madsen, D.B., Becerra-Valdivia, L., Higham, T., Sisson, D.A. et al. (2019) Late Upper Paleolithic occupation at Cooper's Ferry, Idaho, USA, ~16,000 years ago. *Science*, 365: 891–897.

Erlandson, J.M., Braje, T.J., Gill, K.M. and Graham, M.H. (2015) Ecology of the Kelp Highway: did marine resources facilitate human dispersal From Northeast Asia to the Americas? *Journal of Island and Coastal Archaeology*, 10: 392–411, doi:10.1080/15564894.2014.1001923.

Erlandson, J.M., Graham, M.H., Borque, B.J., Corbett, D., Estes, J.A. and Steneck, R.S. (2007) The Kelp Highway Hypothesis: marine ecology, the coastal migration theory, and the peopling of the Americas. *Journal of Island and Coastal Archaeology*, 2(2): 161–174, doi:10.1080/15564890701628612.

Faught, M.K. (2017) Where was the PaleoAmerind standstill? *Quaternary International*, 444: 10–18.

Goebel, T., Slobodin, S.B. and Waters, M.R. (2010) New dates from Ushki-1, Kamchatka confirm 13,000 cal bp age for earliest Paleolithic occupation. *Journal of Archaeological Science*, 37: 2640–2649.

Graf, K. (2009) "The good, the bad, and the ugly": evaluating the radiocarbon chronology of the middle and late Upper Paleolithic in the Enisei River valley, south-central Siberia. *Journal of Archaeological Science*, 36: 694–707.

Graf, K.E. (2013) Siberian odyssey. In K.E. Graf, C.V. Ketron and M.R. Waters (eds) *Paleoamerican Odyssey*, College Station: Texas A&M University Press, pp. 65–80.

Graf, K. (2015) Modern human response to the Last Glacial Maximum. In Y. Kaifu, M. Izuho, T. Goebel, H. Sato and A. Ono (eds) *Emergence and Diversity of Modern Human Behavior in Paleolithic Asia*, College Station: Texas A&M University Press, pp. 506–531.

Graf, K.E. and Buvit, I. (2017) Human dispersal from Siberia to Beringia: assessing a Beringian Standstill in light of the archaeological evidence. *Current Anthropology*, 58 (Supplement 17): S583–S603.

Grebennikov, A.V., Popov, V.K., Glascock, M.D., Speakman, R.J., Kuzmin, Y.V. and Ptashinsky, A.V. (2010) Obsidian provenance studies on Kamchatka Peninsula (Far Eastern Russia): 2003–9 Results. In Y.V. Kuzmin and M.D. Glascock (eds) *Crossing the Straits: Prehistoric Obsidian Source Exploitation in the North Pacific Rim*, British Archaeological Reports (Int. Series), S2152: 89–120.

Kuzmin, Y.V. and Keates, S.G. (2005) Dates are not just data: Paleolithic settlement patterns in Siberia derived from radiocarbon records. *American Antiquity*, 70(4): 773–789.

Montenegro, I., Hetherington, R., Eby, M. and Weaver, A.J. (2006) Modelling pre-historic transoceanic crossings into the Americas. *Quaternary Science Reviews*, 25: 1323–1338.

Raghavan, M., Skoglund, P., Graf, K.E., Metspalu, M., Albrechtsen, A. et al. (2014) Upper Palaeolithic Siberian genome reveals dual ancestry of Native Americans. *Nature*, 505: 87–91.

Sikora, M., Pitulko, V.V., Sousa, V.C., Allentoft, M.E., Vinner, L. et al. (2019) The population history of northeastern Siberia since the Pleistocene. *Nature*, doi:10.1038/s41586-019-1279-z.

Skoglund, P., Mallick, S., Cátira Bortolini, M., Chennagiri, N., Hünemeier, T., Petzl-Erler, M.L., Salzano, F.M., Patterson, N. and Reich, D. (2015) Genetic evidence for two founding populations of the Americas. *Nature*, 325: 104–110.

Slobodin, S. (1999) Northeast Asia in the Late Pleistocene and Early Holocene. *World Archaeology*, 30(3): 484–502, doi:10.1080/00438243.1999.9980425.

Stanford, D.J. and Bradley, B.A. (2012) *Across Atlantic Ice: The Origins of the America's Clovis Culture*. Berkeley and Los Angeles: University of California Press.

Yanshina, O.V. and Kuzmin, Y.V. (2010) The earliest evidence of human settlement in the Kurile Islands (Russian Far East): the Yankito Site Cluster, Iturup Island. *Journal of Island & Coastal Archaeology*, 5(1): 179–184, doi:10.1080/15564891003663927.

12

HOW, WHEN AND WHY DID OUR SPECIES SUCCEED IN COLONISING ASIA?

I end this book by returning to four issues I raised at the beginning: our status as an invasive species; how often and when we dispersed across Asia; the modernity debate and origins of "modern" behaviour; and most contentiously, why we were so successful as a colonising species.

Humans as an invasive species

We humans have been incredibly successful as a colonising, or invasive, species. By 50,000 years ago, humans had dispersed across southern Asia as far as mainland Southeast Asia, Wallacea, and had even reached Australia. By 40,000 years ago, we had dispersed across Central Asia, Siberia, Mongolia and reached north China as well as the Arctic Ocean. By 30,000 years ago, humans were present from Tasmania to the Arctic Ocean, from the Atlantic to the Pacific, and were even living on the Tibetan Plateau. The story of colonisation continues in the late glacial with the colonisation of the Americas, and in the Holocene, the Pacific. Yet this outline still tells only part of the story: we also colonised regions that had never felt a hominin footprint, such as Australia (Chapter 7) and the Japanese islands (Chapter 11), and new environments such as rainforests (Chapters 5 and 6) and the Arctic (Chapter 9).

In addition, as an invasive species we eliminated all indigenous inhabitants: Neandertals and Denisovans in the Palaearctic regions of Europe and Asia (albeit with some absorption of their genes); *Homo luzonensis* in the Philippines; the hobbit *H. floresiensis* on Flores (Chapter 7) and probably *H. erectus* in Southeast Asia and south China (Chapters 6 and 10). This last feature makes us unique as a hominin: after four million years of diversity, the hominin lineage was reduced to just one species after our species left Africa. To quote again John Shea (2011: 28), "'Plays well with others' is not something one is likely to see on *H. sapiens*' evolutionary report card". As Pat Shipman (2015) has pointed out, we are the ultimate invasive species.

It is hard not to feel sorry for *Homo luzonensis* and *H. floresiensis*. Island species that have been isolated for a long time are particularly vulnerable to an invasive predator. These hominins were small, probably had only a rudimentary technology, and above all, were naïve (Chapter 1) in that they had no prior experience of *Homo sapiens*. They were in the equivalent position of the dodo on Mauritius or Stellar's sea cow on the Commander Islands in the Bering Sea. The dodo became extinct within a century, and the Stellar sea cow in only 27 years.[1] I am not suggesting that the island hominins were killed and eaten (although some may have been); for this decimation, it would have been enough to disrupt their territories and social networks, to dominate their resource base and to drive them into marginal habitats. Tasmania might provide a better example than the dodo or Stellar's sea cow: there, the impact of whalers, disease, British convicts and farmers in the 19th century resulted in the extinction of pure-blood aborigines by 1905.[2] *H. luzonensis* probably survived human contact longer than the hobbit (likely extinct ca. 50,000 years ago) simply because Luzon is a much larger island (ca. 110,000 sq km) than Flores (13,000 sq km) and had more places in which to retreat.

How many dispersals?

The colonisation of Asia was emphatically not a single event. Instead, its basic geography (Chapter 3) dictated that dispersal had to take place north or south of the deserts and mountains of central Asia, the Himalayas and the Tibetan Plateau. The southern dispersal from Arabia to India, Southeast Asia and Wallacea (Chapters 4–7) was easier for a species that originated in Africa in that winter temperatures were generally above freezing and (at least in the Oriental Realm) plant foods were available year-round. As indicated in Chapter 3, clothing was either unnecessary for physiological reasons, or could have been made from plant leaves or fibres. On current evidence, this southern dispersal began in Arabia during or before the last interglacial, and had reached southeast Asia by 60–70 ka BP, and Australia by 50–55 ka BP. The northern dispersal (Chapters 8–11) began after 50,000 years ago and was harder because of the severity of the winter months, as well as the presence of Neandertals who would have been an effective rival. It was only when humans had developed long-distance social networks, effective over-wintering strategies and warm, wind-proof winter clothing from hide or fur that they could successfully colonise these harsh environments.

Although there were two main dispersals across Asia, each would have had numerous sub-divisions. Not all dispersals were successful in the sense that they left a genetic legacy that survives to the present. We saw in Chapter 8 that *H. sapiens* was present in the Levant by 177,000–194,000 years ago, and perhaps even in Greece before 200,000 years ago. Humans could have taken advantage of numerous "windows of opportunity" to enter Arabia before the last interglacial

(Chapter 4), and some of these dispersal events may even have introduced the Middle Palaeolithic technology to India (Chapter 5), although we still need skeletal evidence to show our species in south Asia before 100,000 years ago. It is most unlikely that the colonisation of Central Asia, Siberia, Mongolia, north China and the Arctic was a single event (Chapters 9–11), given the vastness of the landscapes, and the inherent variability and severity of the climate, and there must have been numerous failures, local extinctions and contractions. As I stated in my previous book, the early hominin settlement of Asia is a repeated theme of regional expansion and contraction, colonisation and abandonment, integration and isolation as rainfall and temperature increased or decreased (Dennell 2009).

The modernity (non) debate

I discussed the modernity debate, and arguments about the origins of "modern" behaviour in Chapter 2, and explained why the criteria for recognising it were in general unsatisfactory and in any case not likely to be applicable to Asia. In reviewing the Asian evidence in the main part of the book (Chapters 4–11), I have not once used the phrase "modern human behaviour" because I do not find the term useful in an Asian context. When considering the Asian evidence, it is much more useful to consider where people were and what they were doing than to mine their material culture for indications of symbolism or some other claimed attribute of "modernity". As examples, if before 35,000–40,000 years ago people were building boats that could sail across open sea from Wallacea to Australia and make return voyages, or sail to Okinawa, and also to offshore islands of Japan to obtain obsidian that was then exchanged over hundreds of kilometres; or substitute wood with ivory with which to hunt woolly mammoth inside the Arctic Circle; live on the Tibetan Plateau at over 4,000 metres above sea level; produce an effective microlithic technology in India, detoxify otherwise toxic plants and hunt monkeys that lived in the tree canopy in Borneo; catch deep-sea fish off Timor; produce paintings of animals in Borneo and Sulawesi; or maintain extensive social networks over hundreds of kilometres across Mongolia when hunting gazelle and horse, then surely their behaviour is "modern". In an Asian context, the keynote of "modernity" is the human skill in colonising new environments. As pointed out many years ago (Davidson and Noble 1992), human behaviour was "modern" when and because humans reached Australia 55–50,000 years ago, and we have been "modern" ever since. If I had to define "modernity" in an Asia context, I would suggest that it is shown primarily by the ability to survive in environments that had never previously been occupied. One might also (cynically) add that a further important indicator of "modern" human behaviour is an inability to tolerate competition from potentially rival species.

Why were we so successful?

The hardest question to answer is why we were so successful as a colonising species. Numerous explanations have been offered. As discussed in Chapter 1,

Africanists have often highlighted the importance of symbolism in that it demonstrates our ability to project abstract concepts of social status, kinship and community. Unfortunately, there is little evidence for this in Asia, but the absence of evidence does not of course mean that it was absent. Some researchers emphasise the importance of climatic factors in facilitating or impeding colonisation. This is undoubtedly true, but it is hard to attribute the colonisation of Asia, with its incredible diversity of environments, solely to climatic benevolence. Technology clearly played an important part in opening up new environments or increasing the prospects of human survival. Developments in hunting technology were important, whether stone-tipped projectile points, spears with replaceable tips, composite tools of microliths and wooden or bone handles, and fishing hooks. Other potential game-changers were the development of effective clothing in northern latitudes, or effective resins for making composite tools, or perhaps the domestication of wolves (although opinion remains divided on this). Social developments may also have been crucial. The emergence of gendered subsistence, with more clearly defined roles for females and older children who were not necessarily engaged in hunting large mammals would have made a more efficient use of labour; we can add to that our lower calorific requirements than Neandertals, and especially our ability to form large and effective social networks as a form of ecological buffering that was particularly essential in northern latitudes (see Chapters 9–11). What comes across perhaps most from the Asian record is the sheer adaptability of humans, or our "adaptive plasticity" as a "generalised specialist" (Roberts and Stewart 2018) in colonising almost every type of habitat.

It is most unlikely that there is a single reason for the success of our species; more likely, much of the success resulted from the cumulative advantage of small incremental gains, whereby humans were slightly more successful, adaptable, innovative and even luckier more often than Neandertals or Denisovans. I would, however, emphasise three dominant reasons for our success as a coloniser and adaptability: demography, diet and physiology.

Demography and the importance of numbers

The first reason for our success was demographic in that our species had the advantage of numbers over its rivals. We began as an African species, and then colonised the Oriental Realms of South and Southeast Asia, and so by 60,000 years had occupied the southern half of the Asian landmass. These regions of high biodiversity would have supported larger populations than the arid and semi-arid regions of central Asia. In contrast, Neandertals remained in the Palearctic Realm and their population numbers and densities would have been lower (see below). Put simply, there were always likely to be more of us than them.

There are two reasons why population numbers and densities are important. One is that Neandertal populations were vulnerable to being swamped by larger number of *H. sapiens*. The other is that population size affects innovation. According to Bouquet-Appel and Degioanni (2013: 206), "In any population, the production of innovations depends not only on its cognitive biological capacities but also on its demographic size". Neandertals may have been trapped: population growth depends upon technological innovation, but technological innovation requires high population levels to be successful,[3] and Neandertal populations were simply too small. Additionally, because population densities were low, they would have formed "loose" networks (Chapter 1), with little connectivity. Several researchers have emphasised the importance of group "connectedness" in cultural transmission and change (e.g. Malinsky-Buller and Hovers 2019; Hovers and Belfer-Cohen 2013).

Under those constraints, Neandertals were especially vulnerable to disruption by the appearance of a closely related species that was also a skilled predator but reliant on a wider range of hunting techniques, able to hunt more selectively and safely with throwing spears and perhaps bow and arrow, able to access a wider range of resources and capable of maintaining extensive social networks with a high degree of connectivity. With that type of competition, Neandertals would have been vulnerable to habitat fragmentation and loss of habitat (Chapter 3, Figure 3.11). For our species, there were also numerous potential entry points into the Neandertal world. Although the Levant is usually singled out as the main region from which humans dispersed into Europe and continental Asia (Chapter 8), there would have been others such as southern Iran and Pakistan (Chapter 4) into Neandertal territory of Central Asia and Siberia, and Southeast Asia (Chapter 6) for dispersal into China.

The final story of Neandertals in Asia is probably one of local extinction as well as assimilation. Local extinction was probably a persistent feature of living a high-risk lifestyle in an unstable environment, punctuated by frequent Heinrich Events and other downturns (Chapter 3). We know that some degree of assimilation occurred through their genetic legacy in modern non-Africans, and also from the skeletal record of Early Upper Palaeolithic populations (Trinkaus 2007). Recent genetic evidence from Denisovans has been able to demonstrate a young girl who was the daughter of a Neandertal mother and Denisovan father (Chapter 9); with luck and skill, we may find similar evidence for a child born of a Neandertal and a *H. sapiens*. I suggested earlier (Chapter 1) that assimilation may have primarily involved Neandertal females and human males, and it will be interesting to see if that was the case.

Dietary factors

In the prologue to the northern dispersal, I mentioned that Neandertals were a top predator that relied heavily on a meat-based diet, whereas humans were omnivorous. One reason why this was advantageous for humans was that pregnant females and infants probably consumed a greater diversity of essential nutrients such as vitamins A, C and E,[4] whereas the lack of these in Neandertal diet meant that "Neandertal women probably had high incidences of abortions, miscarriages, and stillbirths resulting in high fetal-to-infant mortality" (Hockett and Haws 2005: 30). Consequently, higher rates of infant survival and lower rates of maternal mortality in human populations would allow their populations to grow faster than those of Neandertals. An additional factor may have been that humans were better at dealing with shortages when the main resources failed: with a more diverse diet, they had a wider range of fall-back options (Hockett 2012).

Physiological factors

Two factors are especially relevant to the colonisation of Asia by our species: a rounder brain and a longer childhood.

Brain shape matters more than brain size: the advantages of a rounder brain

It is often mentioned that Neandertals had brains at least as large as those of humans, but in this case, size isn't everything. At the beginning of this book, I mentioned that the braincase of early *H. sapiens* (as at Jebel Irhoud, Morocco; see Figure 2.1) was elongate, whereas humans today have a globular one. The reason why this transformation in brain shape is important is that it occurs during perinatal development. Gunz and colleagues (2012) argue that our globular cranium develops early in infancy and is unique to us, and might have had positive effects on how our brains develop. According to Neubauer et al. (2018), this development is linked to changes in the parietal and the cerebellum:[5] "Parietal areas are involved in orientation, attention, perception of stimuli, sensorimotor transformations underlying planning, visuospatial integration, imagery, self-awareness, working and long-term memory, numerical processing, and tool use" and the cerebellum is associated "not only with motor-related functions like the coordination of movements and balance but also with spatial processing, working memory, language, social cognition, and affective processing". In the first 90 days of life, the cerebellum grows at the fastest rate of all parts of the brain. Significantly, these long-term changes in brain shape parallel the emergence of "modern" behaviour in the archaeological record. (Neubauer et al. 2018: 4–5). Judging by the evidence presented in Chapters 4–11 from all parts of Asia, this re-organisation of the brain and its consequences on behaviour were complete by 50,000–55,000 years ago, i.e. when humans reached Australia and were beginning their dispersal across continental Asia.

I would combine the advantages of a rounder brain with the parallel development of a longer childhood.

The advantage of a long childhood

In Chapter 2, I also mentioned that the child's mandible from Jebel Irhoud shows that it developed at the same rate as a living human and therefore at a slower rate than earlier types of hominin and possibly its Neandertal contemporaries. The reason why this is important is that it indicates a longer and slower rate of development in childhood, which increased the length of childhood dependency upon its parents, but also allowed for the acquisition of complex skills whilst the brain is still developing (Neubauer and Hublin 2012). Several dental specialists have argued that our dental development was slower than that of Neandertals from examination of mandibles from Gibraltar (Dean et al. 1986), Scladina, Belgium (Smith et al. 2007), Lakonis, Greece (Smith et al. 2009) and Obi Rahmat, Uzbekistan (see Chapter 9) (Smith et al. 2011). Rozzi and Bermudez de Castro (2004) suggest that this fast rate of dental development and childhood growth among Neandertals was linked to higher mortality rates amongst adults and the pressures on rapid development of the young to help fuel high calorie intake and a high metabolic rate. Not all researchers agree: analyses of mandibles from La Chaise, France (Macchiarelli et al. 2006), and Roc de Marsac (Bayle et al. 2009) have been claimed to indicate development rates within the range of living humans. Interestingly, in one of the few comparisons of Neandertals with Palaeolithic *H. sapiens*, Bayle et al. (2010) concluded that the *H. sapiens* individual from Lagar Velho, Portugal, differed in dental development from late glacial and Mesolithic individuals but was similar to that of Neandertals, so a clear dichotomy between humans and Neandertals may not be justified for the Early Upper Palaeolithic. Others plead caution in that we need to know more of human variation in dental development (e.g. Guatelli-Steinburg 2009), and Smith (2013) cautions that dental maturation rates are only one aspect of an individual's life history. To complicate matters further, Bermudez de Castro and colleagues (2010) suggest that the modern pattern of dental development was already present a million years ago at Atapuerca, Spain – in which case, Neandertals would also have shared it.

The importance of an early childhood phase of development

Humans differ from extant apes in having a period of early childhood that occurs between infancy and middle, or juvenile, childhood (see Figure 12.1; and Thompson and Nelson 2011). Infancy ends when the child is weaned, or begins to eat solid foods. In the juvenile, middle childhood phase, brain growth is almost complete, the first permanent molars have erupted and individuals are not totally dependent upon their parent(s) but have not yet reached puberty. Juvenile childhood ends with the onset of sexual maturity. Thompson and Nelson (2011) define early childhood as the time when high rates of brain growth occur but only moderate increase in body size, when the child's dentition and digestive system is not yet fully able to process adult foods, and when the child is still dependent on its parent(s). It is hard to assign a precise age range to early childhood, but it is roughly three to six years depending on child and circumstance.

The reason why this stage of development is so important is that this is when a child learns social skills of how to interact with its age group, older children and adults, and also to learn how adults behave by observation and imitation. What is perhaps more important about this phase is that it is when a child begins to develop its skills in what I call the three "i's" of imagination, ingenuity and inventiveness: to invent stories and imaginary worlds, and to experiment with what might be possible, and create different ways of doing things.[6] It is those key

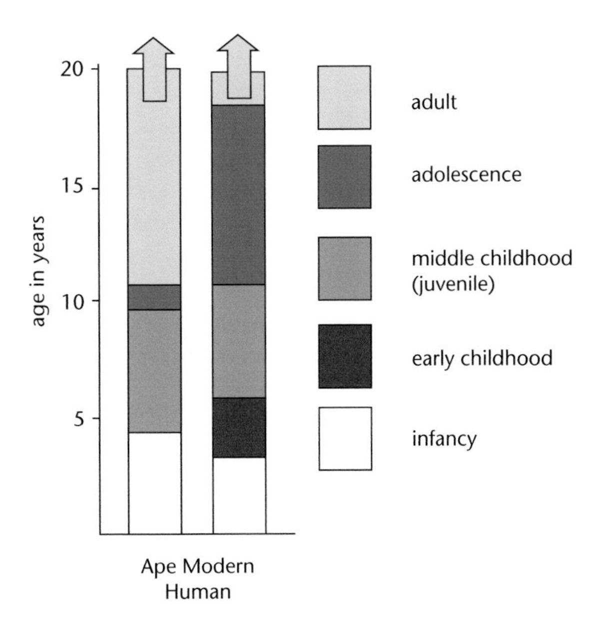

FIGURE 12.1 Developmental rates in apes and humans

Note the childhood phase and extended adolescence that characterise modern humans but not apes.

Source: Thompson and Nelson 2011, Fig. 1.

TABLE 12.1 Examples from Palaeolithic Asia of imagination, ingenuity and inventiveness that are unique to our species and not found among the previous inhabitants

Example	*Where*
Building boats, return voyaging	Wallacea, Japanese islands
Detoxifying poisonous plants	Borneo
Living in rainforests	Borneo, Sri Lanka
Living in the Arctic	Yana, RHS
Bow and arrow	Japan
Pit traps	Japan
Fishing deep-water species	Timor
Microlithic composite technology	India
Figurative art	Borneo, Sulawesi
Substituting ivory for wood	Arctic
Traps, snares, nets for small game	Arctic, Levant
Footwear	Tianyuandong
Eyed needles	Siberia
Operating long-distance exchange networks	The Arctic, Mongolia

skills of imagination, ingenuity and inventiveness that can pay dividends when used in adolescence and adulthood. As examples, the following table lists some examples of ingenuity, inventiveness and imagination in the early colonisation of Asia by our species that were not found in Neandertals or other indigenous inhabitants of Asia. (I omit here later Palaeolithic innovations such as ceramics and pressure flaking.)

In summary, we succeeded in colonising all parts of Asia and eliminating rival species such as Neandertals because we were more numerous, had a more diverse diet that increased the survival rates of mothers and infants, and combined a longer period of childhood development with a re-organised brain that was cognitively more powerful in inventive, imaginative and ingenuous at colonising new environments.

A final perspective

Palaeolithic archaeology began in western Europe in the early 19th century because of a happy coincidence of numerous caves with Palaeolithic deposits and a large number of enterprising (and usually privately funded) researchers who investigated them. Those same deposits also contained the bones of extinct animals, and from that association of bones and stones the deep antiquity of humankind was established by the 1860s as well as the discipline of palaeontology. Since then, the Palaeolithic and Pleistocene of western Europe has become the most intensely researched part of the Palaeolithic world, with by far the largest number of investigators. Palaeolithic archaeology across most of Asia started later: as early as 1860 in British India, but only in the early 20th century in regions such as the Levant, Siberia and China, and not in a concerted manner until the 1960s in areas such as Iran, Japan and Wallacea. A Eurocentric view of the Palaeolithic of Eurasia is easily understood when we note how and where the investigation of our deep past began. At the same time, we should note that the main expansion of our species out of Africa took place before humans had entered the western peninsula of Eurasia: by 50,000 years ago, our species had already colonised Arabia, South and Southeast Asia, Wallacea and had reached Australia; and by 40,000 years ago when humans were entering western Europe, they were already in Iran, Siberia, Mongolia, the Arctic and China and were about to colonise the Tibetan Plateau and the main Japanese islands. At present we have a stream of new information from these regions. As Asia continues its development, this stream is likely to become a river and even a flood, and it will be western Europe that is seen as peripheral, and not the great landmass of Asia.

Notes

1 The dodo was first noticed on Mauritius in 1598 and was last seen in 1662; Stellar's sea cow was discovered by Europeans on the Commander Islands in the Bering Sea in 1741 and was extinct in 1768.

2 Truganni (1812–1876) is often cited as the last full-blooded Tasmanian but Fanny Cochrane Smith (1834–1905) may have that dubious honour.

3 This is known as the "Boserup trap", named after Esther Boserup (1910–1999), a Danish economist. Put briefly, she argued that population growth depends upon technological innovation, but technological innovation requires high population to be successful. This was explained in her classic book in 1965: *The Conditions of Agricultural Growth the economics of agrarian change under population pressure.*
4 These help (among other functions) to protect the immune system. Vitamin A is found in many sources, especially liver and some vegetables; vitamin C is found mainly in plants, and vitamin E in many nuts.
5 The parietal lobe is located between the occipital lobe at the back of the brain and the frontal lobe, and above the temporal lobe. The cerebellum is found at the base of the brain.
6 This section is heavily influenced by watching how my son developed.

References

Bayle, P., Braga, J., Mazurier, A. and Macchiarelli, R. (2009) Dental developmental pattern of the Neanderthal child from Roc de Marsal: a high-resolution 3D analysis. *Journal of Human Evolution*, 56: 66–75.

Bayle, P., Macchiarelli, R., Trinkaus, E., Duarte, C., Mazuriere, A. and Zilhão, J. (2010) Dental maturational sequence and dental tissue proportions in the early Upper Paleolithic child from Abrigo do Lagar Velho, Portugal. *Proceedings of the National Academy of Sciences USA*, 107(4): 1338–1342.

Berger, T.D. and Trinkaus, E. (1995) Patterns of trauma among Neandertals. *Journal of Archaeological Science*, 22: 841–852.

Bermúdez de Castro. J.-M., Martinón-Torres, M., Prado, L., Gómez-Robles, A., Rosell, J., López-Polín, L., Arsuaga, J.L. and Carbonell, E. (2010) New immature hominin fossil from European Lower Pleistocene shows the earliest evidence of a modern human dental development pattern. *Proceedings of the National Academy of Sciences USA*, 107(26): 1739–11744.

Bocquet-Appel, J.-P. and Degioanni, A. (2013) Neanderthal demographic estimates. *Current Anthropology*, 54(Supplement 8): S202–S213.

Boserup, E. (1965) *The Conditions of Agricultural Growth: The Economics of Agrarian Change Under Population Pressure*. London: Allen & Unwin.

Davidson, I. and Noble, W. (1992) Why the first colonisation of the Australian region is the earliest evidence of modern human behaviour. *Archaeology in Oceania*, 27(3): 135–142.

Dean, M.C., Stringer, C.B. and Bromage, T.G. (1986) Age at death of the Neanderthal child from Devil's Tower, Gibraltar and the implications for studies of general growth and development in Neanderthals. *American Journal of Physical. Anthropology*, 70: 301–309.

Dennell, R.W. (2009) *The Palaeolithic Settlement of Asia*. Cambridge: Cambridge University Press.

Guatelli-Steinberg, D. (2009) Recent studies of dental development in Neandertals: implications for Neandertal life histories. *Evolutionary Anthropology*, 18: 9–20.

Gunz, P., Neubauer, S., Golovanova, L., Doronichev, V., Maureille, B. and Hublin, J.-J. (2012) A uniquely modern human pattern of endocranial development. Insights from a new cranial reconstruction of the Neandertal newborn from Mezmaiskaya. *Journal of Human Evolution*, 62: 300–313.

Hockett, B. (2012) The consequences of Middle Paleolithic diets on pregnant Neanderthal women. *Quaternary International*, 264: 78–82.

Hockett, B. and Haws, J.A. (2005) Nutritional ecology and the human demography of Neandertal extinction. *Quaternary International*, 137: 21–34.

Hovers, E. and Belfer-Cohen, A. (2013) On variability and complexity: lessons from the Levantine Middle Paleolithic record. *Current Anthropology*, 54(Supplement 8): S337–S357.

Macchiarelli, R., Bondioli, L., Debenath, A., Mazurier, A., Tournepiche, J.-F., Birch, W., Dean, C. (2006) How Neanderthal molar teeth grew. *Nature*, 444: 748–751.

Malinsky-Buller, A. and Hovers, E. (2019) One size does not fit all: Group size and the late middle Pleistocene prehistoric archive. *Journal of Human Evolution*, 127: 118–132.

Neubauer, S. and Hublin, J.-J. (2012) The evolution of human brain development. *Evolutionary Biology*, 39: 568–586, doi:10.1007/s11692-011-9156-1.

Neubauer, S., Hublin, J.-J. and Gunz, P. (2018) The evolution of modern human brain shape. *Science Advances*, 4(1): eaao5961.

Roberts, P. and Stewart, B.A. (2018) Defining the 'generalist specialist' niche for Pleistocene *Homo sapiens*. *Nature Human Behaviour*, www.nature.com/nathumbehav.

Rozzi, F.V.R. and Bermudez de Castro, J.-M. (2004) Surprisingly rapid growth in Neanderthals. *Nature*, 428: 936–939.

Shea, J.J. (2011) *Homo sapiens* is as *Homo sapiens* was: behavioral variability versus "Behavioral Modernity" in Paleolithic Archaeology. *Current Anthropology*, 52(1): 1–35.

Shipman, P. (2015) *The Invaders: How Humans and Their Dogs Drove Neandertals to Extinction*. Cambridge, Mass.: Harvard University Press.

Smith, T.M. (2013) Teeth and human life-history evolution. *Annual Review of Anthropology*, 42: 191–208.

Smith, T.M., Harvati, K., Olejniczak, A.J., Reid, D.J., Hublin, J.-J. and Panagopoulou, E. (2009) Brief communication: dental development and enamel thickness in the Lakonis Neanderthal molar. *American Journal of Physical Anthropology*, 138: 112–118.

Smith, T.M., Reid, D.J., Olejniczak, A.J., Bailey, S., Glantz, M., Viola, B. and Hublin, J.-J. (2011) Dental development and age at death of a Middle Paleolithic juvenile hominin from Obi-Rakhmat Grotto, Uzbekistan. In S. Condemi and G.-C. Weniger (eds) *Continuity and Discontinuity in the Peopling of Europe: One Hundred Fifty Years of Neanderthal Study*, Springer: Vertebrate Paleobiology and Paleoanthropology, pp. 155–163, doi:10.1007/978-94-007-0492-3_13.

Smith, T.M., Toussaint, M., Reid, D.J., Olejniczak, A.J. and Hublin, J.-J. (2007) Rapid dental development in a Middle Paleolithic Belgian Neanderthal. *Proceedings of the National Academy of Sciences USA*, 104(51): 20220–20225.

Thompson, J.L. and Nelson, A.J. (2011) Middle childhood and modern human origins. *Human Nature*, 22: 249–280.

Trinkaus, E. (2007) European early modern humans and the fate of the Neandertals. *Proceedings of the National Academy of Sciences USA*, 104(18): 7367–7372.

White, M., Pettitt, P. and Schreeve, D. (2016) Shoot first, ask questions later: interpretative narratives of Neanderthal hunting. *Quaternary Science Reviews*, 140: 1–20.

GENERAL INDEX

Entries in **bold** are in figures or tables

Acheulo-Yabrudian assemblages 210
aDNA (ancient DNA) 251, 254, 275, 281, 297, 336
African monsoon 48, 78, 84, 220
Afro-Arabian Realm 51, 71, 203, 235
Ahmarian (Early Upper Palaeolithic) assemblages 213, 215, 217, 220, **222**, **221**, 224
Alborz mountains, Iran 224, 225, 227, 232, 234
Altai Mountains, Russia 11, 201, 248–251, 252–262, **265**, 267, 268, 274–275, 296, 299, 322
amber 8, 320
Anatolia/Anatolian Plateau, Turkey 54, 223, 224, 236
Angara river, Siberia 248, 261, 317
anthraxolite 8, 320
Arabian Peninsula 17, 38, 48, 51, 58, 71, 78–86, 100, 115, 218, 220, 236
Arabian-Persian Gulf **52**, 224
Arctic 1, 8, 17, 21, 48, 52, 53, 56, 60–61, 67, 127, 195, 196, 201, **265**, 317–322, 336, 346, **352**
assimilation 14, **15**, 16, 307, 308, 350
A-T (Aira-Tanzawa) tuff, Japan 328
Aterian lithic assemblages 28, 32, 35–36

Bab el Mandab Strait 58, 78
Baradostian assemblages 231–233, **235**, 236, 250
barrier(s) 8, 32, **34**, 38, 53, 54–58, 59, 73, 98, 124, 125, 135, 167, 168, 173, 224, 234
beads 181, 182, 195, 197, 256, 260, **263**, 298, 307; bone beads 263; ivory beads 318, 320; ostrich eggshell beads 8, 249, **265**, 267, 270, **271**, 274, 275, 299, **302**, 303; shell beads 40, 180, 211, 215, 217, 223, 335; stone beads 298
Beringia 53, 56, **319**, 336–337, 339
Borneo 72–74, 135–140, 143, 146, 149–151, 154–157

cave art 40, 42, 74, 156–157, 183
Central Asia 1, 4, 21, 48, 54, 55, 58, 59, 67, 116, 127, 195, 227, 236, 248–252, 258–260, 275, 284, 299, 346, 348, 350
childhood length 351–354
Cis-Baikalia, Siberia 261, 262
clothing 17, 54, 55, 60–62, 72, 169, 197, 200, 248, 261, 290, 307, 347, 349
coastal dispersal 11, 13–**14**, 115, 121
colonisation 1–24, 32, 53, 55, 59, 63, 65, 67, 73–75, 77, 86, 106, 115, 126
composite tools 119, 349
cordage 17, 40, 179, 197
core areas 9, 222
corridor(s) 8, 32–35, 38, 53, 56–58, 59, 73, 77,78, 80, 86, 92, 94, 97–98, **109**, 135, 137–138, 140, 141, 151, 157, 167, 173, 218, 220, 234, 248,249, 268, 291–293, 308, 338, 339

Dasht-i-Kavir (Sand Desert), Iran 55, 98, 234
Dasht-i-Lut (Salt Desert), Iran 55, 98, 234
detoxifying/detoxification 72, 155, 156, 199, 348, **353**
disease **4**, 16–17, 124, 196, 347
dispersal 1, 9–10, 20, 21, 23, 24, 28, 38, 51, 52, 54–55, 58, 59, 61, 63, **66**, 67
dodo 347, 354; *see also* extinction
domestic dogs/wolves 197, 322–323
dust storms 227, 249, 268
dzuds 268

Early Upper Palaeolithic (EUP) 213, 215, 217, 220, 248–249, 267–268, 270, **271**, 328, 330
East Asian monsoon 48, 136, 137, 274, **283**, 284, 289, 290
Egbert, Ksar Akil, Lebanon **206**, **209**, 213, 214
eggshell beads 8, 249, 270, 274, 275, 299, 303
emergent islands 174–175, 184
EPAS1 305, 306
Ethelruda, Ksar Akil, Lebanon **206**, **209**, 213, 214
extinction 4, 20, 21, 31, 33, 42, 63, 65, 77, 118, 141, **142**, 153, 172, 197, 220, 222, 322, 332, 347, 348, 350; *see also* dodo; Stellar's sea cow

failed dispersal 12, 67, 100, 200
fish 74, 84, 90, 143, 155, 172, 174, 182, 183, 198, 199, 233, 234, **235**, 298, 327, 336, 337, 348, **353**
fish hooks 179, 183, 335
Flores, Indonesia 2, **16**, 63, 74, 141, 142, **167**, 172–173, **177**, 181, 184, 187, 346, 347
Florisbad, South Africa 29, 30, 32
food storage 54, 200, 248–249, 275
footwear 290, 297, **353**
frost bite 59

green Arabia 58, 72, 78, 80–84, 85, 86, 94, 95, 100
green Iranian Plateau 95, 234, 235
green Sahara 32–35, 36, 38, 72, 95
green Thar Desert 95
ground edge axes 324, 325, 330

habitat 1, 5, 8, 9, 24, 92, 112, 137, 139, 140
habitat disruption 63–65, 67
habitat disruption, fragmentation, loss 33, 58, 63–65, 67, 141, 225, 350
habitat loss 58, 63–65, 67, 225, 350
Heinrich Events 51, 53, 54, 220, **221**, **226**, 267, **285**, 350
hides/hide working 40, 61, **62**, 72, 197, 200, 217, 261, 347
high-altitude apoxia 305, 306; *see also* EPAS1
Hindu Kush 55, 249
Hippopotamus amphibius 80, 81, 90
Hoabinhian 307
Hokkaido, Japan 53, 327, 331, 333, 338, 339
Homo erectus 22, 169, 172, 173, 294, 295, 298, 346
Homo floresiensis 2, 16, 63, 74, 169, 171, 172, 173, **177**
Homo heidelbergensis 31, 113, 295, 306
Homo luzonensis 169, 170–171
Homo naledi 31
Homo neanderthalensis 260, 296
Homo sapiens (skeletal evidence only) 28–31, 32, 37, 71, 73, 84–85, 87, 112–115, 125, 144–151, 152, 157, 171, 196, 204–214, 254, 257, 259, 260, 274, 293–295, 297–298, 307, 326, 334, 352
hypothermia 59–61

immigration 10, 98, 115, 141, 185, 196, 206, 212, 223, 281, 282, 289, 290–308
Indian monsoon 78, 84, 114, 225, 305
indigenous populations/species 1, 2–6, 10, 11–17, 18, 19, 21, 23, 24, 42, 63, 74, 98, 113, 118, 126, 127, 149, 195, 196, 199, 200, 209, 260, 281, 307–308, 326, 346, 354
Inner Mongolia 282, 286, 287, **288**, 297, 299, 325
Intertropical Convergence Zone (ITCZ) 78, 84, 218
invasive species 1, 2, 14, **15**, 19–20, 24, 63, 65, 144, 260, 346–347; *see also* naïve faunas
Island Southeast Asia (ISEA) 58, 59, 63, 65
IUP (Initial Upper Palaeolithic) 195–196, 213–217, 220, 223, 248, 256–257, 262, 264, **266**, 267, 270, **271–273**, 299, 303
ivory 39, 156, **265**, **266**, 318, 320, **321**, 348, **353**; *see also* mammoth

Java, Indonesia 73, 135, **136**–140, 148–149, 151–152, 184
jump dispersal 11, **12**, **13**, 24, 37, 337

kelp highway 143, 337–338
Kermanshah valley, Iran 227, 233, 236
Khorramabad valley, Iran 227, 232
Kurishio Current 333

Lake Tana, Ethiopia 35
Lake Urmia, Iran 224
Lake Van, Turkey 225
Lake Zeribar, Iran 224
last glacial maximum (LGM) 53, 84, **96**, 135, **138**, 143, **153**, **186**, 225, 227, 287, 304, 327, **328**, **329**; *see also* MIS 2
Late Upper Palaeolithic (LUP) assemblages 329
Levallois-Mousterian assemblages 35, 37, 215
Levantine Aurignacian assemblages 215, **221**, 224
Levantine Mousterian 210
Loess Plateau, China 282, 284, 286, **287**, 290, 292

mammoth 60, 142, 235, **255**, 260, 267, 282, 287, 290, 295, 317–322, 332, 339, 348
Manchuria 290, 297
Marine isotope (MIS) stage: MIS 2 (= LGM) 49, **50**, 53, 54, 67, 84, 110, 137, **220**, 225, 284, 286, **288**, 319; MIS 3 34, 35, 49, **50**, 52, 58, 59, 74, 80, 90, 92, 93, 94, 95, 99, 110, 142, **220**, 223, 227, 229, 234, 261, **265**, 286, 287, 288, 299, 303, 319, 333; MIS 4 49, **50**, **52**, 54, 58, 59, 74, 77, 84, 92, 93, 94, 99, 100, 110, 118, 137, **139**, 141, 142, 144, 148, 149, 212, 220, 225, 284, 286, 288, 293, 294, 306; MIS 5 34, 49, 58, 59, 72, 77, 80, **81**, 86, 87, 90, 92, 99, 108, **109**, 110, 116, 119, **139**, 141, 199, 210, 212, 220, 222, 227, 286, **288**, 293; MIS 5a 35, 80, 86, 90, 91, 92, 95, 284, 286; MIS 5b 59, 142, 293; MIS 5c 35, 80, 87, 90, 95, 98; MIS 5d 59, 142, 293, 296; MIS 5e 34, 35, 38; MIS 6 33, 35, 49, **50**, 84, 94, 98, 116, 118, **205**, 211, 284, 288, **289**; MIS 7 38, 71, 77, 80, 86, 91, 94, 210; MIS 8 210; MIS 9 38, 71, 80, 81, 85, 86, 94, 100; MIS 11 34, 80, 85, 86, 115, 116
maritime 167, 173–175, 183–185, 326
metapopulation(s) 6–10, **12**, 13–16, 23, 24, 31, 32, 56, 64, 65, **66**, 307
microblade(s) **120**, 259, 260, 270, 299, 303–304, 306, 324, 325, 326, 329, 331, 338
microliths 41, 73, 87, 349
microliths, Indian 99, **111**, **114**, 115, 119–122, **125**, 127, 348, **353**
Middle Palaeolithic 72, 325, 348
Middle Palaeolithic, Arabia 77, 80, **82**, **83**, 84–95
Middle Palaeolithic, Europe 36
Middle Palaeolithic, India 73, 99, **114**, 115–119, 127
Middle Palaeolithic, Iran/Pakistan 73, 95–98, 225, 229
Middle Palaeolithic, Levant 210–213, 215, 224
Middle Palaeolithic, Mongolia 268
Middle Palaeolithic, North Africa 28, 35, 37, 38
Middle Palaeolithic, Zagros 229, **235**, 236
Middle Palaeolithic assemblages, Central Asia and Siberia 195, 196, 249–251, 256–261, 262, 266
Middle Stone Age (MSA) 32, 36, 38, 39–41, 88, 89
Middle Upper Palaeolithic (MUP) assemblages 267
"modern behavior" 38–42, 346, 348, 351
monsoon 48, 50–51, 68, 78, **82**, 84, 98, 108, 112, 114, 122, 136, 218, 220, 225, **226**
Mousterian assemblages 23, 35, 36, 98, 199, 229, 232, **235**, 249, 259, 299, 306, 325, 339
Mu Us Desert, China 284
Mudawwara depression, Jordan 80

naïve faunas 13, 19–21, 23, 332, 347
Neandertal extinction 63, 220, 322
Neandertal(s) 1–4, 10, 11, 14, **15**, **16**, 18, 23, 28, **29**, 36, 39, 40–41, 54, 57, 63, 72, 77, 84, 92, 97, 98, 100, 113, 119, 186, 195–201, 203, **205–208**, 209, 210–214, 217, 220, 222, 227, 228, **230**, 233, 234, 236, 248, 249, 251, 254, 256, 257, 259–261, 275, 281, 297, 299, 303, 305–307, 318, 320, 322, 346, 347, 349–354
needles 128, 197, 260, 303, 320, **353**
Nefud Desert, Saudi Arabia 78, 80, 81, 86, 90, **91**
network **7**, 8, 10, 23, 56, 65, 67, 196, 197, 200, 223, 249, 266, 274, 275, 303, 307, 320, 323, 327, 336, 337, 347, 348–350, **353**
New Siberian Islands, Arctic Ocean 318, 320
Ngandong Fauna 151, 152, 158

Ngandong hominins **136**, **145**, 151–152, 157
Nubian lithic assemblages 35, 36, 87, **88**, 99, 100

obsidian 8, 323, **324**, 327, 333, 338, 348
ochre 8, 40, 41, 183, 197, 211, 258, 263, 296, 303
orangutans 108, 135, 138, 140, 141, 143, 148, 149, 152–157
Oriental Realm/Fauna 51, 52, 67, 72, 74, 106–183, 282, **283**, 290, 293–295, 308, 347, 349
ornaments 35, 41, 169, 195, 197, 211, 215, 217, 223, 249, 258, 259, 260, 263, **264–266**, 267, 298, 303, 307, 320

palaeodemes 6–10, 31, **66**, 307; *see also* metapopulations
PalaeoHonshu 323, 326, 327, 328, 332, 333, 339
palaeolakes 34, 80, **81**, **82**, 84, 90–92, **96**, 286
PalaeoSakhalin-Hokkaido (PSHK) 327
Palawan, the Philippines 135, **136**, 143, 153–154, 157, 169, 174, 187
Palearctic Fauna 4, 61, 235, 287, 306, 327
Palearctic Realm 4, 51, 52, **62**, 67, 74, 122, 195, 196, 201, 203, 235, 248, 281, 282, **283**, 295–305, 308, 349
Pamir mountains 54, 55, 249
Panthera sp. 84, **230**, **255**, 267
pendants 39, 182, 217, 258, 263, **265**, 298
perforated pendants 249, 275
perforated teeth 256, **257**, 260, 320, 298
Philippines 135, **136**, 137, **146**, 166, 168–171, 184, 187, 346
pit traps 325, 330, 331, 332, **353**
plant foods 17, 72, 155, 199, 200, 223, 249, 282, 318, 322, 347
population structure 6–9, 31–32
pre-Aurignacian lithic assemblages 35, 37

Qinling Mountains 282, 283, 287, 290, 306, 308

rainforest(s) 1, 8, 11, 17, 54, 58, 59, 73, 292, 307, 346, **353**; southeast Asia 135, 137–141, 148, 149, 151–153, 157, 158, 291, 292, 307; Sri Lanka 106–**110**, 123–127, 199
range expansion/extension 4, 6, 17, 18, 19, 72, 94, 108, 212
range shift **5**, **6**
recolonisation 21
refugia 53–54, 77, 84, 92–94, **111**, 117, 118
refugia, glacial 4, **66**, 144, 290, 306
refugia, interglacial 290
replacement 15, **16**, 59, 65, 153, 198, 215, 281, 290
Roustamian assemblages, Iran 232
Ryuku islands 327, 332, 333–335, 337, 339

Sahul **52**, 74, 136, 166, **167**, 183–186
sailing *see* maritime
Selenga river, Siberia/Mongolia 249, 268
shellfish 179, 182, 183, 198, 337
Sinai Peninsula 78, 86, 87, 90, 218, 220
small game 19, 40, 198, 199, 210, 217, **353**
snow lines 225, 227
source-sink model 13, 65–67, 144, 289, 290
speleothems 35, 80, **82**, 218, 220, **221**, **222**, 225, **226**, 284, **285**, 294
Stellar's sea cow; *see also* extinction 347, 354
Sulawesi, Indonesia 74, 135, 156, **167**, 171–172, **177**, 181, 183, 184, 187, 348, **353**
Sumatra, Indonesia 71, 73, 74, 117, 135, **136**, 137, **138**, 139, 140, 143, 149, 151, 152, 157, 158, 176, **230**, 295
Sunda Shelf **52**, 53, 59, 65, 67, 73, 74, 134–157, 166, 169, **170**, 176, 183
symbolism 40, 195, 197, 258, 348, 349
Syria 48, 78, 203, **206**, 210, 211, 215, 223, 224

Taiwan 169, 184, 296, 327, 333, 339
Taklamakan desert, China 49, 55
Talaud islands, Indonesia 175, **177**, **179**, 182
Thar Desert, India 17, 51, 58, 71, 72, 77, 98–100, 108, 113, 115, 117, 121, 236
Tibetan Plateau 54, 55, **109**, 195, 200, 201, 282, 296, 299, 304–305, 306, 346, 347, 348, 354
Tien Shan mountains 55, 249
Timor, Indonesia 74, **167**, 176, 179, 182, 184, 187, 335, 348
Timor-Leste **167**, 174, **177**, **178**, 179, 180, **181**, **354**
Toba, Sumatra, Indonesia 143, 157, 158
Toba super-eruption 73, **114**, 115, 117–118, 119, 127, 128
Transbaikalia, Siberia 248, 249, 261–268, 275, 326
trapezoids 330, **331**
traps, snares or nets 155, 199, 223, 322, 325, **353**

Tsushima Strait 323, 327, 328, 333

Uzbekistan 236, 248, 251, 352

vigilant faunas 13, 19, **20**; *see also* naïve faunas

Wallace Line 166, 187
Wallacea 51, 67, 72, 74, 135, 144, 156, 166–184, 337, 346, 348, **353**, 354
wind chill 60–61, 254
Wrangel Island, Arctic Ocean 142, 322

Yangtse Valley 290, 295, 306
Yemeni/Asir Highlands 78, 84, **96**
Youngest Toba Tuff (YTT) 117, 118

Zagros Mountains 48, 55, 56, 58, 94, 95, **96**, **97**, 98, 201, 224–225, 227, **228**, **229**, 236

SITE INDEX

Entries in **bold** are in figures or tables

Afontova-Gora, Siberia 336
Aman Kutan cave, Uzbekistan **252**
Amir Temir, Uzbekistan **252**
Amud cave, Israel **205**, 208, 211–212
Anghilak cave, Uzbekistan **250**, 251
Anui-3, Altai Mtns., Siberia 259, 260
Apidima, Greece 38, 85, 206
Attirampakkam, India **107**, 115

Bamburi 1, India 116
Batadomba lena, Sri Lanka **107**, **111**, 113, **114**, 120, 125
Baz rock shelter, Syria 224
Besitun cave, Iran 227, **228**, 229, **235**
Bhimbetka, India **107**, **114**, 116
Billaspurgam, India 110
Blombos, South Africa 40
Boh Dambang, Cambodia **136**, 141
Boker Tachit, Israel 195, 215
Boodie Cave, Australia 185, **186**
Braholo, Java **136**, 154
Broken Hill, Zambia 31
Bui Ceri Uato, Timor-Leste **167**, **178**
Bunge Toll 1885, Arctic Siberia 318
Byzovaia, Russia 318, 339

Callao Cave, Luzon, the Philippines 74, **136**, 170–171, 176
Carpenter's Gap, Australia **178**, 185, **186**
Chagryskaya Cave, Altai Mountains 254
Chah-i-Jam, Iran 225, **228**, 235
Chikhen-2, Mongolia **262**, **272**
Cooper's Ferry, Idaho, USA 338
Cueva de los Aviones, Spain 40

Daeo, Halmahera, Indonesia **167**, **177**
Dali, China **283**, 295
Darra-i-Kur, Afghanistan 113, 251
Dederiyeh cave, Syria **206**, **209**, 210–212
Denisova cave, Altai Mtns., Russia 251, 254, 256–261, **265**, **266**, 267–268, 275, 296, 298
Denisovan hominin/aDNA 1, 6, 11, 14, 15, 18, 195, 196, 199–209, 248, 251, 254, 256–257, 260–261, 275, 281, 296–297, 299, 303, 305, 306–308, 318, 346, 349–350
Denisovan lithic variant 259
Devils Lair, Australia 185
Dörölj, Mongolia **262**, **271**, 274
Duiktai, Siberia 336
Duoi U'Oi, Vietnam **136**, 141, **145**
Dushan cave, China 306

'Ein Qashish, Israel **205**
El Kowm, Syria 224
Eliye Springs, Kenya 30
Eshkaft-i-Gavi, Iran **96**, **228**, 299

Eyasi, Tanzania 30
Fa Hien Lena, Sri Lanka **107**, **111**, 113, **114**, 120, 125, **126**
Fuyan/Daoxian, China **283**, 293, 294

Galsanri, South Korea 324
Gar Arjeneh cave, Iran **96**, 227, **228**, 231, 232–233, **235**
Garm Roud, Iran **228**, 232
Ghad-i-Barmeh, Iran **228**
Ghar-i-Boof cave, Iran **96**, **228**, **229**, 236
Golo Cave, Gebe Island, Indonesia **167**, **177**, 181, 182
Gorham's Cave, Gibraltar 40
Goyet, Belgium 297, 322
Grotte de Contrebandiers, Morocco 36
Gunang Cave, South Korea 326

Hatnora, India **107**, 112–115
Haua Fteah, Libya 35, 37
Hazar Merd, Iraq 227, 233, **235**
Herto, Ethiopia 30
Hoti Cave, Oman 80, **82**
Houmian, Iran **228**, **235**
Howieson's Poort, South Africa 73, 115, 119, 121
Huanglong, China **283**, 294
Hulu cave, China **285**
Hummal, Syria 224

Idemaruyama, Japan 330, 333
Ifri n'Ammar, Morocco 35
Ivane valley, New Guinea **167**, 185, **186**
Iwo Eleru, Nigeria 30–31

Jahrom, Iran 95, **96**, **228**
Jebel Faya, UAE 85, 96
Jebel Irhoud, Morocco 28, **29**, **30**, 31, 32, 38, 39, 206, 351
Jebel Katefeh (JFK-1), Saudi Arabia 91
Jebel Qattar 1, Saudi Arabia 86, **91**
Jebel Umm Sanman (JSM-1), Saudi Arabia 92
Jeongok-ri (Chongnokni), South Korea 324
Jerimalai, Timor-Leste **167**, 176, **177**, 178, **179**, 180, 182, 183
Jinnuishan cave, China 283, 295
Jinsitai, China **283**, 299, 306, 307, 325
Jubbah, Saudi Arabia **85**, 90, 91
Jwalapuram, India 90, **107**, 113, **114**, **120**, 121

Kaldar cave, Iran **229**, 232, 233
Kalimantan, Borneo, Indonesia 156, 169, 172
Kalinga, Luzon, the Philippines 169
Kamenka, Transbaikalia, Siberia **262**, 263, **264–266**, 267, 270
Kamishirataki 2, Hokkaido, Japan 338
Kana, India **107**, 120, 127
Kapthurin, Kenya 32
Kara Bom, Altai Mountains 250, 257, 258–260, **262**, **265**, **266**, 268, 270
Kara Kamar, Afghanistan 227
Karaca cave, Turkey 225, **226**
Karim Shahir, Iraq 233, **235**
Katoati, India 98, **99**, **107**
Kebara cave, Israel **205**, **208**, 211, 212, 215, 217
Kharganyn Gol 5, Mongolia 268, 270, **272**, 274
Khotyk, Transbaikalia, Siberia **262**, 263, **264**, **265**, **266**, 270
Khudji, Tajikistan **250**, 251, **252**
Kitulgala lena, Sri Lanka **107**, 125
Kobeh cave, Iran **96**, 227, **228**, 233, **235**
Koshidake, Japan; *see also* obsidian 333
Kostienki, Ukraine 322, 336
Kozushima island, Japan; *see also* obsidian 333
Krapina, Croatia 40
Ksar Akil, Lebanon **206**, **209**, 213, 215, 217, 223
Kulbulak, Uzbekistan 249, **250**, **252**
Kunji cave, Iran **96**, 227, **228**
Kuturbulak, Uzbekistan **252**

Laili, Timor-Leste **167**, 176, **178–179**, 182
Lang Rongrien, Thailand **136**, 141, 154
Lang Trang, Vietnam **145**
Leang Barugayya 2, Sulawesi, Indonesia 183
Leang Bulu Bettue, Sulawesi, Indonesia 183
Leang Burung, Sulawesi **167**, **177**
Leang Gurung 2, Sulawesi, Indonesia 183
Leang Sakapao, Sulawesi, Indonesia **167**, **177**, 183
Leang Serru, Salebabu island, Indonesia 181
Leang Timpuseng, Sulawesi, Indonesia **167**, **177**, 183
Lene Hara, Timor-Leste **167**, 176, **177**, 178, **180**, 182
Liang Bua, Flores, Indonesia **74**, **172–173**, **177**
Lida Ajer, Sumatra, Indonesia **136**, **145**, 149, 151, 152, 157, 176, 295
Lingjing; *see also* Xuchang, China 296

Liujiang cave, China **283**, 294
Longlin cave, China 306
Longtanshan cave, China **283**, **294**
Lua Manggetek, Timor **167**, **178**
Lua Meko, Timor **167**, **178**
Lubang Jeriji Saléh, Borneo 156
Luna cave, China 294

Ma'anshan, Guizhou Province, China **283**, 307
Maastricht-Belvedere, the Netherlands 40
Madjedbebe, Australia 74, **178**, 185, **186**, 188
Mahadebbera, India **107**, 120, 127
Makarovo, Cisbaikalia, Siberia **262**
Malaia Syia, Cis-Baikalia 262
Maloyalomanskaya, Altai Mtns., Siberia 260, **266**
Mal'ta, Siberia 336
Maludong cave, China 306
Mammontovaya Kurya, Arctic Siberia **320**
Mamony, Cisbaikalia, Siberia **265**
Manot cave, Israel **205**, **209**, 211, 213–214, 215, 217–218, 220, **221**
Matja Kuru 2, Timor-Leste **177–179**, 180, **181**
Mehtakheri, India **107**, 120
Menindee, Australia 185, **186**
Milanggouwan, China 284
Military Hospital, Cisbaikalia, Siberia **265**
Mirak, Iran 95, 225, **228**, 235
Misliya Cave, Israel 38, 84, 86, **205**, **207**, 210
Moh Khiew, Thailand **136**, 141, **145**, 148
Mughr el-Hamamah, Jordan Valley 215, 217
Mukalla cave, Yemen 80
Mundafan, Saudi Arabia 80, **82**, 84, **85**, 90

Narwala Gabarnmang, Australia 185, **186**
Ngaloba, Tanzania 30
Ngandong, Java, Indonesia **136**, **145**, 151–152, 157, 158
Niah cave, Borneo, Malaysia 72, 74, **136**, **146**, 149–151, 154–157, 182, 199
Nihewan basin, China 295
Nwya Devu, Tibetan Plateau, China 304–305

Obi Rahmat cave, Uzbekistan 251, **252**, 352
Ogzi Kichik cave, Tajikistan **252**
Okinawa, Ryuku Islands, Japan 327, 333–335, 339
Okladnikov Cave, Altai Mountains **3**, 251, 254, 259
Orkhon valley, Mongolia 270; Orkhon-1 **262**, **272**; Orkhon-7 **262**, **272**
Orsang, India 98,113
Otsu-bobata site, Ryuku Islands, Japan 332

Pa Sangar, Iran **235**
Palegawra cave, Iraq 233, **235**
Patpara, India 114, 116
Penghu Channel specimen **283**, 296
Pereselencheskyi Punct, Cisbaikalia, Siberia **265**
Pinnacle Point, South Africa 40, 117
Podzvonkaya, Transbaikalia **262**, 263, **264–266**, 267, 270
Pokrovka, Siberia 254
Predmosti, Czech Republic 322
Punung, Java, Indonesia **136**, **145**, 148, 151, 152, 157
Puritjarra, Australia 13, **186**

Qafzeh cave, Isael 30, **96**, **205**, **207**, 211–213, 217
Qal'e Kord cave, Iran 225
Qaleh Bozi 2 cave, Iran 96, 97, **228**, 233, **235**
Qesem cave, Israel 40, 196, 209

Rangjeonggol cave, South Korea 326
Razboinichya, Altai Mtns., Russia 322, 323
Riwi, Australia **167**, **178**, 185, **186**
Rub'al Khali, Saudi Arabia 49, 78, 80, 84, 90
Ryuku Islands, Japan 327, 333, **334**, 337, 339

Sakitari cave, Okinawa, Japan 334, **335**
Salkhit, Mongolia 268, 274, **283**
Samarkandskaya, Uzbekistan 249, **250**, **252**
Sambungmacan, Java 152
Sanbao cave, China 225, **226**
Schapovo, Cisbaikalia, Siberia **265**
Selungir, Kyrgyzstan **252**
Shaamar loess-palaeosol section **262**, 270, **274**
Shanidar cave, Iraq **96**, 98, 199, 227, **228**, **229**, 231, 233, **235**, 236
Shi'bat Dihya, Wadi Durdud, Yemen **82**, 93
Shugnou, Tajikistan **250**
Shuidonggou (SDG) site cluster 273, 274, 283, 299, 307; Shuidonggou (SDG) 1

299, **300**, **301**, **302**, **303**; Shuidonggou (SDG) 2 299, **300**, **302**; Shuidonggou (SDG) 7 299; Shuidonggou (SDG) 9 229; Shuidonggou (SDG) 12 229, 303
Sibiryachikha variant, Altai Mtns., Siberia 259
Sima de los Huesos, Atapuerca, Spain 254
Site 55, Riwat, Pakistan **107**, **114**, 122, **123**, **124**, 127
SK site (Sopochnaya Karga), Russia 318, **320**
Skuhl cave, Israel **205**, **207**, 211, 212, 217
Song Gupuh, Java **136**, 151, 154
Soreq cave, Israel **221**, 225, **226**
South Temple Canyon, China 283, 299
Strashnaya Cave, Altai Mountains 254, 257–259, 260, **265**
Sunghir, Russia 337
Swan Point, Alaska, USA 338

Tabon cave complex, Palawan, the Philippines **136**, **146**, **153**, 154
Tabun cave, Israel **205**, **207–208**, 209–212
Talepu, Sulawesi, Indonesia 171–172
Tam Pa Ling (TPL), Laos **136**, **144–147**, 154, 157, 176, **283**, 295
Taramsa Hill, Egypt 87, 213
Teshik Tash, Uzbekistan **3**, **250**, 251, **252**, 254
Tianyuandong, China 268, **283**, 297–298, 303, 305–307, 353
T'is al Ghadah, Saudi Arabia 86
Tolbaga, Siberia, Transbaikalia, Siberia 262, 263, **264**, **265**, 267
Tolbor Valley, Mongolia **262**, 268, **269**, 270, 272, **283**; Tolbor 4, Mongolia **265**, 270, **271**; Tolbor 15, Mongolia **265**, **271**; Tolbor 16, Mongolia 270, **271**, **273**, 274; Tolbor 21, Mongolia 270, **271**
Tron Bon Lei, Alor island, Indonesia **167**, 176, **177**, **179**, 182, 183
Tsagaan-Agui, Mongolia **262**
Tsatsyn Ereg-2, Mongolia **262**

Uai Bobo 2, Timor Leste **167**
Üçağızlı, Turkey 215, **216**, 217, 223, 234
Umm el Tlel, Syria 215, 224
Upper Cave, Zhoukoudian, China 283, 298, 306
Ushbulak-1, Kazakhstan **250**
Ushki, Sakhalin, Russia 338
Ust '-Ishim, Siberia 254, **257**, 268
Ust-Kanskaya Cave, Altai Mtns., Siberia 259, **265**
Ust-Karakol, Altai Mtns., Siberia **259**, 260, **265**
Ust-Kova, Cisbaikalia, Siberia **265**

Varvarina Gora, Transbaikalia, Siberia 262, 263, **264**, **265**, 267

Wadi Sannur, Egypt 35
Wadi Surdud, Yemen 92–93, 94
Wajak, Java, Indonesia **136**, **145**, 149, 158
Warratyi, Australia 185, **186**
Warwasi cave, Iran **96**, 227, **228**, 231–233, **235**
Weinan section, China 286
Wezmeh cave, Iran **228**, 229, **230**, 233
Willandra Lakes, Australia 151, **186**
Wusta, Saudi Arabia 84

Xiahe, Tibetan Plateau, China **283**, 296–297, 305
Xiaodong, Yunnan Province, China **283**, 307
Xifeng section, Loess Plateau, China 284
Xuchang, China **283**, **296**, 306
Xujiayao, China **283**, 295, 296, 306

Yafteh cave, Iran **96**, 227, **228**, 229, **231**, 232–234, **235**
Yamachita-cho, Okinawa, Japan 334
Yana, Arctic Siberia 8, **60**, **265**, **319**, 320, **321**, 336–337, 338, 353
Yonaguni island, Ryuku Islands 333, 339
Yonggok (Ryonggok) cave, North Korea 326
Yongho-Dong, South Korea **324**

Zaozer'e, Russia 318
Zarzi cave, Iraq 227
Zhirendong, China 293–294
Zuttiyeh cave, Israel 209